T3-BOI-710

Haken · Wolf

Molecular Physics and Elements of Quantum Chemistry

Hermann Haken Hans Christoph Wolf

Molecular Physics and Elements of Quantum Chemistry

Introduction to Experiments and Theory

Translated by William D. Brewer

With 261 Figures and 43 Tables

Springer

Professor Dr. Dr. h. c. *Hermann Haken*
Institut für Theoretische Physik, Universität Stuttgart, Pfaffenwaldring 57
D-70550 Stuttgart, Germany

Professor Dr. *Hans Christoph Wolf*
Physikalisches Institut, Universität Stuttgart, Pfaffenwaldring 57
D-70550 Stuttgart, Germany

Translator:
Professor Dr. *William D. Brewer*
Freie Universität Berlin, Fachbereich Physik, Arnimallee 14
D-14195 Berlin, Germany

The front cover shows Figs. 5.1c, 19.3, and 4.18.

Title of the German original edition:
H. Haken, H. C. Wolf: *Molekülphysik und Quantenchemie*
Einführung in die experimentellen und theoretischen Grundlagen
(Zweite, verbesserte und erweiterte Auflage)

ISBN 3-540-58363-7 Springer-Verlag Berlin Heidelberg New York
ISBN 0-387-58363-7 Springer-Verlag New York Berlin Heidelberg

Library of Congress Cataloging-in-Publication Data. Haken, H. [Molekülphysik und Quantenchemie. English] Molecular physics and elements of quantum chemistry: introduction to experiments and theory / Hermann Haken, Hans Christoph Wolf; translated by William D. Brewer. p. cm. Includes bibliographical references and index. ISBN 3-540-58363-7 1. Molecules. 2. Chemical bonds. 3. Molecular spectroscopy. 4. Quantum chemistry. I. Wolf, H. C. (Hans Christoph), 1929- . II. Title. QC173.3.H35 1995 539'.6–dc20 94-47315

Printed in Germany

Typesetting: Data conversion by Kurt Mattes, Heidelberg
Offset printing and bookbinding: Druckhaus Beltz, Hemsbach

SPIN: 10126997 56/3144 - 5 4 3 2 1 0 - Printed on acid-free paper

Preface

This textbook is intended for use by students of physics, physical chemistry, and theoretical chemistry. The reader is presumed to have a basic knowledge of atomic and quantum physics at the level provided, for example, by the first few chapters in our book *The Physics of Atoms and Quanta*. The *student of physics* will find here material which should be included in the basic education of every physicist. This book should furthermore allow students to acquire an appreciation of the breadth and variety within the field of molecular physics and its future as a fascinating area of research.

For the *student of chemistry*, the concepts introduced in this book will provide a theoretical framework for his or her field of study. With the help of these concepts, it is at least in principle possible to reduce the enormous body of empirical chemical knowledge to a few fundamental rules: those of quantum mechanics. In addition, modern physical methods whose fundamentals are introduced here are becoming increasingly important in chemistry and now represent indispensable tools for the chemist. As examples, we might mention the structural analysis of complex organic compounds, spectroscopic investigation of very rapid reaction processes or, as a practical application, the remote detection of pollutants in the air.

The present textbook concerns itself with two inseparably connected themes: chemical bonding and the physical properties of molecules. Both have become understandable through quantum mechanics, which had its first successes in the elucidation of atomic structure. While the question of chemical bonding is mainly connected with the ground state of the electrons and its energy as a function of the internuclear separation of the bonded atoms, an explanation of other physical properties of molecules generally requires consideration of excited states. These can refer both to the electronic motions and to those of the nuclei.

The theoretical investigation of these themes thus requires the methods of quantum mechanics, and their experimental study is based on spectroscopic methods, in which electromagnetic waves over a wide spectral range serve as probes. In this way, it becomes possible to obtain information on the structure of a molecule, on its electronic wavefunctions and on its rotations and vibrations. We include here the theoretical and experimental determination of binding energies and the energies of excited states. In the theoretical treatment, we shall meet not only concepts familiar from atomic physics, but also quite new ones, among them the Hartree-Fock approximation, the Born-Oppenheimer approximation, and the use of symmetry properties in group theory. These ideas likewise form the basis of the quantum theory of solids, which is thus intimately connected to molecular physics.

In spite of the central importance held by the combination of molecular physics and quantum chemistry, there previously has been no textbook with the aim we have set for the present one. That fact, along with the extremely positive reception of our introductory text *The Physics of Atoms and Quanta* by students, teachers and reviewers, has stimulated us to write this book. We have based it on lecture courses given over the past years at the University of Stuttgart. We have again taken pains to present the material in a clear and understandable form and in a systematic order, treating problems from both an experimental and from a theoretical point of view and illustrating the close connection between theory and experiment.

Anyone who has been concerned with molecular physics and quantum chemistry will know that we are dealing here with practically limitless fields of study. An important, indeed central task for us was therefore the choice of the material to be treated. In making this choice, we have tried to emphasise the basic and typical aspects wherever possible. We hope to have succeeded in providing an overview of this important and fascinating area of research, which will allow the student to gain access to deeper aspects through study of the published literature. For those who wish to delve deeper into the great variety of research topics, we have provided a list of literature sources at the end of the book. There, the reader will also find literature in the area of reaction dynamics, which is presently experiencing a period of rapid development, but could not be included in this book for reasons of internal consistency. In addition, we give some glimpses into rather new developments such as research on photosynthesis, the physics of supramolecular functional units, and molecular microelectronics.

The book is thus intended to continue to fulfil a dual purpose: on the one hand to give an introduction to the well-established fundamentals of the field of molecular physics, and, on the other, to lead the reader to the newest developments in research.

This text is a translation of the second German edition of *Molekülphysik and Quantenchemie*. We wish to thank Prof. W. D. Brewer for the excellent translation and the most valuable suggestions he made for the improvement of the book.

We thank our colleagues and those students who have made a number of useful suggestions for improvements. In particular, we should like to thank here all those colleagues who have helped to improve the book by providing figures containing their recent research results. The reader is specifically referred to the corresponding literature citations given in the figure captions. We should also mention that this text makes reference to our previous book, *The Physics of Atoms and Quanta*, which is always cited in this book as I.

Last but not least we wish to thank Springer-Verlag, and in particular Dr. H. J. Kölsch and C.-D. Bachem for their always excellent cooperation.

Stuttgart, January 1995 *H. Haken* and *H.C. Wolf*

Contents

List of the Most Important Symbols Used in this Book

a	Hyperfine Coupling Constant (ESR)
	Einstein Coefficient
a_k^+, a_k	Creation and Annihilaton Operators for Fermi Particles (7.50) ff.
A	One-Dimensional Irreducible Representation
$\boldsymbol{A}$	Vector Potential
b	Einstein Coefficient
b_k^+, b_k	Creation and Annihilation Operators for Bose Particles (7.47) ff.
$\boldsymbol{B}$	Magnetic Field Strength
	Magnetic Flux Density
B	Rotational Constant (9.13)
	One-Dimensional Irreducible Representation
B_k^+, B_k	Creation and Annihilation Operators for Vibrational Quanta (11.131) ff.
c	Velocity of Light in Vacuum
c, C	Concentration
c_i	Expansion Coefficient
C, C_ϕ, C_n	Rotation Operators (Rotation of $2\pi/n$)
$\bar{C}_n$	Helicity Operator
d	Electronic State in an Atom
D	Determinant
	Fine Structure Constant (19.35)
	Centrifugal Stretching Coefficient (9.25)
D, D_e, D_0	Dissociation Energy
e	Elementary Charge
$\boldsymbol{e}$	Unit Vector
E	Energy
	Fine Structure Constant (ESR)
	Identity Operator
$\boldsymbol{E}$	Electric Field Strength
$\bar{E}$	Energy Expectation Value
E_{el}	Electronic Energy
$E_{\mathrm{kin}}, E_{\mathrm{pot}}$	Kinetic or Potential Energy
E_{rot}	Rotational Energy
$E_{\mathrm{vib}}, E_{\mathrm{V}}$	Vibrational Energy
ΔE	Energy Difference
f	Oscillator Strength
	Number of Degrees of Freedom
	Electronic State in an Atom
F	Vibrational Term

$F_{l,m}$	Spherical Harmonic Functions
g	g-Factor (Magnetic)
G	Rotational Term
h	Order of a Group
$\hbar = h/2\pi$	Planck's Constant (h = Planck's Quantum of Action)
H	Hamilton Function, Hamiltonian Operator
$H_{k,k}$	Matrix Element of the Hamiltonian Operator (7.16)
i	Imaginary Unit
i	Inversion Operator
I	Intensity
J	Rotational Quantum Number
	Spin-Spin Coupling Constant (NMR)
k	Boltzmann Constant
	Spring Constant, Force Constant
	Component of a Wavevector, Integer
$\boldsymbol{k}$	Wavevector
l	Mean Free Path
	Angular Momentum Quantum Number
$\boldsymbol{L}$	Angular Momentum
	Angular Momentum Operator
L_{l+m}	Laguerre Polynomial
$L_{\pm}$	Creation or Annihilation Operator for the z-Component of Angular Momentum
m	Mass, Magnetic Quantum Number
$\boldsymbol{m}$	Magnetic Moment
m_0	Rest Mass of the Electron
m_r	Reduced Mass
M	Magnetic Quantum Number
	Molecular Mass
n	Index of Refraction
	Principal Quantum Number
n_i	Number of Times the i-th Irreducible Representation Occurs in a Reducible Representation (6.47)
n_λ	Number of Quanta in the State λ
N	Number Density (Number per Unit Volume)
$\boldsymbol{N}$	Angular Momentum of Molecular Rotation
N_A	Avogadro's Number
p	Pressure
	Electronic State in an Atom
	Linear Momentum, Momentum Operator
$\boldsymbol{p}$	Electric Dipole Moment
	Linear Momentum, Momentum Operator
$p_{\mu,\kappa}$	Momentum Matrix Element (16.113)
$\overline{p}$	Expectation Value of Momentum
P	Momentum, Projection Operator (6.58)
$\boldsymbol{P}$	Momentum Operator
P_l^0	Legendre Polynomial
P_l^m	($m \neq 0$) Associated Legendre Function
Q	Class of a Group

r	Radial Distance, Particularly of Electrons
$\boldsymbol{r}$	Radius Vector
R	Distance of Nuclei
	Ideal Gas Constant
	Generalised Group Operation
$\hat{R}$	Reducible Representation Matrix
R_e	Equilibrium Distance or Bond Length
S	Overlap Integral (4.43)
	Spin Quantum Number
$\boldsymbol{S}$	Resultant Spin
	Spin Operator
S_m	Rotation-Inversion Operator
$S_m(j)$	$(m = \pm 1/2)$ Spin Function
S_+	Raising Operator for the z-Component of Total Spin
T	Temperature
	Electronic Term
T_1, T_2	Relaxation Times
v	Velocity
	Vibration Quantum Number
V	Potential
	Potential Energy
	Volume
$w_{\mu.\kappa}$	Transition Probability per Second
W	Energy
	Total Transition Probability
x_e	Anharmonicity Constant
$\overline{x}$	Expectation Value of Position (4.16)
Z	Nuclear Charge, Number of Initial States
α	Absorption Coefficient
	Polarisability
	Function of Moments of Inertia (11.72)
	Spin Function
	Angle
β	Hyperpolarisability
	Magnetic Polarisability (3.36)
	Optical Polarisability (3.14)
	Function of Moments of Inertia (11.72)
	Expansion Function with Respect to Time
	Spin Function
γ	Magnetogyric Ratio
Γ	Representation of a Group
	Linewidth
δ	Chemical Shift (NMR)
$\delta(x)$	Dirac Delta Function
δ_{ij}	Kronecker Delta Symbol
Δ	Difference Symbol

ε	Dielectric Constant
	Extinction Coefficient
	Infinitesimal Parameter
η	Quantum Yield
θ	Spherical Polar Coordinate
Θ	Inertial Tensor, Moment of Inertia (11.52)
$\boldsymbol{\Theta}_{\mu\kappa}$	Transition Dipole Matrix Element (16.120) ff.
λ	Quantum Number of Orbital Angular Momentum (13.2)
	Eigenvalue of a Determinant; Index which Distinguishes Plane Waves with Different Wavevectors k_λ
Λ	Total Orbital Angular Momentum (13.4)
μ	Transition Matrix Element (15.6)
	Magnetic Moment
	Permeability Constant
ξ_i	Displacement from the Rest Position
π	(Orbital) Molecular Orbital (Linear Combination, in particular of p_z-Functions)
$\varrho(E)$	Energy Density
ϱ	Density
	Spin Density (ESR)
σ	Inversion Operator, Spin Matrices
	Diamagnetic Shielding Factor (NMR)
$\overline{\sigma}$	Inversion-Translation Operator
$\sum$	Summation Symbol
Σ	Molecular Term Symbol
ϕ	Wavefunction, Spherical Polar Coordinate
Φ	Wavefunction
χ	Wavefunction (Especially Oscillator Functions)
$\chi(R)$, $\chi_1(R)$	Character of R in a Reducible or Irreducible Representation
ψ	Wavefunction
Ψ	Wavefunction of Several Electrons
ω	Circular Frequency $2\pi\nu$
$\boldsymbol{\Omega}$	Total Electronic Angular Momentum
Ω	Solid Angle
∇	Nabla Operator
∇^2	Laplace Operator

1. Introduction

1.1 What is a Molecule?

When two or more atoms combine to form a new unit, that new particle is termed a *molecule*. The name is derived from the Latin word *molecula*, meaning "small mass". A molecule is the smallest unit of a chemical compound which still exhibits all its properties, just as we have seen the atom to be the smallest unit of a chemical element. A molecule may be decomposed by chemical means into its component parts, i.e. into atoms. The great variety of materials found in the world of matter is a result of the enormous variety of possible combinations in which molecules may be constructed out of the relatively few types of atoms in the Periodic Table of elements.

The simplest molecules are diatomic and homonuclear; that is, they are made up of two atoms of the same type, such as H_2, N_2, or O_2. In these cases, one should imagine the electron distribution as shown in Fig. 1.1 (upper part): there are electrons which belong equally to both atoms, and they form the chemical bond. The next simplest group is that of diatomic molecules containing two different atoms, so-called heteronuclear molecules, such as LiF, HCl, or CuO; see Fig. 1.1 (lower part). In these molecules, in addition to chemical bonding by shared electrons, which is termed homopolar or covalent bonding, another bonding mechanism is important: heteropolar or ionic bonding.

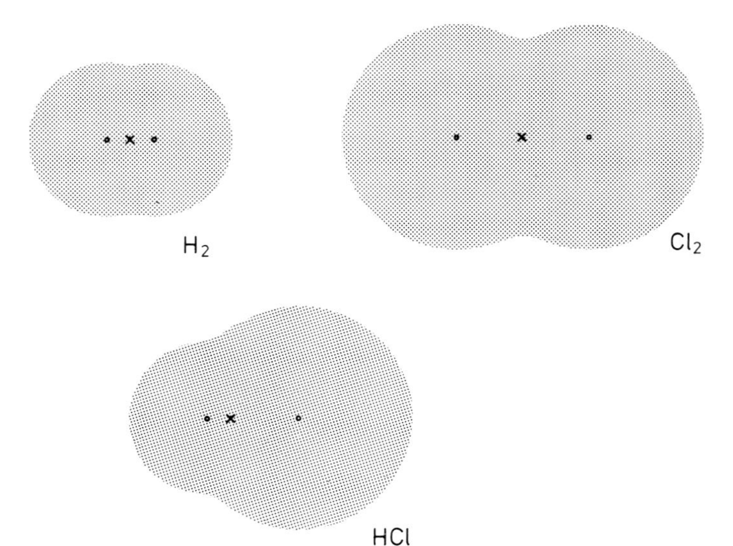

Fig. 1.1. Electron distributions in the small molecules H_2, Cl_2, and HCl, shown schematically. The nuclear separations are 0.74 Å in H_2, 1.27 Å in HCl, and 1.99 Å in Cl_2

We shall explain some of the basic concepts of molecular physics at this point by using as an example the molecule NaCl (in the gas phase). Figure 1.2 shows the potential energy of the system sodium + chlorine as a function of the distance between the atomic nuclei. At large internuclear distances, the interaction between a neutral sodium and a chlorine atom is quite weak and the potential energy of the interaction is thus nearly zero; a slight attractive interaction can, however, be caused by the weak mutual polarisation of the electronic charge clouds. If we bring the neutral atoms close together, at a distance of ca. 0.6 nm a repulsive interaction occurs. This fact can be used to define the size of the atoms, as discussed in more detail in I. (We denote the book *The Physics of Atoms and Quanta*, by H. Haken and H.C. Wolf, as I. We assume knowledge of the atomic physics treated in that book and will refer to it repeatedly in the following.)

At an internuclear distance of 1.2 nm, however, the state in which an electron from the sodium atom passes onto the chlorine atom becomes more energetically favored, and the system Na^+/Cl^- is thus formed by charge transfer. When the distance is further decreased, the effective interaction potential becomes practically the same as the attractive Coulomb potential between the two ions. An equilibrium state is finally reached at a distance of 0.25 nm, due to the competition between this attractive potential and the repulsion of the nuclei and the closed electronic shells of the ions; the repulsion dominates at still smaller distances. This equilibrium distance, together with the electron distribution corresponding to it, determine the size of the molecule.

Continuing through molecules containing several atoms, such as H_2O (water), NH_3 (ammonia), or C_6H_6 (benzene), with 3, 4, or 12 atoms, respectively, we come to large molecules such as chlorophyll or crown ethers, and finally to macromolecules and polymers such as polyacetylene, which contain many thousands of atoms and whose dimensions are no longer measured in nanometers, but instead may be nearly in the micrometer range. Finally, biomolecules such as the giant molecules of deoxyribosenucleic acids (DNA), which are responsible for carrying genetic information (see Sect. 20.6), or molecular functional units such as the protein complex of the reaction centre for bacterial photosynthesis (cf. the schematic representation in Fig. 1.3), are also objects of study in molecular physics. These molecules will be treated in later sections of this book, in particular in Chap. 20.

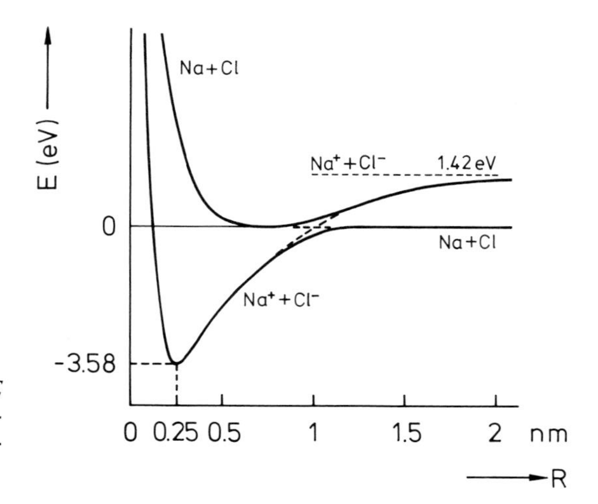

Fig. 1.2. The potential energy E for NaCl and Na^+Cl^- as a function of their internuclear distance R, in the gas phase

The last example already belongs among the *supramolecular* structures, giant molecules or functional units, whose significance for biological processes has become increasingly clear over the past years. When molecules of the same type, or even different molecules, group together to make still larger units, they form molecular clusters and finally solids.

1.2 Goals and Methods

Why does the molecule H_2 exist, but not (under normal conditions) the molecule H_3? Why is NH_3 tetrahedral, but benzene planar? What forces hold molecules together?

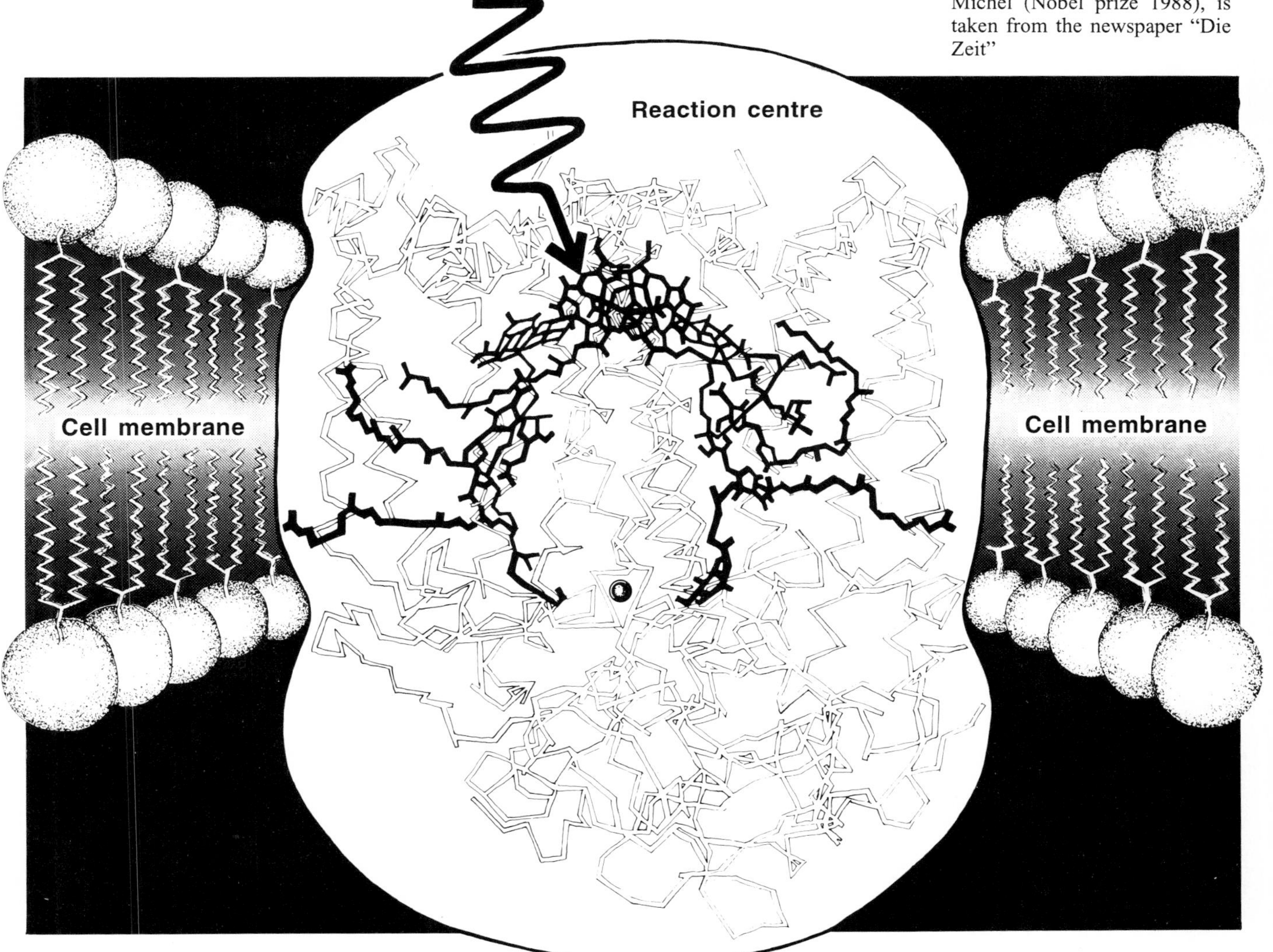

Fig. 1.3. The reaction centre for bacterial photosynthesis as a molecular functional unit. This schematic drawing shows the photoactive molecules, which are embedded in a larger protein unit. The latter is in turn embedded in a cell membrane. Light absorption by the central chlorophyll dimer is the first step in the charge separation which sets off the chemical processes of photosynthesis. This topic will be treated further in Sect. 20.7. The picture, based on the X-ray structure analysis by Deisenhofer, Huber, and Michel (Nobel prize 1988), is taken from the newspaper "Die Zeit"

How large are molecules, and what electrical and magnetic properties do they have? Why does the optical spectrum of a molecule have orders of magnitude more spectral lines than that of an atom? These are some of the questions which can be answered more or less simply when we begin to treat the physics of molecules.

The goal of *molecular physics* is to learn about and to understand the structure, the chemical bonding, and the physical properties of molecules in all their variety. From this basis, one would then like to derive an understanding of the function, the reactions, and the effects of molecules in physical, chemical, and biological systems.

The incomparably greater variety of molecules as compared to atoms has as a consequence that one cannot obtain a basic understanding of all the other molecules by considering the simplest one, as is possible in atomic physics beginning with hydrogen. In the physical investigation of molecules, spectroscopic methods play a special role, as they do in atomic physics as well. However, many more spectroscopic methods are required, in particular because in molecules, unlike atoms, there are more internal degrees of freedom such as rotations and vibrations. In the following, it will become clear just how varied and numerous are the methods of investigation which are used in molecular physics.

We shall see the importance of microwave and infra-red spectroscopies, and how fine details of molecular structure can be uncovered with the techniques of magnetic resonance spectroscopy of electrons and nuclei. We will, however, also gain access to the wide experience on which chemical methods are based, the various calculational techniques of quantum chemistry, and a great variety of experimental methods, beginning with structure determination using X-ray or neutron scattering, mass spectrometry, and photoelectron spectroscopy.

The goal of *quantum chemistry* is to make available the tools with which the electron distribution in molecules, their chemical bonding, and their excited states may be calculated. Its boundary with molecular physics can of course not be defined sharply.

1.3 Historical Remarks

The first precise ideas about molecules resulted from the observation of quantitative relationships in chemical processes. The concept of the molecule was introduced in 1811 by the Italian physicist *Avogadro* in connection with the hypothesis which bears his name, according to which equal volumes of different ideal gases at the same temperature and pressure contain equal numbers of atoms or molecules. This allowed a simple explanation of the *law of constant and multiple proportions* for the weights and volumes of gaseous reactants in chemical reactions. These laws and hypotheses are likewise found at the beginning of atomic physics; they are treated in Sect. 1.2 of I and will not be repeated here.

The investigation of the behaviour of gases as a function of pressure, volume, and temperature in the course of the 19th century led to the *kinetic theory of gases*, a theoretical model in which molecules, as real particles, permit the explanation of the properties of gases and, in a wider sense, of matter in general. On this basis, *Loschmidt* in the year 1865 made the first calculations of the size of molecules, which within his error limits are still valid today.

In the second half of the 19th century, many chemists (we mention here only *Kékulé*, the discoverer of the structure of benzene) made the attempt to obtain information about the atomic and geometric structure of molecules using data from chemical reactions. With the advent of modern atomic and quantum physics in the 20th century, an effort has also been made to gain an exact understanding of *chemical bonding*. Following the pioneering work of *Kossel* on heteropolar and of *Lewis* and *Langmuir* on homopolar bonding (1915–1920), *Hund, Heitler, London* and others after 1927 laid the foundations of a quantitative quantum theory of chemical bonding, and thus of quantum chemistry. Since then, a multitude of researchers have contributed to the increasing degree of refinement of these theoretical ideas in numerous research papers.

The various instrumental and experimental advances which have allowed an increasingly detailed analysis of the physical properties of countless molecules will be treated in those chapters of this book which deal with the respective methods. It can be most readily verified that such experimental methods tell us much about molecular structure – to be sure indirectly, but yet precisely – if we can view the molecules themselves. Using the methods of X-ray scattering and interference, this becomes possible with high accuracy when sufficiently large, periodically recurring units can be simultaneously investigated, i.e. with single crystals. An example is discussed in the next chapter in connection with the determination of the sizes of molecules; see Fig. 2.2. With the modern techniques of transmission electron microscopy (Fig. 1.4), and in particular using the recently-developed scanning tunnel microscope (Fig. 1.5), it is now possible to obtain images of individual molecules. The existence of molecules and an understanding of their physical properties have long ceased to be simply hypotheses: instead, they are established experimental results and form the basis for our understanding of many structures and processes not only in chemistry, but also in many other fields such as biology, materials science, and technology.

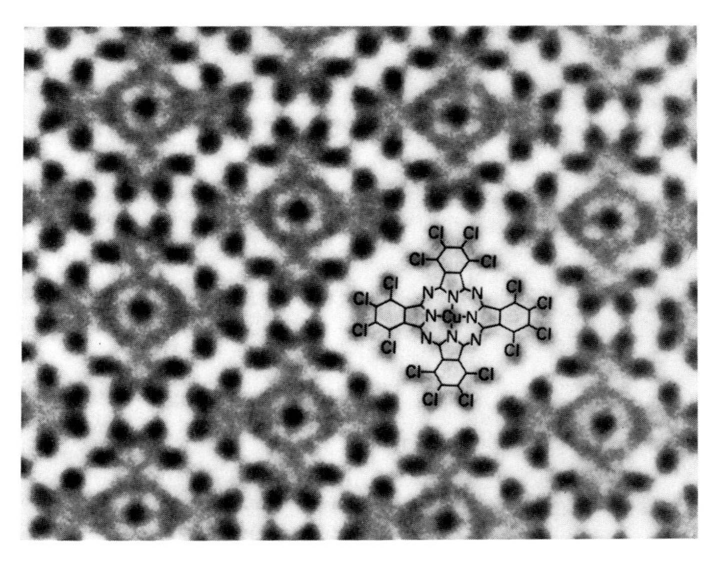

Fig. 1.4. An electron microscope image of hexadecachloro copper phthalocyanine molecules. The molecules form a thin, oriented layer on an alkali halide crystal which serves as substrate. The image was made with a high-resolution 500 kV transmission electron microscope and was processed using special image-enhancement methods. The central copper atoms and the 16 peripheral chlorine atoms may be most clearly recognised. (This picture was kindly provided by Prof. N. Uyeda of Kyoto University)

1.4 The Significance of Molecular Physics and Quantum Chemistry for Other Fields

Molecular physics and quantum chemistry provide the connecting link between our knowledge of atomic structure and our efforts to gain a comprehension of the physical and biological world. They form the basis for a deeper understanding of chemical phenomena and for knowledge of the countless known and possible molecules, their physical properties and their interactions. They lead us to an understanding of microscopic forces and bonding structures, of the electrical, magnetic, and mechanical properties of crystals and other materials used in science and technology. They provide us with the fundamentals needed to understand the biological world: growth, reproduction, and perception; metabolic processes, photosynthesis in plants, and all of the basic processes of organic life. In short, all living things become comprehensible only if we understand the molecular structures which underlie them, the molecules which are actively and passively involved in life-processes, together with their functions and their interactions.

Small molecules such as H_2 or HCl are particularly suitable as examples to introduce important principles, theoretical treatments and experimental methods. These small molecules will therefore assume an important place in this book, due to their relative simplicity and clarity. In the following chapters, we shall learn about a number of methods and concepts using as examples small molecules in the gas phase. In the process, however, we must not forget how varied, and correspondingly complex, the world of molecules as a whole is. To a greater degree than in atomic physics, we will have to consider the multiplicity of phenomena in our material world, the details and not just the basic principles, in order to gain an understanding of molecular physics. The next chapters aim to give an idea of this multiplicity of detail.

Fig. 1.5. An image of benzene molecules made with a scanning tunnel microscope. The benzene was evaporated onto a Rhenium (111) surface together with CO molecules, which serve to anchor the larger molecules and are themselves nearly invisible. As a result of the substrate-molecule interaction, we see in the picture partially localised states which make the benzene molecules appear to have a reduced (threefold) symmetry; what is seen are thus not the individual C atoms, but rather molecular orbitals. (From H. Ohtani, R.J. Wilson, S. Chiang, and C.M. Mate, Phys. Rev. Lett. **60**, 2398 (1988); picture provided by R.J. Wilson)

In the foreground of our considerations will be the individual molecule: isolated molecules in a gas. In contrast to atoms, molecules have internal degrees of freedom involving motions of the component atomic nuclei, which give rise to rotations and to vibrations. We shall discover spectroscopy to be the most important method for elucidating molecular structure, just as in atomic physics; but in the case of molecules, the microwave and infra-red regions of the spectrum, where rotational and vibrational excitations are found, will occupy much more of our attention.

The interactions of molecules with each other and with other types of molecules finally will lead us to the physics of fluids, to solid state physics, and to the physical and structural fundamentals of biology. In this book, we shall restrict the treatment of those fields to the basic knowledge which is required for understanding the molecules themselves. Conversely, we shall learn a much greater amount about methods and results which are essential for an understanding of phenomena in the above fields. Our goal, here as in I, will be to begin with observations and experimental results, and from them, to work out the basic principles of molecular physics and quantum chemistry. This book thus does not intend to provide specialised knowledge directly, but rather to smooth the way for the reader to gain access to the enormous body of technical literature.

2. Mechanical Properties of Molecules, Their Size and Mass

Only in recent years and in particularly favorable cases has it become possible to directly generate images of molecules. In order to determine their sizes, masses, and shapes, there are however numerous less direct but older and simpler methods which date back even to the field of classical physics. Such methods are the subject of the following sections.

2.1 Molecular Sizes

If by the "size" of a molecule we mean the spatial extent of its electronic shells, rather than the internuclear distances of its component atoms, then we can start from rather simple considerations in order to determine the size of a small molecule containing only a few atoms. Following *Avogadro*, we know that 1 mole of an ideal gas at standard conditions occupies a volume of $22.4 \cdot 10^{-3}$ m^3 and contains N_A molecules, where N_A is Avogadro's number, $6.02205 \cdot 10^{23}$ mol^{-1}. When we condense the gas to a liquid or a solid, its volume will decrease by a factor of about 1000. If we now assume that the molecules just touch each other in the condensed phase, then from the above data we calculate the order of magnitude of the molecular radii to be 10^{-10} m, i.e. 0.1 nm or 1 Å. In a similar manner, starting with the density ϱ of a liquid, we can calculate the volume occupied by its individual molecules if we assume that they are spherically close-packed or if we know the packing, i.e. their spatial arrangement.

Additional, more precise methods for determining molecular sizes based on macroscopic measurements are the same as those which we have already met in atomic physics. We shall repeat them only briefly here:

- From determinations of the pV isotherms of real gases and using Van der Waals' equation of state for the pressure p and the volume V:

$$\left(p + \frac{a}{V^2}\right)(V - b) = RT , \tag{2.1}$$

  (where T is the absolute temperature, R the ideal gas constant, and p and V refer to one mole), we can obtain numerical values for the quantity b, the covolume. In the framework of the kinetic theory of gases, it is equal to 4 times the actual volume of the molecules. Van der Waals' equation of state must be used instead of the ideal gas equation when the interactions between the particles (a/V^2) and their finite volumes (b) are taken into account. Table 2.1 contains measured values of b and the molecular diameters calculated from them for several gases.
- From measurements of so-called transport properties such as diffusion (transport of mass), viscosity (transport of momentum), or thermal conductivity (transport of energy), one

Table 2.1. Measured values for the covolume b in Van der Waals' equation of state (2.1), in units of liter mol^{-1}, and the molecular diameters d in Å calculated from them, for several gas molecules. After Barrow

Molecule	b	d
H_2	0.0266	2.76
H_2O	0.0237	2.66
NH_3	0.0371	3.09
CH_4	0.0428	3.24
O_2	0.0318	2.93
N_2	0.0391	3.14
CO	0.0399	3.16
CO_2	0.0427	3.24
C_6H_6	0.155	4.50

obtains the mean free path l of molecules in the gas and from it, their diameters, in the following way:
For the viscosity or internal friction of a gas, we have

$$\eta = \frac{1}{3}\varrho l \sqrt{\overline{v^2}} \tag{2.2}$$

(ϱ = density, $\overline{v^2}$ = mean squared velocity of the molecules). We know that the molecules do not have a single velocity, but instead obey the Maxwell-Boltzmann distribution.
With the equation:

$$p = \frac{1}{3}\varrho \overline{v^2}$$

for the gas pressure p, (2.2) can be modified in terms of directly measurable quantities. Substituting, we obtain

$$l = \eta \sqrt{\frac{3}{p\varrho}}. \tag{2.3}$$

Thus, by determining the pressure, density, and viscosity of a gas, we can calculate its mean free path.

– Another method makes use of the thermal conductivity. For the thermal conductivity λ, we find:

$$\lambda = \frac{1}{3}N\frac{C_V}{N_A}l\sqrt{\overline{v^2}} \tag{2.4}$$

(N is the number density of the molecules, C_V is the molar heat capacity at constant volume, and N_A is Avogadro's number).

We see that a low thermal conductivity is typical of gases with molecules of large mass, since then $\overline{v^2}$ is small at a given temperature.

From the mean free path l, we obtain the interaction cross section and thus the size of the molecules, as indicated in Sect. 2.4 of I. We find:

$$l = \frac{1}{\sqrt{2}\pi N d^2}, \tag{2.5}$$

(where N is again the number density of the molecules and d is their diameter assuming a circular cross section). Some data obtained in this manner are collected in Table 2.2.

For N_2 (nitrogen) under standard conditions, it is found that $N = 2.7 \cdot 10^{25}$ m^{-3}, $l = 0.6 \cdot 10^{-7}$ m, and from this the diameter of the molecules is $d = 3.8 \cdot 10^{-10}$ m. For the mean time between two collisions of the molecules, we find, using

$$\tau\sqrt{\overline{v^2}} = l, \qquad \text{the value} \qquad \tau = 1.2 \cdot 10^{-10}\text{s}.$$

All of the methods mentioned treat the molecule in the simplest approximation as a sphere. To determine the true form and shape of molecules, more sophisticated physical methods are needed.

Methods involving interference of scattered X-rays or electron beams, which were also mentioned in I, permit the determination of the molecular spacing in solids, and therefore

Table 2.2. Diameter d (in Å) of some small molecules derived from gas-kinetic interaction cross sections

Molecule	d
H_2	2.3
O_2	3.0
CO_2	3.4
C_2H_6	3.8

of the molecular sizes including an anisotropy of the molecules, i.e. when they deviate from spherical shapes. See I, Sect. 2.4, for further details. For these methods, one also needs single crystal samples or at least solids having a certain degree of long-range order. When the molecules are in a disordered environment, for example in a liquid or a glass, one may obtain less clear interference patterns due to short-range order present in the glass or the liquid. Short-range order means that particular intermolecular distances occur with especially high probabilities.

The distances between atoms within a molecule, i.e. between the component atoms of the molecule, can be determined through interference of electron beams diffracted by the molecules. For this purpose, the intensity distribution in the electron diffraction pattern must be measured. Making the assumptions that each atom within the molecule acts as an independent scattering centre, and that the phase differences in the scattered radiation depend only on the interatomic distances, one can derive values for the characteristic internuclear separations in molecules, as illustrated in Fig. 2.1.

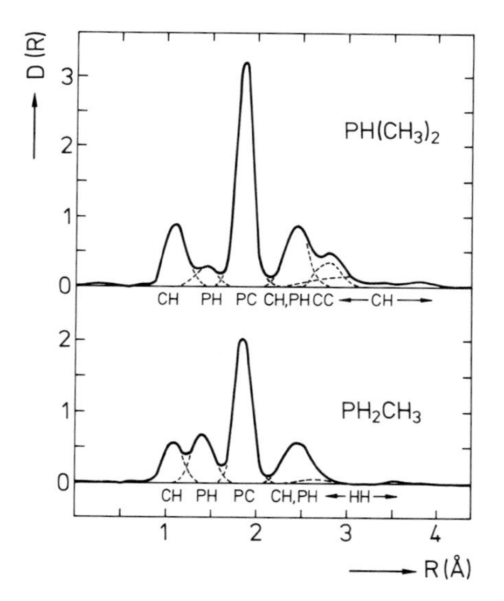

Fig. 2.1. Radial distribution functions D describing the electron density as a function of the bond length R between atomic nuclei in the molecules $PH(CH_3)_2$ and PH_2CH_3, obtained from electron diffraction patterns. The maxima in the distribution functions can be correlated with the internuclear distances indicated. [After Bartell, J. Chem. Phys. **32**, 832 (1960)]

If one wishes to measure the precise electron density distribution of a molecule, and thereby obtain more information from X-ray interference patterns of single crystals besides just the crystal structure and the distances between the molecular centres of gravity, then the relative intensities of the interference maxima must be precisely measured. The scattering of X-rays by a crystal is essentially determined by the three-dimensional charge distribution of its electrons, which can be reconstructed from the measured intensities of the interference patterns using Fourier synthesis. One thus obtains maps of the electron density distribution in molecules, such as the one shown in Fig. 2.2. Electrons directed at the crystal are also scattered by the electronic shells of its component atoms or molecules and can likewise be used to obtain electron density maps. Electron diffraction is, to be sure, applicable only to thin film samples, owing to the shallow penetration depth ("information depth") of the electron beams.

The case of neutron diffraction is quite different. Since neutrons are scattered primarily by nuclei and, when present, by magnetic moments, neutron diffraction can be used to determine

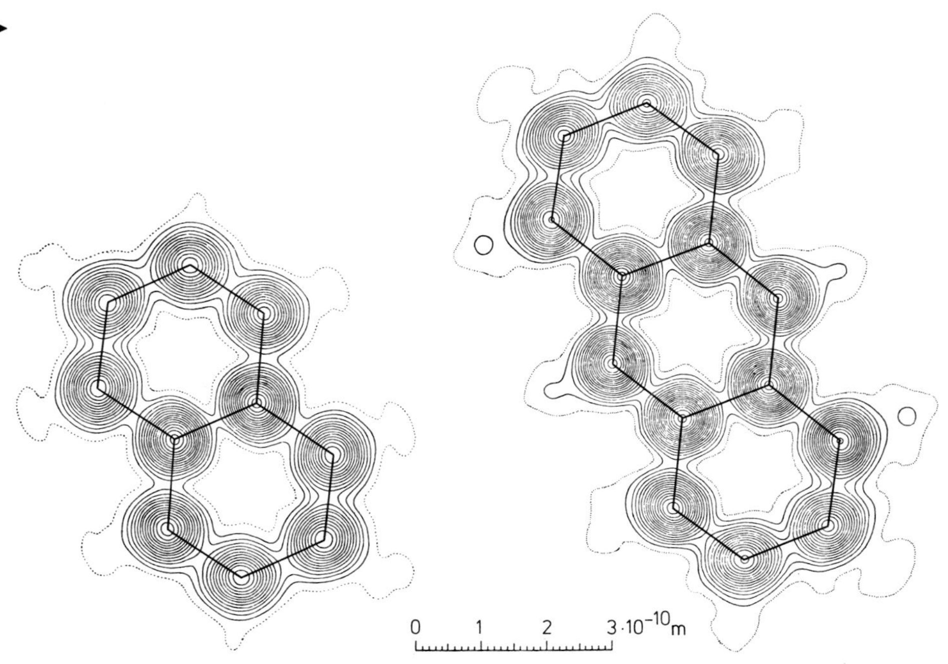

Fig. 2.2. Cross sections through the molecular plane of napthalene (left) and anthracene. The contour lines representing the electron density are drawn at a spacing of 1/2 electron per Å^3; the outermost, dashed line just corresponds to this unit. (After J.M. Robertson, Organic Crystals and Molecules, Cornell University Press (1953))

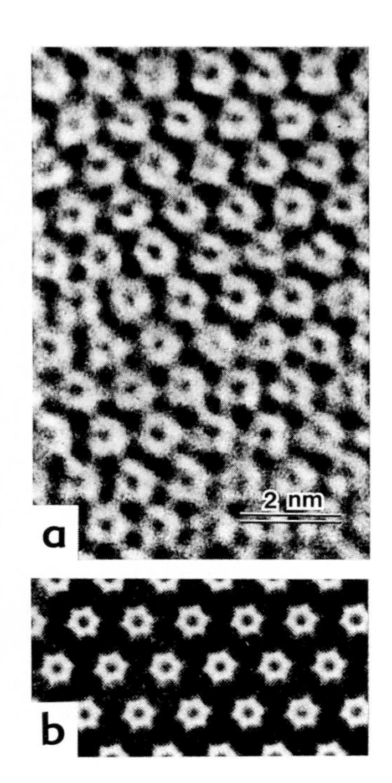

Fig. 2.3. **(a)** A transmission electron microscope image of a thin C_{60} crystal in the (111) direction. The spatial resolving power of the apparatus was 0.17 nm. The picture was taken using the HREM (high resolution electron microscopy) method and a special image processing technique. **(b)** For comparison: the calculated image from a crystal having a thickness corresponding to 2 unit cells (4.9 nm). (From S. Wang and P.R. Busek, Chem. Phys. Lett. **182**, (1991), with the kind permission of the authors)

directly the structure of the nuclear framework of a molecule. The electronic structure can, in contrast, be investigated only to a limited extent by using neutrons.

A microscopic image of molecules can be obtained with the electron microscope. The spatial resolution of transmission electron microscopes has become so high in recent decades that structures in the range of 1 to 2 Å can be imaged. An example is shown in Fig. 1.4, Chap. 1. A further example is given in Fig. 2.3, which shows an image of a thin fullerene (C_{60}) crystal taken with a high-resolution transmission electron microscope. The nearly spherical C_{60} molecules can be readily recognised (see also Fig. 4.18) in their densely-packed arrangement. Although imaging of molecules using the field emission microscope (see Fig. 2.14 in I) has as yet attained no great practical significance, the scanning tunnel microscope (STM), developed in the years following 1982, promises to become an important tool for the identification, imaging, and perhaps even for the electrical manipulation of individual molecules.

Since the introduction of the STM by Binnig and Rohrer in 1982 and its further development, it has become possible to obtain detailed images of surfaces at atomic or molecular resolution. In the original, simplest version, employing a constant tunnelling current, the STM functions are indicated schematically in Fig. 2.4.

A probe electrode having an extremely thin point is brought so close to a conducting surface that a low operating voltage (mV to V) gives rise to a measurable current between the probe and the surface without a direct contact; the current is due to the tunnel effect (cf. I, Sect. 23.3). This so-called tunnel current depends very strongly on the distance from the probe to the surface. The probe is now scanned over the surface, varying its distance z with the aid of a feedback circuit in such a way as to keep the tunnel current constant. An image of the surface is then obtained by plotting the distance z as a function of the surface coordinates x and y. This is shown schematically in the lower part of Fig. 2.4. With this type

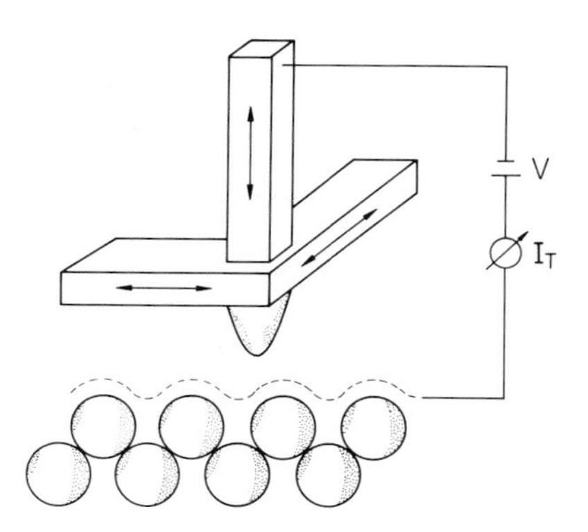

Fig. 2.4. A schematic representation of a scanning tunnel microscope. The tunnel current I_T between the surface being imaged and the probe electrode, which has the form of an extremely thin point, is plotted as a function of the surface spatial coordinates x,y using the distance z of the probe from the surface

of microscope, it is also possible to image individual molecules adsorbed onto a surface. An example at molecular resolution is shown in Fig. 1.5.

A further development of the STM is the force microscope. Here, the quantity directly measured is not the tunnel current, but rather the force between the probe and the substrate surface; it can thus be used even with insulating substrates. With the aid of such scanning microscopes, the structures of molecules and their arrangements on surfaces can be made visible. The recrystallisation of molecules on surfaces can also be followed as a function of time. An additional example of a molecular image made with the STM is shown in Fig. 2.5.

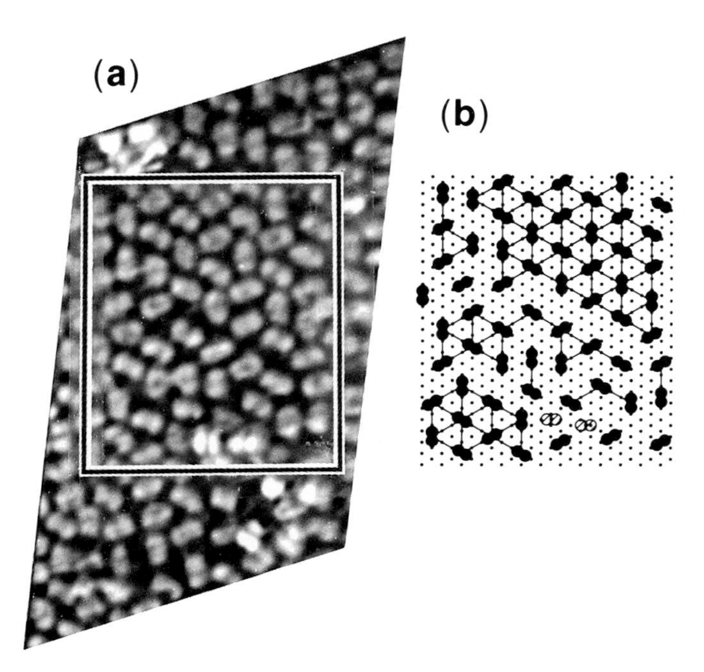

Fig. 2.5. (a) An STM picture of napthalene molecules on a Pt (111) substrate. **(b)** A schematic representation of the orientation of napthalene molecules on the Pt (111) surface. [From V.M. Hallmark, S. Chiang, J.K. Brown, and Ch. Wöll, Phys. Rev. Lett. **66**, 48 (1991). A review was given by J. Frommer, Angew. Chem. **104**, 1325 (1992)]

A quite different method of determining the sizes of molecules can be applied to molecular layers. Long-chain hydrocarbon molecules which carry a water-soluble (hydrophilic) group at one end, while the opposite end is hydrophobic, can spread out to form monomolecular layers on a water surface. This was first shown by the housewife *Agnes Pockels* in 1891.

The technique was developed further by *Lord Rayleigh*, whom she informed of it, and later in particular by *Langmuir*. He was able to show that the molecules can be compressed up to a well-defined smallest distance on the water surface, so that they touch each other in equilibrium. From the molecular mass and density, one can then determine the number of molecules per unit surface area. From this, a numerical value for the cross-sectional area of the molecules can be calculated. This method can of course be applied only to molecules having a very special structure. Details are given in Fig. 2.6.

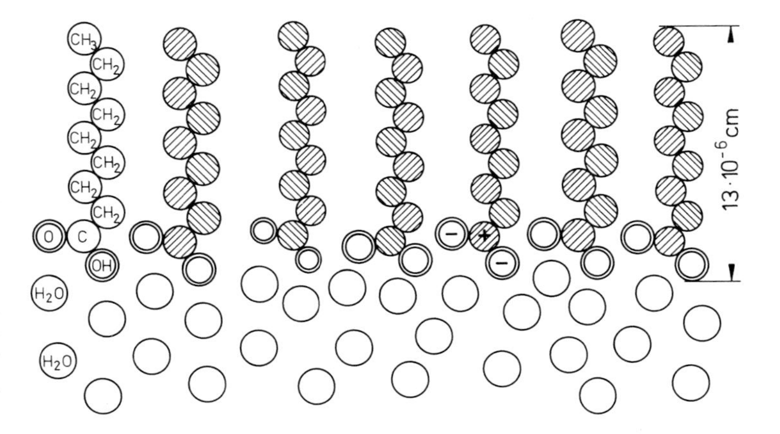

Fig. 2.6. Schematic representation of the arrangement of fatty acid molecules on an aqueous surface. The water molecules are indicated by ◯, hydrophilic oxygen or hydroxyl groups by ◎, and hydrophobic carbon atoms or CH_2 groups by ⊘. More information is given in Sect. 20.7

We should mention here that these monomolecular layers, so-called Langmuir-Blodgett films, have been the subject of considerable renewed interest in recent years. They can, for example, be transferred from the water onto substrates, and several layers can be placed one atop the other. Thus, the behaviour and interactions between individual molecules in structures of low dimensionality or at precisely defined distances and relative positions from one another can be studied. These layers are also used as models for biological membranes. It is a goal of present-day research to build artificial molecular functional units from such ordered layers; cf. Sect. 20.7.

The methods discussed above yield rather precise values for the sizes of molecules in a relatively simple manner. As we shall see in later chapters, there is a whole series of spectroscopic techniques with which one can obtain considerably more detailed knowledge about the structure of a molecule, the spatial arrangement and extent of its components, its nuclear framework, and the effective radius of its electronic shells.

In any case, when speaking of "size", we must define the physical property which we are considering. This is illustrated in Fig. 2.7. If, for example, we wish to determine the size of a molecule by measuring collision cross sections, we can define either the distance of closest approach of the collision partners, d_T, or else the distance d_0 at which the electronic shells of the collision partners detectably overlap, or finally the distance d_{Min} at which the interaction energy E takes on its minimum value, to be the molecular size. In this process, we must keep in mind that molecules are not "hard", but rather are more or less strongly deformed during the collision, as is indicated in Fig. 2.7 in defining the distance d_T. T stands for temperature, since the molecules have a mean energy kT in the collisions. The electronic wavefunctions are also not sharply bounded. It should therefore not be surprising that the measured values of molecular size differ according to the method of measurement; for example, for the H_2

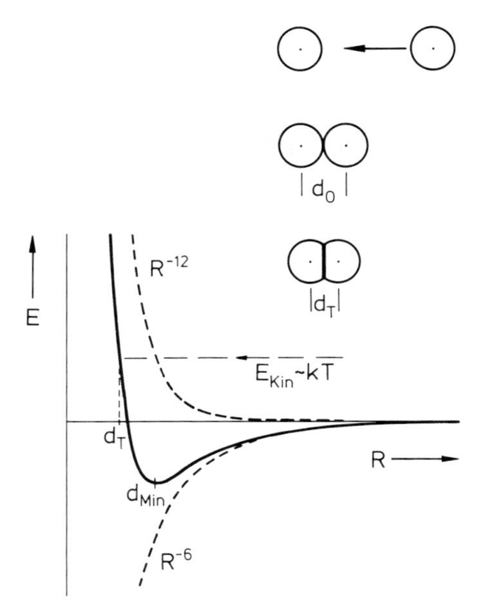

Fig. 2.7. Defining the "size" of a molecule: one can distinguish between d_0, the distance at which two colliding molecules detectably touch; d_T, the closest distance of approach which is attained in collisions at a kinetic energy kT; and d_{Min}, the distance which corresponds to a minimum in the interaction potential. Here, we mean the interaction potential between two neutral molecules, not to be confused with the intramolecular potential. The interaction typically follows an R^{-6} law for the attractive part and R^{-12} for the repulsive part

molecule, we find the numerical values (in Å) 2.47 from the viscosity, 2.81 from the Van der Waals equation, and $R_e = 0.74$ Å from spectroscopic data for the equilibrium distance of the centres of gravity of the two H nuclei in H_2.

2.2 The Shapes of Molecules

Molecules are spherical only in rare cases. In order to investigate their spatial structures, one has to determine both the arrangement of their nuclear frameworks and also the distributions and extensions of their electronic shells. This is illustrated further in Fig. 2.8 by two simple examples.

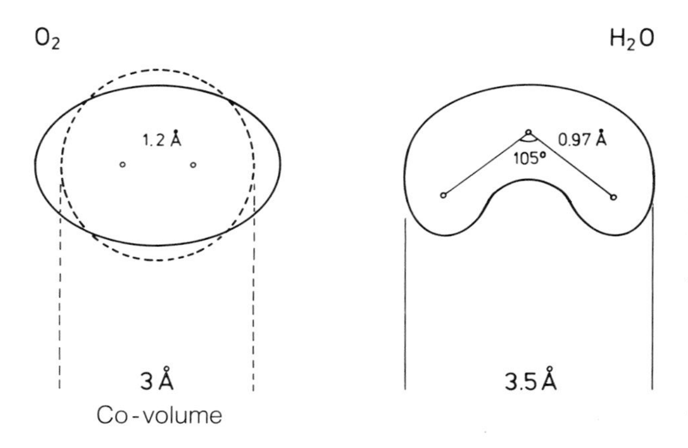

Fig. 2.8. As a rule, molecular contours deviate from a spherical form. For example, here we show the molecules O_2 and H_2O. In addition to the bond lengths and the bond angles, the spatial extension of the molecular electronic shells is an important measurable quantity

The nuclear framework, i.e. the bond lengths of the atomic nuclei which make up the molecule and their relative orientations to one another, can be determined very precisely. Aside from X-ray, electron and neutron diffraction, spectroscopic methods such as infrared absorption spectroscopy and nuclear magnetic resonance (NMR) are required for this determination; they will be treated in detail later in this book.

We list some small molecules here as examples:

diatomic, homonuclear	H_2	H−H	bond length 0.74 Å
	I_2	I−I	bond length 2.66 Å
	O_2	O−O	bond length 1.20 Å
diatomic, heteronuclear	HCl	H−Cl	bond length 1.28 Å
triatomic, symmetric–linear	CO_2	O−C−O	bond length 1.15 Å
triatomic, bent	H_2O	O H H	bond length 0.97 Å ∡105°
tetratomic, symmetric pyramidal	NH_3	N H H H	NH bond length 1.01 Å
pentatomic, tetrahedral	CH_4	H C H H H	CH bond length 1.09 Å
polyatomic hydrocarbons, paraffines	C_2H_6 ethane	H H H−C−C−H H H	C−C bond length 1.55 Å C−H bond length 1.09 Å ∡(H−C) = 109.5° ∡(HCH) = 111.5°
aromatics	C_6H_6 benzene	H H C C H C C H C C H H	C−H bond length 1.08 Å C−C bond length 1.39 Å
biological macromolecules	DNA	double helix 10^5-10^6 atoms	200 Å long

The precision with which these data can be derived from an analysis of electron and X-ray diffraction on ordered structures is very great. Internuclear distances can be quoted with certainty to a precision of ± 0.01 Å and angles to $\pm 1°$. As we mentioned above, the boundaries of the electronic shells of molecules are not precisely defined, since the electron density falls off continuously with increasing distance from the nuclei; however, surfaces of constant electron density (contour surfaces) can be defined, and thus regions of minimal electron density can be located, upon which precise bond length determinations can be based. If surfaces of constant spatial electron density are cut by a plane, their intersections with the plane (of the drawing) yield electron density contour lines. Adding the structure of the nuclear framework, when it is known, produces pictures of molecules like that shown in Fig. 2.9.

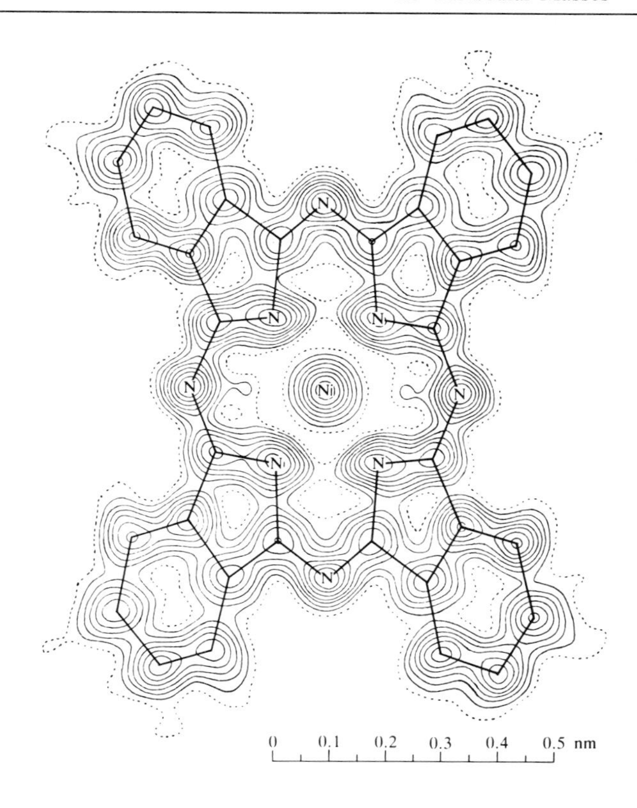

Fig. 2.9. An electron density diagram of the nickel phthalocyanine molecule. As in Fig. 2.2, the H atoms are not visible, since they are poorly detected by X-ray diffraction methods compared to atoms with higher electron densities. The contour lines represent the electron density. Their interval corresponds to a density difference of one electron per Å^2, and the dashed lines represent an absolute density of 1 electron per Å^2. The lines around the central Ni atom have a contour interval corresponding to 5 electrons per Å^2. After Robertson

2.3 Molecular Masses

The *mass* of a molecule, like that of an atom, can be most readily determined by weighing it. One mole of a substance, i.e. 22.4 l of gas under standard conditions of temperature and pressure, contains $N_A = 6.022 \cdot 10^{23}$ molecules. From the mass of a mole, one can therefore determine the mass of a molecule by dividing by the number of molecules, i.e. Avogadro's number.

A particularly important method of determining molecular masses is *mass spectroscopy*, making use of the deflection of beams of charged molecules by electric and magnetic fields. The basic principles of this method are described in I, Sect. 3.2. In atomic physics, mass spectroscopy is used for the precise determination of atomic masses and for the investigation of isotopic mixtures; in molecular physics, it can be used in addition for analysis and for the determination of molecular structures. Electron bombardment can be employed to decompose many molecules into fragments. By investigating the nature of the fragments using mass spectroscopy, one can obtain information on the structure of the original molecule by attempting to reconstruct it from the fragments, like a puzzle. An example is shown in Fig. 2.10.

Other methods are especially important in the case of biological macromolecules. For example, from the radial distribution of molecules in an ultracentrifuge, one can determine their masses. When the size of the molecules becomes comparable to the wavelength of scattered light, the angular distribution of the light intensity gives information on the shape and size of the scatterers and thus indirectly on their masses. Light scattering is caused by

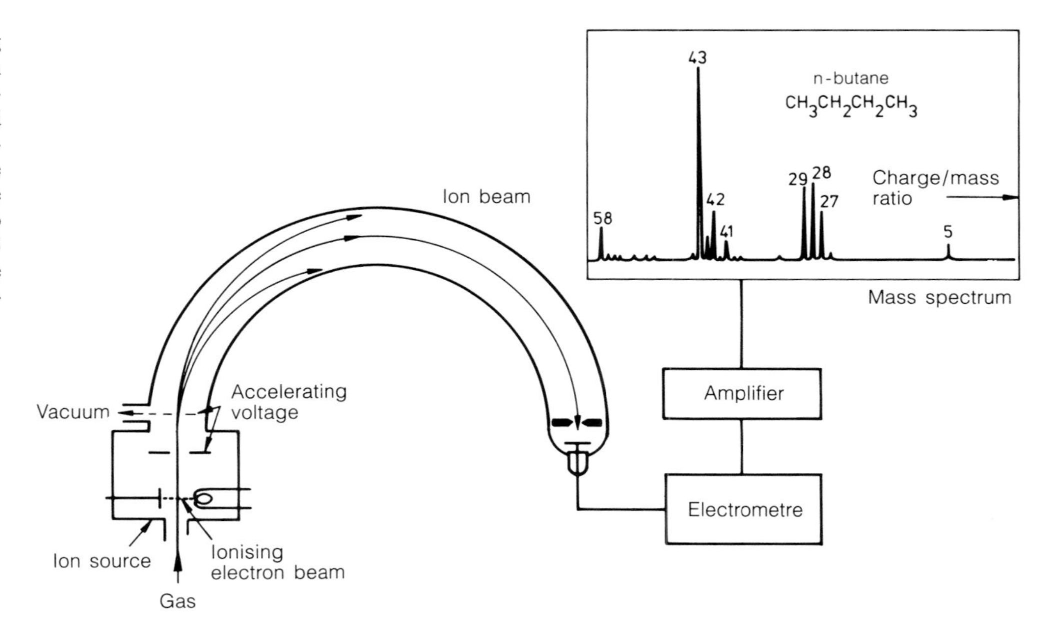

Fig. 2.10. A schematic drawing of a mass spectrometer, which functions by means of electromagnetic deflection of ionised molecular fragments. As an example, the mass spectrum of the butane molecule is shown in the inset. Maxima corresponding to fragments of masses between 5 and 58 can be recognised; we will not discuss their detailed interpretation here. After *Barrow*

different parts of the molecule and these different scattered rays can interfere, giving rise to an angular distribution of the scattered radiation which no longer corresponds to simple Rayleigh scattering. The principle is illustrated in Fig. 2.11.

Using the methods which are referred to as small-angle X-ray and neutron scattering (SAXS, SANS), a measurement or at least an estimate of the spatial extent of larger molecules is often possible.

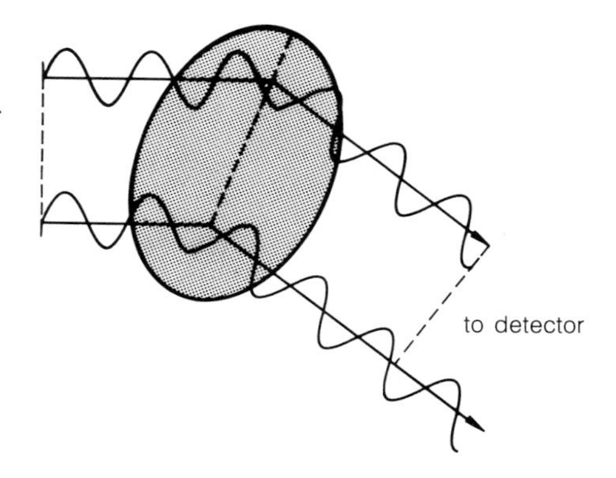

Fig. 2.11. Light which is scattered from different parts of larger molecules can interfere, leading to an intensity distribution of the scattered light which differs from that for Rayleigh scattering. From it, information on the size and shape of the scattering molecules can be obtained. This method is not very specific, but is experimentally relatively simple

In the case of macromolecules, the methods mentioned above can fail for several reasons, in particular when it is desired to investigate the shape, size, and mass of the molecules in their natural environment, i.e. frequently in the liquid phase. On the one hand, the size and shape of the molecules can change with changing surroundings; on the other, the methods are based to some extent on isolating the molecules from their environment. In these cases, other methods can be applied, such as osmosis through membranes, the equilibrium or velocity distribution of sedimentation in the gravitational field of the Earth or in the centrifugal field of an ultracentrifuge, the transport of molecules under the influence of an electric field in paper or in a gel, called electrophoresis, or filtration through micropores. These methods, which are applied also to biologically active molecules, will not be discussed in detail here.

2.4 Momentum, Specific Heat, Kinetic Energy

The *momentum* and *kinetic energy* of molecules were derived in the 19th century by applying the atomic hypothesis to thermodynamic properties of gases.

The mean kinetic energy of molecules in a gas is given by the expression

$$\overline{E}_{\text{kin}} = \frac{m}{2}\overline{v^2}, \tag{2.6}$$

where $\overline{v^2}$ is again the mean squared velocity of the molecules in the gas, and m is their mass.

For the pressure p we have, from elementary thermodynamics,

$$p = \frac{2}{3}N\overline{E}_{\text{kin}} \tag{2.7}$$

(N = particles/unit volume).

Because of the equation of state of an ideal gas,

$$pV = nRT, \tag{2.8}$$

where n is the number of moles of the gas, V its volume, R the gas constant, and T the absolute temperature, it then follows for the individual molecules that

$$\overline{E}_{\text{kin}} = \frac{3}{2}kT, \tag{2.9}$$

with $k = R/N_{\text{A}}$ = Boltzmann constant; and since the number of degrees of freedom of translational motion is 3, for the energy per degree of freedom f we find:

$$\overline{E}_{\text{kin}\,f} = \frac{1}{2}kT. \tag{2.10}$$

For the total energy of a mole of particles, we then have

$$\overline{E}_{\text{mole}} = \frac{3}{2}RT \tag{2.11}$$

and for the specific heat at constant volume

$$C_V = \frac{dE}{dT} = \frac{3}{2}R \tag{2.12}$$

and at constant pressure

$$C_p = C_V + R = \frac{5}{2}R.$$

With monoatomic gases, these values are in fact found in measurements; for molecular gases, higher values are measured. This is due to the fact that molecules, in contrast to atoms, have additional degrees of freedom, which are associated with rotational and vibrational motions, and that these motions also contribute to the specific heat. The rotational degrees of freedom each contribute $\frac{1}{2}kT$ to C_V. In general, a molecule has three rotational degrees of freedom corresponding to rotations around the three principal body axes, i.e. the axes of the ellipsoid of the moment of inertia. In the case of a linear molecule, all the mass points lie on a line, and the moment of inertia around the corresponding axis vanishes; there are then only two rotational degrees of freedom. We thus find for the specific heat of di- or triatomic molecules, respectively, initially ignoring quantum effects, the following formulas:

$$C_V = \frac{5}{2}R \quad \text{or} \quad 3R; \quad C_p = \frac{7}{2}R \quad \text{or} \quad 4R. \tag{2.13}$$

In addition, internal vibrations can be excited in a molecule. They contribute one degree of freedom for a diatomic molecule, three for a triatomic molecule, and $3n - 6$ for a molecule with n atoms. The number of these degrees of freedom and thus of the normal modes (cf. Chap. 10) can be calculated in the following way: each atom contributes three degrees of freedom of motion; for n atoms, there are $3n$ degrees of freedom. Three of these correspond to the translational motion of the centre of gravity of the whole molecule and three to rotations. A molecule containing n atoms thus has $3n - 6$ vibrational degrees of freedom. This formula is valid for $n \geq 3$. In a diatomic molecule, owing to its two rotational degrees of freedom, there is exactly one vibrational degree of freedom.

The mean thermal energy per degree of freedom is twice as large as for translation and rotation, since in the case of vibrations, both kinetic and potential energy must be taken into account. The specific heats of polyatomic molecules are correspondingly larger at temperatures at which the vibrations can be thermally excited.

In all these considerations, it must be remembered that the vibrational and rotational states in molecules are quantised. The energy quanta have different magnitudes, depending on the molecular structure, and are generally smaller for rotations than for vibrations. They can be thermally excited only when the thermal energy kT is sufficiently large in comparison to the quantum energy $h\nu$. One thus observes a temperature-dependent specific heat C_V or C_p for molecules, as indicated schematically in Fig. 2.12. At very low temperatures, only the translational degrees of freedom contribute to the specific heat and one measures the value $C_V = \frac{3}{2}R$. With increasing temperature, the additional degrees of freedom of the rotation are excited, and at still higher temperatures, those of molecular vibrations also contribute to the measured C_V. One can thus draw conclusions about the number and state of motion of the atoms in a molecule even from measurements of its specific heat.

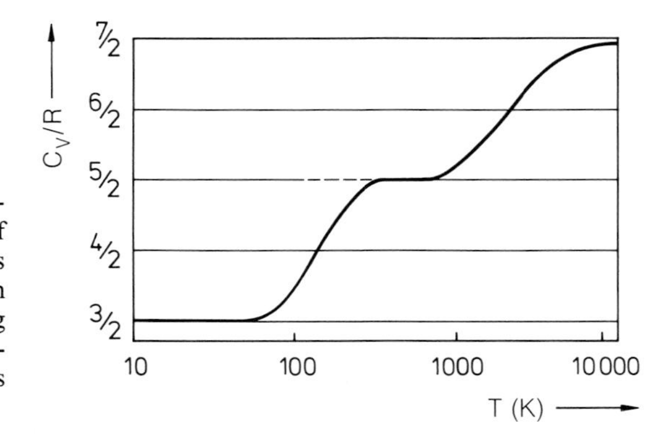

Fig. 2.12. The temperature dependence of the specific heat of a gas. The curve corresponds approximately to the hydrogen molecule, H_2. With decreasing temperature, the degrees of freedom of vibrations and rotations are "frozen in" in two steps

3. Molecules in Electric and Magnetic Fields

Macroscopic materials properties such as the dielectric constant ε and the permeability μ are determined by the electric and magnetic characteristics of the basic building blocks of matter. We show in Sects. 3.1 through 3.4 how the electric properties of molecules can be investigated by measuring ε and the index of refraction, n. Sections 3.5 through 3.8 give the corresponding information about magnetic moments and polarisabilities from determinations of the magnetic susceptibility.

3.1 Dielectric Properties

Molecules are in general electrically neutral. However, they can possess an *electric dipole moment* $\boldsymbol{p}$ (and other, higher moments such as a quadrupole moment), and their electrical *polarisability* is generally anisotropic. In this section, we will show how information about the electrical characteristics of molecules can be obtained from measurements of *macroscopic materials properties*, particularly in the presence of electric fields. The most readily accessible quantity here is the dielectric constant ε; it is most simply determined by measuring the capacitance of a condensor with and without a dielectric consisting of the material under study. The ratio of the two measured values is the dielectric constant. The present section concerns itself with the definition of the dielectric constant and with its explanation on a molecular basis.

For the quantitative description of electric fields, we require two concepts from electromagnetic theory:

- The electric *field strength* $\boldsymbol{E}$. It is derived from the force which acts on a test charge in an electric field.
- The electric *displacement* $\boldsymbol{D}$. It is defined by the surface influence charge density produced on a sample in a field.

In a medium with the dielectric constant ε, the displacement $\boldsymbol{D}_m$ is given by

$$\boldsymbol{D}_m = \varepsilon\varepsilon_0 \boldsymbol{E} \tag{3.1}$$

with

$$\varepsilon_0 = 8.85 \cdot 10^{-12} \frac{\mathrm{As}}{\mathrm{Vm}}.$$

(The notation ε_r is also used, where r stands for 'relative', instead of the quantity denoted above by ε; the product $\varepsilon_\mathrm{r}\varepsilon_0$ is then called ε.)

The dimensionless dielectric constant ε is a scalar quantity in isotropic materials and a tensor in anisotropic materials. ε is always larger than 1 in matter. In dielectric materials, the numerical value of ε is only slightly greater than 1 and is nearly independent of the temperature. In paraelectric materials, ε can be much greater than 1 and decreases with increasing temperature. As we shall see in the following, paraelectric materials consist of molecules which have permanent electric dipole moments. In dielectric materials, the dipole moment is *induced* by an applied electric field.

Some values of ε for dielectric and paraelectric materials are given in Table 3.1.

Table 3.1. Numerical values of ε (under standard conditions). The materials in the left column are dielectric, the others are paraelectric

He	1.00007	H_2O	78.54	LiF	9.27
H_2	1.00027	Ethanol	24.30	AgBr	31.1
N_2	1.00058	Benzene	2.27	NH_4Cl	6.96

In addition, the electric *polarisation* $\boldsymbol{P}$ is also defined by means of the equation

$$\boldsymbol{P} = \boldsymbol{D}_m - \boldsymbol{D} \quad \text{or} \quad \boldsymbol{D}_m = \varepsilon_0 \boldsymbol{E} + \boldsymbol{P}\,, \tag{3.2}$$

where $\boldsymbol{D}_m$ is the displacement *in the material* and $\boldsymbol{D}$ that in vacuum.

$\boldsymbol{P}$ measures the *contribution of the material* to the electric displacement and has the dimensions and the intuitive meaning of an electric dipole moment per unit volume.

From (3.1) and (3.2), it follows that

$$\boldsymbol{P} = (\varepsilon - 1)\varepsilon_0 \boldsymbol{E} = \chi \varepsilon_0 \boldsymbol{E}\,. \tag{3.3}$$

The quantity $\varepsilon - 1$ is also referred to as the dielectric *susceptibility* χ.

The polarisation can be explained on a *molecular basis*. It is the sum of the dipole moments $\boldsymbol{p}$ of the N molecules in the volume V. We thus have

$$\boldsymbol{P} = \frac{1}{V}\sum_{i=1}^{N} \boldsymbol{p}_i = \boldsymbol{p}\,'N\,, \tag{3.4}$$

where $\boldsymbol{p}'$ denotes the contribution which, averaged over space, each molecular dipole moment makes to the polarisation $\boldsymbol{P}$. In the case of complete alignment of all the dipole moments parallel to the field, we find $\boldsymbol{P} = N\boldsymbol{p}$. In (3.4), it should be considered that the number density N (number of molecules per unit volume) depends on Avogadro's number N_A (number of molecules in a mole of substance) through the equation $N = N_A(\varrho/M)$, where ϱ is the density and M the molar mass of the substance. Thus, according to (3.3) and (3.4), a relation exists between the *macroscopic quantity measured*, ε, and the molecular property *dipole moment* $\boldsymbol{p}$.

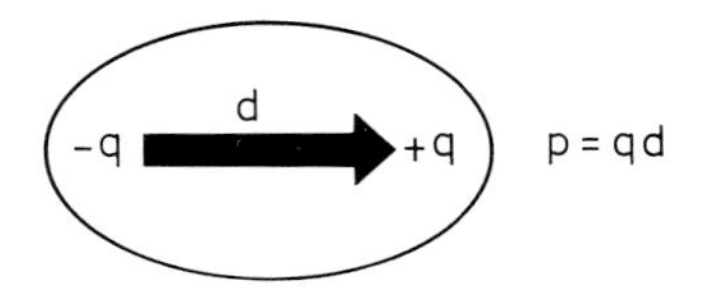

Fig. 3.1. The electric dipole moment of two charges $+q$ and $-q$ at a distance d is equal to $\boldsymbol{p} = q\boldsymbol{d}$; its direction points from the negative towards the positive charge

We refer to an *electric dipole moment* of a molecule when the centres of charge of its positive and negative charges do not coincide. For example, two point charges $+q$ and $-q$ at a distance d (Fig. 3.1) have the dipole moment

$$\boldsymbol{p} = q\boldsymbol{d} \quad [\mathrm{As\,m}]\,. \tag{3.5}$$

The *vector* of the dipole moment points from the negative to the positive charge. In addition to the unit [As m], the unit *Debye* (D) is also used, with $1\,\mathrm{D} = 3.336 \cdot 10^{-30}$ As m. Two elementary charges at a distance of 1 Å = 10^{-10} m have a dipole moment of $1.6 \cdot 10^{-29}$ As m = 4.8 D. This is the order of magnitude of molecular dipole moments. Molecules with finite dipole moments are called *polar*. Polar molecules such as HCl or NaCl have a *permanent dipole moment* $\boldsymbol{p}$, which in the case of predominantly ionic bonding can even be calculated quite accurately as the product of charge times bond length. The dipole moment of HCl is 1.08 D and that of H_2O is 1.85 D. We shall discuss polar molecules in more detail in Sect. 3.3; however, we first treat nonpolar molecules in the following Sect. 3.2.

3.2 Nonpolar Molecules

Symmetric molecules such as H_2, O_2, N_2, or CCl_4 are nonpolar, i.e. they have no permanent dipole moments which remain even when $\boldsymbol{E} = 0$. They can, however, have an *induced dipole moment* in a field $\boldsymbol{E} \neq 0$. For this dipole moment $\boldsymbol{p}_{\text{ind}}$, induced by polarisation in the external field, we have:

$$\boldsymbol{p}_{\text{ind}} = \alpha \boldsymbol{E}_{\text{loc}} , \qquad \begin{aligned} \alpha &= \text{polarisability} , \\ \boldsymbol{E}_{\text{loc}} &= \text{field strength at the molecule} . \end{aligned} \tag{3.6}$$

The *polarisability* α is a measure of the ease of displacement of the positive charge relative to the negative charge in the molecule, and is thus an important molecular property. The resulting polarisation is called the *displacement polarisation*. It is useful to distinguish two cases:

- When the induced dipole moment results from a displacement of the electronic clouds relative to the positive nuclear charges, we speak of an *electronic polarisation*;
- When, in contrast, a displacement of massive positive ions relative to massive negative ions occurs, we speak of *ionic polarisation*.

The polarisability α is thus the sum of an electronic and an ionic contribution, $\alpha = \alpha_{\text{el}} + \alpha_{\text{ion}}$.

When we refer to the polarisability α, then strictly speaking we mean the polarisability averaged over all directions in the molecule, $\overline{\alpha}$. In reality, for all molecules excepting those having spherical symmetry, α depends on the direction of the effective field $\boldsymbol{E}$ relative to the molecular axes; α is thus a tensor. If the anisotropy of the polarisability is known, it can be used to draw conclusions about the structure of the molecule. The anisotropy can be measured using polarised light, by aligning the molecules and measuring the dielectric constant ε in the direction of the molecular axis. This type of alignment can be produced, for example, by an applied electric field. The double refraction exhibited by some gases and liquids in an applied electric field, referred to as the electro-optical Kerr effect, is based on this kind of alignment. Another possibility for aligning the molecules and measuring their polarisabilities in different molecular directions is to insert them into a crystal lattice. A typical result, e.g. for the CO molecule, is $\alpha_{\parallel}/\varepsilon_0 = 5.3 \cdot 10^{-24}\,\text{cm}^3$ along the molecular axis and $\alpha_{\perp}/\varepsilon_0 = 1.8 \cdot 10^{-24}\,\text{cm}^3$ perpendicular to the axis.

In *strong electric fields*, as found for example in laser beams, there are in addition to the linear term in (3.6) also *nonlinear terms* which must be considered; they are proportional to the second, third, or higher powers of $\boldsymbol{E}_{\text{loc}}$. In practice, the most important term is the quadratic one, proportional to $\boldsymbol{E}_{\text{loc}}^2$. The coefficient β in the term $\beta \boldsymbol{E}_{\text{loc}}^2$ is called the *hyperpolarisability*.

The *dimensions* of α are, from (3.6), [As m^2 V^{-1}]. The dimensions of α', defined as the quotient α/ε_0, are simpler: they are those of a volume. For molecules having axial symmetry, it is sufficient to determine two values of the polarisability, perpendicular and parallel to the molecular axis. The (electronic) polarisability is an indication of how strongly the electron distribution in the molecule is deformed by an applied electric field. When the molecule contains heavy atoms in which some of the electrons are farther apart from their nuclei, then the electron distribution is less rigidly connected to the nuclei and the electronic polarisability is correspondingly large.

Some numerical values for the polarisabilities of simple molecules are given in Table 3.2.

Table 3.2. Polarisabilities α/ε_0, in 10^{-10}m^3

	$\overline{\alpha}/\varepsilon_0$	$\alpha_\perp/\varepsilon_0$	$\alpha_\parallel/\varepsilon_0$
H_2	0.79	0.61	0.85
O_2	1.60		
Cl_2		3.2	6.6
C_6H_6	10.3	6.7	12.8
H_2O	1.44		
CCl_4	10.5		

In *gases* at moderate pressures, the molecules do not mutually influence each other. The total polarisation $\boldsymbol{P}$ of the observed volume can thus be calculated using (3.4), as the sum of the polarisations of all the molecules within the volume. We then find the following expression for the polarisation resulting from the induced moments of molecules having a number density N, assuming complete alignment of their moments by the applied field:

$$\boldsymbol{P} = N\boldsymbol{p}_{\text{ind}} = N\alpha\boldsymbol{E}_{\text{loc}} \, . \tag{3.7}$$

With

$$N = \frac{N_A \varrho}{M} \tag{3.7a}$$

(ϱ = density, M = molecular mass),

we obtain for the displacement polarisation

$$\boldsymbol{P} = \frac{N_A \varrho}{M}\alpha\boldsymbol{E} \, . \tag{3.8}$$

In the case of dilute gases, the local field $\boldsymbol{E}_{\text{loc}}$ at the position of each molecule is naturally equal to the applied field $\boldsymbol{E}$.

From (3.8) and (3.3), it follows that

$$\varepsilon = 1 + \frac{N_A \varrho}{M\varepsilon_0}\alpha \, . \tag{3.9}$$

We thus obtain the *polarisability* α of the molecules by measuring the dielectric constant ε.

In a dielectric of *greater density*, one has to take into account the fact that the local field $\boldsymbol{E}_{\text{loc}}$ is not equal to the applied field $\boldsymbol{E}$. In the neighbourhood of a molecule under consideration, there are other molecules whose charge distributions give a contribution to the local field. This must be allowed for in computations; cf. Fig. 3.2. For the local field, we have:

$$\boldsymbol{E}_{\text{loc}} = \boldsymbol{E} + N\frac{\boldsymbol{P}}{\varepsilon_0} , \tag{3.10}$$

where here, N refers not to the particle number density, but rather to the depolarizing factor.

The *depolarizing factor* defined above depends on the shape of the sample and can be calculated for a given shape.

Following Lorentz, the field inside a spherical cavity in a dielectric can be calculated using a depolarizing factor $N = 1/3$, and thus

$$\boldsymbol{E}_{\text{loc}} = \boldsymbol{E} + \frac{1}{3}\frac{\boldsymbol{P}}{\varepsilon_0} . \tag{3.11}$$

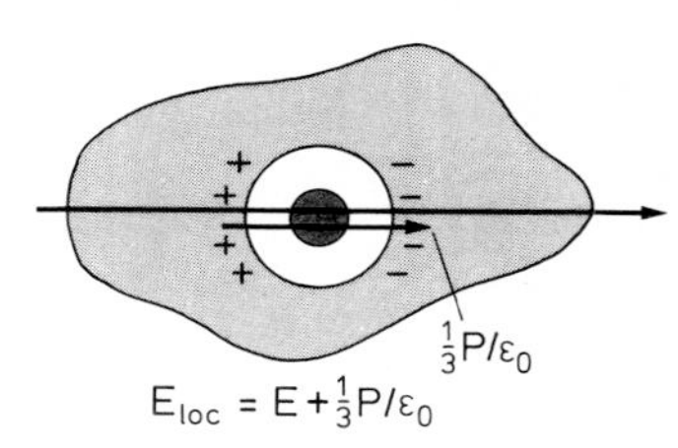

Fig. 3.2. The definition of the local field $\boldsymbol{E}_{\text{loc}}$: in a dielectric medium, the applied field $\boldsymbol{E}$ is augmented by the field resulting from the induced surface charges. This Lorentz field, assuming a spherical cavity, is equal to $P/3\varepsilon_0$

Then, from (3.7), we have

$$\boldsymbol{p}_{\text{ind}} = \frac{\boldsymbol{P}}{N} = \frac{\boldsymbol{P}M}{N_{\text{A}}\varrho} = \alpha\left(\boldsymbol{E} + \frac{1}{3}\frac{\boldsymbol{P}}{\varepsilon_0}\right) . \tag{3.12}$$

Using (3.3), we eliminate $\boldsymbol{E}$ from (3.12) to yield

$$\frac{\boldsymbol{P}M}{N_{\text{A}}\varrho} = \alpha\left(\frac{\boldsymbol{P}}{\varepsilon_0(\varepsilon-1)} + \frac{1}{3}\frac{\boldsymbol{P}}{\varepsilon_0}\right) = \frac{\boldsymbol{P}(\varepsilon+2)}{3\varepsilon_0(\varepsilon-1)}\alpha$$

and we obtain:

$$\frac{\varepsilon-1}{\varepsilon+2}\frac{M}{\varrho} = \frac{1}{3}\frac{N_{\text{A}}}{\varepsilon_0}\alpha \equiv \boldsymbol{P}_{\text{Mol}} . \tag{3.13}$$

This is the *Clausius-Mosotti equation*. It defines the molar polarisation $\boldsymbol{P}_{\text{Mol}}$ and connects the *macroscopic* measurable quantities ε, M, and ϱ with the *molecular* quantity α.

So far, we have considered only the polarisation in a static $\boldsymbol{E}$ field. We now make some remarks concerning the behaviour of molecules in an alternating field; in particular, the electric field in a light beam is relevant. The applied electric field oscillates at the frequency ν and would reverse the polarisation of matter in the field at that frequency. In general, this succeeds for the displacement polarisation up to frequencies corresponding to the infra-red range, and the contribution of the polarisability to the polarisation remains constant. At higher frequencies, it is necessary to distinguish between the electronic and the ionic polarisations. For the latter, the time required to reverse the polarisation is typically about equal to the period of a molecular vibration. The ionic contribution to the displacement polarisation therefore vanishes when the frequency of the light increases from the infra-red to the visible range, i.e. when it becomes greater than the important molecular vibration frequencies. The nuclei in the molecules, and their charge distributions, have too much inertia to follow the rapidly reversing field of the polarizing light beam at higher frequencies. At the frequencies of visible light, only the less massive electrons can follow the alternating field, leading to reversal of the polarisation; thus only the electronic part of the displacement polarisation contributes at these frequencies.

From the Maxwell relation $\varepsilon\mu = n^2$ (μ = permeability constant, n = index of refraction), considering that for many molecules, $\mu \approx 1$ and therefore $n = \sqrt{\varepsilon}$, it follows from (3.13) that the *Lorentz-Lorenz equation*:

$$\frac{n^2-1}{n^2+2}\frac{M}{\varrho}=\frac{1}{3\varepsilon_0}N_A\beta \equiv R_{\text{Mol}} \tag{3.14}$$

holds. R_M is the *molar refraction*. The optical polarisability β (not to be confused with the hyperpolarisability introduced above!) is the polarisability at the frequencies of visible or ultraviolet light. It differs, as explained above, from the *static polarisability* α, and depends on the frequency of the light. This frequency dependence is called *dispersion*. An example: the index of refraction, n, of water at 20° C has the value 1.340 for $\lambda = 434$ nm and $n = 1.331$ for $\lambda = 656$ nm.

3.3 Polar Molecules

The *displacement polarisation* which we have discussed so far, and the values of ε and $\boldsymbol{P}$ to which it gives rise, are only slightly or not at all dependent on the temperature. In contrast, there are many materials in which ε and $\boldsymbol{P}$ decrease strongly with increasing temperature. The explanation depends on the concept of *orientation polarisation*, which is to be distinguished from displacement polarisation. While the latter as discussed above is induced by an applied electric field, an orientation polarisation occurs in materials whose molecules have permanent electric dipole moments, $\boldsymbol{p}_p$ (*Debye*, 1912). Such molecules are termed polar, and materials containing them are called paraelectric. The orientation polarisation is based on the alignment of permanent dipoles by an applied electric field. It should be mentioned that the permanent dipole moments are usually much larger than induced moments; some *numerical values* are given in Table 3.3.

For *comparison*, we can calculate the induced dipole moment $\boldsymbol{p}_{\text{ind}}$ in a field $\boldsymbol{E} = 10^5$ V/cm, using a polarisability $\alpha' = 10^{-24}\text{cm}^3$, typical of nonpolar molecules, i.e. $\alpha\varepsilon_0 = 10^{-40}$ As m²/V; we find $\boldsymbol{p}_{\text{ind}} = \alpha\varepsilon_0\boldsymbol{E} = 10^{-33}$ As m, i.e. 3 orders of magnitude smaller than the typical permanent dipole moments as shown in Table 3.3.

Table 3.3. Permanent dipole moments p_p in 10^{-30} As m (1 D = $3.3356 \cdot 10^{-30}$ As m)

Molecule	p_p	Molecule	p_p
HF	6.0	H_2	0
HCl	3.44	H_2O	6.17
HBr	2.64	CH_3OH	5.71
CO	0.4	KF	24.4
CO_2	0	KCl	34.7
NH_3	4.97	KBr	35.1
C_6H_6	0		

A glance at Table 3.3 shows that the measurement of molecular *permanent dipole moments* can allow the determination of important *structural data*: while for the CO_2 molecule, one observes a zero dipole moment and thus concludes that the molecule is linear, O–C–O, the nonvanishing dipole moment of the water molecule indicates a bent structure for H_2O.

Thus, the displacement polarisation [see (3.7)]

$$\boldsymbol{P}_{\text{ind}} = \frac{\sum \boldsymbol{p}_{\text{ind}}}{V} \tag{3.15}$$

is independent of or only slightly dependent on the temperature, and at least in part follows an applied ac field up to very high frequencies (those of UV light!) owing to the small inertia of the displaced electrons; it thus makes a contribution to the index of refraction n.

In contrast, the *orientation polarisation*

$$\boldsymbol{P}_{\mathrm{or}} = \frac{\sum \boldsymbol{p}_{\mathrm{p}}}{V} = N\boldsymbol{p}_{\mathrm{p}}' \tag{3.16}$$

is dependent on the temperature (and on the frequency of the applied field). The alignment of the permanent dipoles $\boldsymbol{p}_{\mathrm{p}}$ in an electric field $\boldsymbol{E}$ is the result of a competition between the orientation energy $W_{\mathrm{or}} = -\boldsymbol{p}_{\mathrm{p}} \cdot \boldsymbol{E}$, which tends to produce a complete alignment of the dipoles parallel to the field, and the thermal energy $W_{\mathrm{th}} \approx kT$, which tends to randomise the directions of the dipoles in the applied field. As a result of this competition, each dipole contributes only $\boldsymbol{p}' < \boldsymbol{p}$ to the total polarisation, averaged over time.

Due to this competition, an equilbrium is reached which nearly corresponds to a Boltzmann distribution. The calculation (*Langevin*, 1900) gives the following result for the mean value of $\cos\theta$ at higher temperatures, $kT \gg \boldsymbol{p}_{\mathrm{p}} \cdot \boldsymbol{E} = pE\cos\theta$, where θ is the angle between the directions of $\boldsymbol{p}$ and $\boldsymbol{E}$ and the interaction between the dipoles themselves can be neglected:

$$\overline{\cos\theta} = \frac{p_{\mathrm{p}}E}{3kT} \quad \text{and} \quad \boldsymbol{P}_{\mathrm{or}} = N\frac{p_{\mathrm{p}}^2 E}{3kT} \ . \tag{3.17}$$

This is known as *Curie's law*; it was first derived in this form for temperature dependent paramagnetism. It states that the orientation polarisation is proportional to the reciprocal of the absolute temperature.

(A less approximate calculation for the mean value of $\boldsymbol{p}$ yields the equation

$$\boldsymbol{p}' = p\,\overline{\cos\theta} = p\,L\left(\frac{pE}{kT}\right)$$

with the Langevin function

$$L(x) = \frac{\mathrm{e}^x + \mathrm{e}^{-x}}{\mathrm{e}^x - \mathrm{e}^{-x}} - \frac{1}{x} = \coth x - \frac{1}{x} \ .$$

At room temperature, $kT \approx 5 \cdot 10^{-21}$ Ws and the orientation energy W_{or} of the dipoles in a field of $E = 10^5$ V/cm is about 10^{-24} Ws. The condition $pE/kT \ll 1$ is thus fulfilled and the function L can be expanded in a series which is terminated after the first term. This yields $p' = p^2E/3kT$, the so-called high-temperature approximation.)

Now that we know the contribution of permanent dipoles to the polarisation, we should like to calculate the dielectric constant of a dilute system (with $\varepsilon - 1 \ll 1$), by adding together the displacement polarisation and the orientation polarisation to give an overall polarisation. We refer to (3.3), (3.9) and (3.17) and obtain

$$\varepsilon = 1 + N\left(\alpha + \frac{p_{\mathrm{p}}^2}{3\varepsilon_0 kT}\right) = 1 + \chi \ . \tag{3.18}$$

When the interaction of the dipoles among themselves can no longer be neglected, i.e. especially in condensed phases, then instead of the Clausius-Mosotti equation, the *Debye equation* holds:

$$\frac{\varepsilon - 1}{\varepsilon + 2}\frac{M}{\varrho} = \frac{1}{3\varepsilon_0}N_A\left(\alpha + \frac{p_p^2}{3kT}\right) \equiv P_{Mol} . \tag{3.19}$$

Experimentally, α and $\boldsymbol{p}_p$ are determined from a measurement of ε as a function of the temperature. When the *molar polarisation* P_{Mol} is plotted against $1/T$, from (3.18) one finds a straight line. Its slope yields p and its intercept gives α. For nonpolar molecules, the slope is zero. Figure 3.3 shows some experimental data. For gases, one finds $(\varepsilon - 1) = 1...10 \cdot 10^{-3}$; for liquid water at room temperature, ε=81.

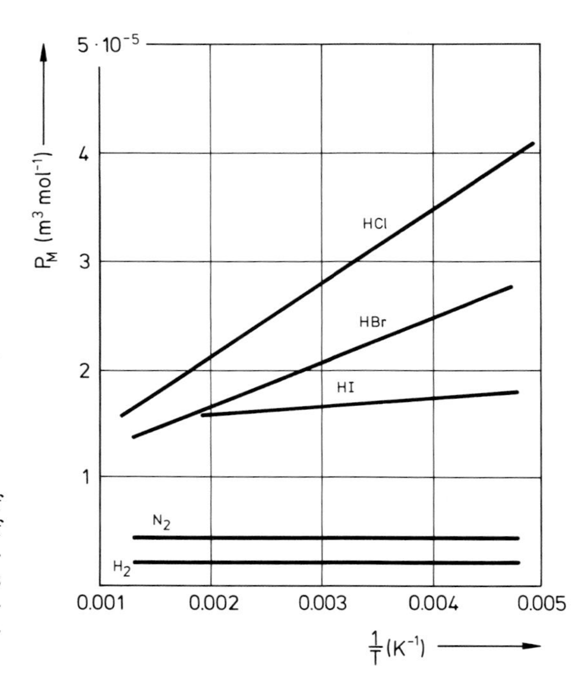

Fig. 3.3. Molar polarisation of some gases as a function of the temperature, for the determination of dipole moments and polarisabilities from measurements of the dielectric constant ε

The orientation polarisation is produced against the inertial mass of the whole molecule and thus, already at lower frequencies than for the displacement polarisation, it cannot follow a rapidly-changing ac field. This is because not only must the outer electrons move with the field relative to the atomic cores, or the atomic cores relative to one another within the molecule, but rather the whole molecule has to reorient at the frequency of the applied ac field. Assuming that a typical time for molecular rotation in a liquid is about 10^{-12} s, then the molecules can no longer follow the field reversals at frequencies above about 10^{11} s^{-1} (in the microwave range). The Debye equation (3.19) is then replaced by the Clausius-Mosotti equation (3.13).

3.4 Index of Refraction, Dispersion

In ac fields at high frequencies, for example in a light beam, one usually measures the index of refraction n instead of the dielectric constant ε. According to Maxwell, they are related by $n = \sqrt{\varepsilon\mu}$ (μ = permeability); for $\mu = 1$, we have $n = \sqrt{\varepsilon}$.

The frequency dependence of ε or of n reflects the different contributions to the polarisation, the displacement and the orientation. In the frequency range of visible light, as mentioned above, only the electronic displacement polarisation is present.

The frequency dependence of ε or of n in the optical range, called the dispersion, can be calculated to a good approximation using a simple model in which the molecules are treated as damped harmonic oscillators having an eigenfrequency ω_0, a mass m, and a damping constant γ. The displacement x of the oscillator from its zero point, multiplied by the elementary charge e, then represents the dipole moment of the molecule. The $\boldsymbol{E}$-field of the light oscillates with the circular frequency ω. We then obtain the oscillator equation:

$$m\ddot{x} + \gamma\dot{x} + m\omega_0^2 x = eE_0 e^{i\omega t} . \tag{3.20}$$

A stationary solution of this equation is

$$x(t) = X e^{i\omega t} \tag{3.21}$$

with

$$X = \frac{eE_0}{m(\omega_0^2 - \omega^2) + i\gamma\omega} . \tag{3.22}$$

This complex expression can be rewritten as the sum of a real and an imaginary part:

$$X = \left(\frac{em(\omega_0^2 - \omega^2)}{m^2(\omega_0^2 - \omega^2)^2 + \gamma^2\omega^2} - i\frac{e\gamma\omega}{m^2(\omega_0^2 - \omega^2)^2 + \gamma^2\omega^2} \right) E_0 \tag{3.23}$$

or

$$X = X' - iX'' . \tag{3.24}$$

A corresponding solution holds for the dipole moment $p = ex$ (charge e and separation x) and thus, according to (3.9) and (3.7a),

$$\varepsilon = 1 + \frac{N}{\varepsilon_0}\alpha = 1 + \frac{N}{\varepsilon_0}\frac{ex}{E_0} , \tag{3.25}$$

$$\varepsilon = \varepsilon' - i\varepsilon'' , \tag{3.26}$$

i.e. we obtain a complex dielectric constant, where ε' and ε'' are given by the real and the imaginary parts of the parenthesis in (3.23), respectively.

The real and imaginary parts of ε are related to one another by the so-called *Kramers-Kronig* relations. Losses (absorption, ε'') and refraction (dispersion, ε') thus are connected. There exists for example no loss-free material with a large dispersion. Since ε is complex, the Maxwell relation requires the index of refraction also to be complex; we obtain:

$$\tilde{n} \equiv \sqrt{\varepsilon' - i\varepsilon''} = n + ik .$$

The real quantities n and k are shown in Fig. 3.4.

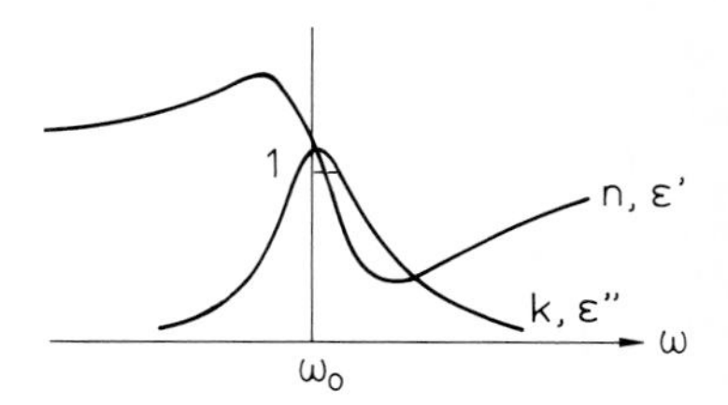

Fig. 3.4. Real and imaginary parts of the index of refraction due to the displacement polarisation in the neighbourhood of a resonance, for a damped oscillator

As can be seen, this model may be extended almost intact to give a quantum-mechanical description; then each molecule must be described by a whole set of oscillators. Which of these oscillators is active at a given excitation frequency depends on whether its frequency is near to the corresponding eigenfrequency.

The dielectric constant ε is the sum of contributions from the displacement polarisation and the orientation polarisation. The following relation holds:

$$\varepsilon = 1 + \chi_{el} + \chi_{ion} + \chi_{or} ,$$

where χ_{or} denotes the contribution of the orientation polarisation to the susceptibility, etc. This contribution is often denoted as χ_{dip} (for dipolar). The frequency dependence of this contribution is not described by a calculation similar to that given above for χ_{el} and χ_{ion}; instead, it must be treated as a relaxation process. χ_{or} decreases with increasing frequency, because a certain time is required for the reorientation of the molecular dipoles in the ac field: the relaxation time. In Fig. 3.5, the overall frequency dependence of ε is shown schematically. Measured values of the dielectric constant ε and the absorption coefficient k in the low frequency range are shown in Fig. 3.6 for a particular molecule, namely water, H_2O. In this frequency range, the orientation polarisation is dominant.

Starting with the static value $\varepsilon = 81$, which hardly changes up to a frequency of about 10^{10} Hz, we pass through frequencies where first the molecular vibrations and then the electronic clouds can no longer follow the excitation field, giving finally $n = 1.33$, corresponding to $\varepsilon = 1.76$, for visible light.

As we have seen, the quantity ε is complex and strongly dependent on the measurement frequency in certain ranges. For this reason, many different experimental methods, in addition to simple capacity measurements in a condensor, must be applied to determine it over a wide frequency range. Other methods use for example the index of refraction of electromagnetic waves, absorption and reflection in all spectral ranges, or the polarisation of scattered radiation.

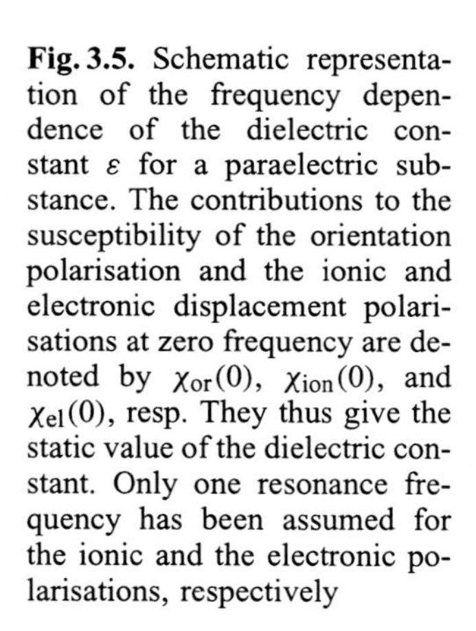

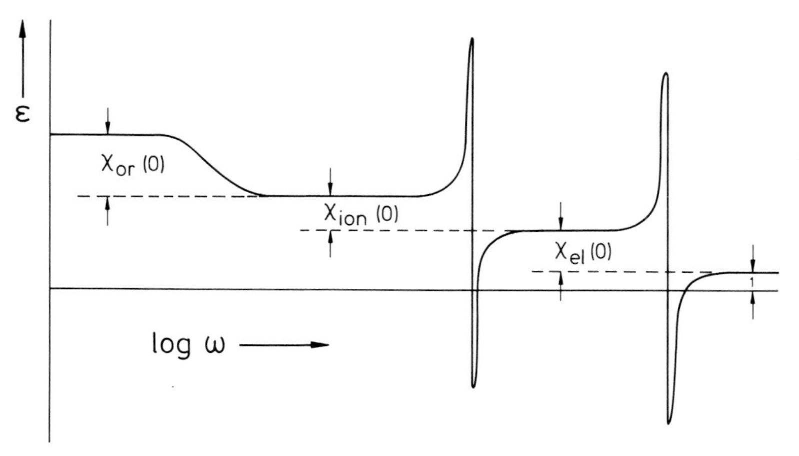

Fig. 3.5. Schematic representation of the frequency dependence of the dielectric constant ε for a paraelectric substance. The contributions to the susceptibility of the orientation polarisation and the ionic and electronic displacement polarisations at zero frequency are denoted by $\chi_{or}(0)$, $\chi_{ion}(0)$, and $\chi_{el}(0)$, resp. They thus give the static value of the dielectric constant. Only one resonance frequency has been assumed for the ionic and the electronic polarisations, respectively

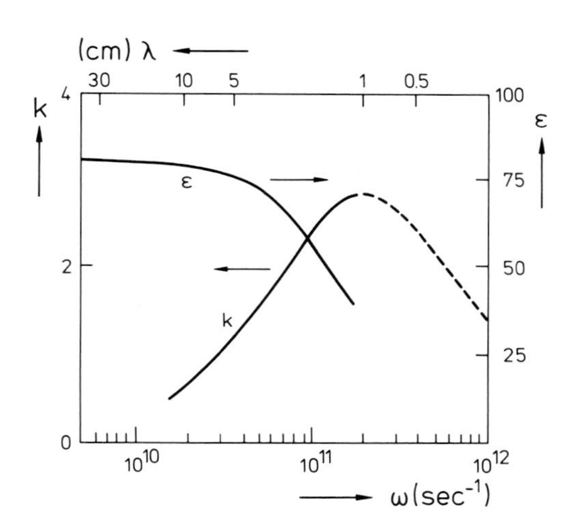

Fig. 3.6. Frequency dependence of the dielectric constant ε and the absorption coefficient k for water. In the relatively low frequency range shown here, the dispersion is dominated by the orientation polarisation. With increasing frequency, the water dipoles can no longer follow the oscillating field

3.5 The Anisotropy of the Polarisability

To complete the discussion of the behaviour of molecules in an electric field, we should mention the fact that up to now, we have for simplicity's sake practically ignored *anisotropies.* Only spherically symmetrical molecules such as for example CCl_4 have an isotropic polarisability, i.e. one which has the same value for all angles between the electric field and the molecular axes. In general, as mentioned in Sect. 3.2, the polarisability of a molecule is anisotropic; this means that the quantities ε and n vary depending upon the orientation of the molecule relative to a measuring field – for example, the direction of polarisation of a light beam. From a knowledge of the anisotropy of the polarisability, one can thus obtain information about the shape of the molecules. In gases and liquids, the rapid molecular motions cause an averaging over all possible orientations of the molecules relative to the $\boldsymbol{E}$-vector of the light. If one wishes to measure the anisotropy directly, the molecules must be oriented, for example by substituting them into a molecular crystal. The measured dielectric constants ε of such crystals may then show a strong anisotropy.

Another possibility is the *electro-optical Kerr effect,* discovered in 1875. This term is applied to the observation that many molecular substances exhibit double refraction in strong electric fields. This comes about in the following way: in an electric field, the molecules tend to align themselves in such a manner that their dipole moments are parallel to the field. If the molecules are anisotropic with respect to α, the relation between α and ε or the index of refraction n gives rise to a difference between n for light with its electric field vector parallel to the direction of the applied field and for light whose electric field oscillates perpendicular to the applied field.

A further important consequence of anisotropic polarisability is the *depolarisation of light scattered by molecules* due to an anisotropy or to motion of the molecules. To illustrate this point, Fig. 3.7 shows the angular distribution of scattered radiation from a spherically symmetric molecule for unpolarised and for polarised incident light. This is the angular distribution of a Hertzian dipole oscillator. If the molecule is no longer spherically symmetric, or if it moves during the scattering process, deviations from this angular distribution occur. Polarised incident light is depolarised more strongly as the electronic shells of the scattering molecule become more asymmetric; very long or very flat molecules exhibit a high degree

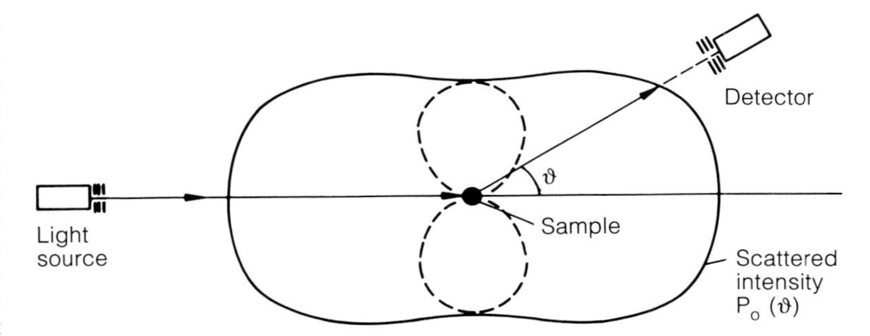

Fig. 3.7. Rayleigh scattering depends in a characteristic way on the scattering angle θ. The diagram shows the spatial distribution (in the plane) of the light intensity scattered by an isotropic, spherical sample. The full curve holds for unpolarised light, the dashed curve is for polarised incident light. This angular distribution diagram is for a spherically symmetric molecule. It may also change when the scattering particles move; thus, one can investigate motional processes of molecules or of functional groups in molecules by measuring the anisotropy of Rayleigh scattering

of depolarisation. Thus, one can obtain more information about the structure and motions of molecules.

Finally, we mention here the optical activity of some organic molecules. This refers to the difference in their indices of refraction for left and right circularly polarised light, i.e. *circular dichroism*. It is caused by the asymmetric arrangement of the atoms in the molecule. Particularly in the case of large molecules, one can learn something about the asymmetry of the electron density in the molecule from this effect.

3.6 Molecules in Magnetic Fields, Basic Concepts and Definitions

The macroscopic magnetic properties of matter are measured collectively by determining the materials constant μ, the permeability. The following relation holds:

$$\mu = \frac{\text{magnetic flux density } B_m \text{ in matter}}{\text{magnetic flux density } B \text{ without matter}} . \tag{3.27}$$

A derived quantity is the magnetic polarisation, which is a measure of the contribution due to the matter in the sample:

$$\boldsymbol{J} = \boldsymbol{B}_m - \boldsymbol{B} = (\mu - 1)\mu_0 \boldsymbol{H} \tag{3.28}$$

($\boldsymbol{H}$ is the magnetic field strength).

$\boldsymbol{J}$ can be defined by the expression

$$\boldsymbol{J} = \mu_0 \frac{\text{magnetic moment } \boldsymbol{M}}{\text{Volume}} . \tag{3.29}$$

μ_0 is the so-called magnetic field constant or permeability constant of vacuum, with the numerical value $\mu_0 = 1.256 \cdot 10^{-6}$ $\mathrm{VsA^{-1}m^{-1}}$, which is defined by the proportionality of the flux density to the magnetic field strength, $\boldsymbol{B} = \mu_0 \boldsymbol{H}$ in vacuum, or

$$\boldsymbol{B}_m = \mu \mu_0 \boldsymbol{H} \tag{3.30}$$

in matter.

In addition, the magnetic *susceptibility*

$$\kappa = \mu - 1 \tag{3.31}$$

is used; μ and κ are dimensionless number quantities.

Materials with $\kappa < 0, \mu < 1$ are called *diamagnetic*. In such materials, the atoms or molecules have no permanent magnetic moments. Materials with $\kappa > 0, \mu > 1$ are *paramagnetic*; here, the atoms or molecules have permanent moments, which can be oriented by an applied magnetic field.

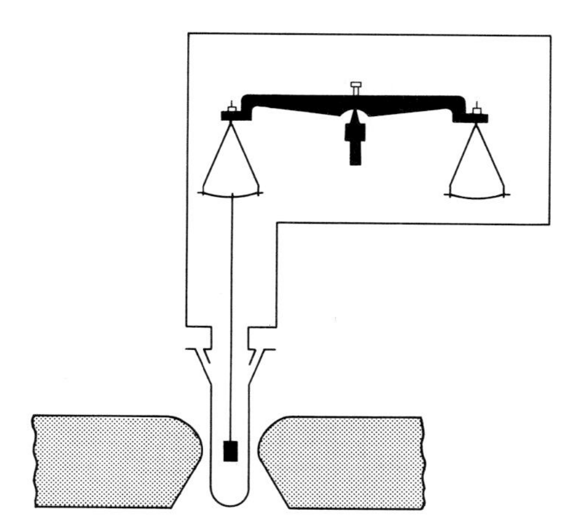

Fig. 3.8. A magnetic balance or Faraday balance. An inhomogeneous magnetic field exerts an attractive force on a paramagnetic sample and a repulsive force on a diamagnetic sample. Magnetic susceptibilties can be determined in such an apparatus

The magnetic susceptibility of a sample can be determined by measuring, for example, the force it experiences in an inhomogeneous magnetic field (*Faraday balance*, Fig. 3.8); or alternatively by measuring the inductance of a coil in which the sample has been placed. A modern method for paramagnetic materials is electron spin resonance (ESR), which will be treated in Chap. 19. Some values of the susceptibility are shown in Table 3.4.

Table 3.4. Magnetic susceptibilities at room temperature

Diamagnetic Materials		Paramagnetic Materials	
H_2	$-0.002 \cdot 10^{-6}$	O_2	$1.86 \cdot 10^{-6}$
H_2O	$-9.0 \cdot 10^{-6}$	O_2 liquid	$3620 \cdot 10^{-6}$
NaCl	$-13.9 \cdot 10^{-6}$	$Dy_2(SO_4)_3 \cdot 8H_2O$	$6320003 \cdot 10^{-6}$
Cu	$-7.4 \cdot 10^{-6}$	Al	$21.2 \cdot 10^{-6}$
Bi	$-153 \cdot 10^{-6}$	Cu^{++}	$264 \cdot 10^{-6}$

The macroscopic materials properties μ and κ can be explained, measured, and calculated in terms of the microscopic properties of the molecules involved, just like the electrical quantity ε and in an analogous manner. Conversely, from a measurement of the macroscopic quantities, the magnetic properties of the molecules can be derived. This will be demonstrated in the following. An understanding of these materials properties is important for understanding molecular structure and for chemistry.

A para- or diamagnetic object having a volume V experiences a magnetic polarisation or magnetisation in a magnetic field $\boldsymbol{B}$, given by the following expression:

$$\boldsymbol{J} = (\mu - 1)\boldsymbol{B}. \tag{3.32}$$

It thus acquires a magnetic moment $\boldsymbol{M}$ parallel to the direction of the applied field; this moment per unit volume is given by

$$\boldsymbol{J} = \frac{\mu_0 \boldsymbol{M}}{V}. \tag{3.33}$$

In a molecular picture, the moment $\boldsymbol{M}$ is interpreted as the sum of the time-averaged contributions $\boldsymbol{m}\,'$ from the n molecules, i.e.

$$\boldsymbol{J} = \mu_0 \boldsymbol{m}\,' \frac{n}{V} = \mu_0 \boldsymbol{m}\,' N. \tag{3.34}$$

From (3.32) and (3.34) it follows for $\boldsymbol{m}\,'$ that:

$$\boldsymbol{m}\,' = \frac{\boldsymbol{J}}{\mu_0 N} = \frac{\boldsymbol{B}(\mu - 1)}{\mu_0 N}. \tag{3.35}$$

We now define a molecular property, the *magnetic polarisability* β:

$$\beta = \frac{\boldsymbol{m}\,'}{\boldsymbol{B}} = \frac{\mu - 1}{\mu_0 N} = \frac{\kappa}{\mu_0 N}. \tag{3.36}$$

This is reasonable, since experimentally, μ is a materials constant independent of $\boldsymbol{B}$.

In condensed matter, the applied magnetic flux density may differ from the flux density which acts at the site of a molecule within the sample. This has to be taken into account appropriately.

β has the dimensions [$\mathrm{Am^4/Vs}$], and the product $\beta\mu_0$ has the dimensions [$\mathrm{m^3}$]. From a measurement of the susceptibility κ, one thus obtains using (3.31), (3.35), and (3.36) the molecular quantity β.

3.7 Diamagnetic Molecules

The electronic shells of most molecules possess no permanent magnetic moments. They have an even number of electrons whose angular momenta add to zero; they thus lack magnetic moments and are diamagnetic. Like all materials, however, these molecules acquire an induced magnetic moment $\boldsymbol{m}_{\mathrm{ind}}$ in an applied magnetic field $\boldsymbol{B}$, which, according to Lenz's rule, is opposed to the inducing field, i.e. is negative. This diamagnetic contribution to the magnetisation has only a slight temperature dependence. From (3.35) we find

$$\boldsymbol{m}_{\mathrm{ind}} = \frac{\boldsymbol{B}(\mu - 1)}{\mu_0 N}, \tag{3.37}$$

and

$$\beta = \frac{\boldsymbol{m}\,'_{\mathrm{ind}}}{\boldsymbol{B}}. \tag{3.38}$$

As a numerical example, we consider the diamagnetic hydrogen molecule, H_2, for which a determination of μ yields

$$\mu_0\beta = -3 \cdot 10^{-36}\mathrm{m}^3\,, \quad \beta = -3 \cdot 10^{-30}\,\frac{\mathrm{Am}^4}{\mathrm{Vs}}.$$

In a laboratory field $B = 1$ Vs/m^2, the induced magnetic moment of each molecule is then equal to

$$m'_{\text{ind}} = -3 \cdot 10^{-30} \frac{\text{Am}^4}{\text{Vs}} \cdot 1\ \text{Vs/m}^2 = -3 \cdot 10^{-30}\text{Am}^2.$$

This numerical value is small compared to the Bohr magneton, $\mu_B = 9.27 \cdot 10^{-24}\text{Am}^2$.

The values of induced magnetic moments are always much less than the Bohr magneton μ_B, the unit of atomic magnetism, and thus are small compared to the permanent magnetic moments of atoms or molecules.

The magnetic polarisability of non-spherically symmetrical molecules is in general anisotropic. For example, in the benzene molecule the measured values perpendicular and parallel to the plane of the molecule are

$$\mu_0\beta_\perp = -152 \cdot 10^{-36}\text{m}^3 \quad \text{and} \quad \mu_0\beta_\parallel = -62 \cdot 10^{-36}\text{m}^3.$$

The anisotropy is in this case intuitively understandable: the π-electrons can react to an inducing magnetic field more easily in the plane of the molecule than perpendicular to it, producing a current loop in the plane. The anisotropy of the magnetic polarisability in benzene or in other molecules with aromatic ring systems is an important indication of the delocalisation of the π-electrons along chains of conjugated double bonds (see also Sect. 18.3). Diamagnetism is indeed based upon the production of molecular eddy currents by a change in the external magnetic flux. The diamagnetic susceptibility is therefore greater when the electronic mobility along closed loops perpendicular to the applied magnetic field is larger.

3.8 Paramagnetic Molecules

As already mentioned, there are also molecules with *permanent magnetic dipole moments*. Examples are molecules of the gases O_2 and S_2 (cf. Sect. 13.3). Their electronic ground states are *triplet* states having total spins of $S = 1$. Also in this class are the so-called radicals, i.e. molecules with unpaired electronic spins ($S = 1/2$), or organic molecules in metastable triplet states ($S = 1$) (cf. Fig. 15.1). A consideration of how the paramagnetism of a molecule results from its spin and orbital functions will be given in later chapters of this book.

For paramagnetic molecules, one observes in contrast to diamagnetic substances large and positive values of the permeability, which increase with decreasing temperature. Experimentally, a proportionality to $1/T$ is usually observed. This paramagnetic behaviour can be understood in a quite analagous manner to the orientation polarisation in electric fields: it results from a competition between the aligning tendency of the applied field $\boldsymbol{B}$, with its orientational energy $W_{\text{or}} = \boldsymbol{m}_{\text{p}} \cdot \boldsymbol{B}$, and the thermal motions of the molecules, whose energy $W_{\text{th}} = kT$ tends towards a randomisation of the molecular orientation.

Without an applied field, the directions of the permanent moments $\boldsymbol{m}_{\text{p}}$ are randomly distributed, and the vector sum of the moments is, as a time and spatial average, equal to zero as a result of the thermal motions of the molecules. In an applied field, a preferred direction is defined and each molecular moment makes a contribution to the time-averaged magnetisation $\boldsymbol{M}$.

The contribution $\boldsymbol{m}^*$ of an individual molecule with a permanent moment $\boldsymbol{m}_\mathrm{p}$ to the macroscopic moment $\boldsymbol{M}$ can be written as

$$m^* = x m_\mathrm{p} \,, \tag{3.39}$$

where the index p is used here to indicate that permanent magnetic moments are meant. The factor x, which is in general small, can be readily calculated by analogy to the procedure used for the electronic orientation polarisation in Sect. 3.3. In sufficiently dilute systems, in which the interactions between the molecules may be neglected, we find as a good approximation

$$x \approx \frac{1}{3} \frac{\boldsymbol{m}_\mathrm{p} \cdot \boldsymbol{B}_m}{kT} \,. \tag{3.40}$$

Making use of the relations (3.35), (3.36), (3.39), and (3.40), we find after simple rearrangements the paramagnetic contribution to the magnetic polarisation:

$$\boldsymbol{J} = \mu_0 \frac{\sum \boldsymbol{m}^*}{V} = \frac{1}{3} \frac{m_\mathrm{p}^2 \mu_0 N \boldsymbol{B}}{kT} \,, \tag{3.41}$$

and

$$\kappa = \frac{1}{3} \frac{m_\mathrm{p}^2 \mu_0 N}{kT} \,. \tag{3.42}$$

This is *Curie's Law*, which describes the temperature dependence of paramagnetism.

Using (3.41) and (3.36) we obtain the following relation:

$$m_\mathrm{p} = \sqrt{\beta 3 kT} \,, \tag{3.43}$$

which in turn allows us to calculate the permanent moment of a paramagnetic molecule by making use of the magnetic polarisability β obtained from (3.36). For O_2, the measured value

$$\beta = 5.5 \cdot 10^{-26} \frac{\mathrm{Am}^4}{\mathrm{Vs}}$$

at $T = 300$ K leads, using (3.43), to

$$m_\mathrm{p} = 2.58 \cdot 10^{-23} \mathrm{Am}^2$$

for the magnetic moment. This value is of the order of magnitude of the Bohr magneton μ_B. Other magnetic moments determined in this manner are, for example, $1.70 \cdot 10^{-23}\,\mathrm{Am}^2$ for the NO molecule and $4.92 \cdot 10^{-23}\,\mathrm{Am}^2$ for the iron ion Fe^{+++}.

The overall susceptibility of a substance is given by the sum of the diamagnetic contribution and the paramagnetic contribution, when the latter is present.

We thus have

$$\begin{aligned} \mu &= \mu_\mathrm{dia} + \mu_\mathrm{para} \\ &= 1 + N \left(\beta_\mathrm{dia} + \frac{\mu_0 m_\mathrm{p}^2}{3kT} \right) \,. \end{aligned} \tag{3.44}$$

The quantities β_dia and $\boldsymbol{m}_\mathrm{p}$ are found by plotting the measured values of κ or μ against $1/T$, as we have already seen in the case of the electrical properties of matter in Sect. 3.3.

There are many molecules which are diamagnetic in their ground states but which have paramagnetic electronically excited states. Particularly important and interesting are the triplet states of many organic molecules. We shall have more to say on this topic, especially in Sect. 15.3 and in Chap. 19.

At low temperatures, in certain materials, a preferred parallel or antiparallel ordering of the spins and thus of the magnetic moments of the molecules is observed even in the absence of an applied magnetic field, i.e. spontaneously. This is termed ferromagnetism or antiferromagnetism. In the case of molecular substances, the latter is more common; i.e. the paramagnetic molecules order at low temperatures with their spins in pairs having antiparallel orientation.

In the chapters up to now, we have met with a number of the most important basic quantities for molecular physics, mainly from the experimental point of view. We now turn in the following four chapters to the theory of chemical bonding. Chapters 4 and 5 are of general interest, while Chaps. 6 and 7 contain more extended theoretical approaches and may be skipped over in a first reading of this book.

4. Introduction to the Theory of Chemical Bonding

In this chapter, we begin by reviewing the most important concepts of quantum mechanics and then discuss the difference between heteropolar and homopolar bonding. In the following sections, we treat the hydrogen molecule-ion and the hydrogen molecule, using the latter to illustrate various important theoretical methods. Finally, we turn to the topic of hybridisation, which is particularly significant for the carbon compounds.

4.1 A Brief Review of Quantum Mechanics

Classical physics failed to explain even the structure of the atom. Consider, for example, the hydrogen atom, in which one electron orbits around the nucleus. The (charged) electron behaves as an oscillating dipole and would, according to classical electrodynamics, continuously radiate away energy, so that it must fall into the nucleus after a short time. Furthermore, the appearance of discrete spectra is unexplainable. Particular difficulties occur in the attempt to explain chemical bonding; we will treat this topic in more detail in the next section. Molecular physics can clearly not get along without quantum mechanics. We therefore start with a brief review of the basic concepts of quantum mechanics, keeping the hydrogen atom in mind as a concrete example. For a more thorough treatment, we refer the reader to I, Chaps. 9 and 10.

We assume the atomic nucleus to be infinitely massive, so that we need consider only the electron's degrees of freedom. Its energy is given by

$$E = E_{\text{kin}} + E_{\text{pot}} , \tag{4.1}$$

where the kinetic energy may be written as

$$E_{\text{kin}} = \frac{m_0}{2} \boldsymbol{v}^2 ; \tag{4.2}$$

m_0 is here the mass of the electron, and $\boldsymbol{v}$ is its velocity. In order to arrive at the correct starting point for a quantum-mechanical treatment, we replace the velocity $\boldsymbol{v}$ by the canonically conjugate variable $\boldsymbol{p}$, the momentum, according to:

$$m_0 \boldsymbol{v} = \boldsymbol{p} , \tag{4.3}$$

so that we can write the kinetic energy in the form

$$E_{\text{kin}} = \frac{1}{2m_0} \boldsymbol{p}^2 . \tag{4.4}$$

The potential energy can be given as a position-dependent potential:

$$E_{\text{pot}} = V(\boldsymbol{r}) \tag{4.5}$$

where $\boldsymbol{r} = (x, y, z)$. The energy expression (4.1) can then be written as a Hamilton function

$$H = \frac{1}{2m_0}\boldsymbol{p}^2 + V(\boldsymbol{r}) . \tag{4.6}$$

This expression is the starting point for the quantisation. According to Jordan's rule we must replace the momentum $\boldsymbol{p}$ by a momentum operator:

$$p_x = \frac{\hbar}{\mathrm{i}}\frac{\partial}{\partial x} , \quad p_y = \frac{\hbar}{\mathrm{i}}\frac{\partial}{\partial y} , \quad p_z = \frac{\hbar}{\mathrm{i}}\frac{\partial}{\partial z} , \tag{4.7}$$

or, in vector notation,

$$\boldsymbol{p} = \frac{\hbar}{\mathrm{i}}\boldsymbol{\nabla} . \tag{4.8}$$

The Hamilton function (4.6) thus becomes the Hamiltonian operator:

$$H = \frac{1}{2m_0}\left(\frac{\hbar}{\mathrm{i}}\nabla\right)^2 + V(\boldsymbol{r}) . \tag{4.9}$$

If we calculate the square of the nabla operator, we obtain the Laplace operator ∇^2, defined by:

$$\nabla^2 = \frac{\partial^2}{\partial x^2} + \frac{\partial^2}{\partial y^2} + \frac{\partial^2}{\partial z^2} . \tag{4.10}$$

We can then finally write the Hamiltonian operator in the form

$$H = -\frac{\hbar^2}{2m_0}\nabla^2 + V(\boldsymbol{r}) . \tag{4.11}$$

Using this operator, we can formulate the time-dependent Schrödinger equation, which contains a time- and position-dependent wavefunction $\psi(\boldsymbol{r}, t)$:

$$H\psi(\boldsymbol{r}, t) = \mathrm{i}\hbar\frac{\partial}{\partial t}\psi(\boldsymbol{r}, t) . \tag{4.12}$$

In many cases, the Hamiltonian is itself not explicitly time-dependent. In such a case, one can simplify the time-dependent Schrödinger equation (4.12) by making the substitution

$$\psi(r, t) = \exp\left(-\frac{\mathrm{i}}{\hbar}Et\right)\psi(\boldsymbol{r}) , \tag{4.13}$$

i.e. by separating out a time-dependent exponential function and leaving the position-dependent function $\psi(\boldsymbol{r})$. Inserting (4.13) into (4.12), differentiating with respect to time and dividing out the exponential function which occurs in (4.13), we obtain the time-independent Schrödinger equation:

$$H\psi = E\psi . \tag{4.14}$$

In solving either (4.12) or (4.14), we must take into account the boundary conditions for ψ, which depend on the position vector $\boldsymbol{r}$. In general, they state that ψ vanishes when $\boldsymbol{r}$

goes to infinity. As can be quite generally shown, the Schrödinger equation (4.14), together with the boundary conditions, yields a set of so-called *eigenvalues* E_ν and corresponding eigenfunctions ψ_ν, where ν is an index denoting the *quantum numbers*. Therefore, in place of (4.14), we could write

$$H\psi_\nu = E_\nu \psi_\nu \,. \tag{4.15}$$

According to the basic postulate of quantum mechanics, the values obtained as the result of a measurement are just those which occur as eigenvalues in (4.15). In measurements of quantities other than the total energy, different values may result from each individual measurement. In this case, the theory can in general predict only *expectation values*, e.g. for position, momentum, kinetic or potential energy. These expectation values are defined by

$$\overline{x} = \int \psi^*(\boldsymbol{r}, t) x \psi(\boldsymbol{r}, t) dV \,, \tag{4.16}$$

$$\overline{p}_x = \int \psi^*(\boldsymbol{r}, t) p_x \psi(\boldsymbol{r}, t) dV \,, \tag{4.17}$$

$$E_{\text{kin}} = \int \psi^*(\boldsymbol{r}, t) \left(-\frac{\hbar^2}{2m_0} \nabla^2 \right) \psi(\boldsymbol{r}, t) dV \,, \tag{4.18}$$

$$E_{\text{pot}} = \int \psi^*(\boldsymbol{r}, t) V(\boldsymbol{r}) \psi(\boldsymbol{r}, t) dV \,. \tag{4.19}$$

The quantities $p_x, x, \ldots$ have now become operators in (4.16)–(4.19). We can use them to construct expressions for additional operators, e.g. for the angular momentum operator, using the relation

$$\boldsymbol{L} = [\boldsymbol{r}, \boldsymbol{p}] \,,$$

or, applying (4.8),

$$\boldsymbol{L} = \left[\boldsymbol{r}, \frac{\hbar}{\mathrm{i}} \nabla \right] \,. \tag{4.20}$$

We now consider the hydrogen atom, or, more generally, an atom having the nuclear charge Z and containing only one electron. It is not our intention here to develop the quantum mechanics of the hydrogen atom in detail; this is done in I, Chap. 10. Instead, we wish only to remind the reader of some basic results. In the case of the hydrogen atom, the Hamiltonian is given explicitly by

$$H = -\frac{\hbar^2}{2m_0} \nabla^2 - \frac{1}{4\pi\varepsilon_0} \frac{Ze^2}{r} \,. \tag{4.21}$$

Since the Hamiltonian depends only on the radius but not on the angles in a spherical polar coordinate system, it is useful to transform (4.21) to spherical polar coordinates using

$$\boldsymbol{r} \rightarrow r, \theta, \phi \,. \tag{4.22}$$

As may be shown, the wavefunction can then be written in the form

$$\psi_{nlm}(\boldsymbol{r}) = R_{nl} P_l^m(\cos\theta) \, \mathrm{e}^{\mathrm{i}m\phi} \,, \tag{4.23}$$

where the indices $n\ l\ m$ refer to *quantum numbers*: n is the principal quantum number, l the angular momentum quantum number, and m the magnetic quantum number. The wavefunction thus can be separated into a radial part R, which depends only on r, and an angular part $P_l^m e^{im\phi}$. The energy is found to be

$$E_n = -\frac{m_0 Z^2 e^4}{2\hbar^2 (4\pi\varepsilon_0)^2} \frac{1}{n^2} \ . \tag{4.24}$$

It thus depends only on the principal quantum number n, which can take on the values 1,2,3.... This characterises the bound states of the atom.

In the following, the angular dependence of ψ is mainly of interest. We therefore remind the reader of the simpler angular momentum states, cf. Fig. 4.1. For $l = 0$, there is one state, which does not depend on angles, i.e. it has spherical symmetry.

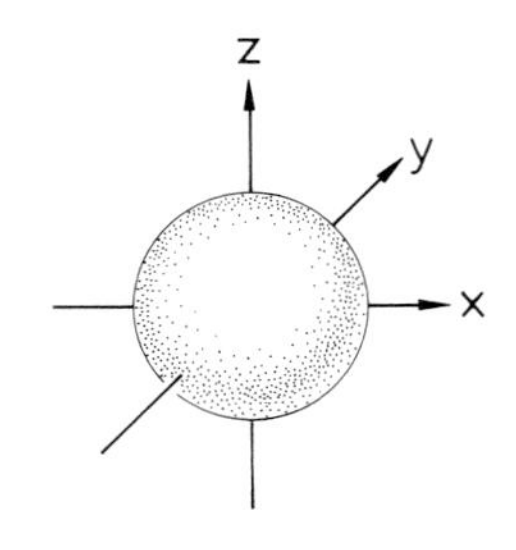

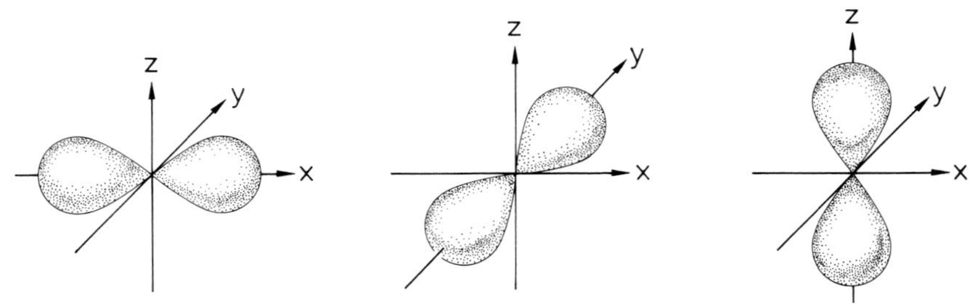

Fig. 4.1. Representation of the angular momentum functions for an s-state (spherically symmetric) and the p-functions (real representation)

We denote the angle-dependent factor in (4.23) by:

$$F_{l,m}(\theta,\phi) \equiv P_l^m(\cos\theta)\, e^{im\phi} \ . \tag{4.25}$$

For $l = 0, 1$, we obtain the following expressions for $F_{l,m}$:

$$l = 0 \quad F_{0,0} = \frac{1}{\sqrt{4\pi}} \tag{4.26}$$

$$l = 1 \quad F_{1,0} = \sqrt{\frac{3}{4\pi}} \cos\theta = \sqrt{\frac{3}{4\pi}} \frac{z}{r} \tag{4.27}$$

$$F_{1,\pm 1} = \pm\sqrt{\frac{3}{8\pi}} \sin\theta\, e^{\pm i\phi} = \pm\sqrt{\frac{3}{8\pi}} \frac{x \pm y}{r} \ , \tag{4.28}$$

where, in the last term of these equations, we have expressed the angular dependence by using Cartesian coordinates x, y, z. The radial function $R_{n,l}$ which occurs in (4.23) has the explicit form

$$R_{n,l} = N_{n,l} e^{-\kappa_n r} r^l L_{n+1}^{2l+1}(2\kappa_n r) , \tag{4.29}$$

where N is a normalisation factor, defined in such a way that:

$$\int_0^\infty R^2 r^2 dr = 1 . \tag{4.30}$$

The constant κ_n is given by the expression:

$$\kappa_n = \frac{1}{n} \frac{m_0 Z e^4}{\hbar^2 4\pi\varepsilon_0} . \tag{4.31}$$

The function L_{n+1}^{2l+1} is defined as a derivative of the Laguerre polynomials L_{n+1}, according to

$$L_{n+1}^{2l+1}(\varrho) = d^{2l+1} L_{n+1}/d\varrho^{2l+1} , \tag{4.32}$$

whereby the Laguerre polynomials themselves can be calculated using a differentiation formula:

$$L_{n+1}(\varrho) = e^{\varrho} d^{n+1}(e^{-\varrho}\varrho^{n+1})/d\varrho^{n+1} . \tag{4.33}$$

In the simplest case, $n = 1$, $l = 0$, we obtain

$$L_1(\varrho) = -\varrho + 1 \tag{4.34}$$

and thus

$$L_1^1 = dL_1/d\varrho = -1 , \tag{4.35}$$

so that $R_{1,0}$ is given by

$$R_{1,0} = N e^{-\kappa_1 r} . \tag{4.36}$$

Some examples are shown in Fig. 4.2.

4.2 Heteropolar and Homopolar Bonding

A theory of chemical bonding must be able to explain why it is possible for certain atoms to form a particular molecule, and it must be able to calculate the binding energy of the molecules formed. Before the development of quantum mechanics, one special type of bonding – heteropolar bonding – seemed to be easily explainable, but the other type – homopolar bonding – could not be understood at all. An example of heteropolar bonding (heteropolar = differently charged) is provided by the *common salt molecule*, NaCl (cf. Fig. 1.2). The formation of its bond can be imagined to take place in two steps: first, an electron is transferred from the Na atom to the Cl atom. The now positively-charged Na^+ ion attracts the negatively-charged Cl^- ion and *vice versa*, owing to the Coulomb force, which thus is responsible for the bonding. Considered more carefully, this explanation is only apparently complete, since it gives no theoretical justification for the electron transfer from Na to Cl. The theoretical basis for this transfer was given only by the quantum theory, according to which it is energetically more favorable for the electron to leave the open shell of the Na atom and to pass to the Cl atom, completing its outermost shell. Thus, to properly explain even heteropolar bonding, we require quantum mechanics.

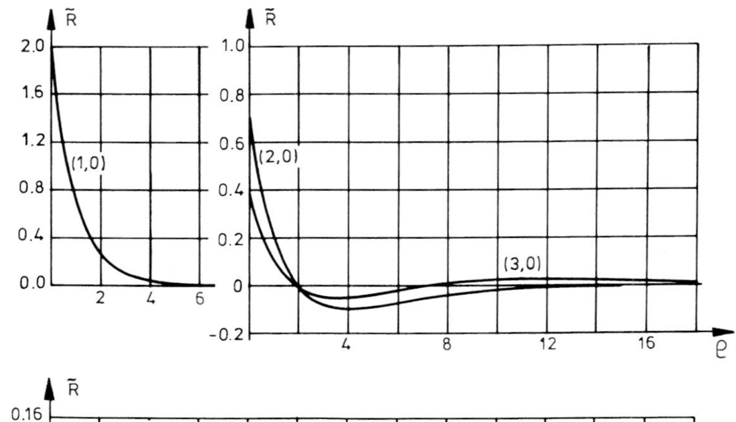

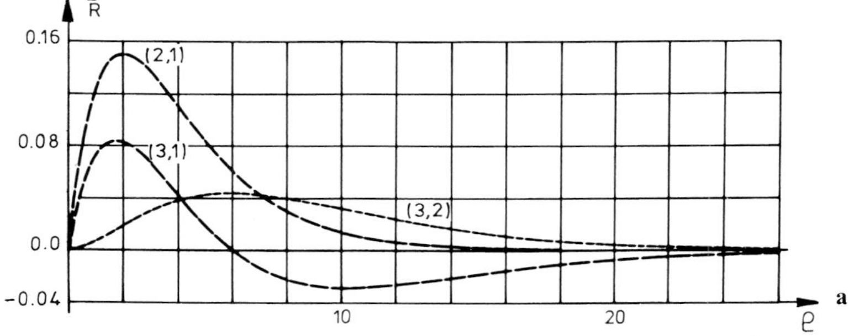

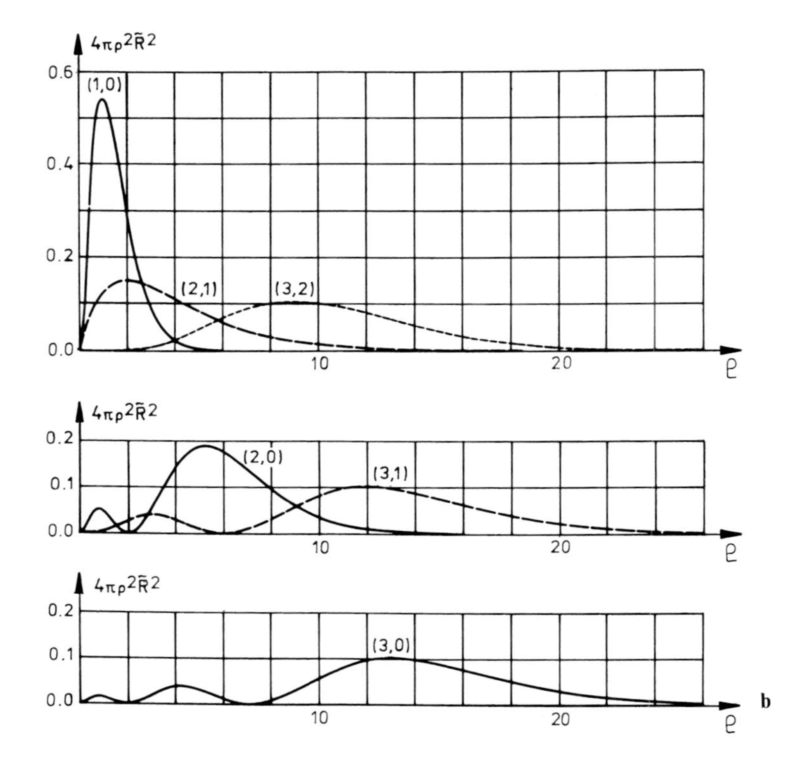

Fig. 4.2. (**a**) The radial part of the wavefunctions $\tilde{R}(\varrho = 2\kappa r) \equiv R(r)$ (4.29) of the H-atom is plotted against the dimensionless coordinate ϱ. The indices (1,0), (2,1),... on the curves correspond to (n, l), where n is the principal quantum number and l the angular momentum quantum number. (**b**) The corresponding probability amplitudes in the radial dimension, i.e. $4\pi\varrho^2\tilde{R}(\varrho)$, are plotted against the dimensionless coordinate ϱ

The question of the explanation of homopolar bonding was even more difficult. How, for example, could a hydrogen molecule, H_2, be formed from two *neutral* H atoms? Here, the quantum theory provided a genuine breakthrough. Its basically new idea can be discussed by using as an example the H_2^+ hydrogen molecule-ion, which corresponds to neutral H_2 from which an electron has been removed. The remaining electron must hold the two protons together. According to quantum mechanics, it can do this (pictorially speaking) by jumping back and forth between the two nuclei, staying for a while near one proton and then for a while near the other. Its probability of occupying the space between the two protons is thus increased; it profits from the Coulomb attraction to both nuclei and can thus compensate for the repulsive Coulomb force between the protons, as long as they do not approach each other too closely. We shall show in Sect. 4.3 that this picture can be precisely defined by calculating the wavefunctions of H_2^+. We will see there how the wave nature of the electron plays a decisive role. The wavefunctions which describe the electron's occupation of the space near the one proton or the other interfere constructively with each other, increasing the probability of finding the electron between the two protons and giving rise to a *bonding* state. A similar picture is found for the hydrogen molecule, H_2 (cf. Fig. 1.1). It is interesting that destructive interference is also possible – the occupation probability is then reduced and even becomes zero along the plane of symmetry between the two nuclei – and an *antibonding* state is produced, which releases the bound H atoms.

Let us now turn to the quantum mechanical calculation.

4.3 The Hydrogen Molecule-Ion, H_2^+

In this section, we start to develop the quantum theory of chemical bonding. The simplest case of chemical bonding is that of the hydrogen molecule-ion, H_2^+. This molecule can be observed as a bound state in a gas discharge in hydrogen atmosphere; in such a discharge, electrons are removed from the hydrogen molecules. The binding energy of H_2^+, identical to its dissociation energy, has been found to be 2.65 eV. Here, we are dealing with two hydrogen nuclei, i.e. protons, but only one electron. The two nuclei are distinguished by using the indices a and b (cf. Fig. 4.3). If they are separated by a very large distance, we can readily imagine that the electron is localised near either the one nucleus or near the other. Its wavefunction is then just like that of the ground state of the hydrogen atom. In the following, we denote the distance of the electron to nucleus a or to nucleus b as r_a or r_b, respectively. If we call the wavefunction of the hydrogen ground state belonging to nucleus a ϕ_a, it must obey the Schrödinger equation

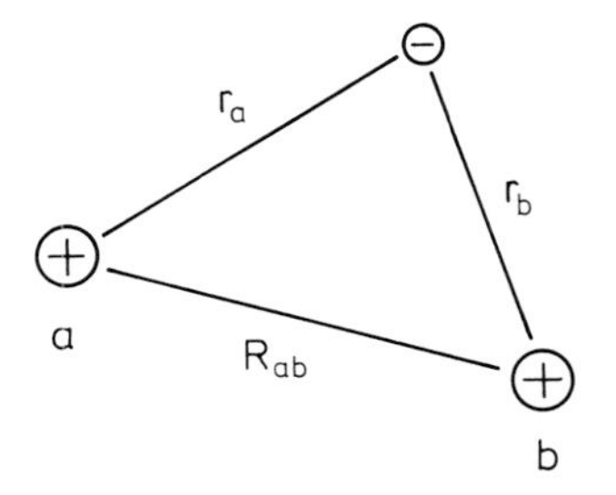

Fig. 4.3. Overview sketch of the hydrogen molecule-ion. The two nuclei (protons) are denoted as a and b, and their separation as R_{ab}. r_a and r_b give the distance of the electron to nucleus a or nucleus b, respectively

$$\underbrace{\left(-\frac{\hbar^2}{2m_0}\nabla^2 - \frac{e^2}{4\pi\varepsilon_0 r_a}\right)}_{H_a} \phi_a(r_a) = E_a^0\, \phi_a(r_a) , \tag{4.37}$$

and a corresponding equation holds for the wavefunction ϕ_b, with the energies E_a^0 and E_b^0 being equal:

$$E_a^0 = E_b^0 = E^0 . \tag{4.38}$$

If we now let the two nuclei approach one another, then the electron, which was originally near nucleus a, for example, will respond to the attractive Coulomb force of nucleus b.

Correspondingly, an electron which was originally near nucleus b will now respond to the Coulomb attraction of nucleus a. We therefore need to write a Schrödinger equation which contains the Coulomb potentials of both nuclei (Fig. 4.4). Furthermore, in order to calculate the total energy, we need to take into account the Coulomb repulsion of the nuclei. If we denote the nuclear separation by R_{ab}, then this additional energy is equal to $e^2/4\pi\varepsilon_0 R_{ab}$.

Since this additional term does not affect the energy of the electron, it simply results in a shift of the energy eigenvalues by a constant amount. We shall initially leave off this constant, and add it back in at the end of the calculation.

These considerations lead us to the Schrödinger equation

$$\left(-\frac{\hbar^2}{2m_0}\nabla^2 - \frac{e^2}{4\pi\varepsilon_0 r_a} - \frac{e^2}{4\pi\varepsilon_0 r_b}\right)\psi = E\,\psi\ , \tag{4.39}$$

in which the wavefunction ψ and the energy E must still be calculated.

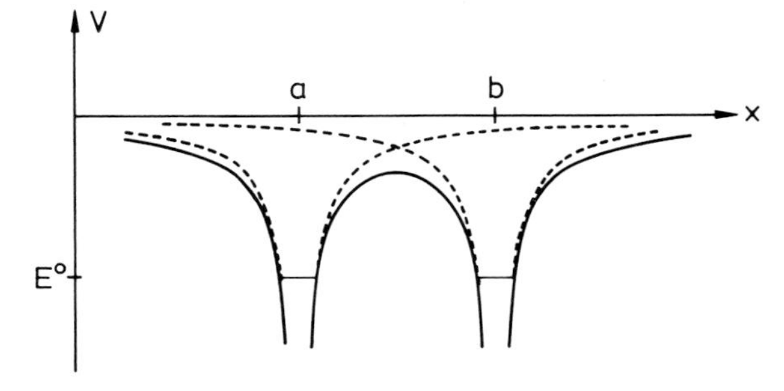

Fig. 4.4. The hydrogen molecule-ion: the potential energy V of the electron due to the Coulomb attraction to the two nuclei a und b is plotted against the x-coordinate. The *dashed curves* show the potential energy of the electron in the field of one nucleus, a or b. The *solid curve* is the total potential energy. The binding energy E^0 of the electron in the field of a single nucleus is also indicated

We now make an approximate determination of the wavefunction ψ. To this end, we make use of an idea borrowed from perturbation theory in the presence of degenerate levels. The electron could, in principle, be found near nucleus a or nucleus b (cf. Fig. 4.5), and would have the same energy in either case; compare (4.37) and (4.38). These two states, ϕ_a and ϕ_b, are thus degenerate in energy. Now, however, the other nucleus also affects the electron and perturbs its energy levels; we can expect that this would lift the degeneracy of the two states. Exactly as in perturbation theory with degeneracy, we take as a trial solution to (4.39) a linear combination of the form:

$$\psi = c_1\phi_a + c_2\phi_b\ , \tag{4.40}$$

where the two coefficients c_1 and c_2 are still to be determined. To calculate them, we proceed in the usual manner: we first insert the trial function (4.40) into (4.39) and obtain

$$\Bigg(\underbrace{-\frac{\hbar^2}{2m_0}\nabla^2 - \frac{e^2}{4\pi\varepsilon_0 r_a}}_{H_a} - \frac{e^2}{4\pi\varepsilon_0 r_b}\Bigg) c_1\phi_a$$

$$+\Bigg(\underbrace{-\frac{\hbar^2}{2m_0}\nabla^2 - \frac{e^2}{4\pi\varepsilon_0 r_b}}_{H_b} - \frac{e^2}{4\pi\varepsilon_0 r_a}\Bigg) c_2\phi_b = E(c_1\phi_a + c_2\phi_b)\ . \tag{4.41}$$

In the two large parentheses in (4.41), we have collected the terms in such a way that the operator H_a acts on ϕ_a and the operator H_b on ϕ_b. We can now refer to (4.37) and the corresponding equation for ϕ_b to simplify these expressions, by putting for example $E_a^0\phi_a$ in place of $H_a\phi_a$ and correspondingly for $H_b\phi_b$.

If we now bring the right-hand side of (4.41) to the left, we obtain

$$\Bigg(\underbrace{E^0 - E}_{\Delta E} - \frac{e^2}{4\pi\varepsilon_0 r_b}\Bigg) c_1\phi_a + \Bigg(\underbrace{E^0 - E}_{\Delta E} - \frac{e^2}{4\pi\varepsilon_0 r_a}\Bigg) c_2\phi_b = 0 \,. \tag{4.42}$$

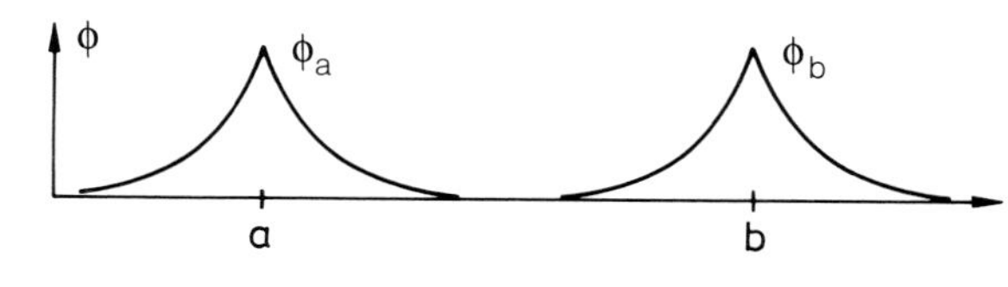

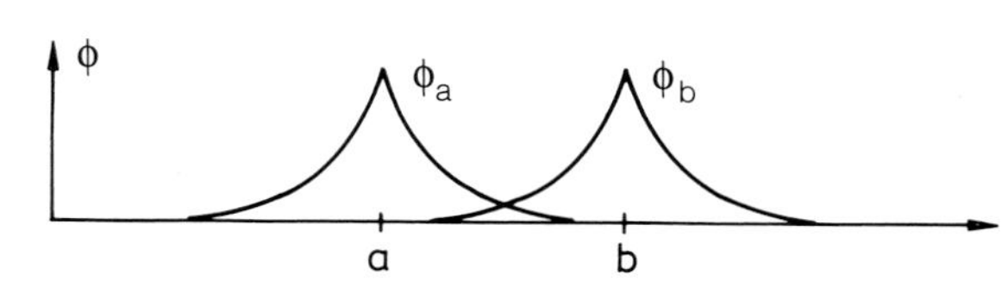

Fig. 4.5. (*Upper part*) The wavefunction ϕ_a of the electron when it is localised in the field of nucleus a, and the corresponding wavefunction ϕ_b of the electron near nucleus b. (*Lower part*) When the internuclear spacing between a and b is decreased, the two wavefunctions ϕ_a and ϕ_b begin to overlap in the central region

Although ϕ_a and ϕ_b are functions of the position coordinates, the coefficients c_1 and c_2 are assumed to be position-independent. In order to find a position-independent equation for the c's, we multiply (4.42) by ϕ_a^* or ϕ_b^*, as accustomed from perturbation theory, and integrate over the electronic coordinates. In the following, we assume that the functions ϕ_a and ϕ_b are real, which is the case for the ground state wavefunction of hydrogen. We have to keep in mind that the functions ϕ_a and ϕ_b are not orthogonal, i.e. that the integral

$$\int \phi_a\phi_b dV = S \tag{4.43}$$

is not equal to zero. If we multiply (4.42) by ϕ_a and then integrate over electronic coordinates, we obtain expressions which have the form of matrix elements, namely the integrals:

$$\int \phi_a(r_a)\left(-\frac{e^2}{4\pi\varepsilon_0 r_b}\right)\phi_a(r_a)\,dV = C \,, \tag{4.44}$$

$$\int \phi_a(r_a)\left(-\frac{e^2}{4\pi\varepsilon_0 r_a}\right)\phi_b(r_b)\,dV = D \,, \tag{4.45}$$

which we denote by the letters C and D. The meaning of the first integral becomes immediately apparent if we recall that $-e\phi_a^2$ is the charge density of the electron; (4.44) is then nothing other than the *Coulomb interaction energy* between the electronic charge density and the nuclear charge e (compare Fig. 4.6). In the integral (4.45), in contrast, instead of the electronic charge density, the expression $-e\phi_a\phi_b$ occurs. This means that the electron in a sense spends part of its time in state ϕ_a and the rest in state ϕ_b, or in other words, that there is an exchange between the two states. The product $\phi_a\phi_b$ is therefore called the exchange density and integrals in which such products are found are termed *exchange integrals* (cf. Fig. 4.7). These integrals express an effect which is specific to quantum theory. If we had multiplied (4.42) by ϕ_b instead of ϕ_a and integrated, we would have found expressions quite similar to (4.44) and (4.45), with only a permutation of the indices a and b. Since, however, the problem is completely symmetric with respect to these indices, the new integrals would have the same values as the original ones.

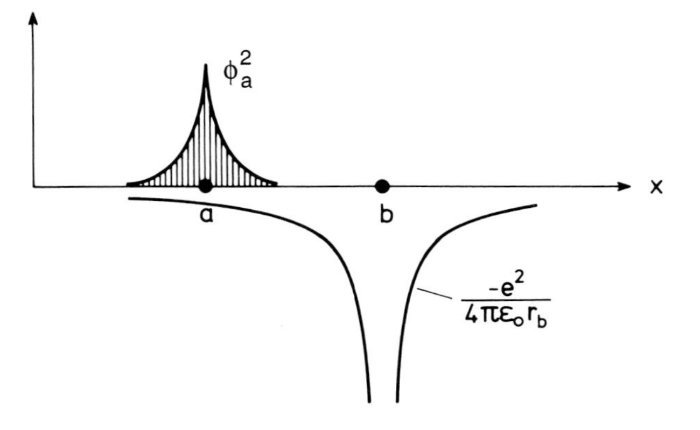

Fig. 4.6. An intuitive picture of the integral (4.44), which gives the Coulomb interaction energy of an electron cloud having the probability distribution ϕ_a^2 in the Coulomb field of a nucleus. The charge density distribution ϕ_a^2 (*shaded region*) is plotted along with the potential energy (*solid curve*) of a point charge in the Coulomb field of nucleus b. In calculating the integral, at each point in space the value of ϕ_a^2 is multiplied by the value of $-e^2/4\pi\varepsilon_0 r_b$ at the same point, and the products are then integrated over all space

Collecting all the terms obtained through multiplying by ϕ_a and integrating, we find that (4.42) has become the following equation:

$$(\Delta E + C)\, c_1 + (\Delta E\ S + D)\, c_2 = 0 \, , \tag{4.46}$$

and correspondingly after multiplication of (4.42) by ϕ_b and integration, we obtain the equation:

$$(\Delta E\ S + D)\, c_1 + (\Delta E + C)\, c_2 = 0 \, . \tag{4.47}$$

These are two simple algebraic equations for the unknown coefficients c_1 and c_2. In order that the equations have a non-trivial solution, the determinant of their coefficients must vanish, i.e.

$$(\Delta E + C)^2 - (\Delta E\ S + D)^2 = 0 \, . \tag{4.48}$$

This is a quadratic equation for the energy shift ΔE, which in the present case can be solved quite simply by bringing the second term in (4.48) to the right-hand side and taking the square root of both sides:

$$(\Delta E + C) = \pm(\Delta E\ S + D) \, . \tag{4.49}$$

The two possible signs, $\pm$, occur because of taking the square root. Inserting (4.49) into (4.46) or (4.47), we obtain immediately for the upper sign

$$c_2 = -c_1 \equiv -c \, . \tag{4.50}$$

In this case, the total wavefunction is given by

$$\psi = c(\phi_a - \phi_b) \, . \tag{4.51}$$

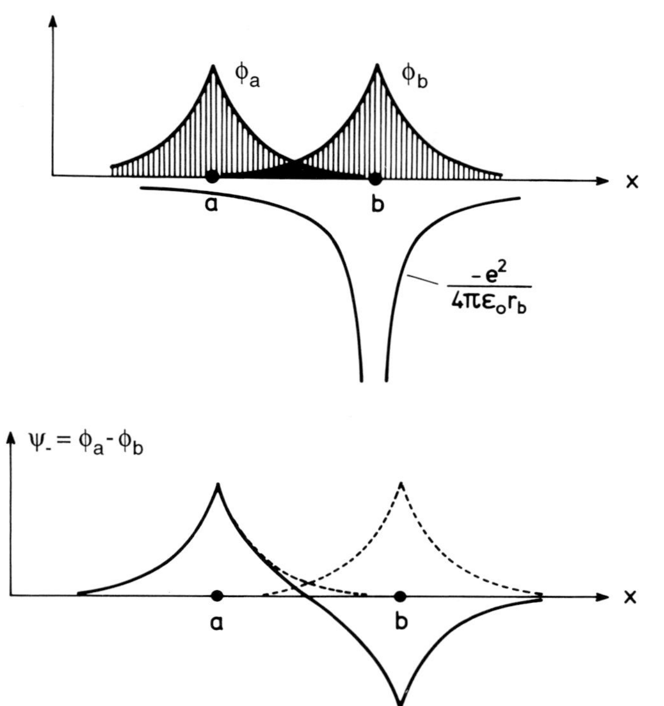

Fig. 4.7. An intuitive picture of the meaning of the integral (4.45). The three functions ϕ_a, ϕ_b, and $-e^2/4\pi\varepsilon_0 r_b$ which occur in the integral are plotted. The integral contains the product of these three functions, which is non-zero only where the two wavefunctions ϕ_a and ϕ_b overlap; this is the region *shaded heavily* in the figure. The integral is obtained by taking the functional values of ϕ_a, ϕ_b, and $-e^2/4\pi\varepsilon_0 r_b$ at each point in space, multiplying them, and then integrating this product over all space

Fig. 4.8. The antisymmetric wavefunction ψ_- is formed by taking the difference of ϕ_a and ϕ_b. Its occupation probability can be seen to vanish in the plane of symmetry between the two nuclei

The constant c is fixed by the normalisation of the total wavefunction ψ. The corresponding wavefunction is represented in Fig. 4.8. If we take the lower sign in (4.49), we obtain $c_2 = c_1 = c$ for the coefficients and thus for the total wavefunction:

$$\psi = c(\phi_a + \phi_b) \ . \tag{4.52}$$

(compare Fig. 4.9). Using (4.49), we can calculate the energies corresponding to (4.51) and (4.52), setting $E = E^0 - \Delta E$.

The *antisymmetric* wavefunction has the electronic energy

$$E = E^0 + \frac{C - D}{1 - S} \tag{4.53a}$$

and the *symmetric* wavefunction corresponds to the energy

$$E = E^0 + \frac{C + D}{1 + S} \ . \tag{4.53b}$$

As can be seen by considering Figs. 4.6 and 4.7, the quantities S, C, and D depend on the internuclear distance, whereby $0 < S \leq 1$ and $C, D < 0$. If the nuclei are allowed to approach one another, the electronic energy splits into two terms according to (5.53a) and (5.53b). In order to decide whether bonding occurs via the electron, we must still add the Coulomb repulsion energy between the protons, $e^2/4\pi\varepsilon_0 R_{ab}$, to (5.53a) or (5.53b). Furthermore, we must compare the energy at a finite internuclear separation R_{ab} with that at infinite separation, where C and D are zero. We thus have to examine

$$E_{\text{binding}} = \frac{C \pm D}{1 \pm S} + \frac{e^2}{4\pi\varepsilon_0 R_{ab}} \,. \tag{4.54}$$

As shown by numerical calculation, the overlap integral S hardly changes the result, so that we can leave it out of our further discussion.

Let us first consider the behaviour of C as a function of the internuclear distance R_{ab}. If R_{ab} is large compared to the spatial extent of the wavefunction ϕ_a (or ϕ_b), then C is practically equal to the potential energy E_{pot} of a point charge in the potential of the other nucleus, i.e. equal to $-e^2/4\pi\varepsilon_0 R_{ab}$. For large distances R_{ab}, C and the last term in (4.54) thus compensate each other. However, for small distances $R_{ab} \to 0$, the last term in (4.54) becomes infinite, while C approaches a (negative) finite value. This can be seen directly from (4.44), since for $R_{ab} \to 0$, the distance r_a becomes equal to r_b and (4.44) then becomes the same as the expectation value of the potential energy in the hydrogen atom, which as is well known is finite. The sum $C + e^2/4\pi\varepsilon_0 R_{ab}$ is thus positive and there is no bond formation.

The final decisive factor in the question of bond formation is thus D (4.45), which contains the exchange density. For $R_{ab} \to 0$, ϕ_b and ϕ_a become identical, so that D and C are the same and D cannot compensate the effect of $e^2/4\pi\varepsilon_0 R_{ab}$. If R_{ab} is now allowed to increase, then both $e^2/4\pi\varepsilon_0 R_{ab}$ and D, which have opposite signs, decrease in magnitude. A numerical calculation shows that in a certain region, E_{binding} becomes negative (cf. Fig. 4.10). The corresponding state is termed a *bonding state*. Conversely, no bonding occurs in the state (4.51); it represents a non-bonding or "antibonding" state.

As must be clear from our discussion, the bonding effect is based entirely upon the occurrence of the exchange density $\phi_a\phi_b$ in D. The bonding of the hydrogen molecule-ion is thus a typically quantum-mechanical phenomenon. Nevertheless, one can form an intuitive picture of the bonding and non-bonding effects.

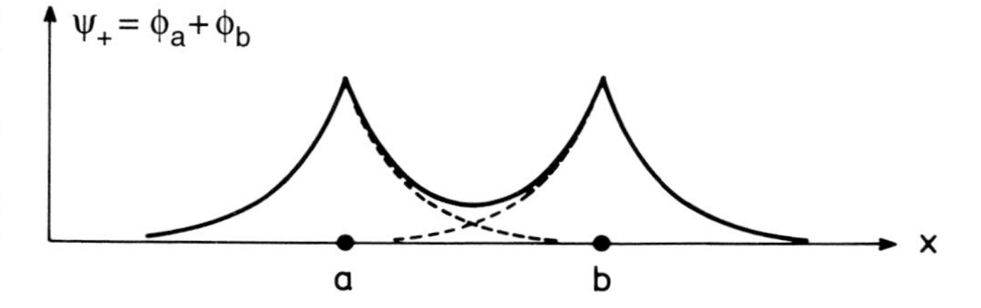

Fig. 4.9. The symmetric wavefunction ψ_+ is formed by adding the wavefunctions ϕ_a and ϕ_b. Due to the overlap between ϕ_a and ϕ_b, the occupation probability in the region between the two nuclei is increased

As may be seen from Fig. 4.9, the occupation probability of the electron in the region between the two nuclei in the bonding state is relatively high. It can thus profit from the Coulomb attraction of both nuclei, lowering the potential energy of the whole system. In the non-bonding state (Fig. 4.8), the occupation probability for the electron between the two nuclei is low; in the centre, it is in fact zero. This means that the electron is affected by the attractive force of practically one nucleus only.

For the decrease in energy of the hydrogen molecule-ion as compared to the hydrogen atom, the above calculation gives the result 1.7 eV; the experimental value is 2.65 eV. Our trial wavefunction thus indeed gives a bound state, but it is weakly bound compared to what is found experimentally. An improvement in first order can be obtained by using the trial wavefunction

$$\psi = c\,(\mathrm{e}^{-\alpha r_a/a_0} + \mathrm{e}^{-\alpha r_b/a_0}) \,,$$

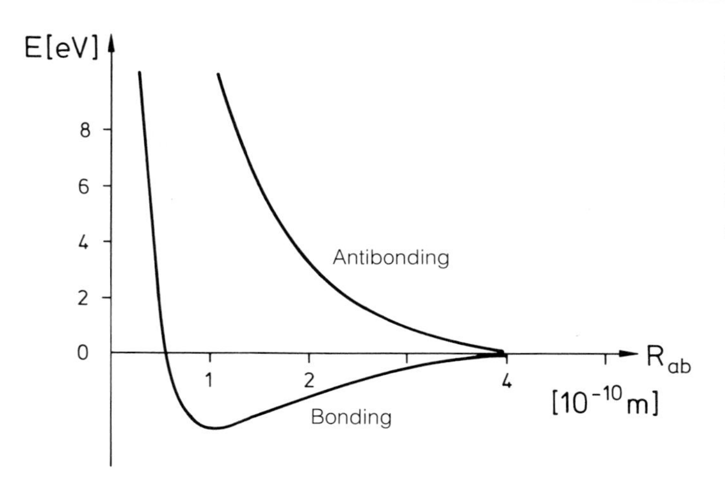

Fig. 4.10. The energy E of the hydrogen molecule-ion including the mutual Coulomb repulsion of the nuclei. The energy curves are plotted against the internuclear separation R_{ab} for the bonding and the antibonding states

where a_0 is the first Bohr radius and α is a variational parameter. In the energy minimum, it is found that $\alpha = 1.24$, i.e. the effective Bohr radius a_0/α is reduced. The result of this reduction is that the electron cloud perpendicular to the bonding axis is more strongly concentrated in the region between the nuclei; the Coulomb interaction between the electron and the nuclei is thus intensified. This interpretation is supported by the precise numerical solution of (4.39).

4.4 The Hydrogen Molecule, H_2

4.4.1 The Variational Principle

We now turn to the problem of chemical bonding when more than one electron participates in bond formation. However, before we consider in detail the simplest example, i.e. the hydrogen molecule H_2, we make some preliminary remarks which are of fundamental importance for other problems in quantum mechanics, also.

We shall often encounter the task of solving a Schrödinger equation

$$H\Psi = E\Psi \ , \tag{4.55}$$

which will frequently turn out not to be possible in closed form. In addition to the method of perturbation theory, which we have already discussed, there is a fundamentally different and very important approach based on the variational principle. In order to explain it, we suppose the Schrödinger equation (4.55) to have been multiplied by Ψ^* and integrated over all of the coordinates on which Ψ depends. We then obtain

$$E = \frac{\int \Psi^* H \Psi dV_1 \dots dV_n}{\int \Psi^* \Psi dV_1 \dots dV_n} \ . \tag{4.56}$$

Here, n is the number of electrons, while dV_j, $j = 1, \dots, n$ is a volume element referring to the j-th electron for the integration over its coordinates.

Since the Hamiltonian H is the operator belonging to the total energy of the system, expression (4.56) is just the expectation value of the total energy, which in the present case is identical with the energy eigenvalue of the Schrödinger equation. What would happen, though, if for Ψ we used some arbitrary wavefunction instead of a solution of the Schrödinger equation? Then (4.56) still has the dimensions of an energy, but it is not necessarily equal to the correct eigenvalue of the Schrödinger equation which we are seeking. Applying mathematics, one can at this point prove an extremely important relation: if we in fact do not use a true eigenfunction of the ground state of the system for Ψ, but rather some other wavefunction, then its corresponding energy expectation value will always be *larger* than the eigenvalue of a solution to (4.55). In this sense, we can give a criterion for how well we have approached the true eigenfunction: the lower the calculated expectation value (4.56), the better the trial wavefunction used to obtain it.

We shall use this criterion repeatedly later on. Now, however, we want to set out to determine the wavefunctions and the energy of the hydrogen molecule in the ground state, at least approximately. In choosing a suitable approximate wavefunction, our physical intuition will play an essential role. Depending on which aspects of the physical problem are emphasised, we will arrive at different approaches, which are known by the names of their original authors: the Heitler-London and the Hund-Mullikan-Bloch methods. In addition to these approaches, we will meet up with improvements such as the so-called covalent-ionic resonance (Sect. 4.4.3), and also a wavefunction which includes all the others described as special cases, and thus opens the way to a first general treatment of the many-electron problem in molecules (Sect. 4.4.5).

4.4.2 The Heitler-London Method

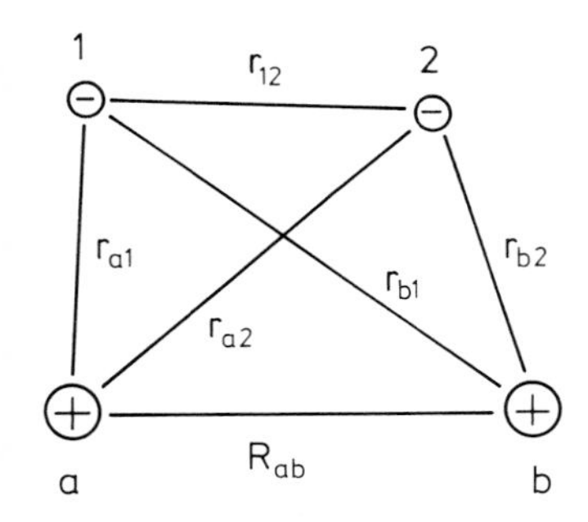

Fig. 4.11. An overview sketch of the hydrogen molecule. The two nuclei are denoted by the indices a and b, the two electrons by 1 and 2. The internuclear, interelectronic, and electron-nuclear distances with their respective notations are shown in the figure

The two atomic nuclei (protons) are distinguished by the indices a and b, and the two electrons by the indices 1 and 2. Due to the fact that the Coulomb force acts between all four particles, we need to introduce the corresponding distances, which are defined in Fig. 4.11. In order to write down the Hamiltonian, we recall the energy balance from classical physics. We are dealing with the kinetic energies of electron 1 and electron 2, and with the various contributions to the Coulomb interaction energy. We first translate the classical expression for the kinetic energy into quantum-mechanical terms! If $\boldsymbol{p}_1$ and $\boldsymbol{p}_2$ are the momenta of electrons 1 and 2, then the (classical) kinetic energy is given by

$$E_{\text{kin}} = \frac{1}{2m_0}\boldsymbol{p}_1^2 + \frac{1}{2m_0}\boldsymbol{p}_2^2 . \tag{4.57}$$

We now need to convert $\boldsymbol{p}_1$ and $\boldsymbol{p}_2$ to quantum-mechanical operators using the rule (4.7); in the process, we must add the indices 1 and 2 to the spatial coordinates. We thus obtain

$$p_{x1} = \frac{\hbar}{\mathrm{i}}\frac{\partial}{\partial x_1} , \quad p_{y1} = \frac{\hbar}{\mathrm{i}}\frac{\partial}{\partial y_1} , \quad p_{z1} = \frac{\hbar}{\mathrm{i}}\frac{\partial}{\partial z_1} , \tag{4.58}$$

$$p_{x2} = \frac{\hbar}{\mathrm{i}}\frac{\partial}{\partial x_2} , \quad p_{y2} = \frac{\hbar}{\mathrm{i}}\frac{\partial}{\partial y_2} , \quad p_{z2} = \frac{\hbar}{\mathrm{i}}\frac{\partial}{\partial z_2} , \tag{4.59}$$

or, using the nabla operator,

$$\boldsymbol{p}_1 = \frac{\hbar}{\mathrm{i}}\nabla_1 , \; \boldsymbol{p}_2 = \frac{\hbar}{\mathrm{i}}\nabla_2 . \tag{4.60}$$

For the kinetic energy operator, we then obtain

$$H_{\mathrm{kin}} = -\frac{\hbar^2}{2m_0}\nabla_1^2 - \frac{\hbar^2}{2m_0}\nabla_2^2 . \tag{4.61}$$

The square of the nabla operator can once again be expressed as the Laplace operator:

$$\nabla_1^2 = \frac{\partial^2}{\partial x_1^2} + \frac{\partial^2}{\partial y_1^2} + \frac{\partial^2}{\partial z_1^2} \tag{4.62}$$

and correspondingly for the index 2. Adding the various contributions to the Coulomb interaction energy to the kinetic energy operator (4.61), we obtain for the Hamiltonian

$$\begin{aligned} H = & \underbrace{-\frac{\hbar^2}{2m_0}\nabla_1^2 - \frac{e^2}{4\pi\varepsilon_0 r_{a1}}}_{H_1} \underbrace{-\frac{\hbar^2}{2m_0}\nabla_1^2 - \frac{e^2}{4\pi\varepsilon_0 r_{b2}}}_{H_2} \\ & - \frac{e^2}{4\pi\varepsilon_0 r_{b1}} - \frac{e^2}{4\pi\varepsilon_0 r_{a2}} + \frac{e^2}{4\pi\varepsilon_0 R_{ab}} + \frac{e^2}{4\pi\varepsilon_0 r_{12}} . \end{aligned} \tag{4.63}$$

We again assume that the nuclei are infinitely massive. Our task is now to solve the Schrödinger equation

$$H\Psi(r_1, r_2) = E\Psi(r_1, r_2) \tag{4.64}$$

with the Hamiltonian (4.63). If the nuclei were infinitely far apart, it would be sufficient to consider them separately, i.e. to solve the equations

$$\left(-\frac{\hbar^2}{2m_0}\nabla_1^2 - \frac{e^2}{4\pi\varepsilon_0 r_{a1}}\right)\phi_a(r_1) = E_0\phi_a(r_1) , \tag{4.65}$$

$$\left(-\frac{\hbar^2}{2m_0}\nabla_2^2 - \frac{e^2}{4\pi\varepsilon_0 r_{b2}}\right)\phi_b(r_2) = E_0\phi_b(r_2) . \tag{4.66}$$

However, we are dealing here with a two-electron problem; accordingly, we must take the Pauli exclusion principle into account, i.e. we have to consider the fact that electrons have a *spin*. If the two hydrogen atoms did not influence each other, we could immediately write down the overall wavefunction using the wavefunctions ϕ_a and ϕ_b which occur in (4.65) and (4.66). As we can see by insertion into a Schrödinger equation with $H = H_1 + H_2$, a solution would be:

$$\phi_a(r_1)\phi_b(r_2) . \tag{4.67}$$

In order to take the existence of spin into account, we have to multiply this trial solution by appropriate spin functions. The reader who is not familiar with the spin formalism should not be disturbed at this point, as we need only a few properties of the spin functions and will then be able to dispense with them completely during the further course of the calculation.

We denote the function referring to an electron with spin 'up' by α. (This type of spin wavefunction was denoted in I, Sect. 14.2.2 as $\phi_\uparrow$.) If we are dealing with electron 1, we call the wavefunction $\alpha(1)$. If both electrons have their spins in the same direction ("up"), then our wavefunction becomes

$$\phi_a(r_1)\phi_b(r_2)\alpha(1)\alpha(2) . \tag{4.68}$$

This function, however, does not obey the Pauli principle, which states in its mathematical formulation that a wavefunction must be antisymmetric in all the coordinates of the electrons (i.e. spatial and spin coordinates). In other words, when we exchange the indices 1 with the indices 2 everywhere, the wavefunction must change its sign. The wavefunction (4.68) does not have this property; however, the following wavefunction *does* have it:

$$\Psi = \phi_a(r_1)\alpha(1)\phi_b(r_2)\alpha(2) - \phi_a(r_2)\alpha(2)\phi_b(r_1)\alpha(1) . \tag{4.69}$$

If we factor out the spin functions $\alpha(1)$ and $\alpha(2)$, the wavefunction assumes the simple form

$$\Psi = \alpha(1)\alpha(2)\underbrace{[\phi_a(r_1)\phi_b(r_2) - \phi_a(r_2)\phi_b(r_1)]}_{\Psi_\mathrm{u}} , \tag{4.70}$$

i.e. it is the product of a spin function and a spatial wavefunction. (In quantum mechanics, wavefunctions which are symmetric with respect to exchange of the electronic spatial coordinates are termed *gerade*, abbreviated "g", from the German for "even"; antisymmetric wavefunctions are denoted by a "u", for *ungerade* = "odd".)

Looking forward to an important general approach for representing many-electron wavefunctions, we write (4.69) in a still different form. It may be represented as a determinant:

$$D = \begin{vmatrix} \phi_a(r_1)\alpha(1) & \phi_a(r_2)\alpha(2) \\ \phi_b(r_1)\alpha(1) & \phi_b(r_2)\alpha(2) \end{vmatrix} . \tag{4.71}$$

If we calculate this determinant following the usual rule

D = product of the main diagonal
− product of the secondary diagonal,

then we obtain just the expression (4.69). The determinant has a clearly apparent structure: the rows refer to the states a and b, and the columns refer to the numbers 1 and 2 of the two electrons.

Although (4.70) refers to two electrons whose spins are *parallel* and directed upwards, we can also construct wavefunctions for electrons with *parallel* spins which point *downwards*. We denote the spin function of a single electron whose spin is in the downwards state by β; then the total wavefunction becomes

$$\Psi = \beta(1)\beta(2)\Psi_\mathrm{u} . \tag{4.72}$$

For completeness, we also give the third wavefunction, belonging to the substate of the "triplet" state in which the spins are parallel. This state has its z-component of the total spin equal to zero, and is given by

$$\Psi = \frac{1}{\sqrt{2}}[\alpha(1)\beta(2) + \alpha(2)\beta(1)]\,\Psi_\mathrm{u} . \tag{4.73}$$

As the following calculation shows, the wavefunction Ψ does not belong to the state which is lowest in energy, since its spins are parallel. We need to find a wavefunction whose spins, in contrast, are *antiparallel*, i.e. one in which the one electron is described by a "spin up" function α and the other by a "spin down" function β. Here, expanding on (4.68), there are a number of possibilities. One of them is:

$$\phi_a(r_1)\phi_b(r_2)\alpha(1)\beta(2) . \tag{4.74}$$

Other functions can be found by starting with (4.74) and exchanging the coordinates r_1 and r_2 or the arguments of α or β, i.e. 1 and 2, or by exchanging everything at the same time. None of these combinations is antisymmetric as it stands. We therefore will attempt to find a combination of (4.74) with some of these other possible trial functions which is antisymmetric and which can be written as the product of a spin part and a spatial part, similarly to (4.70). This is in fact possible, as one discovers after some trial and error, and leads to the wavefunction

$$\Psi = \underbrace{[\phi_a(r_1)\phi_b(r_2) + \phi_a(r_2)\phi_b(r_1)]}_{\Psi_g} [\alpha(1)\beta(2) - \alpha(2)\beta(1)] . \tag{4.75}$$

The spin function is clearly antisymmetric here, while the spatial function Ψ_g is symmetric. If we exchange the spatial and spin coordinates of the two electrons simultaneously, this overall wavefunction changes sign: it is antisymmetric, in agreement with the Pauli exclusion principle.

The spin functions were here only a means of establishing the required symmetry of the total wavefunction. Since, however, no operators occur in the Hamiltonian of the Schrödinger equation (4.64) which act in any way upon the electronic spins, we can treat the spin functions just as a number when inserting (4.70) and (4.75) into that equation, and can divide them out from both sides. The resulting equation contains only the spatial functions Ψ_g or Ψ_u. This means that in the approximation to which we are calculating here, the interaction of the spins with one another (the spin-spin interactions) and of the spins with the spatial functions (spin-orbit interactions) are not taken into account. From now on, we concern ourselves only with the functions Ψ_g and Ψ_u and compute the energy expectation values belonging to these wavefunctions.

Following the basic idea of Heitler and London, we take these wavefunctions Ψ_g and Ψ_u as trial solutions of the Schrödinger equation with the Hamiltonian (4.63), which contains all the Coulomb interactions between the electrons and the protons, and imagine that we can then approximate the exact energy by applying (4.56). We thus have the task of calculating the energy eigenvalues for these wavefunctions. This calculation is not difficult, but it requires some patience.

As a first effort towards the calculation of the eigenvalues, we consider the normalisation integral which occurs in the denominator of (4.56). It has the form:

$$\begin{aligned} &\iint |\Psi(r_1, r_2)|^2 \, dV_1 dV_2 \\ &\quad = \iint [\phi_a(r_1)\phi_b(r_2) \pm \phi_a(r_2)\phi_b(r_1)]^* \\ &\quad\quad \cdot [\phi_a(r_1)\phi_b(r_2) \pm \phi_b(r_2)\phi_a(r_1)] \, dV_1 dV_2 . \end{aligned} \tag{4.76}$$

After multiplying out all the terms (and assuming that ϕ_a and ϕ_b are real), we obtain

$$\int \phi_a^2\, dV_1 \int \phi_b^2\, dV_2 + \int \phi_a^2\, dV_2 \int \phi_b^2\, dV_1$$
$$\pm \int \phi_a(r_1)\phi_b(r_1)\, dV_1 \int \phi_a(r_2)\phi_b(r_2)\, dV_2$$
$$\pm \int \phi_a(r_2)\phi_b(r_2)\, dV_2 \int \phi_a(r_1)\phi_b(r_1)\, dV_1 \,. \tag{4.77}$$

As a result of the normalisation of the wavefunctions ϕ_a and ϕ_b, the first two expressions can be reduced to:

$$\int \phi_a^2\, dV_1 = \int \phi_b^2\, dV_2 = 1 \,, \tag{4.78}$$

while the remaining two expressions are squares of the overlap integral

$$\int \phi_a(r_1)\phi_b(r_1)\, dV = S \,. \tag{4.79}$$

We can thus write the normalisation integral (4.76) in the simple form

$$2(1 \pm S^2) \,. \tag{4.80}$$

In evaluating the numerator of the energy expectation value (4.56), we encounter, analogously to (4.77), altogether four expressions, which occur in pairs of equivalent terms.

We begin with the expression

$$\iint \phi_a(r_1)\phi_b(r_2) \left\{ H_1 + H_2 - \frac{e^2}{4\pi\varepsilon_0 r_{b1}} - \frac{e^2}{4\pi\varepsilon_0 r_{a2}} + \frac{e^2}{4\pi\varepsilon_0 R_{ab}} + \frac{e^2}{4\pi\varepsilon_0 r_{12}} \right\}$$
$$\cdot\, \phi_a(r_1)\phi_b(r_2)\, dV_1 dV_2 \,. \tag{4.81}$$

Since the Hamiltonian H_1 in (4.81) acts only on ϕ_a, we can use the fact that ϕ_a obeys the Schrödinger equation (4.65) in our further calculations. Applying the same considerations to H_2, we can simplify (4.81) to the form:

$$\iint \phi_a(r_1)^2 \phi_b(r_2)^2$$
$$\cdot \Big\{ \underbrace{2E_0}_{1)} - \underbrace{\frac{e^2}{4\pi\varepsilon_0 r_{b1}}}_{2)} - \underbrace{\frac{e^2}{4\pi\varepsilon_0 r_{a2}}}_{3)} + \underbrace{\frac{e^2}{4\pi\varepsilon_0 R_{ab}}}_{4)} + \underbrace{\frac{e^2}{4\pi\varepsilon_0 r_{12}}}_{5)} \Big\}\, dV_1 dV_2 \,. \tag{4.82}$$

For what follows, it is useful to examine the meaning of the terms in (4.82) individually.

1) Owing to the normalisation of the wavefunctions ϕ_a and ϕ_b, the expression

$$\iint \phi_a(r_1)^2 \phi_b(r_2)^2 2E_0\, dV_1 dV_2$$

reduces to

$$2E_0 \,, \tag{4.83}$$

i.e. the energy of the two hydrogen atoms at infinite distance from each other.

2) The expression

$$\int \phi_a(r_1)^2 \left(-\frac{e^2}{4\pi\varepsilon_0 r_{b1}}\right) dV_1 = C < 0 \tag{4.84}$$

represents the Coulomb interaction energy of nucleus b with electron 1 in state a.

3) The integral

$$\int \phi_b(r_2)^2 \left(-\frac{e^2}{4\pi\varepsilon_0 r_{a2}}\right) dV_2 = C < 0 \tag{4.85}$$

is the Coulomb interaction energy of electron 2 in state b in the field of nucleus a. From the symmetry of the problem, it follows that the two integrals 2) and 3) are equal.

4) Owing to the normalisation of the wavefunctions ϕ_a and ϕ_b, the expression

$$\iint \phi_a(r_1)^2 \phi_b(r_2)^2 \frac{e^2}{4\pi\varepsilon_0 R_{ab}}\, dV_1 dV_2$$

reduces to

$$\frac{e^2}{4\pi\varepsilon_0 R_{ab}}\,. \tag{4.86}$$

This is the Coulomb repulsion energy of the two nuclei.

5) The integral

$$\iint \phi_a(r_1)^2 \phi_b(r_2)^2 \frac{e^2}{4\pi\varepsilon_0 r_{12}}\, dV_1 dV_2 = E_{\mathrm{RI}} \tag{4.87}$$

represents the repulsive Coulomb interaction energy of the two electrons.

Adding up the contributions (4.83) through (4.87) we obtain a contribution to the energy expectation value of (4.81) (which we abbreviate as $\hat{E}$)

$$\hat{E} = 2E_0 + 2C + E_{\mathrm{RI}} + \frac{e^2}{4\pi\varepsilon_0 R_{ab}}\,. \tag{4.88}$$

This is, however, still not the final result, since on inserting the wavefunctions Ψ_{g} or Ψ_{u} into the expression (4.56) for the energy eigenvalue, we also obtain exchange terms of the form

$$\pm \iint \phi_b(r_1)\phi_a(r_2)\{\ldots\}\phi_b(r_2)\phi_a(r_1)\, dV_1 dV_2\,, \tag{4.89}$$

where the expression in curly brackets, $\{\ldots\}$, is the same as in (4.81). Explicitly written out, (4.89) thus becomes

$$\pm \iint \phi_b(r_1)\phi_a(r_2)\phi_a(r_1)\phi_b(r_2) \cdot \Big\{\underbrace{2E_0}_{1)} \underbrace{- \frac{e^2}{4\pi\varepsilon_0 r_{b1}}}_{2)} \underbrace{- \frac{e^2}{4\pi\varepsilon_0 r_{a2}}}_{3)} \underbrace{+ \frac{e^2}{4\pi\varepsilon_0 R_{ab}}}_{4)} \underbrace{+ \frac{e^2}{4\pi\varepsilon_0 r_{12}}}_{5)}\Big\}\, dV_1 dV_2\,. \tag{4.90}$$

The various terms have the following forms and meanings:

1) The expression

$$\iint \phi_b(r_1)\phi_a(r_2)(\pm 2E_0)\phi_a(r_1)\phi_b(r_2)\, dV_1 dV_2$$

reduces on applying the definition (4.79) of the overlap integral S to

$$\pm 2E_0 S^2 \, . \tag{4.91}$$

This is the energy of the two separated hydrogen atoms multiplied by the square of the overlap integral S.

2) The exchange integral

$$\pm \underbrace{\int \phi_a(r_2)\phi_b(r_2)\, dV_2}_{S} \underbrace{\int \phi_b(r_1)\left(-\frac{e^2}{4\pi\varepsilon_0 r_{b1}}\right)\phi_a(r_1)\, dV_1}_{D} \tag{4.92}$$

is the product of the overlap integral S and the one-electron exchange integral D [compare (4.45)].

3) The exchange integral

$$\pm \iint \phi_b(r_1)\phi_a(r_2)\left(-\frac{e^2}{4\pi\varepsilon_0 r_{a2}}\right)\phi_a(r_1)\phi_b(r_2)\, dV_1 dV_2$$

reduces in exact analogy with (4.92) to

$$\pm SD \, . \tag{4.93}$$

4) The exchange integral

$$\pm \iint \phi_b(r_1)\phi_a(r_2)\left(\frac{e^2}{4\pi\varepsilon_0 R_{ab}}\right)\phi_a(r_1)\phi_b(r_2)\, dV_1 dV_2$$

reduces directly to

$$\pm S^2 \frac{e^2}{4\pi\varepsilon_0 R_{ab}} \, , \tag{4.94}$$

i.e. to the square of the overlap integral S multiplied by the Coulomb interaction energy between the two nuclei.

5) The exchange integral

$$\pm \iint \phi_b(r_1)\phi_a(r_2)\frac{e^2}{4\pi\varepsilon_0 r_{12}}\phi_a(r_1)\phi_b(r_2)\, dV_1 dV_2 = \pm E_{\mathrm{CE}} \tag{4.95}$$

represents the Coulomb interaction energy between the two electrons, but computed using not the normal charge density, but rather the *exchange density*. This integral is therefore referred to as the Coulomb-exchange interaction.

The total contribution of (4.91–4.95), which we abbreviate as $\tilde{E}$, is then given by

$$\tilde{E} = \pm 2E_0 S^2 \pm 2DS \pm E_{\mathrm{CE}} \pm \frac{e^2}{4\pi\varepsilon_0 R_{ab}} S^2 \, . \tag{4.96}$$

We now recall our original task, which was to compute the numerator of (4.56), using the wavefunctions Ψ_g and Ψ_u. If we multiply all the functions within Ψ_g or Ψ_u, respectively, by each other, then we obtain (as already pointed out) contributions of the type (4.81) twice, and contributions of the type (4.89) twice. Finally, we have to divide the whole thing by the normalisation integral. We then obtain for the total energy of the hydrogen molecule the following expression:

$$E_{g,u} = 2\frac{\hat{E} \pm \tilde{E}}{\iint |\Psi|^2 dV_1 dV_2}\,, \tag{4.97}$$

where the upper or lower sign applies in the energies $\hat{E}$ and $\tilde{E}$, according to whether the wavefunction Ψ_g or Ψ_u was used:

$$E_g = 2E_0 + \frac{2C + E_{RI}}{1 + S^2} + \frac{2DS + E_{CE}}{1 + S^2} + \frac{e^2}{4\pi\varepsilon_0 R_{ab}}\,, \tag{4.98}$$

$$E_u = 2E_0 + \frac{2C + E_{RI}}{1 - S^2} - \frac{2DS + E_{CE}}{1 - S^2} + \frac{e^2}{4\pi\varepsilon_0 R_{ab}}\,. \tag{4.99}$$

In order to determine whether or not chemical bonding occurs, we must test whether E_g or E_u is lower than the energy of the two infinitely separated H atoms, given by $2E_0$. Various effects are in competition here, as we can see on closer examination of the individual terms in e.g. (4.98). Thus, C, the potential energy of an electron in the Coulomb field of the opposite proton, is negative [cf. (4.84)], while the Coulomb interaction energy between the two electrons, E_{RI}, is positive. Furthermore, the last term in (4.98), which describes the Coulomb repulsion of the protons for each other, is also positive. In addition, there are the typically quantum-mechanical effects represented by the exchange interactions, which can be summarised in

$$K = 2DS + E_{CE}\,. \tag{4.100}$$

While DS is negative, the Coulomb-exchange interaction between the electrons, E_{CE}, is found to be positive. Whether or not chemical bonding finally comes about thus depends on the numerical values of the individual integrals.

It is not our purpose here to deal with the numerical evaluation of the integrals in detail. This evaluation reveals that the overall contribution of the exchange integrals (4.100) is negative; this makes the energy corresponding to the even (g) wavefunction lower than that of the odd (u) wavefunction. Furthermore, for the even wavefunction Ψ_g, the nett effect of the various Coulomb interactions is to yield an energy lower than that of two free hydrogen atoms. This state is therefore referred to as the bonding state. The lowering of the energy is – in addition to the effects of exchange (4.100) – due to the fact that the electrons can both occupy the region between the two nuclei simultaneously and thus can profit from the attractive Coulomb potential of both protons, in such a way as to compensate the repulsive potential between the electrons themselves and between the nuclei. This is similar to the case of H_2^+ discussed earlier. The energy lowering depends on the distance between the nuclei; an energy minimum is found for a particular internuclear distance (Fig. 4.12). As can be seen in the figure, the odd wavefunction Ψ_u does not lead to an energy lowering; for this reason, the corresponding state is called the antibonding (or non-bonding) state.

The dissociation energy, which is equal to the difference between the minimum energy at the equilibrium nuclear distance (bond length) and the energy at a distance $R_{ab} = \infty$,

Fig. 4.12. The binding energy of the hydrogen molecule as a function of the internuclear distance R_{ab}, taking the repulsive Coulomb interaction of the nuclei into account. (*Lower curve*) The electron spins are antiparallel. (*Upper curve*) The electron spins are parallel

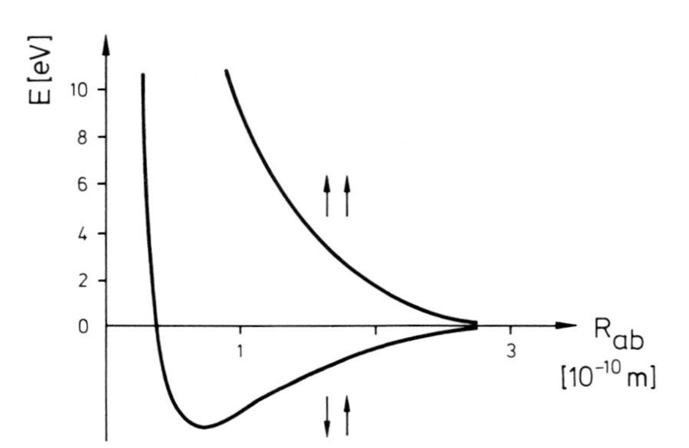

is found from a calculation based on the wavefunction given above to be 3.14 eV. The observed binding energy, which is equal to the dissociation energy, is, in contrast, 4.48 eV; however, it should be remembered that the nuclei themselves make a contribution through their kinetic energy. If this contribution, which was neglected in our calculation where we assumed the nuclear masses to be infinite, is subtracted, we arrive at a binding energy of 4.75 eV. We see that there is still a considerable difference between the calculated and the measured binding energies. This means that the wavefunctions of the Heitler-London model are still a very rough approximation. Although they show us that the bonding in the hydrogen molecule can be understood theoretically, they can give only a rough approach to the form of the true wavefunctions. In order to improve the wavefunctions, some additional effects must be taken into account; we shall discuss here one of the most typical, which is called covalent-ionic resonance.

4.4.3 Covalent-Ionic Resonance

In the previous section, we used as a wavefunction for the two electrons in the hydrogen molecule one in which the first electron spends its time for the most part near one nucleus, while the second electron is near the opposite nucleus. In this case, which is termed "covalent", the wavefunction has the form

$$\Psi_{\text{cov}} = N[\phi_a(r_1)\phi_b(r_2) + \phi_a(r_2)\phi_b(r_1)] \, , \tag{4.101}$$

where N is a normalisation factor.

It is of course possible, at least with a certain probability, that *both* electrons are on one of the hydrogen atoms; the wavefunction is then of the form:

$$\phi_a(r_1)\phi_a(r_2) \, . \tag{4.102}$$

Since the two nuclei are equivalent, both electrons could just as well be near nucleus b, which would correspond to the wavefunction

$$\phi_b(r_1)\phi_b(r_2) \, . \tag{4.103}$$

The functions (4.102) and (4.103) describe states in which there is a negatively charged hydrogen ion present. They are therefore referred to as "ionic" states. The states represented by (4.102) and (4.103) are energetically degenerate, and so we must form a linear combination to obtain the overall wavefunction. We do this in a symmetric form:

$$\Psi_{\text{ion}} = N'[\phi_a(r_1)\phi_a(r_2) + \phi_b(r_1)\phi_b(r_2)] , \qquad (4.104)$$

so that (4.104) has the same symmetry as (4.101). Now we must expect that nature does not choose exclusively the wavefunction (4.101) nor the wavefunction (4.104), since the electrons repel each other to some extent but can also be near the same nucleus some of the time. Both situations are possible, and thus according to the basic rules of quantum mechanics, the most realistic wavefunction should be constructed as a linear combination of the two possible states, (4.101) and (4.104):

$$\Psi = \Psi_{\text{cov}} + c\,\Psi_{\text{ion}} , \qquad (4.105)$$

where the constant c represents a variable parameter, which must be adjusted so as to minimise the energy expectation value belonging to the wavefunction (4.105).

4.4.4 The Hund-Mullikan-Bloch Theory of Bonding in Hydrogen

Along with the Heitler-London method, which we have described above, a second method is often used in molecular physics; in general, it does not give such good results for the total binding energy as the Heitler-London method, but it does allow the spatial probability distribution of the electrons to be more closely delineated. This is particularly important for spectroscopic investigations of molecules, since in such work, usually only one electronic state undergoes a change and it is just this change which one wishes to describe theoretically.

In this method, one at first ignores the fact that two electrons are present. Instead, we consider the motion of a single electron in the field of the two nuclei or, in other words, we begin with the solution of the hydrogen molecule-ion problem. We examined this solution in Sect. 4.3; it has the form:

$$\psi_{\text{g}}(\boldsymbol{r}) = N[\phi_a(\boldsymbol{r}) + \phi_b(\boldsymbol{r})] . \qquad (4.106)$$

The idea is now to place *both* of the electrons of the hydrogen molecule into the state (4.106). To solve the Schrödinger equation with the Hamiltonian (4.63) for the two electrons, we therefore take as trial wavefunction

$$\Psi(\boldsymbol{R}_1, \boldsymbol{R}_2) = \psi_{\text{g}}(\boldsymbol{r}_1)\psi_{\text{g}}(\boldsymbol{r}_2) \cdot \text{spin function} , \qquad (4.107)$$

where $\boldsymbol{R}_1$ and $\boldsymbol{R}_2$ include both the spatial coordinates $\boldsymbol{r}_1$ and $\boldsymbol{r}_2$ and the spin coordinates. We shall concentrate our attention here on the case of antiparallel spins, so that the spin function is antisymmetric and has the form

$$\text{spin function} = \frac{1}{\sqrt{2}}[\alpha(1)\beta(2) - \alpha(2)\beta(1)] . \qquad (4.108)$$

The total wavefunction (4.107) is clearly antisymmetric with respect to the spatial and spin coordinates of the electrons. Using the trial function (4.107), the expectation value of the total energy can again be computed. It is found to be higher in energy than that of the Heitler-London method, i.e. not as realistic. The method we have just described is called the LCAO method, for *Linear Combination of Atomic Orbitals*. Such a linear combination, e.g. (4.106), represents the wavefunction of a single electron in a molecule and is therefore termed a *Molecular Orbital* (MO).

This method can be extended to more complex molecules, as we shall see later. However, it requires some modifications for many molecules, and we shall treat the most important and most characteristic of them in this book.

4.4.5 Comparison of the Wavefunctions

In later chapters, we will be concerned with finding suitable trial wavefunctions for molecules containing more than two electrons. We therefore now compare the different trial functions for the hydrogen molecule in its ground state with the electronic spins antiparallel. For the sake of clarity, we leave off the normalisation factor of the functions Ψ_g on the right-hand side of the following equations, since we are interested only in the structure of the wavefunctions. The trial functions are then given by:

Heitler-London

$$\Psi_g = [\phi_a(1)\phi_b(2) + \phi_a(2)\phi_b(1)] \tag{4.109}$$

Heitler-London + ionic

$$\Psi_g = [\phi_a(1)\phi_b(2) + \phi_a(2)\phi_b(1)] + c[\phi_a(1)\phi_a(2) + \phi_b(1)\phi_b(2)] \tag{4.110}$$

Hund-Mullikan-Bloch

$$\Psi_g = [\phi_a(1) + \phi_b(1)]\,[\phi_a(2) + \phi_b(2)]\;. \tag{4.111}$$

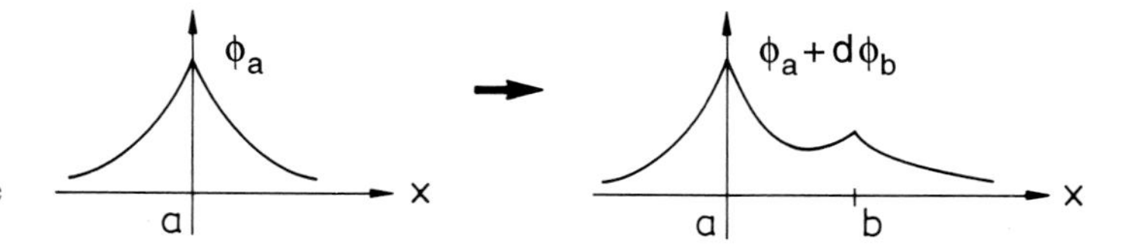

Fig. 4.13. A visualisation of the substitution (4.112)

We will now show that all these trial functions, (4.109–4.111), are special cases of a more general wavefunction, which we construct in this section. In the process, we mix into the wavefunction which originally referred to atom a a portion of the wavefunction from atom b and *vice versa* for the wavefunction originally referring to atom b. We thus make the substitution (see Fig. 4.13):

$$\phi_a \to \phi_a + d\phi_b\;, \qquad \phi_b \to \phi_b + d\phi_a\;, \tag{4.112}$$

where d is a constant coefficient, with $d \leq 1$.

We thereby define a new wavefunction according to

$$\begin{aligned}\Psi_g(1,2) = {} & [\phi_a(1) + d\phi_b(1)]\,[\phi_b(2) + d\phi_a(2)] \\ & + [\phi_a(2) + d\phi_b(2)]\,[\phi_b(1) + d\phi_a(1)]\;.\end{aligned} \tag{4.113}$$

This can be transformed by a simple calculation into:

$$\begin{aligned}\Psi_g(1,2) = {} & (1+d^2)[\phi_a(1)\phi_b(2) + \phi_a(2)\phi_b(1)] \\ & + 2d[\phi_a(1)\phi_a(2) + \phi_b(1)\phi_b(2)]\;.\end{aligned} \tag{4.114}$$

If we now set $d = 0$, then we obtain the Heitler-London trial wavefunction, (4.109). On the other hand, $d = 1$ yields the Hund-Mullikan-Bloch trial function, (4.111). If we factor out $(1+d^2)$ from the right-hand side of (4.114) and put it into the common normalisation constant, a comparison between (4.114) and (4.110) gives the result

$$\frac{2d}{1+d^2} = c\,. \tag{4.115}$$

In other words, the trial function (4.110), which contained an improvement to the original Heitler-London function through the addition of an ionic part, is also included as a special case in (4.113). The trial function (4.113) can be improved still further by including the wavefunctions of *excited* atomic states in the linear combination of (4.112). These considerations show us a first, important way towards formulating the wavefunctions for molecules with many electrons.

4.5 Hybridisation

An important case which is of particular interest for organic chemistry is that of *hybridisation*. In considering it, we also for the first time deal with atoms containing more than one electron. In forming molecules, the electrons in the inner, closed atomic shells are not strongly influenced; chemical bonding occurs via the outer electrons (valence electrons), which are more weakly bound to their atomic nuclei. In the carbon atom, two of the six electrons are in the $1s$ orbital, two in the $2s$ orbital, and two are distributed among the three orbitals $2p_x$, $2p_y$, and $2p_z$. The l degeneracy of the $n=2$ shell, which was found to hold in the hydrogen atom, is lifted here. However, the 4 eV energy splitting between the $2s$ and $2p$ states is not very large, and there is in fact an excited state of the carbon atom in which an electron from the $2s$ state has made a transition into the $2p$ state. In this case, the states $2s$, $2p_x$, $2p_y$, and $2p_z$ each contain one electron. Let us now consider these singly-occupied states carefully while we allow external forces to act on an electron by bringing a hydrogen atom close to the carbon atom. These external forces can, so to speak, compensate the energy difference which still remains between the $2s$ and the $2p$ states, making them practically degenerate in energy.

As we know from perturbation theory in the presence of degeneracy, in such a case we have to take linear combinations of the old functions, which were degenerate. For example, instead of the $2s$- and $2p$-functions, we construct two new functions having the form:

$$\begin{aligned}\psi_+ &= \psi_s + \psi_{p_x}\\ \psi_- &= \psi_s - \psi_{p_x}\,.\end{aligned} \tag{4.116}$$

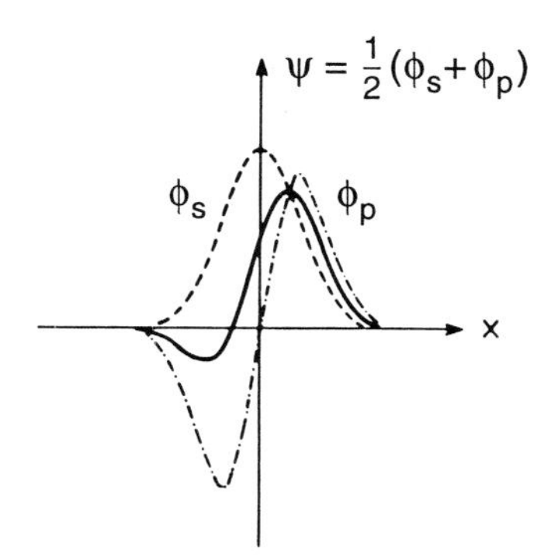

Fig. 4.14. The shape of the wavefunctions in the case of diagonal hybridisation. The s-function ϕ_s (*dashed curve*) and the p-function ϕ_p (*dot-dashed curve*) as well as the function which results from their superposition (*solid curve*) are plotted against the distance from the nucleus. The figure clearly shows how the centre of gravity of the wavefunctions shifts to the right on superposing the two functions ϕ_s and ϕ_p

Linear combinations of this type can shift the centre of gravity of the electronic charge clouds relative to that of the s-function (see Fig. 4.14). Exactly this phenomenon occurs in hybridisation.

Let us consider several types of hybridisation, beginning with the most well-known case, that of methane, CH_4, where the carbon atom is surrounded by four hydrogen atoms. Experimentally, it is known that the carbon atom sits at the centre of a tetrahedron with the four hydrogen atoms at its vertices (Fig. 4.15). Interestingly, the four degenerate wavefunctions of the $n=2$ shell in the carbon atom can be used to form four linear combinations whose centres of gravity are shifted precisely towards the four vertices of a tetrahedron. If we remember that the wavefunctions of the p states have the form $f(r)x$, $f(r)y$, and $f(r)z$, then it becomes clear that the following linear combinations produce the shifts in the charge centres of gravity described above (tetrahedral configuration):

$$\begin{aligned}\psi_1 &= \tfrac{1}{2}(\psi_s + \psi_{p_x} + \psi_{p_y} + \psi_{p_z}) ,\\ \psi_2 &= \tfrac{1}{2}(\psi_s + \psi_{p_x} - \psi_{p_y} - \psi_{p_z}) ,\\ \psi_3 &= \tfrac{1}{2}(\psi_s - \psi_{p_x} + \psi_{p_y} - \psi_{p_z}) ,\\ \psi_4 &= \tfrac{1}{2}(\psi_s - \psi_{p_x} - \psi_{p_y} + \psi_{p_z}) .\end{aligned} \tag{4.117}$$

These wavefunctions are mutually orthogonal *in the quantum-mechanical sense*, as one can readily verify by inserting the ψ_j for $j = 1, \ldots, 4$ into $\int \psi_j^*(\boldsymbol{r})\psi_k(\boldsymbol{r})dV$ and using the orthogonality of the ψ_s, ψ_{p_x}, ψ_{p_y}, and ψ_{p_z} functions. This type of orthogonality is not to be confused with orthogonality of the spatial orientation! Using these new linear combinations, (4.117), we can "tune" the electrons of the carbon atom to the tetrahedral environment. Each one of the four wavefunctions in (4.117) can now form a chemical bond with the corresponding hydrogen atom (Fig. 4.15).

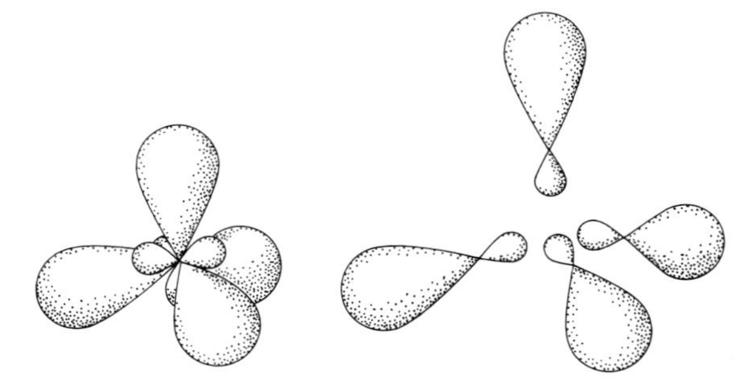

Fig. 4.15. (*Left*) The electron density distribution of the four orbitals in tetrahedrally hybridised carbon. (*Right*) An exploded view of the hybrid orbitals

Taking as an example the direction of vertex 1, we denote the carbon hybrid wavefunction ψ_1 in (4.117) more precisely as $\psi_{\mathrm{C}1}$, and that of the hydrogen atom at this vertex as $\psi_{\mathrm{H}1}$. Similarly to the case of the hydrogen molecule, we now generate a wavefunction for each of the two electrons involved in the bond formation; these take the following form, according to the LCAO prescription:

$$\psi(\boldsymbol{r}) = \psi_{\mathrm{C}1}(\boldsymbol{r}) + c\psi_{\mathrm{H}1}(\boldsymbol{r}) . \tag{4.118}$$

Owing to the difference between the carbon atom and the hydrogen atom, the constant coefficient c will always be $\neq 1$ (in contrast to the hydrogen molecule), and it must be determined by applying the variational method.

In the present case, we have oriented our considerations to the experimental finding that the four hydrogen atoms are located at the vertices of a tetrahedron. One could now be tempted to ask the question as to whether the wavefunctions (4.117) are initially present and the hydrogen atoms then locate themselves at the vertices of the tetrahedron thus defined, or conversely the hydrogen atoms first move to the vertices of a tetrahedron and thereby cause the carbon wavefunctions to generate corresponding hybrid orbitals. From the quantum-mechanical point of view, such speculations are pointless. The positions of the hydrogen atoms and the orientation of the hybrid wavefunctions are mutually consistent. The overall configuration is adopted by the CH_4 molecule in such a way as to minimise the total energy.

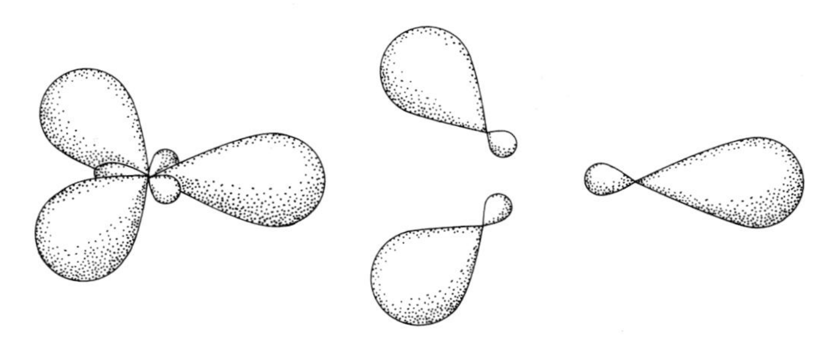

Fig. 4.16. (*Left*) The density distribution of the three orbitals in the case of trigonal hybridisation of carbon. (*Right*) An exploded view of the orbitals

The tetrahedral hybridisation just discussed, i.e. an arrangement of the wavefunctions resulting in tetrahedral symmetry, is not the only type of hybridisation possible for the carbon atom. We have already mentioned a second type, *diagonal* hybridisation, which is expressed in the wavefunctions (4.116) (see Fig. 4.14).

For carbon, still a third type of hybridisation is possible, the *trigonal* configuration, in which the s-, p_x-, and p_y-wavefunctions hybridise as suitable linear combinations to yield hybrid orbitals in three preferred directions within a plane. In order to give the reader an impression of how such hybrid orbitals are written, we show them explicitly (Fig. 4.16):

$$\begin{aligned} \psi_1 &= \sqrt{\tfrac{1}{3}}(\psi_s + \sqrt{2}\psi_{p_x}) \,, \\ \psi_2 &= \sqrt{\tfrac{1}{3}}(\psi_s + \sqrt{\tfrac{3}{2}}\psi_{p_y} - \sqrt{\tfrac{1}{2}}\psi_{p_x}) \,, \\ \psi_3 &= \sqrt{\tfrac{1}{3}}(\psi_s - \sqrt{\tfrac{3}{2}}\psi_{p_y} - \sqrt{\tfrac{1}{2}}\psi_{p_x}) \,. \end{aligned} \tag{4.119}$$

These wavefunctions are also mutually orthogonal in the quantum-mechanical sense.

Clearly, in generating these three hybrid wavefunctions, no use is made of the fourth original carbon wavefunction, $2p_z$. It plays an additional role in bonding, as we shall see directly. We consider the case of *ethene*, C_2H_4. Here, two carbon atoms take on the trigonal configuration. The hydrogen-carbon bonds are again formed by wavefunctions of the type given in (4.118), where for ψ_{C1} we insert, e.g. ψ_2 from (4.119). One carbon-carbon bond is formed by the first of these wavefunctions, with each carbon atom contributing one electron. However, the electrons occupying the p_z-orbitals are still left over. These remaining atomic orbitals form linear combinations, in analogy to the hydrogen molecule in the Hund-Mullikan-Bloch model, giving rise to an additional carbon-carbon bond. We thus have a case of double bond formation between the two carbon atoms (Fig. 4.17). This configuration is referred to as sp^2 or trigonal hybridisation.

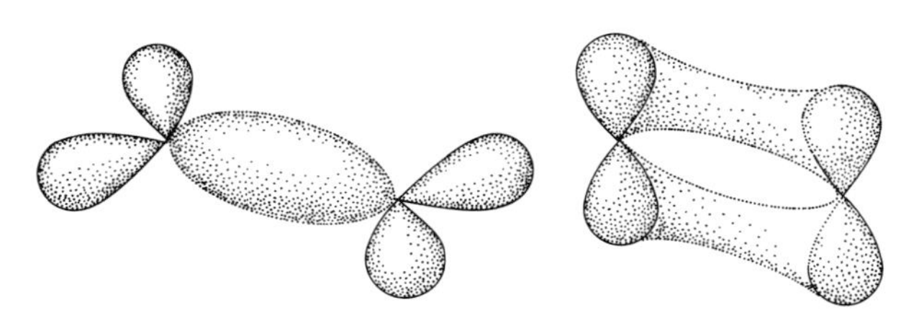

Fig. 4.17. The electron density distribution of the hybrid orbitals of the carbon atom in ethene, C_2H_4. (*Left*) The two carbon atoms are located at the two opposite nodes, and each takes on a trigonal configuration together with the corresponding hydrogen atoms. (*Right*) The perpendicularly-oriented p_z functions of the two carbon atoms form an additional carbon-carbon bond

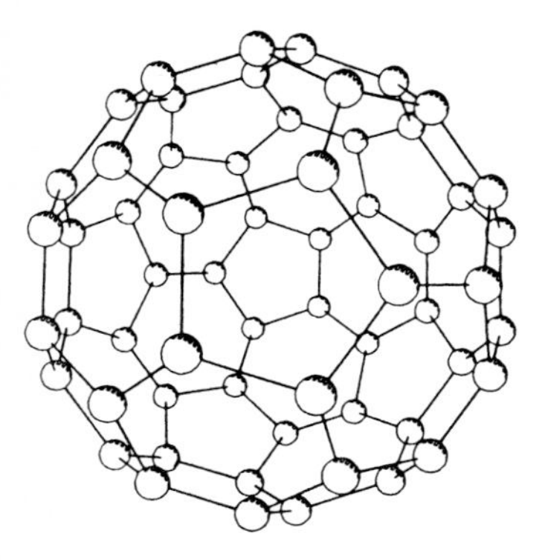

Fig. 4.18. The structure of the C_{60} molecule, "Buckminster-Fullerene", discovered in a molecular beam. [After H.W. Kroto, J.R. Heath, S.C. O'Brien, R.F. Curl, and R.E. Smalley, Nature **318**, 162 (1985)]. It can also be produced by vapourising graphite in a helium atmosphere. [See W. Krätschmer, K. Fostiropoulos, and D.R. Hoffmann, Chem. Phys. Lett. **170**, 167 (1990)]

An especially elegant example of trigonal hybridisation is provided by the "Buckminster-Fullerene" molecule, C_{60}, known for short as "fullerene", which was discovered in 1985. This molecule has attracted considerable attention because of its properties, which are quite unusual in a variety of ways. It consists of 12 pentagonal and 20 hexagonal units, i.e. altogether 32 rings, and has the shape of a soccer ball with a diameter of roughly 7 Å; see Fig. 4.18. As in benzene, the p-orbitals which extend outside the spherical surface of the molecule are not localised and their electrons can move as π-electrons throughout the molecule. C_{60} can form various compounds, such as $C_{60}H_{60}$. In addition to C_{60}, other molecules of the C_n structure have been identified, with n varying from 32 up to several hundred. These molecules can also act as cages, in which other atoms can be trapped, or in which different C_n molecules can be enclosed in a multiple-shell structure, like Russian dolls.

5. Symmetries and Symmetry Operations: A First Overview

In this chapter, we cover the fundamentals and theoretical approaches which we will need for – among other things – determining the wavefunctions and the energies of the π-electrons in benzene. A second example will be the ethene molecule.

5.1 Fundamental Concepts

Symmetries and symmetry operations play a still more important role in molecular physics than they do in the quantum theory of atoms. In the present section, we will cast an initial glance at this topic, and will then directly apply some of the knowledge we have gained. In Chap. 6, we shall again treat the subject of symmetries and symmetry operations systematically and in more detail.

In molecular physics, it is generally important to know the geometry of the molecule of interest from experimental studies before attempting a theoretical treatment. We will have the task of calculating the wavefunctions, or also the possible vibrational motions of the nuclei, taking this observed symmetry into account. We can draw on the example of the benzene molecule as a starting point for our considerations (Fig. 5.1a). It is planar and has the shape of an equilateral hexagon, i.e. if we rotate the molecule through an angle of 60° about an axis perpendicular to its plane, it remains unchanged. Another example is provided by H_2O, which remains unchanged if it is rotated through an angle of 180° about an axis perpendicular to its plane (see Fig. 5.2). NH_3 is symmetric with respect to rotations of 120° (Fig. 5.3). The ICl_4^- ion is planar and is unchanged by a rotation of 90° (Fig. 5.4), while all linear molecules, such as HCN (Fig. 5.5), are symmetric with respect to a rotation about the common internuclear axis through any arbitrary angle ϕ.

Making use of these examples, we discuss more precisely just what is meant by symmetry and symmetry operations. For this purpose, we first carry out a little thought experiment: we imagine that in H_2O, the initially quite identical hydrogen nuclei are distinguishable, and

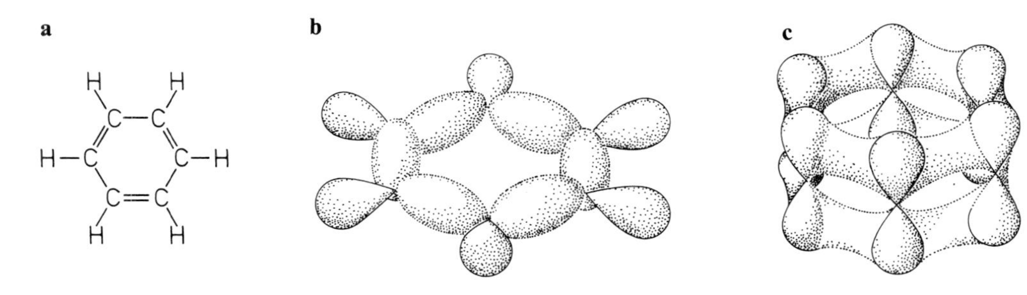

Fig. 5.1a–c. Benzene, C_6H_6. (**a**) structure formula; (**b**) charge density of the σ-electrons; (**c**) charge density of the π-electrons

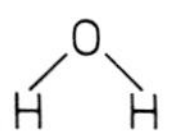

Fig. 5.2. H_2O

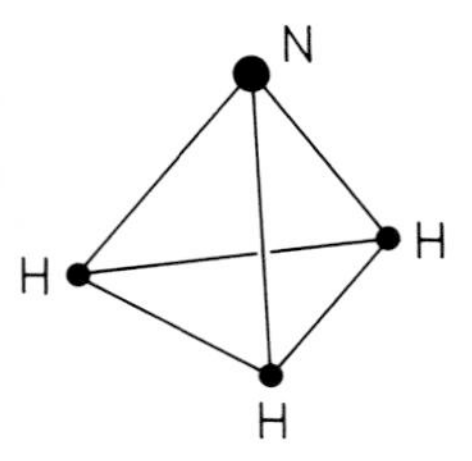

Fig. 5.3. NH_3

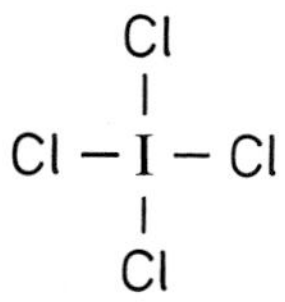

Cl
|
Cl − I − Cl
|
Cl

Fig. 5.4. ICl_4

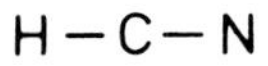

H − C − N

Fig. 5.5. HCN

then we rotate the water molecule in such a way that the two protons exchange places. We then make the protons again indistinguishable. After the rotation of the molecule, one can thus no longer see that it had been rotated at all. In the course of such a rotation through an angle ϕ, the coordinates of the individual atoms in the molecule are of course changed. Using the standard notation for molecules, we denote the rotation as C. In order to specify the angle ϕ through which the rotation took place, we can put it as an index on C: C_ϕ. We shall use that notation occasionally in this section. However, it is more usual to choose the index as the number n, which tells us how many times a rotation must be repeated until the original state is again restored; in other words, $n\phi = 2\pi$. For example, if $\phi = 60°$ (or in radians, $\phi = \pi/3$), we find $n = 6$. In the benzene molecule, the rotational symmetry can thus be described as C_6.

We now consider the effect of a rotation on the Cartesian coordinates. They can be written compactly in terms of the position vector

$$\boldsymbol{r} = \begin{pmatrix} x \\ y \\ z \end{pmatrix} . \tag{5.1}$$

A rotation through the angle ϕ corresponds to a new position vector $\boldsymbol{r}'$. The relation between $\boldsymbol{r}$ and $\boldsymbol{r}'$ is then given by:

$$\boldsymbol{r}' = C_\phi \boldsymbol{r} , \tag{5.2}$$

where C_ϕ means: carry out a rotation of $\boldsymbol{r}$ through the angle ϕ. As we know from elementary mathematics, the primed and unprimed coordinate systems are related by the equations:

$$\begin{aligned} x' &= x \cos\phi + y \sin\phi , \\ y' &= -x \sin\phi + y \cos\phi , \\ z' &= z . \end{aligned} \tag{5.3}$$

In order to keep the notation simple, in the following we will leave off the angle ϕ or the number n as index to C:

$$C_\phi \to C . \tag{5.4}$$

Since the distance from the origin remains constant in a rotation, we can immediately write down the relation

$$Cr^2 = r'^2 = r^2 , \tag{5.5}$$

i.e. we could also write

$$r' = r . \tag{5.6}$$

The rotation operation can now be applied to the coordinates of any particle we wish; not only to the protons in hydrogen, but also to, e.g. the electron in a hydrogen atom. The application of the rotation operator C to the wavefunctions $\psi(\boldsymbol{r})$ of the hydrogen atom then means simply that we rotate the coordinates $\boldsymbol{r}$, i.e. the following relation holds:

$$C\psi(\boldsymbol{r}) = \psi(C\boldsymbol{r}) = \psi(\boldsymbol{r}') . \tag{5.7}$$

Let us consider how the wavefunctions transform under the rotation C according to (5.7). We begin with the $1s$-function of hydrogen, which has the form

$$\psi(\boldsymbol{r}) = N\mathrm{e}^{-r/r_0} \tag{5.8}$$

(cf. Fig. 4.1), where N is a normalisation constant. According to the definition (5.7), and taking the relation (5.6) into account, we obtain

$$C\psi(\boldsymbol{r}) = N\mathrm{e}^{-r'/r_0} = N\mathrm{e}^{-r/r_0} \ . \tag{5.9}$$

Under rotation, the wavefunction of the hydrogen atom in the $1s$-state thus remains unchanged, or in other words, it is *invariant with respect to a rotation* C.

Let us see as a preparation for future use how the p functions transform; they can be represented using either real or complex functions. Starting with the real representation, we associate the wavefunction

$$\psi_{p_x} = xf(r) \tag{5.10}$$

with a "dumbbell" lying along the x-direction (compare Fig. 4.1). The function $f(r)$, which depends only on the radius r, can be written in the form

$$f(r) = N\mathrm{e}^{-r/r_0} \ , \tag{5.11}$$

but any distance-dependent function, e.g. the radial functions for larger values of the principal quantum number, would also meet the requirements of which we make use in the following. The other two dumbbells are given by

$$\psi_{p_y} = yf(r) \tag{5.12}$$

and

$$\psi_{p_z} = zf(r) \ . \tag{5.13}$$

Consider now what happens when we let the rotation operator C for a rotation about the z-axis act on these wavefunctions according to the general definition (5.7). We obtain:

$$\begin{aligned} C\psi_{p_x} &= x'f(r') = \cos\phi xf(r) + \sin\phi yf(r) = \cos\phi\psi_{p_x} + \sin\phi\psi_{p_y} \ , \\ C\psi_{p_y} &= y'f(r') = -\sin\phi xf(r) + \cos\phi yf(r) = -\sin\phi\psi_{p_x} + \cos\phi\psi_{p_y} \ , \\ C\psi_{p_z} &= \psi_{p_z} \ . \end{aligned} \tag{5.14}$$

In (5.14), the first step (from left to right) was carried out according to (5.7), the second according to (5.3), and for the third, we made use of the definitions (5.10) and (5.12). Application of the rotation operation C thus transforms the wavefunctions ψ_{p_x}, ψ_{p_y}, and ψ_{p_z} among themselves. This already shows us the tip of the iceberg of a general truth which we shall meet again in a much more general context. We note in this connection that the p-functions of the hydrogen atom just referred to all belong to the same energy. As we shall show generally later on, wavefunctions which belong to the same energy are transformed into linear combinations of the same set of wavefunctions by symmetry operations. The question will also arise as to whether there are not simple cases where a wavefunction is transformed into *itself* on application of a symmetry operation. This in fact is true in the present case, if instead of the real representation of the p-state wavefunctions we use certain linear combinations of them. These are complex and are eigenfunctions of the operator for the z-component of angular momentum. They are given by

$$\psi_{\pm} = \psi_{p_x} \pm \mathrm{i}\psi_{p_y} = N(x \pm \mathrm{i}y)\,\mathrm{e}^{-r/r_0} \ , \tag{5.15}$$

where N is again a normalisation constant. The form $x \pm \mathrm{i}y$ can be treated as a complex variable in the complex plane, and we introduce the usual polar coordinates for it:

$$x + \mathrm{i}y = r\mathrm{e}^{\mathrm{i}\phi} . \tag{5.16}$$

Then (5.15) becomes

$$\psi_{\pm} = Nr\mathrm{e}^{-r/r_0}\mathrm{e}^{\pm\mathrm{i}\phi} . \tag{5.17}$$

In the complex plane, a rotation through an angle ϕ_0 means that the original angle ϕ is to be replaced by $\phi + \phi_0$. We thus obtain

$$C_{\phi_0}\mathrm{e}^{\mathrm{i}\phi} = \mathrm{e}^{\mathrm{i}(\phi+\phi_0)} \tag{5.18}$$

and therefore

$$C_{\phi_0}\psi_{+} = \mathrm{e}^{\mathrm{i}\phi_0}\psi_{+} , \tag{5.19}$$

and, correspondingly,

$$C_{\phi_0}\psi_{-} = \mathrm{e}^{-\mathrm{i}\phi_0}\psi_{-} . \tag{5.20}$$

The relations (5.19) and (5.20) naturally mean that the application of the rotation operator leaves the functions ψ_+ and ψ_- unchanged aside from a constant factor $\mathrm{e}^{\mathrm{i}\phi_0}$ or $\mathrm{e}^{-\mathrm{i}\phi_0}$.

5.2 Application to Benzene: the π-Electron Wavefunctions by the Hückel Method

As is known experimentally, the benzene molecule, C_6H_6, is planar: the H atoms lie in the same plane as the C atoms, which are joined to form a hexagonal ring (cf. Fig. 5.1a). If we look at a particular carbon atom, we find that it has a trigonal arrangement for the bonds to the two neighbouring C atoms and the H atom which extends outside the ring. Just as in ethene (see Sect. 4.4), we see that each carbon atom has one p_z orbital, containing one electron, left over after forming the trigonal hybrid orbitals. All such p_z orbitals in the 6 different carbon atoms are energetically equivalent; an electron could, in principle, occupy any one of these states. Let us now recall the basic approach of the LCAO method, i.e. the method of linear combinations of atomic orbitals (cf. Sect. 4.4.4). It requires us first to search for the wavefunction of each single electron in the field of all the atomic cores, i.e. here in the field of all 6 carbon atoms. In principle we are dealing here with a generalisation of the hydrogen molecule problem; however, an electron can now be spread over six atoms instead of over two.

We suppose that all the orbitals of the carbon atoms which lie in the molecular plane, i.e. the $1s$ orbitals and the hybrid orbitals made up of the $2s$, $2p_x$, and $2p_y$ atomic orbitals, have been filled with electrons of lower energies. These hybrid orbitals in benzene are referred to as σ orbitals (Fig. 5.1b). (We shall have more to say about the notation for orbitals in Chap. 13.) Similarly to the case of ethene, some electronic wavefunctions remain: those derived from the $2p_z$ states; they extend outwards perpendicular to the molecular plane and are localised on the individual carbon atoms. We can assume that the electrons which the carbon atoms contain in these orbitals move independently of each other in the fields of the carbon atomic cores, including the already-occupied σ orbitals. We will justify this assumption in detail later on; for the moment, we have the task of determining the wavefunction of an electron in a field which is symmetric with respect to rotations of 60° about an axis perpendicular to the molecular plane. We apply the Hund-Mullikan-Bloch method just as in the case of the hydrogen molecule: we represent the wavefunction we are seeking as a linear combination of wavefunctions located on the carbon atoms, more precisely the $2p_z$ wavefunctions. The molecular orbitals which result are called π orbitals. In order to make use of our symmetry considerations, we first investigate the behaviour of such a function belonging to the carbon atom with index j. According to Fig. 5.6, we can represent this function as

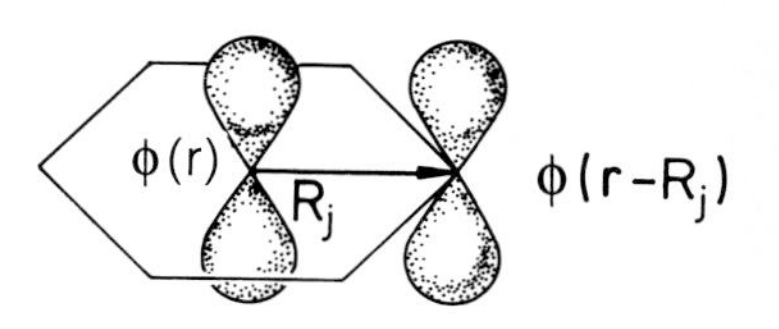

Fig. 5.6. The p_z function $\phi(\boldsymbol{r})$ is transformed into $\phi(\boldsymbol{r}-\boldsymbol{R}_j)$ by a translation through the vector $\boldsymbol{R}_j$

$$\phi_j(\boldsymbol{r}) = \phi(\boldsymbol{r}-\boldsymbol{R}_j) , \tag{5.21}$$

where for concreteness we will keep in mind a representation of the function (5.13). When we carry out a rotation through an angle of 60° (see Fig. 5.7), we obtain the result

$$C_6\phi_j(\boldsymbol{r}) = \phi_j(C_6\boldsymbol{r}) = \phi(C_6\boldsymbol{r}-\boldsymbol{R}_j) , \tag{5.22}$$

where we have used the definition (5.21). As a result of the symmetry of the problem, the vector $\boldsymbol{R}_j$, which points from the centre of the molecule towards the nucleus of carbon atom j, can be interpreted as a rotated vector which was produced from the vector $\boldsymbol{R}_{j-1}$ by a rotation through 60° (Fig. 5.7):

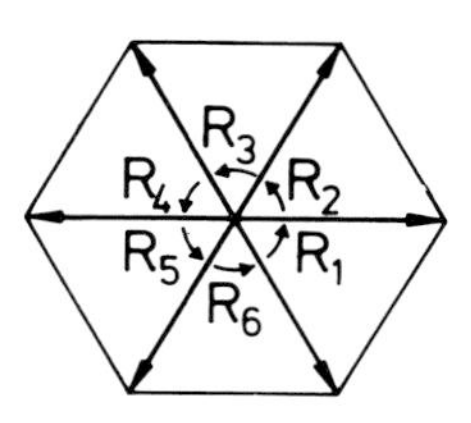

Fig. 5.7. On rotation through 60° (*small arrows*), the vectors $\boldsymbol{R}_j$ are transformed into one another

$$C_6\boldsymbol{R}_{j-1} = \boldsymbol{R}_j . \tag{5.23}$$

Then instead of (5.22), we can write

$$C_6\phi_j(\boldsymbol{r}) = \phi(C_6\boldsymbol{r} - C_6\boldsymbol{R}_{j-1}) . \tag{5.24}$$

Next, we can factor out the operator C_6 from the parenthesis in (5.24), yielding:

$$C_6\phi_j(\boldsymbol{r}) = \phi[C_6(\boldsymbol{r}-\boldsymbol{R}_{j-1})] . \tag{5.25}$$

Now, the z-direction is not influenced at all by a rotation about the z-axis, and furthermore the distance $\boldsymbol{r}-\boldsymbol{R}_{j-1}$ remains unchanged by a rotation. Therefore, (5.25) is equivalent to:

$$C_6\phi_j(\boldsymbol{r}) = \phi(\boldsymbol{r}-\boldsymbol{R}_{j-1}) . \tag{5.26}$$

Thus, with the aid of some mathematical reformulations, we have obtained the result that a p_z-wavefunction on one carbon atom is transformed by a 60° rotation of the molecule into the corresponding wavefunction on an adjacent carbon atom:

$$C_6\phi_j(\boldsymbol{r}) = \phi_{j-1}(\boldsymbol{r}) . \tag{5.27}$$

Following these elementary preparations, we shall now see how useful symmetry considerations can be in molecular physics. To this end, we consider the wavefunction $\psi(\boldsymbol{r})$ of an electron which moves throughout the whole molecule in its potential field, as mentioned above. The corresponding Schrödinger equation is given by:

$$H(\boldsymbol{r})\psi(\boldsymbol{r}) = E\psi(\boldsymbol{r}) , \tag{5.28}$$

where the Hamiltonian H contains the kinetic energy of the electron and its potential energy in the molecular potential field. A rotation through 60° leaves this Hamiltonian unchanged, i.e. we obtain the relation

$$CH(\boldsymbol{r}) = H(\boldsymbol{r}') = H(\boldsymbol{r}) \, . \tag{5.29}$$

Now, we apply the rotation operation C to both sides of (5.28), yielding:

$$CH(\boldsymbol{r})\psi(\boldsymbol{r}) = EC\psi(\boldsymbol{r}) \, . \tag{5.30}$$

Using (5.29), the operation C on the left side of (5.30) acts only on the wavefunction:

$$H(\boldsymbol{r})C\psi(\boldsymbol{r}) = EC\psi(\boldsymbol{r}) \, . \tag{5.31}$$

If we compare the left sides of (5.30) and (5.31) and remember that they remain valid for any arbitrary wavefunction $\psi(\boldsymbol{r})$, we can by subtraction obtain an operator equation:

$$CH - HC = 0 \, . \tag{5.32}$$

The rotation operator C and the Hamiltonian H thus commute. This is another way of expressing the fact that the Hamiltonian is invariant under the rotation C. It follows from (5.31) that if $\psi(\boldsymbol{r})$ is a solution of the Schrödinger equation, then so is $C\psi(\boldsymbol{r})$.

We now assume for the moment that only a single wavefunction belongs to the energy E, i.e. that the energy level is not degenerate. In such a case, when two apparently different wavefunctions belong to the same energy, then there is a contradiction unless the two wavefunctions are in fact identical aside from a multiplicative constant which we will call λ; we thus obtain the relation:

$$C\psi(\boldsymbol{r}) = \lambda\psi(\boldsymbol{r}) \, . \tag{5.33}$$

Mathematically, one can show that in general, a relation like (5.33) always holds under rotation operations. This is related to the fact that for rotations, there always exists a number M such that an M-fold application of the rotation operation transforms the wavefunction back into itself. Formally, this means that

$$C^M = 1 \, . \tag{5.34}$$

We now use relation (5.33) to determine the coefficients of the LCAO wavefunction in a simple way. We represent ψ as a linear combination of the atomic wavefunctions ϕ_j according to

$$\psi(\boldsymbol{r}) = c_1\phi_1 + c_2\phi_2 + \ldots + c_6\phi_6 \, . \tag{5.35}$$

Inserting (5.35) into (5.33), we obtain

$$\begin{aligned} &c_1C\phi_1(\boldsymbol{r}) + c_2C\phi_2(\boldsymbol{r}) + \ldots + c_6C\phi_6(\boldsymbol{r}) \\ &\quad = \lambda[c_1\phi_1(\boldsymbol{r}) + c_2\phi_2(\boldsymbol{r}) + \ldots + c_6\phi_6(\boldsymbol{r})] \, . \end{aligned} \tag{5.36}$$

However, as we have just seen, the application of a rotation to the wavefunction ϕ_j, produces simply an exchange of the index j of the "base" carbon atom. Using this fact, (5.36) is changed into

$$c_1\phi_6(\boldsymbol{r}) + c_2\phi_1(\boldsymbol{r}) + \ldots + c_6\phi_5(\boldsymbol{r}) = \lambda[c_1\phi_1(\boldsymbol{r}) + \ldots + c_6\phi_6(\boldsymbol{r})] \, . \tag{5.37}$$

Since here the individual wavefunctions ϕ_j are linearly independent of each other, (5.37) can be valid only if the coefficients of the same functions ϕ_j on the left and the right sides of the equation are equal. This leads immediately to the relations

$$\begin{aligned} c_1 &= \lambda c_6 \ , \\ c_2 &= \lambda c_1 \ , \\ c_3 &= \lambda c_2 \ , \\ &\vdots \\ c_6 &= \lambda c_5 \ . \end{aligned} \tag{5.38}$$

To solve them, we take the trial solution

$$c_j = \lambda^j c_0 \ , \tag{5.39}$$

where c_0 is a normalisation constant. If we apply the rotation operation in the case of benzene six times, the molecule is returned to its original state; from this, it follows that

$$\lambda^6 = 1 \ . \tag{5.40}$$

According to the calculational rules for complex numbers, (5.40) has the solution

$$\lambda = \mathrm{e}^{2\pi k \mathrm{i}/6} \ , \tag{5.41}$$

with k an integer which must be chosen according to:

$$k = 0, 1, 2, \dots, 5$$

or

$$k = 0, \pm 1, \pm 2, +3 \ . \tag{5.42}$$

We now insert the result (5.39) together with (5.41) and (5.42) into (5.35) and obtain the explicit form which the wavefunction must take, namely:

$$\psi = c_0 \sum_{j=1}^{6} \mathrm{e}^{2\pi k \mathrm{i} j/6} \phi_j(\boldsymbol{r}) \ . \tag{5.43}$$

This is the wavefunction of the π-electrons of benzene (compare Fig. 5.1c). We have thus succeeded in solving the Schrödinger equation without having to carry out any calculations involving the Hamiltonian operator. Symmetry alone was sufficient to fix the coefficients uniquely, leaving only the normalisation constant c_0 to be determined.

5.3 The Hückel Method Once Again. The Energy of the π-Electrons

As we know, the carbon atom has two electrons in the $1s$ shell ("core electrons") and in addition 4 electrons in the $n = 2$ shell. These four electrons participate in bonding to other atoms and are therefore called valence electrons. We have seen that a distinction is made in the bonding of carbon in the benzene molecule between σ- and π-electrons. The wavefunctions of the σ-electrons are located in the plane of the molecule, while the π-electrons, which originate with the p_z atomic orbitals, are oriented perpendicularly to the molecular plane; it is for them a nodal plane.

We select one of these π-electrons and assume that it moves in the combined potential of the nuclei, the σ and core electrons, and the other π-electrons. The direct interaction of the electrons with each other is thus replaced by an effective potential. As we shall see later, and should already know from atomic physics (cf. I), such a procedure can be justified in the framework of the Hartree-Fock approximation. The Hamiltonian which refers to the π-electrons is then

$$H_\pi^{\text{Hückel}} = \sum_\mu \left[-\frac{\hbar^2}{2m_0} \nabla_\mu^2 + V(\boldsymbol{r}_\mu) \right] , \tag{5.44}$$

where the sum over μ runs from 1–6, enumerating the six π-electrons of the carbon atoms. Equation (5.44) clearly contains a sum of Hamiltonians, each one of which refers to a *single* electron. Therefore, the Schrödinger equation belonging to (5.44) can be solved by finding the wavefunctions of the *individual electrons* as solutions to the Schrödinger equation

$$\left[-\frac{\hbar^2}{2m_0} \nabla^2 + V(\boldsymbol{r}) \right] \psi(\boldsymbol{r}) = E\psi(\boldsymbol{r}) . \tag{5.45}$$

The potential in (5.45) can be decomposed into two parts:

$$V(\boldsymbol{r}) = V_n(\boldsymbol{r}) + V^S(\boldsymbol{r}) \tag{5.46}$$

of which one part, $V_n(\boldsymbol{r})$, is due to the nuclei and the other part, $V^S(\boldsymbol{r})$, to the σ- and π-electrons. Following the prescription of the Hund-Mullikan-Bloch method, we represent the wavefunction of a single electron as a linear combination of atomic wavefunctions, in this case the carbon $2p_z$ wavefunctions, as follows:

$$\psi = \sum_{j=1}^{N} c_j \phi_j(\boldsymbol{r}) . \tag{5.47}$$

The coefficients c_j are still unknown and can be determined with the aid of the variational principle, according to which the left-hand side of

$$\frac{\int \psi^* H \psi \, dV}{\int \psi^* \psi \, dV} = E \tag{5.48}$$

is to be minimised by a suitable choice of the coefficients. Inserting (5.47) into the numerator of (5.48), we obtain

$$\sum_{jj'} c_j^* c_{j'} \cdot \underbrace{\int \phi_j^* H \phi_{j'} \, dV}_{H_{jj'}} , \tag{5.49}$$

where we will use the abbreviation $H_{jj'}$ in what follows. In the same way, we find the denominator of (5.48):

$$\sum_{jj'} c_j^* c_{j'} \cdot \underbrace{\int \phi_j^* \phi_{j'} \, dV}_{S_{jj'}} . \tag{5.50}$$

The energy on the right-hand side of (5.48) is a function of the coefficients, so that we can write

$$E = E(c_1, c_1^*, c_2, c_2^*, \ldots) \,. \tag{5.51}$$

A necessary condition for obtaining a minimum in E is that the derivatives with respect to the coefficients c_j and c_j^* vanish:

$$\frac{\partial E}{\partial c_j} = \frac{\partial E}{\partial c_j^*} = 0 \,. \tag{5.52}$$

For computational reasons, it is more practical to multiply equation (5.48) on both sides by the denominator and consider the resulting expression,

$$\sum_{jj'} c_j^* c_{j'} \cdot H_{jj'} = E \sum_{jj'} c_j^* c_{j'} \cdot S_{jj'} \,. \tag{5.53}$$

We now differentiate this equation with respect to the coefficients c_j^*, obtaining

$$\sum_{j'} H_{jj'} \cdot c_{j'} = E \sum_{j'} c_{j'} \cdot S_{jj'} \,. \tag{5.54}$$

We have already set the derivatives of E with respect to c_j^* equal to zero in (5.54), using (5.52). Equation (5.54) is a system of equations for the coefficients c_j which we can write explicitly in the form

$$\begin{aligned}
(H_{11} - S_{11}E)c_1 + (H_{12} - S_{12}E)c_2 + \quad \ldots \quad &+ (H_{1N} - S_{1N}E)c_N = 0 \,, \\
(H_{21} - S_{21}E)c_1 + (H_{22} - S_{22}E)c_2 + \quad \ldots \quad &+ (H_{2N} - S_{2N}E)c_N = 0 \,, \\
\vdots \qquad\qquad\qquad & \qquad\quad \vdots \\
(H_{N1} - S_{N1}E)c_1 + (H_{N2} - S_{N2}E)c_2 + \quad \ldots \quad &+ (H_{NN} - S_{NN}E)c_N = 0 \,.
\end{aligned} \tag{5.55}$$

Since this is a system of homogeneous equations, the determinant of its coefficients,

$$\begin{vmatrix}
H_{11} - ES_{11} & H_{12} - ES_{12} \ldots & H_{1N} - ES_{1N} \\
\vdots & \vdots & \vdots \\
H_{N1} - ES_{N1} & H_{N2} - ES_{N2} \ldots & H_{NN} - ES_{NN}
\end{vmatrix} = 0 \,, \tag{5.56}$$

must vanish, if we wish to obtain a nontrivial solution. This is clearly not a very simple problem, since we are already dealing with a $6 \cdot 6$ determinant.

Using symmetry considerations, however, one can solve this problem very simply! In the previous section, we saw that the coefficients are known [compare (5.43)]. It is therefore unnecessary to solve the determinant equation (5.56); instead, we can substitute the known coefficients directly into the system (5.55). In this way, we can determine the energy E for the general system (5.55) explicitly. In order to emphasise the essentials, we assume the following simplifications:

$$\begin{aligned}
&S_{jj} = 1, \qquad S_{jj'} = 0, \qquad j \neq j' \\
&H_{jj} = A, \qquad H_{j,j-1} = H_{j,j+1} = B, \qquad \text{otherwise} = 0 \,.
\end{aligned} \tag{5.57}$$

These conditions are equivalent to neglecting the overlap between the wavefunctions and considering interaction energies only within one atom and with next-neighbour atoms. We now insert the simplifications (5.57) and the form of the coefficients c_j [from (5.43)],

$$c_j = c_0 \mathrm{e}^{2\pi \mathrm{i} jk/6} \,, \tag{5.58}$$

into, e.g. the first line of (5.55), thus obtaining

$$\mathrm{e}^{2\pi \mathrm{i}k/6}(A - E) + \mathrm{e}^{2\pi \mathrm{i}2k/6} B + \mathrm{e}^{2\pi \mathrm{i}6k/6} B = 0 \,, \tag{5.59}$$

which can be immediately resolved into the form

$$E = A + B(\mathrm{e}^{2\pi \mathrm{i}k/6} + \mathrm{e}^{-2\pi \mathrm{i}k/6}) \,. \tag{5.60}$$

All the other lines of (5.55) give the same result. Using the real representation, (5.60) can be written as

$$E = A + 2B \cos\left(\frac{2\pi k}{6}\right) \,. \tag{5.61}$$

In this equation, k takes on the values prescribed by (5.42), i.e.

$$k = 0, \pm 1, \pm 2, +3 \,. \tag{5.62}$$

Taking into account the fact that the exchange integral B is negative,

$$B < 0 \,, \tag{5.63}$$

we obtain the term diagram shown in Fig. 5.8 for the π-electrons of benzene. It can be filled with the carbon electrons, starting from the lowest energy level and taking the Pauli exclusion principle into account. The energies shown in Fig. 5.9 are then obtained.

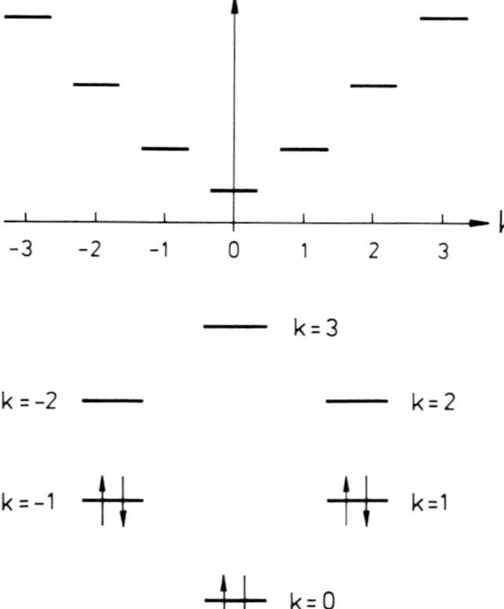

Fig. 5.8. The term diagram for the π-electrons of benzene

Fig. 5.9. The states occupied by the π-electrons of benzene. [Note that $E(-k) = E(k)$]

The use of symmetry considerations brought a considerable simplification compared to traditional theoretical methods in this case. We were able to determine the coefficients explicitly beforehand, without having to solve the system of equations (5.55). In particular, we did not need to calculate the determinant (5.56) and find its eigenvalues explicitly, which otherwise would have been necessary. Furthermore, our calculation using the Hückel method has the advantage that we can also treat excited states according to the term diagram of Fig. 5.8, since their energies are already known from (5.60).

5.4 Slater Determinants

Let us return to the solution of the many-electron problem, for example in the case of benzene. Here, we make use of two pieces of knowledge which we had gained previously: if the Hamiltonian consists of a sum of operators, then – generalising the method used for the hydrogen molecule – the wavefunction of all the electrons may be written as a product of the wavefunctions of individual electrons. In doing this, it is important to take the spin of each electron into account using the spin functions α (spin up) and β (spin down). In order that the overall wavefunction be antisymmetric in the spatial *and* the spin coordinates as required by the Pauli exclusion principle, we use a determinant for the ground state wavefunction, generalising the approach given by (4.71). In this determinant, the counting index of the electrons is the row index and the quantum number of the state occupied by the electron is the column index. The determinant thus has the form:

$$\Psi(1,2,\ldots,6)=\begin{vmatrix} \psi_1(\boldsymbol{r}_1)\alpha(1) & \psi_1(\boldsymbol{r}_1)\beta(1) & \psi_2(\boldsymbol{r}_1)\alpha(1) & \psi_2(\boldsymbol{r}_1)\beta(1)\ldots \\ \psi_1(\boldsymbol{r}_2)\alpha(2) & \psi_1(\boldsymbol{r}_2)\beta(2)\ldots & & \\ \vdots & & & \\ \psi_1(\boldsymbol{r}_6)\alpha(6) & \psi_1(\boldsymbol{r}_6)\beta(6)\ldots & & \end{vmatrix} . \tag{5.64}$$

This expression is called a Slater determinant. Clearly, writing down such determinants is tedious; they are therefore often abbreviated in the form:

$$\Psi(1,2,\ldots,6)=|\psi_1,\overline{\psi}_1,\psi_2,\overline{\psi}_2\ldots\psi_6,\overline{\psi}_6| , \tag{5.65}$$

where the arguments of ψ refer to the electrons and the indices of the wavefunctions ψ to the individual states, and we assume that each wavefunction is occupied by two electrons having antiparallel spins. Equation (5.65) thus yields the determinant (5.64) if we make the following replacements

$$\psi_j \rightarrow \psi_j\alpha , \quad \overline{\psi}_j \rightarrow \psi_j\beta , \tag{5.66}$$

and use the convention that a bar over the wavefunction refers to an electron with its spin down.

5.5 The Ethene Wavefunctions. Parity

As an additional example of the power of symmetry considerations, we treat the ethene molecule (Fig. 5.10). This molecule evidently has a centre of inversion symmetry at the midpoint of the line joining the two C atoms; i.e. if we reverse the signs of all the coordinates, x, y, z becoming $-x, -y, -z$, then the entire molecule remains unchanged. If we subject the wavefunction ψ of a single electron to this mirror operation, and again assume that the

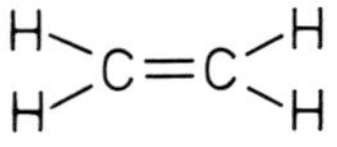

H₂C=CH₂ structure: H H >C=C< H H

Fig. 5.10. Ethene

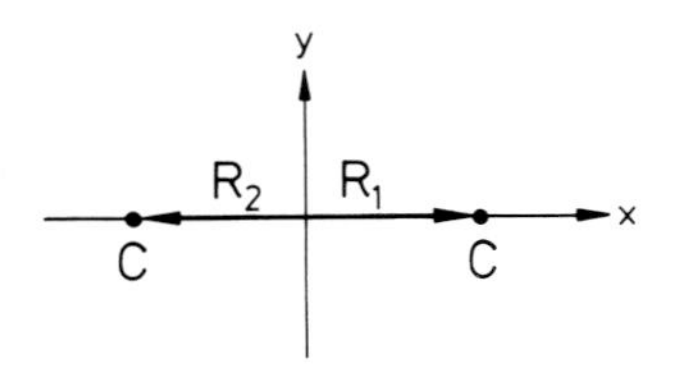

Fig. 5.11. The definitions of $\boldsymbol{R}_1$ and $\boldsymbol{R}_2$

wavefunctions are nondegenerate, we obtain $\psi(\boldsymbol{r}) = \lambda^2\psi(\boldsymbol{r})$. This means that λ can take on only the values +1 or −1. We then find

$$\psi(-\boldsymbol{r}) = \pm\psi(\boldsymbol{r}) , \tag{5.67}$$

or, as one also says, ψ has *even parity* (upper sign) or *odd parity* (lower sign).

The wavefunctions of the individual π-electrons are taken, as in the hydrogen molecule-ion, to be linear combinations of the $2p_z$ wavefunctions of the two carbon atoms (cf. Fig. 5.11):

$$\phi_1(\boldsymbol{r}) = \phi(\boldsymbol{r} - \boldsymbol{R}_1) , \quad \phi_2(\boldsymbol{r}) = \phi(\boldsymbol{r} - \boldsymbol{R}_2) ,$$

$$\psi(\boldsymbol{r}) = c_1\phi_1(\boldsymbol{r}) + c_2\phi_2(\boldsymbol{r}) . \tag{5.68}$$

For the atomic wavefunctions, we have the symmetry properties

$$\phi_1(-\boldsymbol{r}) = -\phi_2(\boldsymbol{r}) \tag{5.69}$$

and

$$\phi_2(-\boldsymbol{r}) = -\phi_1(\boldsymbol{r}) , \tag{5.70}$$

as one can readily see by making use of the explicit representation of ϕ:

$$\phi = Nz\mathrm{e}^{-r/r_0} . \tag{5.71}$$

Inserting (5.68) into (5.67) and making use of the properties (5.69) and (5.70), we obtain

$$c_1\phi_1(-\boldsymbol{r}) + c_2\phi_2(-\boldsymbol{r}) = -c_1\phi_2(\boldsymbol{r}) - c_2\phi_1(\boldsymbol{r}) = \pm c_1\phi_1(\boldsymbol{r}) \pm c_2\phi_2(\boldsymbol{r}) . \tag{5.72}$$

Comparing the coefficients of the same wavefunctions on the left and right-hand sides of (5.72), we find the relations

$$c_1 = \pm c_2 . \tag{5.73}$$

With this relation, we can substitute in the linear equation (5.54) for the coefficients, just as we did in the case of benzene; now, however, we have to deal with only two coefficients. In complete analogy to the calculation for benzene, we obtain

$$E = A \pm B , \tag{5.74}$$

where

$$B < 0 . \tag{5.75}$$

We can see that $c_1 = c_2$ leads to a bonding state and $c_1 = -c_2$ to an antibonding state. The term diagram which results is given in Fig. 5.12.

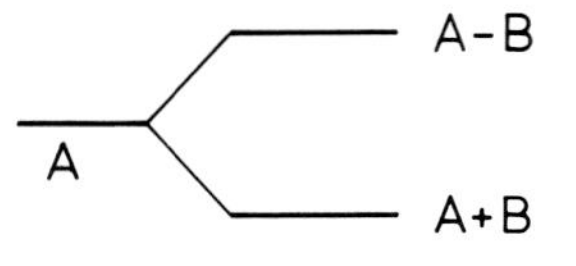

Fig. 5.12. The term diagram of ethene

5.6 Summary

In Chaps. 4 and 5, using concrete examples we have demonstrated some fundamental concepts required for the (at least approximate) calculation of the electronic wavefunctions of molecules, i.e. the molecular orbitals. We can summarise the basic ideas as follows:

1) The wavefunction of *all* the electrons of a molecule is approximated as a product or a determinant containing the wavefunctions of the *individual* electrons.
2) The individual wavefunction (molecular orbital) is constructed as a linear combination from atomic wavefunctions (LCAO method).
3) The coefficients of the LCAO wavefunctions are determined by using symmetry considerations, giving a considerable reduction in the computational effort required.

Clearly, some important questions are raised by the points 1) – 3):

1) Why is it allowable to use the approximation 1)? This question leads us to the Hartree-Fock method and its extensions, which we treat in Chap. 7.
2) and 3) How can we generalise the symmetry considerations? We take up this question in Chap. 6, where we treat molecular symmetries in a quite general way.

In Chaps. 6 and 7, the reader will thus gain a detailed introduction to the modern electron theory of molecules, which will allow him or her to delve into the scientific literature dealing with this subject.

6. Symmetries and Symmetry Operations. A Systematic Approach*

This chapter provides a systematic approach to the application of group theory for the determination of molecular wavefunctions. We treat molecular point groups, the effect of symmetry operators on wavefunctions, and then the basic concepts of the theory of group representations. The method is demonstrated using the explicit example of the H_2O molecule.

6.1 Fundamentals

In the preceding chapter, we saw how we could determine the π-electron orbitals of benzene in an especially elegant way by making use of the rotational symmetry of the molecule. In this chapter, we shall deal systematically with symmetries and symmetry operations, keeping concrete examples of molecules in mind. The symmetry properties of a molecule are characterised by the possible symmetry operations, e.g. rotations. In the course of such a symmetry operation, every point in space is transformed into another point, keeping the lengths of all distances constant. The object before and after the operation is indistinguishable.

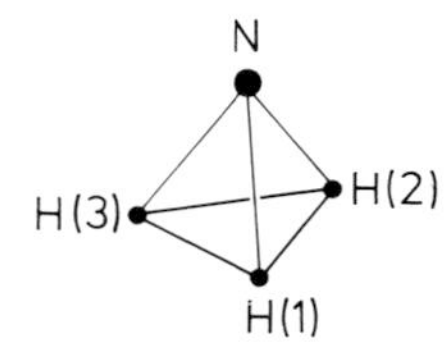

Fig. 6.1. The NH_3 molecule. The numbers 1, 2, and 3 are used to denote the positions of the hydrogen atoms

As a first example, we choose the NH_3 molecule, which can be described as a trigonal pyramid (Fig. 6.1). The three hydrogen atoms are located at the vertices of the equilateral basal triangle, and the nitrogen atom is directly above the centroid of the triangle. If the molecule is rotated about an axis passing through the N atom and the centroid, by an angle of 120° in the positive sense (i.e. counterclockwise as seen from above), then the H atoms exchange their places in the following manner: $H_3 \rightarrow H_1$, $H_1 \rightarrow H_2$, and $H_2 \rightarrow H_3$ (cf. Fig. 6.2). The N atom maintains its position. The state attained after each such rotation is indistinguishable from the original state, since the H atoms are all equivalent. In the course of these operations, neither lengths nor angles within the molecule are changed; the operations can therefore be considered to be symmetry operations. Analogous considerations hold for the reflection operations sketched in Fig. 6.3. The mirror planes are perpendicular to the basal triangle of the molecule and each one contains a bisector of the triangle. Thus, a reflection in the σ_1 plane exchanges the atoms H_2 and H_3, while H_1 and N remain unchanged.

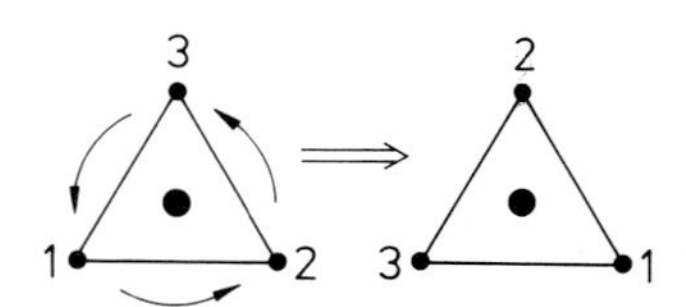

Fig. 6.2. The effect of the symmetry operation C_3, i.e. a rotation of 120° about a vertical axis

Symmetry operations are not to be confused with so-called symmetry elements. In the above example of the NH_3 molecule, the symmetry operation C_3 tells us how to carry out a rotation through 120°. The set of all points which do not change their spatial positions during this symmetry operation form the symmetry element "axis of rotation", which is likewise denoted by C_3. In the case of the reflections in σ_1, σ_2, and σ_3, the symmetry element is the respective mirror plane. A symmetry element is defined as the set of all points on which the symmetry operation is carried out. In the case of elementary or non-composite symmetry operations (Table 6.1), the symmetry element is equivalent to the set of all points which

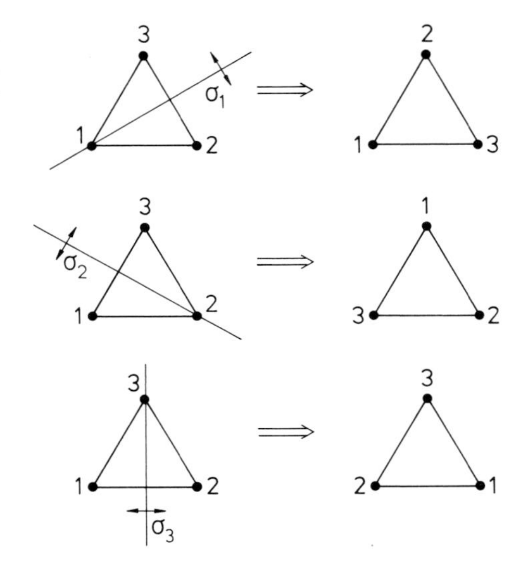

Fig. 6.3. The effects of the symmetry operations σ_1, σ_2, and σ_3, i.e. of the reflections illustrated

remain fixed in space when the symmetry operation is carried out. If at least one point remains invariant in the course of an operation (for higher symmetries: one line or one plane), the operation is referred to as a point symmetry operation. An example is inversion, i, in which the origin of the coordinate system forms the symmetry element and a coordinate vector $\boldsymbol{r}$ is transformed into $-\boldsymbol{r}$. Furthermore, it is useful for mathematical reasons to define the identity E formally as a symmetry operation. In this operation, all points of a three-dimensional object remain unchanged. In the case of polymers with a regular chain conformation or of crystal lattices, two additional symmetry operations can occur, which depend on the periodicity of the molecular chain or the lattice: the screw operation (translation + rotation), and the translation-reflection operation. Table 6.1 lists first the four simple point symmetry operations: the identity E, reflection σ, rotation C, and inversion i; and then the combined point symmetry operations: improper rotation S, translation-reflection $\overline{\sigma}$, and the screw operation $\overline{C}_n$, with their corresponding symmetry elements.

We can gain an intuitive understanding of the individual symmetry operations by considering an equilateral triangle as in Figs. 6.2 through 6.4. The symmetry operations result here in a permutation of the vertex numbers as shown in the figures. We will show with a few examples how a concatenation of two symmetry operations produces a new symmetry operation, which in the case of Fig. 6.4 can in fact be expressed through an operation that was already defined. Looking at Fig. 6.4a, we first carry out the reflection σ_2 and then the rotation C_3. The final product of this composite operation, shown on the right in the figure, could also have been obtained by reflecting the original triangle through the symmetry element σ_1 (Fig. 6.4b). We thus see that the relation $C_3\sigma_2 = \sigma_1$ holds (note that the operations on the left-hand side of this equation are to be read from right to left!). Now, what will happen if we reverse the order of the reflection and the rotation, i.e. first rotate and then reflect, as shown in Fig. 6.5a? The resulting triangle could also be obtained by a reflection through the σ_3 plane (Fig. 6.5b). We thus obtain the operator relation $\sigma_2 C_3 = \sigma_3$. Comparing the result of Fig. 6.4a with that of Fig. 6.5a, we can see that the results of a combined rotation and reflection depend on the order in which the operations are performed.

In other words, symmetry operations do not commute, at least in the present case. Quite generally, one can show that the product of two rotations again is equivalent to a rotation,

Table 6.1. Elementary and composite symmetry operations with the corresponding symmetry elements

Symbol	Symmetry operation	Symmetry element
E	"Identity operation"	Identity
C_n	Rotation through $2\pi/n$	n-fold axis of rotation
σ	Reflection	Mirror plane
i	Inversion (reflection at an inversion centre	Centre of inversion symmetry
S_n	Rotation through $2\pi/n$ followed by reflection (Improper rotation)	n-fold axis of rotary reflection symmetry
$\overline{\sigma}$	Translation-reflection (translation followed by a reflection)	Translation-reflection plane
$\overline{C}_n$	Screw operation (Translation followed by a rotation through $2\pi/n$)	Screw axis

while the product of a reflection followed by a rotation, or a rotation followed by a reflection, is equivalent to a reflection. Two successive reflections can be replaced by a rotation. We can summarise these results in a group operation table, as shown in Table 6.2.

Table 6.2. The multiplication table for the symmetry group C_{3v}. A multiplication BA leads to the new elements listed in the table

	Operation A						
	C_{3v}	E	C_3	C_3^2	σ_1	σ_2	σ_3
Operation B	E	E	C_3	C_3^2	σ_1	σ_2	σ_3
	C_3	C_3	C_3^2	E	σ_3	σ_1	σ_2
	C_3^2	C_3^2	E	C_3	σ_2	σ_3	σ_1
	σ_1	σ_1	σ_2	σ_3	E	C_3	C_3^2
	σ_2	σ_2	σ_3	σ_1	C_3^2	E	C_3
	σ_3	σ_3	σ_1	σ_2	C_3	C_3^2	E

We have thus arrived at the concept of a *group*. A group consists of a set of elementary operations with the following properties: concatenating two operations A and B yields a new operation, which likewise belongs to the group, according to $A\,B = C$. The set of symmetry operations contains an identity operation E which is defined so that $E\,A = A\,E = A$. Every operation A has an inverse operation A^{-1}, with $A\,A^{-1} = E$. It can then be shown that $A^{-1}\,A = E$ also holds. For the operations A, B, and C, an associative law is valid: $(A\,B)\,C = A\,(B\,C)$.

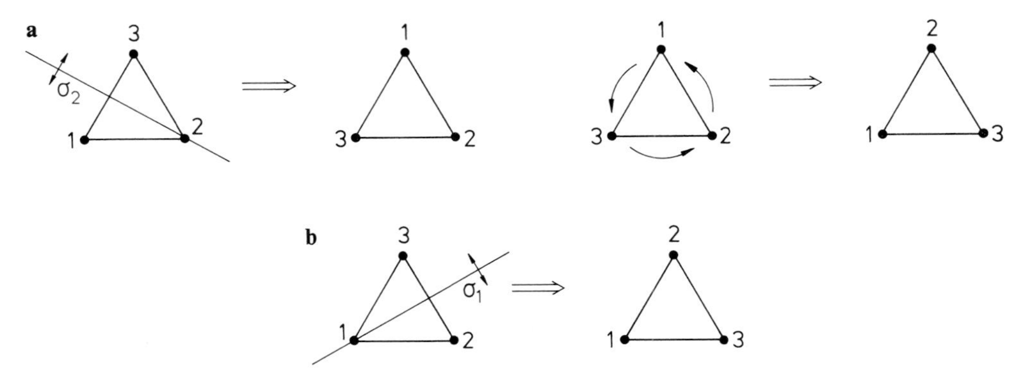

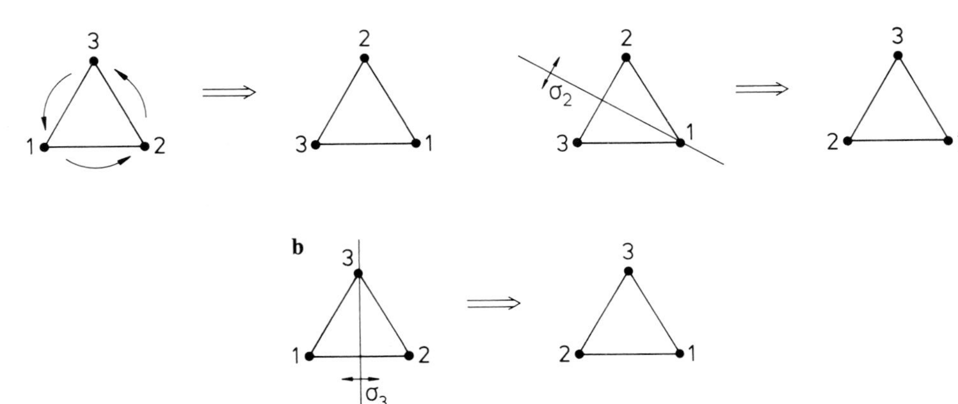

Fig. 6.4. (a) Effect of the symmetry operation σ_2 followed by C_3. **(b)** The same effect as in **(a)** is produced by σ_1

Fig. 6.5. (a) The effects of the symmetry operations C_3 followed by σ_2. **(b)** The same effect as in **(a)** is produced by σ_3

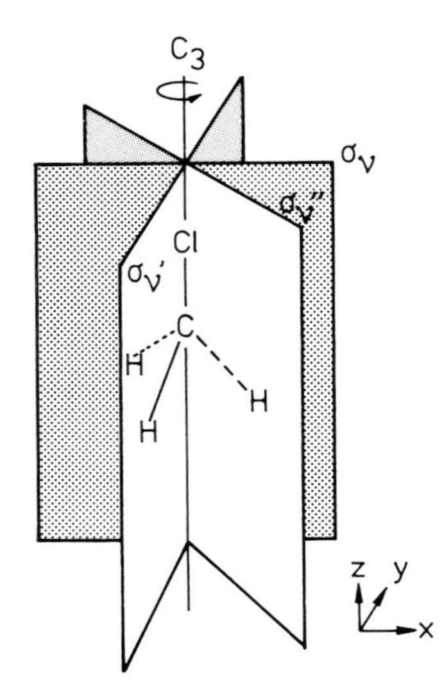

Fig. 6.6. The symmetry elements of the point group C_{3v}. The molecule chloromethane (CH_3Cl) is shown as an example

When the operations all mutually commute, i.e. $A\,B = B\,A$ for all A and B in the group, the group is called *Abelian*. Using the group table, it is easy to verify that the symmetry operations E, C_3, C_3^2, σ_1, σ_2, and σ_3 form a group. Following a notation convention which we will discuss in detail below, this group is called C_{3v}. From the symmetry operations of a molecule which form a group, we can often choose certain operations which among themselves fulfill the conditions for forming a group; these symmetry operations are placed in a subgroup of the original group. The multiplication table in Table 6.2 shows that the operations E, C_3, and C_3^2 form a subgroup of C_{3v}. In addition to NH_3, another example of a molecule with the point group C_{3v} is chloromethane. It is shown in Fig. 6.6, together with the symmetry elements of the group C_{3v}.

6.2 Molecular Point Groups

For the classification of the molecular point groups, we use the notation introduced by *Schönflies*. (Another notation, preferred by crystallographers, is that due to *Hermann-Mauguin*.) In the following section, we have collected all the point groups of molecules. We begin with molecules which allow the smallest possible number of symmetry operations in addition to the identity operation which is a member of all point groups. We then consider molecules with higher degrees of symmetry. Some examples are given in Fig. 6.7.

Molecules without an axis of rotational symmetry belong to the *point groups*

C_1: This point group contains, aside from the identity E, no additional symmetry elements. example: NHFCl.

C_S: The only symmetry element is a mirror plane. Example: NOCl (in Fig. 6.7, first row, left).

C_i: The only symmetry element is the centre of inversion symmetry, i. Example: ClBrHC–CHClBr in the *trans*-conformation.

All the other point groups refer to molecules with axes of rotational symmetry (rotation groups).

C_n: Molecules with an n-fold rotational axis ($n \neq 1$) as their only symmetry element. Examples: H_2O_2 (C_2) and $Cl_3C–CH_3$ (C_3). Linear molecules without a centre of inversion symmetry belong to the rotation group C_∞; they in addition possess an infinite number of mirror planes which intersect in the molecular axis ($C_{\infty v}$).

S_n: Molecules which have as their only symmetry element an axis of rotational-reflection symmetry of even order ($n = 2m$, beginning with $m = 2$). (For an example, see Fig. 6.7.) The point group S_2 contains only the inversion i and the identity operation E; therefore, $S_2 \equiv C_i$.

C_{nh}: Molecules with a rotational axis of order $n > 1$ ($C_{1h} \equiv C_S$) and a (horizontal) mirror plane perpendicular to it. (The term "horizontal" results from the convention that the rotation axis is taken to be vertical.) The $2n$ symmetry operations follow from those of the rotation group C_n and its combination with the reflection σ_h; $S_n = \sigma_h C_n$. If n is an even integer, the molecule contains a centre of inversion symmetry due to $S_2 \equiv i$. Example: butadiene in the planar *trans* conformation, C_{2h}. These molecules are invariant under the following elementary symmetry operations: identity operation E; rotation by 180° about an axis perpendicular to the plane of the image and passing through the centre of gravity of the molecule; reflection in the plane perpendicular to that axis and containing all the atoms of the molecule; and finally inversion about the centre of gravity of the molecule, which is also a centre of inversion symmetry.

C_{nv}: Molecules with a rotational axis and n mirror plane(s), all of which contain the rotational axis. The mirror plane is "vertical", since it contains the axis of rotation; it is denoted as σ_v. In the case $n > 2$, the symmetry operation C_n creates additional equivalent, vertical mirror planes. The symmetry operations of the point group C_{nv} are the rotations about the n-fold axis of rotation and the n reflections in the mirror planes. If n is an even integer, a distinction is made between two different types of mirror planes: every second mirror plane is denoted as σ_v, while the planes between are called σ_d (for dihedral). Examples: H_2CCl_2 (C_{2v}), NH_3 (C_{3v}). The group $C_{\infty v}$ contains linear molecules without a mirror plane perpendicular to the molecular axis (for example OCS); the symmetry operations are: infinitely many rotations about this axis, and just as many reflections in planes containing the molecular axis.

D_n: Molecules with an n-fold rotational axis ($C_n, n \geq 2$) and a twofold rotational axis perpendicular to the principal axis. An example of D_3 is $H_3C–CH_3$, if the CH_3 groups are staggered relative to each other by an angle which must not be a multiple of $\pi/3$.

D_{nh}: This point group contains, in addition to the symmetry elements of the point group D_n, a plane σ_h perpendicular to the principal axis (i.e. horizontal). Combining the rotation operations of the rotation group D_n with the reflections σ_h yields n improper rotations S_n ($S_n = \sigma_h C_n$) and n reflections σ_v ($\sigma_v = C_2\sigma_h$) in addition to the operations of D_n.

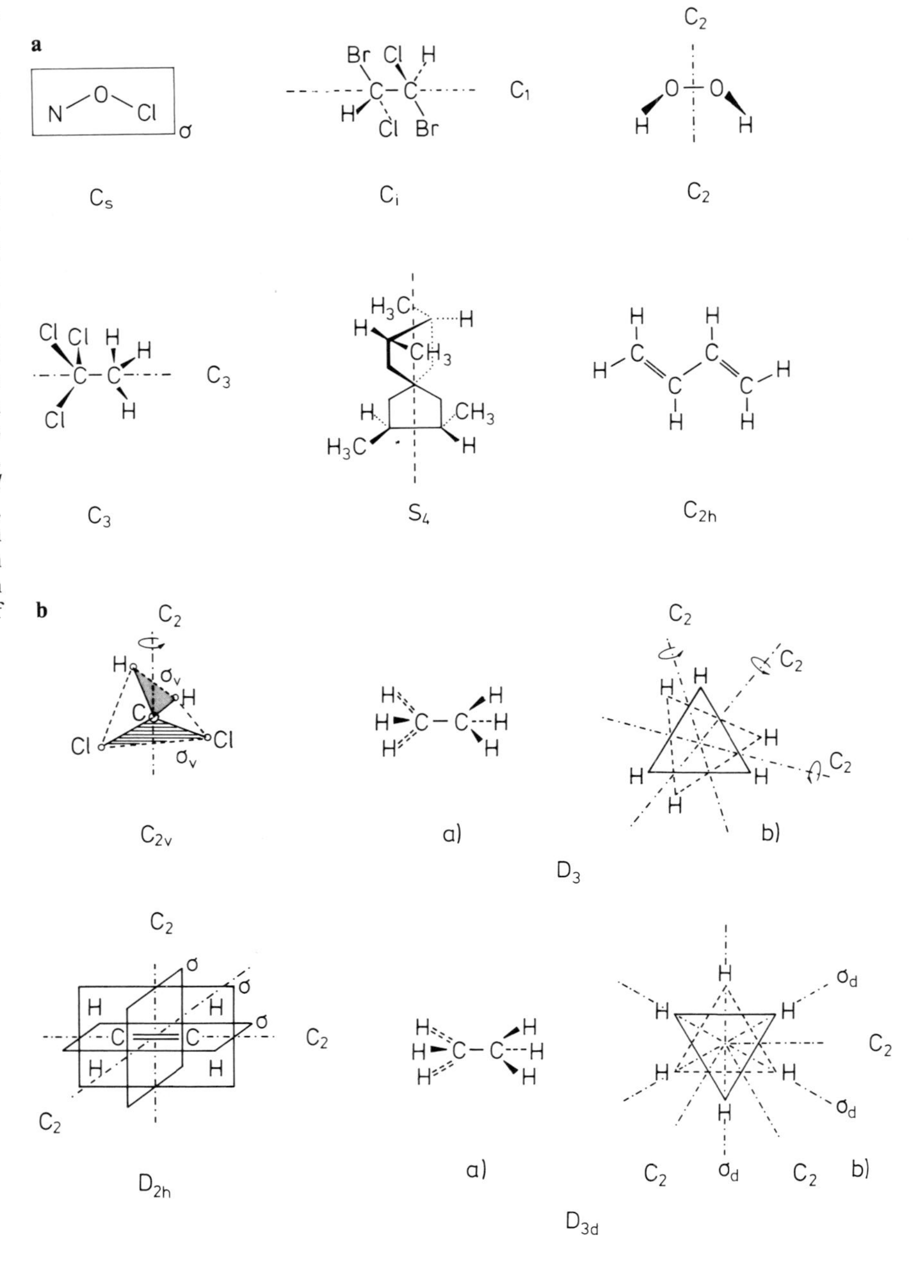

Fig. 6.7a,b. Some examples of point groups. (**a**) *1st row, left*: NOCl, point group C_s; *centre*: ClBrHC–CHClBr in the *trans* conformation, point group C_i; *right*: H_2O_2, point group C_2. *2nd row, left*: Cl_3C–CH_3, point group C_3; *centre*: point group S_4; *right*: butadiene in the planar *trans* conformation, point group C_{2h}. (**b**) *1st row, left*: H_2CCl_2, point group C_{2v}; *centre*: H_3C–CH_3, side view; *right*: ditto, but viewed along the C–C molecular axis. The CH_3 groups make an angle which is *not* a multiple of $\pi/3$; point group D_3. *2nd row, left*: H_2C–CH_2, point group D_{2h}; *centre and right*: as in the 1st row centre, but the CH_3 groups make an angle of $\pi/3$, i.e. the hydrogen atoms of one methyl group fit in the gaps between hydrogens of the other group

If n is even, the n mirror planes are divided into $n/2$ σ_v planes (containing a C_2 axis perpendicular to the principal axis) and $n/2$ σ_d planes (containing the angle bisectors between two C_2 axes perpendicular to the principal axis). Here, when n is even, a centre of inversion symmetry is again present, due to $S_2 \equiv i$.

D_{nd}: This point group contains, in addition to the symmetry operations of the group D_n, n reflections σ_d in planes containing the C_n axis and bisecting the angles between two neighbouring C_2 axes. The combination $C_2\sigma_d = \sigma_n C_{2n} = S_{2n}(C_2 - C_n)$ produces n additional improper rotations S_{2n}^k ($k = 1, 3, \ldots, 2n-1$). This point group has a centre of inversion symmetry when n is an odd number. An example for the point group D_{3d} is H_3C–CH_3, when the H atoms of the two CH_3 groups are offset into the gaps of the opposing group.

We now consider molecules which have more than one symmetry axis of more than twofold symmetry. The most important of these point groups are those which are derived from the equilateral tetrahedron and the regular octahedron (Fig. 6.8). Their pure rotation groups – that is the groups of operations which consist only of rotations about symmetry axes – are denoted by T and O. The symmetry group of the regular octahedron is at the same time that of a cube, since the latter has the same symmetry elements. In addition, a regular octahedron can be inscribed within a cube in such a manner that the sides and the vertices of the cube are equivalent with respect to the octahedron. O is thus the pure rotation group of a cube. An equilateral tetrahedron can also be inscribed within a cube. The vertices of the cube are now, however, no longer equivalent (four of the cube's vertices are now also vertices of the tetrahedron). The equilateral tetrahedron thus has a lower symmetry than that of a cube; T must be a subgroup of O. Figure 6.9 shows the axes of rotation belonging to the groups T and O.

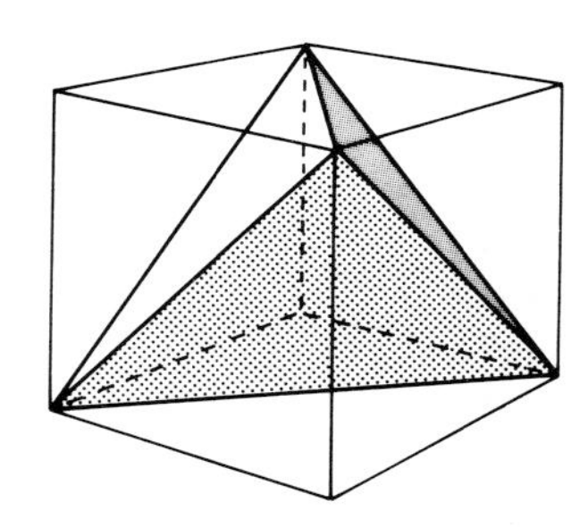

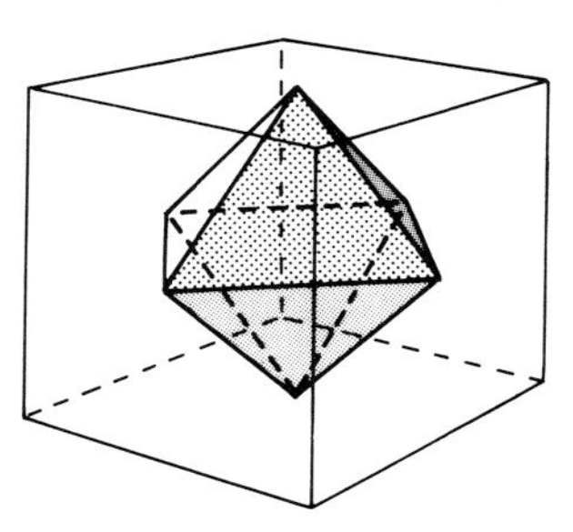

Fig. 6.8. *Upper part*: An equilateral tetrahedron inscribed in a cube; *Lower part*: A regular octahedron in a cube

T: The symmetry elements are the 4 threefold and the 3 twofold axes of the regular tetrahedron, which permit 12 rotational symmetry operations. The 4 C_3 axes of this rotation group pass through the centroid and one vertex each of the tetrahedron. The C_2 axes pass through the midpoints of opposite edges of the tetrahedron.

O: The 3 fourfold, the 4 threefold, and the 6 twofold rotational axes form the symmetry elements of the rotation group of the regular octahedron and allow 24 rotational symmetry operations. The C_4 axes pass through opposite vertices, the C_3 axes through the centroids of opposite faces, and the C_2 axes through the midpoints of opposite edges of the octahedron.

T_d: The full symmetry group of the equilateral tetrahedron consists of the rotational elements of the group T as well as 6 reflections in the σ_d mirror plane and 6 fourfold axes of improper rotation, S_4. The molecules CH_4, P_4, CCl_4, and a number of complex ions with tetrahedral symmetry belong to this point group.

O_h: Adding all 9 mirror planes of the cube to the pure rotation group O, we obtain the important group O_h. Examples of this symmetry are the molecule SF_6, the ion $(PtCl_6)^{2-}$, and numerous octahedral coordination compounds. The mirror planes $3\sigma_h$ and $6\sigma_d$ which belong to the group O_h give rise to the additional symmetry operations $6S_4$, $8S_6$, and i.

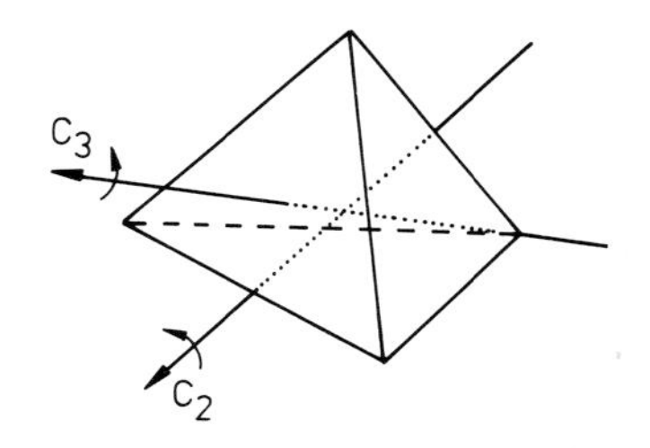

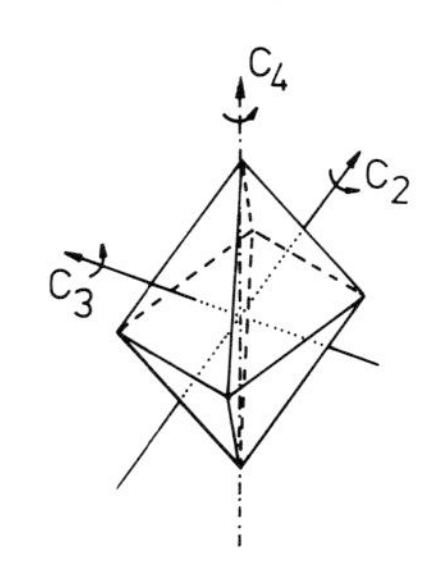

Fig. 6.9. Examples of the rotational axes of the equilateral tetrahedron (rotation group T) (*Upper part*) and of the regular ocathedron (rotation group O) (*Lower part*)

6.3 The Effect of Symmetry Operations on Wavefunctions

In Sect. 5.2, we showed using the example of benzene how a rotation of the coordinate system causes a transformation of the wavefunctions. We now want to expand on what we learned there in two ways:

1) We generalise to the case of arbitrary symmetry operations, not just rotations.
2) The wavefunctions may refer not only to a single electron, but to several.

In order to study the effects of symmetry operations on wavefunctions, let us assume that the set of mutually degenerate wavefunctions

$$\Psi_1, \Psi_2, \ldots, \Psi_M \tag{6.1}$$

all belong to a particular energy eigenvalue of a Schrödinger equation.

We first consider just one single symmetry operation, which we denote by A. Since the Hamiltonian is supposed to be invariant with respect to the transformation A, it must commute with A. However, this means that not only Ψ_1, but also A applied to Ψ_1 is an eigenfunction of the Schrödinger equation belonging to the same energy as the set (6.1) [compare (5.29–32)]. Since (6.1) were supposed to be the only wavefunctions belonging to this energy, it must necessarily be possible to represent $A\Psi_1$ as a linear combination of these wavefunctions, having the form:

$$A\,\Psi_1 = \sum_{m=1}^{M} a_{1m}\Psi_m \; . \tag{6.2}$$

The coefficients a_{1m} in this equation are constants, while the wavefunctions naturally depend upon the electronic coordinates. A relation of the form (6.2) holds not only for Ψ_1, but also for any wavefunction in the set (6.1), so that we obtain

$$A\,\Psi_j = \sum_{m=1}^{M} a_{jm}\Psi_m \; . \tag{6.3}$$

Here, the coefficients a_{jm} depend on the one hand on the index of the wavefunction which occurs on the left, but on the other hand also on the indices of the wavefunctions Ψ_m on the right. In this sense, we can say that the effect of the operator A on Ψ corresponds to the multiplication of the vector (6.1), which is then to be written as a column vector, by a matrix $[a_{jm}]$:

$$A \to [a_{jm}] \; . \tag{6.4}$$

The identity operation, which leaves the vector (6.1) unchanged, is denoted by E.

Let us now see what happens if we let first the operator A and then the operator B act on Ψ_j. We thus investigate the effect of the product BA when it is applied to the wavefunction Ψ_j, whereby in analogy to (6.3) we may assume that

$$B\,\Psi_j = \sum_{m=1}^{M} b_{jm}\Psi_m \tag{6.5}$$

holds.

We first insert the right side of (6.3) into $BA\Psi_j = B(A\Psi_j)$, obtaining

$$BA\,\Psi_j = B(A\,\Psi_j) = B\left(\sum_{m=1}^{M} a_{jm}\Psi_m\right) \; . \tag{6.6}$$

However, since B has nothing to do with the coefficients, but rather acts only on the wavefunction Ψ which follows them, we can write for the right-hand side of (6.6):

$$\sum_{m=1}^{M} a_{jm}B\,\Psi_m \tag{6.7}$$

and then use (6.5):

$$\sum_{m=1}^{M} a_{jm} \sum_{l=1}^{M} b_{ml} \Psi_l \ . \tag{6.8}$$

The two summations over l and m can be exchanged, so that we finally obtain instead of (6.6) the following equation:

$$BA\,\Psi_j = \sum_{l=1}^{M} \underbrace{\left(\sum_{m=1}^{M} a_{jm} b_{ml} \right)}_{c_{jl}} \Psi_l \ . \tag{6.9}$$

Application of the product BA to Ψ_j thus yields again a linear combination of the Ψ_j, with however new coefficients c_{jl}. We can therefore associate a matrix C with the operator product BA,

$$BA \rightarrow [c_{jl}] = C \ , \tag{6.10}$$

where according to (6.9), the coefficients c_{jl} are related to the coefficients a_{jm} and b_{ml} through the equation

$$c_{jl} = \sum_{m=1}^{M} a_{jm} b_{ml} \ . \tag{6.11}$$

This is simply the product rule for the matrices $A' = [a_{jm}]$, $B' = [b_{ml}]$ and $C' = [c_{jl}]$, with $A'B' = C'$. We thus recognise the fundamental concept that the operators $A, B, \ldots$ can be represented by matrices, including the rule for matrix multiplication, but with the operator product BA corresponding to the matrix product $A'B'$, that is in reversed order.

Let us now see what the *inverse* of A does. We first write the expression

$$A^{-1}A\,\Psi_j(r) = A^{-1} \left(\sum_{m=1}^{M} a_{jm} \Psi_m \right) \tag{6.12}$$

and use as a trial function for $A^{-1}\Psi_m$ on the right-hand side of (6.12) the following expression:

$$A^{-1}\Psi_m(r) = \sum_{l=1}^{M} f_{ml} \Psi_l \ . \tag{6.13}$$

Going through the following straightforward steps

$$\Psi_j(r) = \sum_{m=1}^{M} a_{jm} A^{-1} \Psi_l \tag{6.14}$$

$$= \sum_{m=1}^{M} a_{jm} \sum_{l=1}^{M} f_{ml} \Psi_l \tag{6.15}$$

$$= \sum_{m=1}^{M} \sum_{l=1}^{M} a_{jm} f_{ml} \Psi_l \tag{6.16}$$

we obtain

$$\sum_{m=1}^{M} a_{jm} f_{ml} = \delta_{jl} ; \tag{6.17}$$

or, if we collect a_{jm} and f_{ml} into matrices $A' = [a_{jm}]$ and $F' = [f_{ml}]$, we get the equivalent matrix equation

$$A'F' = E' . \tag{6.18}$$

However, this means that F' is none other than the inverse of the matrix A':

$$F' = A'^{-1} . \tag{6.19}$$

The operator A^{-1} is thus associated to the matrix A'^{-1}

Let us summarise: as we saw in Sect. 6.1, the symmetry operations which we denoted as $A, B, C, \ldots$ form a group. Each application of a group element $A, \ldots$ to the set of wavefunctions is associated with a matrix $A', \ldots$, which transforms the wavefunctions among themselves. The product of two group elements corresponds to a matrix product according to the rule

$$BA \to A'B' , \tag{6.20}$$

whereby one has to take care that the order of the corresponding matrices is reversed relative to that of the operators in the product. The inverse of the operation A, i.e. A^{-1}, corresponds to the inverse of the matrix A', i.e. A'^{-1}. Furthermore, the identity operation E naturally corresponds to the unit matrix E'. Finally, as we know from linear algebra, matrices obey an associative law, e.g. $(A'B')C' = A'(B'C')$. We thus can see that all of the properties of the original group of operations A, B, C are to be found in the corresponding matrices $A', B', C', \ldots$. The matrices $A', B', C', \ldots$ themselves form a group; this group of matrices is referred to as a *representation* of the (abstract) group with the elements $A, B, C, \ldots$.

6.4 Similarity Transformations and Reduction of Matrices

We now recall a bit of knowledge which we acquired when considering rotations: we saw that a real representation of the wavefunctions exists such that rotations transform the p-functions into linear combinations [cf. (5.14)], and a complex representation, where rotations simply cause the wavefunctions to be multiplied by a constant factor [cf. (5.19)]. This leads us to the general question as to whether we cannot find a basis set of wavefunctions in the present more complicated case which keeps the number of wavefunctions involved in a transformation A to a minimum. This question will lead us into a few basic mathematical considerations. As the reader will soon see, we are looking for a *similarity transformation*. We consider the matrix C', which possesses an inverse and whose indices k and j run just through the set of indices $1, \ldots, M$ of the wavefunctions. We introduce a new set of wavefunctions χ_k according to

$$\chi_k = \sum_{j=1}^{M} (C'^{-1})_{kj} \Psi_j . \tag{6.21}$$

The inversion of (6.21) is naturally

$$\Psi_j = \sum_{k=1}^{M} C'_{jk}\chi_k \ . \tag{6.22}$$

We now apply the symmetry operation A to χ_k, leading in a simple manner to

$$A\chi_k = \sum_{l=1}^{M}(C'^{-1})_{kl} A\,\Psi_l = \sum_{l=1}^{M}(C'^{-1})_{kl}\sum_{m=1}^{M} a_{lm}\Psi_m \ . \tag{6.23}$$

Now, we express Ψ_m on the right-hand side of (6.23) again in terms of χ_k according to (6.22):

$$A\chi_k = \sum_{l=1}^{M}(C'^{-1})_{kl}\sum_{m=1}^{M} a_{lm}\sum_{j=1}^{M} c_{mj}\chi_j \ . \tag{6.24}$$

Rearranging the sums leads to

$$A\chi_k = \sum_{j=1}^{M}\underbrace{\left[\sum_{l=1}^{M}(C'^{-1})_{kl}\sum_{m=1}^{M} a_{lm}c_{mj}\right]}_{b_{kj}}\chi_j \ . \tag{6.25}$$

In this equation, we have introduced the abbreviation b_{kj}, defined by:

$$b_{kj} = \sum_{l=1}^{M}(C'^{-1})_{kl}\sum_{m=1}^{M} a_{lm}c_{mj} \ . \tag{6.26}$$

The reader who is familiar with matrix algebra will recognise that the right-hand side of (6.26) contains simply a product of matrices. If we collect the elements of b_{kj} into a matrix $\tilde{A}'$, we can rewrite (6.26) in the form

$$\tilde{A}' = C'^{-1}A'C' \ . \tag{6.27}$$

In the language of matrix algebra, the matrix $\tilde{A}'$ is obtained from A' by means of a *similarity transformation*. The group properties remain unaffected by this transformation: if we multiply out the individual elements in (6.27), the products from $C'^{-1}C'$ just yield unity. Now from mathematics, we know that a similarity transformation can change a matrix A' into a simpler form, in which only the elements along the main diagonal and nearby, in the shape of square arrays, are nonzero; cf. Fig. 6.10. This form is called the "Jordan normal form" or "block form". If the powers of A' remain finite, i.e. if $A'^n, n \to \infty$ are all finite, then A' can even be diagonalised. One could be tempted to believe that we can always take the matrix A' to be diagonal.

Unfortunately, a difficulty arises at this point, when namely the general correspondence

$$\begin{aligned} A &\to C'^{-1}A'C' = \tilde{A}' \\ B &\to C'^{-1}B'C' = \tilde{B}' \end{aligned} \tag{6.28}$$

holds. Then, in the case of group elements which do not commute, it can happen that we cannot choose C' in such a way that all the matrices $\tilde{A}', \tilde{B}', \ldots$ are simultaneously diagonalised.

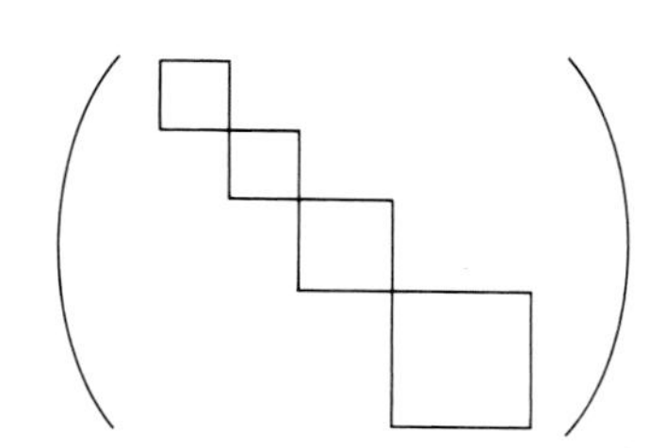

Fig. 6.10. The typical structure of a reduced matrix (block form). Outside the boxes, all the matrix elements are zero

Now there is an important branch of mathematics dealing with the *theory of group representations*, in which it is shown that a minimal representation of A', B', ... can be obtained by applying the similarity transformation (6.28). What does such a representation mean? It means that one can choose the basis set of the wavefunctions χ_k in such a way that on application of *all* the symmetry operations of the group, *only a certain subset* of the χ_k are transformed into each other. In other words, the basis of the χ_k can be decomposed into components. These components naturally have, in general, a much simpler behaviour under transformations than the original Ψ_j. And now comes what is perhaps the most wonderful idea to arise out of the combination of group theory with quantum mechanics: we had assumed a certain basis set Ψ_j or χ_k in our considerations; but the transformation properties of the χ_k do not depend at all any more on the concrete quantum-mechanical problem at hand, but rather only on the underlying symmetry group.

Thus, instead of finding the wavefunctions directly as solutions to the Schrödinger equation, which can be very complicated, it will offen suffice to use group theory to determine what transformation behaviour the basis vectors have in their representations. We can then require this symmetry behaviour of the wavefunctions, just as we did in the case of ethene or of benzene (see Chap. 5). In those cases, we could determine the coefficients for the construction of ψ from atomic wavefunctions uniquely. In the general case, this will not always be possible, but in any event, the number of unknown coefficients can be drastically reduced by using the group properties. The transformation behaviour of basis functions under particular symmetry groups is tabulated in the literature. Treating this topic in detail here would by far overreach the framework of this book; it would become an encyclopedic listing, which would not permit any useful physical insights. For this reason, we shall treat only a few such symmetry properties and their notation as examples.

6.5 Fundamentals of the Theory of Group Representations

6.5.1 The Concept of the Class

For later application, let us learn some fundamental concepts from the theory of group representations. The number of elements in a group is called its *order* and is often denoted by the letter h. Thus, $h = 4$ for the group C_{2h} and $h = 6$ for C_{3v}. Two elements A, B of a group are called *conjugate* to one another if there exists an element C such that

$$B = C^{-1}AC \tag{6.29}$$

holds. If we represent the group operations by matrices, then (6.29) simply denotes a similarity transformation; one thus speaks of a similarity transformation also in the case of abstract group relations such as (6.29). If we multiply (6.29) from the left by C and from the right by C^{-1}, we obtain

$$A = CBC^{-1} \, , \tag{6.30}$$

which means just that the conjugate relationship is reciprocal. A *class* is then defined as the set of all the elements of the group which are conjugate to each other. In order to find out which elements belong to the same class, we have to investigate the similarity transformations. Taking as an example the group C_{3v}, we first choose the element E and go through all the transformations C, obtaining

$$E^{-1}EE = E \,, \tag{6.31}$$

$$C_3^{-1}EC_3 = E \,, \tag{6.32}$$

and corresponding relations, in each of which E appears on the right-hand side, since E multiplied by any element of the group yields that same element. It follows from these relations that E is in a class by itself. Taking as a second example σ_v, we look for elements in the same class as σ_v. We take

$$E^{-1}(\sigma_v E) = \sigma_v \,, \tag{6.33}$$

which follows immediately from the properties of E. In order to verify the next example in the equation

$$C_3^{-1}\sigma_v C_3 = C_3^{-1}\sigma_v' = C_3^2\sigma_v' = \sigma_v'' \,, \tag{6.34}$$

we look at the group table of the group C_{3v} (Table 6.2) and immediately obtain the result in the left-hand side of the equations (6.34). Since $C_3^3 = E$, where we can write the left side as $C_3 C_3^2$, we find $C_3^{-1} = C_3^2$. Finally, we look again at the group table, and can verify the last equation in (6.34). In a similar manner, we obtain the results

$$(C_3^2)^{-1}(\sigma_v C_3^2) = \sigma_v' \,, \tag{6.35}$$

$$\sigma_v^{-1}(\sigma_v\sigma_v) = \sigma_v \,, \tag{6.36}$$

$$\sigma_v'^{-1}\sigma_v\sigma_v' = \sigma_v' \,, \tag{6.37}$$

$$\sigma_v'^{-1}\sigma_v\sigma_v'' = \sigma_v' \,. \tag{6.38}$$

Clearly, the elements σ_v, σ_v' and σ_v'' belong to the same class. If we begin with σ_v' instead of σ_v, we can arrive at σ_v by inverting C_3 and from there to the other class member σ_v''. One can easily convince oneself that such operations never lead outside the class itself. In a similar way, we can show that C_3 and C_3^2 belong to a class. The order of this class is the number of its elements; the class containing σ_v, σ_v' and σ_v'' thus has the order 3, while the order of the class to which C_3 and C_3^2 belong is 2.

6.5.2 The Character of a Representation

A central tool in the theory of group representations is the "character". As we have seen, each element of a group can be associated with a matrix. The term *character* denotes the trace, or in other words the sum of the diagonal elements, of this matrix.

For the matrix

$$A' = \begin{bmatrix} a_{11} & a_{12} & \cdots \\ a_{21} & a_{22} & \cdots \\ \cdots & \cdots & a_{kk} \end{bmatrix} , \tag{6.39}$$

we thus have

$$\text{Character of } A' = \text{Trace}(A') = \sum_{l=1}^{k} a_{ll} \,. \tag{6.40}$$

Let us see how the character of a representation, which initially may very well be reducible, can be determined. We take as an example the N_2H_2 molecule, whose geometric structure is

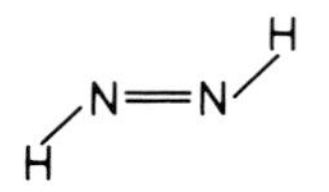

Fig. 6.11. N_2H_2

shown in Fig. 6.11. One can readily convince oneself that the symmetry operations of the group C_{2h} leave this molecule invariant. These are the following operations (cf. Fig. 6.12): the identity operation E, rotation by 180° about an axis which is perpendicular to the plane of the figure and passes through the centre of gravity of the molecule, reflection in a plane perpendicular to this axis and containing the atoms of the molecule, and finally inversion through the centre of gravity, which is simultaneously a centre of inversion symmetry. The group table is given in Table 6.3.

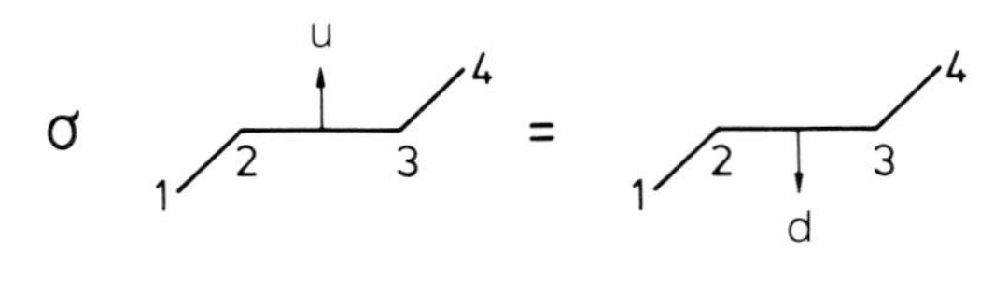

Fig. 6.12. The effect of the symmetry operations σ, C_2, and i on the N_2H_2 molecule. The arrows marked 'u' and 'd' refer to 'up' and 'down'

Table 6.3. Group multiplication table

C_{2h}	E	C_2	σ	i
E	E	C_2	σ	i
C_2	C_2	E	i	σ
σ	σ	i	E	C_2
i	i	σ	C_2	E

We now seek a particular representation by considering the lengths of the N–H bonds in the various positions of the molecule, denoting them as ΔR_1 and ΔR_2 (compare Fig. 6.13). This example also makes it clear that the objects which are operated upon by the symmetry operations may be not only wavefunctions [cf. (6.1)], but also geometric forms. The identity operation E changes nothing in the molecule, so that we immediately obtain the representation

$$E\begin{pmatrix}\Delta R_1\\ \Delta R_2\end{pmatrix}=\begin{pmatrix}1&0\\0&1\end{pmatrix}\begin{pmatrix}\Delta R_1\\ \Delta R_2\end{pmatrix}. \tag{6.41}$$

On rotation about an axis perpendicular to the plane of the figure through the midpoint of the N–N bond, ΔR_1 is transformed to ΔR_2 and ΔR_2 to ΔR_1. We thus obtain

$$C_2\begin{pmatrix}\Delta R_2\\ \Delta R_1\end{pmatrix}=\begin{pmatrix}0&1\\1&0\end{pmatrix}\begin{pmatrix}\Delta R_1\\ \Delta R_2\end{pmatrix}. \tag{6.42}$$

The molecule is invariant under a reflection in a plane which is identical to the plane of Fig. 6.13. For this operation, we find

$$\sigma_h\begin{pmatrix}\Delta R_1\\ \Delta R_2\end{pmatrix}=\begin{pmatrix}1&0\\0&1\end{pmatrix}\begin{pmatrix}\Delta R_1\\ \Delta R_2\end{pmatrix}. \tag{6.43}$$

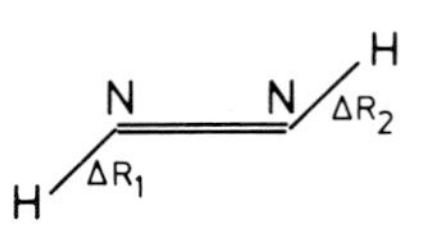

Fig. 6.13. The distances ΔR_1 and ΔR_2 in N_2H_2

Finally we find for the inversion:

$$i\begin{pmatrix}\Delta R_2\\ \Delta R_1\end{pmatrix}=\begin{pmatrix}0&1\\1&0\end{pmatrix}\begin{pmatrix}\Delta R_1\\ \Delta R_2\end{pmatrix}. \tag{6.44}$$

The matrices which occur in (6.41–44) are those special ones which we were seeking, for which we can immediately give the sum of the diagonal elements as the respective characters. We thus arrive at Table 6.4.

As we shall show later, the representation given by the matrices in Table 6.4 is reducible. Mathematically, it is possible to find the irreducible representations systematically and to describe them in character tables. It is found that not only one set of matrices can represent a given group of symmetry operations, but rather that the representation can be realised in various ways, i.e. using various sets of matrices. These different possibilities for representations are formally distinguished by using indexed Greek letters, e.g. Γ_1, Γ_2, etc. We thus obtain the following character table (Table 6.5):

Table 6.4

C_{2h}		Character
E	$=\begin{pmatrix}1 & 0\\0 & 1\end{pmatrix}$	2
C_2	$=\begin{pmatrix}0 & 1\\1 & 0\end{pmatrix}$	0
σ_h	$=\begin{pmatrix}1 & 0\\0 & 1\end{pmatrix}$	2
i	$=\begin{pmatrix}0 & 1\\1 & 0\end{pmatrix}$	0

At the left in the first column are the symmetry operations, followed by their matrix representations; in the second column, the corresponding characters are listed

Table 6.5. Character table for C_{2h}

C_{2h}	E	C_2	σ_h	i
Prelim. Name				
Γ_1	1	1	1	1
Γ_2	1	−1	−1	1
Γ_3	1	1	−1	−1
Γ_4	1	−1	1	−1

In the upper left-hand corner of this table is the symbol for the symmetry group, in the present case C_{2h}. To the right in the same row are the symbols for the group operations, i.e. the identity operation E, rotation about a twofold symmetry axis C_2, reflection in a horizontal plane σ_h, and inversion i. The next row contains the characters belonging to the representation Γ_1 and corresponding to the respective group elements. The following rows contain the characters for the representations $\Gamma_2, \ldots$.

Now, how large is the number of irreducible representations of a group? It is, as can be proven mathematically, equal to the number of classes in the group. If we choose a particular irreducible representation, then the character of all the operations is the same within the same class. This can be readily understood, since the elements within a class differ only by similarity transformations from one another; however, the trace of a matrix is unchanged by a similarity transformation, i.e. the characters remain the same. As can be shown for the group C_{2h}, each of the 4 elements forms a class by itself; there are thus 4 classes, each of which contains a single element.

Table 6.6. Character table for C_{3v}

C_{3v}	E	C_3	C_3^2	σ_v	σ_v'	σ''_v
Γ_1	1	1	1	1	1	1
Γ_2	1	1	1	−1	−1	−1
Γ_3	2	−1	−1	0	0	0

Table 6.7. Character table for C_{3v}

C_{3v}	E	$2C_3$	$3\sigma_v$
Γ_1	1	1	1
Γ_2	1	1	−1
Γ_3	2	−1	0

Let us consider a further example, the character table for the point group C_{3v}. It is given in Table 6.6. As we saw above, C_3 and C_3^2 form a class by themselves, and likewise σ_v, σ_v', and σ''_v. This naturally means, considering what was said above about characters and classes, that the characters of C_3^2 for all representations Γ_1, Γ_2, Γ_3 are the same as those of C_3, as we can see from the character Table 6.6. The same is true of σ_v, σ_v', and σ''_v. For this reason, Table 6.6 contains redundant information; it can be condensed into the more compact form of Table 6.7.

In this latter table, the numbers 2 and 3 in front of C_3 and σ_v indicate how many operations there are in the respective class. We note that E and i are always each in a class by themselves.

At this point, we need an additional concept, the *dimension of an irreducible representation*. This is the dimensionality of the matrices in the representation. Since the character of E is just the number of elements in the main diagonal and thus is equal to the dimension of the corresponding irreducible representation, we can see that the character of E gives the dimension of the representation. In the literature, following a convention introduced by *Mullikan*, somewhat different character tables are often used, as shown in the following example (Table 6.8) for the group C_{3v}.

Table 6.8. Complete character table for C_{3v}

C_{3v}	E	$2C_3$	$3\sigma_v$	Operation	
A_1	1	1	1	z	$x^2 + y^2, z^2$
A_2	1	1	-1	R_z	
E	2	-1	0	$(x, y)(R_x, R_y)$	$(x^2 - y^2, xy)(xz, yz)$

The first row begins with C_{3v}, and all of the group operations are familiar, as is the block of characters which is listed below them; what is new is the notation A_1, A_2, E for the irreducible representations. The E which appears here as a symbol for an irreducible representation is not to be confused with the E which occurs in the first row and denotes a symmetry operation of the group. The letters A and E denote a particular behaviour with respect to symmetry, which we will discuss below. The fifth column, containing z, R_z, indicates which coordinates [here (z) or (R_z)] exhibit the particular symmetry behaviour denoted by the $A_1, \ldots$ beginning the same row. It is thus made clear that, for example, the z coordinate in a Cartesian coordinate system is invariant with respect to the operations of A_1, i.e. the matrix of the transformation reduces to a 1, which is then identical to the character of the representation of the particular symmetry operation. In the last row, x and y thus serve as a basis. Finally, in the last column of Table 6.8, those basis elements are given which can be formed from the squares or from quadratic or bilinear expressions using x, y and z.

6.5.3 The Notation for Irreducible Representations

Let us now explain the reason for the change of notation from Γ to A_1, A_2, etc. The purpose of this change is to show by means of the symbol for a particular representation whether it is one- or multidimensional and which special symmetry properties it has. The letters A and B refer to one-dimensional irreducible representations, with A reserved for representations which are symmetric with respect to rotations about the principal axis and B for those which are antisymmetric. The character of the symmetric representation is +1 and that of the antisymmetric representation is -1. The letters E and T denote two- or three- dimensional representations. The indices g and u are added to A or B when the representation is even (g) or odd (u) with respect to inversion. A prime or double prime is added to the symbol to denote symmetric or antisymmetric behaviour with respect to reflection in the horizontal mirror plane. The indices 1 and 2 are added to A or B when the corresponding representation is symmetric (1) or antisymmetric (2) with respect to the C axis, with the C_2 axis being perpendicular to the principal axis, or, if C_2 is not present, to a vertical mirror plane. The indices 1 and 2 on E and F are complicated and will not be discussed here. This notation is summarised in Table 6.9.

Table 6.9. Notation for irreducible representations

Dimension of the representation	Characters under the operation					Symbols
	E	C_n	i	σ_h	C_2^* or σ_v	
1	1	1				A
	1	-1				B
2	2					E
3	3					T
			1			A_g B_g E_g T_g
			-1			A_u B_u E_u T_u
				1		A' B'
				-1		A'' B''
					1	A_1 B_1
					-1	A_2 B_2

* C_2-axis perpendicular to the principal axis

In order to give the reader an example of the use of this new notation A_g, etc., we give here the character table for the group C_{2h} (Table 6.10):

Table 6.10. Character table for C_{2h}

C_{2h}	E	C_2	i	σ_h		
A_g	1	1	1	1	R_z	x^2, y^2, z^2, xy
B_g	1	-1	1	-1	R_x, R_y	xz, yz
A_u	1	1	-1	-1	z	
B_u	1	-1	-1	1	x, y	

6.5.4 The Reduction of a Representation

An important question is naturally that of how we can reduce or decompose a representation and how we know which irreducible representations are contained in it. The characters help us to answer this question. If, for example, we consider the matrix A shown in Fig. 6.10, we can on the one hand find its character by taking the sum of its diagonal elements. On the other hand, this matrix contains the matrices of the individual irreducible representations, which have their own group characters, and we can see at once that the character of the reducible representation must be equal to the sum of the characters of the irreducible representations which it contains. This is naturally true of each element in the group to which the matrices correspond.

Table 6.11. Decomposition of the characters of a reducible representation

	E	C_2	i	σ_h
A_g	1	1	1	1
B_u	1	-1	-1	1
Sum				
$A_g + B_u$	2	0	0	2

Let us consider the characters which occur in the example of the symmetry operations relating to the lengths ΔR_1, ΔR_2 in the N_2H_2 molecule. These characters, according to Table 6.4, are given by 2,0,0, and 2. The question is now: How can we relate these characters to those of the irreducible representations which are given in Table 6.5? This means that for each group operation, E, C_2, i, and σ_h, a suitable sum of the characters of the representations must be found. We thus arrive at Table 6.11, i.e. precisely the desired combination (2,0,0,2).

As is shown by group theory, and as we shall demonstrate in the following, the decomposition is unique. In addition to finding by trial and error which combinations of the individual irreducible representations lead to the given reducible representation, one can also proceed systematically. As we just pointed out, the character of the reducible representations for each group element is equal to the sum of the characters of the irreducible representations contained in them. This can be expressed by the following formula:

$$\chi(R) = \sum_i n_i \chi_i(R) \, . \tag{6.45}$$

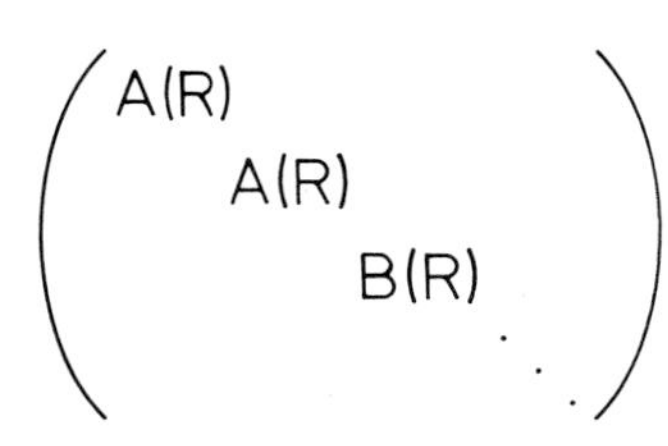

Fig. 6.14. An example of the reduction of a matrix with two irreducible representations

Here, χ is the character of the in general reducible representation which corresponds to the group operation R, where R can be any one of the symmetry operations. On the right-hand side, a sum is taken over the various irreducible representations which are distinguished by an index i, whereby n_i is the number of equivalent irreducible representations, i.e. the number of equivalent blocks in the matrix (see Fig. 6.14).

Equation (6.45) has a certain formal similarity to relations from quantum mechanics, where for example an arbitrary wavefunction ψ can be decomposed into a linear combination of wavefunctions ψ_i. In fact, an orthogonality relation of the form

$$\frac{1}{h} \sum_R \chi_j(R) \chi_i(R) = \delta_{ij} \tag{6.46}$$

holds here also; the sum is to be carried out over all of the symmetry operations R. To be sure, we cannot derive relation (6.46) here. In complete analogy to quantum mechanics, employing (6.45) and (6.46), we can however show how often an irreducible representation i is contained in the reducible representation. To this end, we multiply (6.45) by χ_i and sum over the individual group elements. We thus obtain:

$$n_i = \frac{1}{h} \sum_R \chi_i(R) \chi(R) \, . \tag{6.47}$$

We saw above that the characters of the irreducible representations are the same when they refer to the different elements of a group belonging to the same class. For this reason, it is sufficient to sum over only those elements which belong to different classes, taking into account how many elements are in each class. We thus arrive at the formula

$$n_i = \frac{1}{h} \sum_Q N \chi(R) \chi_i(R) \, . \tag{6.48}$$

Here, the summation is to be carried out over the classes. The meaning of the various quantities in (6.48) is summarised in Table 6.12.

We again consider as an example the group C_{2h} (Table 6.10) and look at the representation Γ_1, which has the characters 2,0,0,2. The order of the group is $h = 4$. Applying formula (6.48), we obtain the following relations:

$$n_{A_g} = \tfrac{1}{4}(\underset{E}{1 \cdot 2 \cdot 1} + \underset{C_2}{1 \cdot 0 \cdot 1} + \underset{i}{1 \cdot 0 \cdot 1} + \underset{\sigma_h}{1 \cdot 2 \cdot 1}) = 1 \tag{6.49}$$

and

$$n_{B_g} = \tfrac{1}{4}\{1 \cdot 2 \cdot 1 + 1 \cdot 0 \cdot (-1) + 1 \cdot 0 \cdot 1 + [1 \cdot 2 \cdot (-1)]\} = 0 \,, \tag{6.50}$$

as can readily be seen. In a corresponding manner, we find

$$n_{A_u} = 0 \tag{6.51}$$

$$n_{B_u} = 1 \,. \tag{6.52}$$

We thus obtain the result that the representation Γ_1 can be decomposed into the representations A_g and B_u. We can take an additional important step: our goal is, finally, to construct electronic wavefunctions (or molecular vibrational functions) which correspond to irreducible representations. These then give the minimal set of functions which are mutually degenerate, i.e. which belong to the same energy.

Table 6.12. The meaning of the quantities occurring in (6.48)

n_i:	Number of times that the i-th irreducible representation occurs in the reducible representation
h:	Order of the group
Q:	Class of the group
N:	Number of operations in the class Q
R:	Group operation
$\chi(R)$:	Character of R in the reducible representation
$\chi_i(R)$:	Character of R in the irreducible representation

6.6 Summary

The method which we have applied in this chapter can be summarised as follows:

Many molecules exhibit symmetries. Under a symmetry operation, the molecule is left unchanged. The symmetry operations form a group, in which the product of two operations is given by the group table. If the symmetry operations are applied to a set of mutually degenerate wavefunctions, these functions undergo a linear transformation among themselves. The transformation coefficients form a matrix, and the group of the symmetry operations can be represented by matrices. By a suitable choice of the basis of the wavefunctions, the matrices can be brought into a simple (block) form: this corresponds to decomposing the representation into its irreducible representations. The characters (traces of each matrix) are a valuable aid to finding the irreducible representations.

If we wish to apply this method to the exact (or more often, approximate) electronic wavefunctions for a particular molecule, the following essentials are sufficient: we need calculate only those wavefunctions which belong to a particular irreducible representation of the symmetry group of the molecule under consideration. Then, e.g. in the LCAO method, we can determine exactly the unknown coefficients or at least reduce their number drastically. We shall demonstrate this using the example of the H_2O molecule.

6.7 An Example: The H_2O Molecule

In this section, we want to derive the one-electron wavefunctions of the H_2O molecule. We use the molecular orbital method, where the orbitals ψ are represented as linear combinations of atomic wavefunctions ϕ_j:

$$\psi = \sum_j c_j \phi_j \ . \tag{6.53}$$

The coefficients c_j are to be determined with the aid of the variational method which we have already used in Sect. 5.3; we remind the reader of the formula

$$\frac{\int \psi^* H \psi dV}{\int \psi^* \psi dV} = \text{Min!} \ . \tag{6.54}$$

If we insert (6.53) into (6.54) and assume the atomic orbitals to be practically orthogonal to each other, we obtain a result with which we are already familiar:

$$\sum_j (H_{ij} - E_\alpha \delta_{ij})\, c_j^{(\alpha)} = 0 \ . \tag{6.55}$$

As we know, this is a set of homogeneous linear equations, which has a nontrivial solution only when the determinant of the coefficients is zero. This condition fixes a set of eigenvalues which are identical to the energy values E, as well as the corresponding wavefunctions.

We now make use of the basic ideas of Sect. 5.2, where we saw that the coefficients c_j could be entirely or partially determined by using group-theoretical considerations, without the need to solve the generally complicated equations (6.55). As a concrete example, we consider the water molecule, H_2O. Our goal is to determine the molecular orbitals ψ in such a way that they correspond to the irreducible representations of the symmetry operations of the molecule, in this case H_2O. To this end, we undertake the following steps:

1) We determine the symmetry group of the molecule.
2) We choose the atomic orbitals from which the molecular orbitals are to be constructed according to (6.53).
3) The atomic orbitals are used as a basis set, from which a representation of the symmetry group is generated. The details of this process will become clear in the following.
4) The representation obtained in step 3) is then decomposed into its irreducible representations. We thus obtain the possible linear combinations of atomic orbitals which can be used to form the molecular orbitals.

The H_2O molecule is shown in Fig. 6.15, which also indicates its various symmetry elements. The molecule clearly may be placed into a Cartesian coordinate system in such a way that the H atoms lie in the plane spanned by the x- and z-axes. The xz-plane is a symmetry plane (mirror plane), on which the reflection symmetry operation σ_v' can be carried out. The yz-plane, perpendicular to the xz-plane, is likewise a plane of symmetry on which the operation denoted by σ_v is performed. An additional symmetry element is the z-axis, around which a twofold rotation, transforming the H atoms into one another, can be carried out. All together, these symmetry operations yield the symmetry group C_{2v}. Its multiplication table is given as Table 6.13.

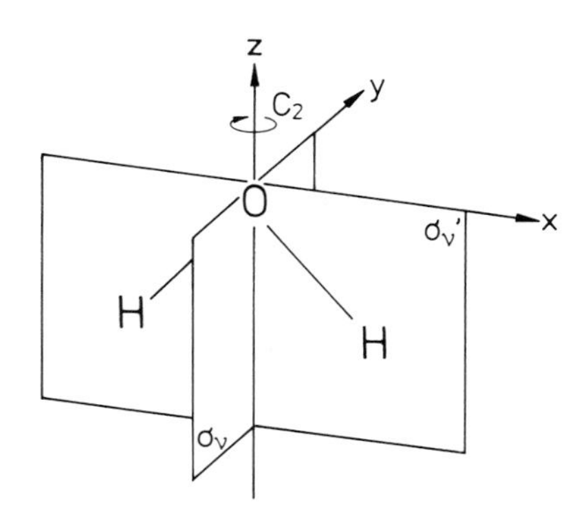

Fig. 6.15. H_2O with its symmetry elements

The following properties can readily be derived from this multiplication table: the group is a commutative group (Abelian group), i.e. each pair of elements A and B obeys a multiplication rule $AB = BA$. From this property it follows immediately that each element is in a class by itself, and since there are 4 elements, there must be 4 classes. These 4 classes correspond to 4 irreducible representations which are all inequivalent. It can then be seen that each irreducible representation is one dimensional. The corresponding character table, which can be derived using mathematical methods, is shown by Table 6.14.

Table 6.13. Group multiplication table for C_{2v}

C_{2v}	E	C_2	σ_v	σ_v'
E	E	C_2	σ_v	σ_v'
C_2	C_2	E	σ_v'	σ_v
σ_v	σ_v	σ_v'	E	C_2
σ_v'	σ_v'	σ_v	C_2	E

Table 6.14. Character table for C_{2v}

C_{2v}	E	C_2	σ_v	σ_v'		
A_1	1	1	1	1	z	x^2, y^2, z
A_2	1	1	-1	-1	R_z	xy
B_1	1	-1	1	-1	x, R_y	xz
B_2	1	-1	-1	1	y, R_x	yz

We now have to consider which atomic orbitals we will choose as a basis. Since the hydrogen atoms are in their ground states before forming chemical bonds to the oxygen, and since it requires a considerable excitation energy to raise them to the first excited state with principal quantum number $n = 2$, it seems reasonable to use $1s$ orbitals for the wavefunctions which are contributed by the hydrogen atoms. In the case of the oxygen atom, the $1s$ functions form a *closed shell*, which practically does not participate in bond formation. For this reason, we use the wavefunctions of the next shell; these are the $2s$ and $2p$ orbitals. We then have the following wavefunctions as a basis set of atomic orbitals ϕ_j: $s_1, s_2, 2s, 2p_x, 2p_y$, and $2p_z$ (compare Fig. 6.16). (More precisely, we should write here e.g. ϕ_{s_1} instead of s_1, etc.) We now wish to decompose the matrices of the representation according to the basis set of these five wavefunctions. As the calculation shows, however, it suffices to consider separately the wavefunctions s_1 and s_2, which come from the two hydrogen atoms, and those which come from the oxygen atom, i.e. $2s, 2p_x, 2p_y$, and $2p_z$. It may thus be shown that the matrices of the representation can be decomposed into blocks corresponding to H and O (cf. Fig. 6.17).

Let us look at the behaviour of the wavefunctions of the hydrogen atoms more closely. They form a basis s_1, s_2. These functions, as we have already seen in the case of the hydrogen

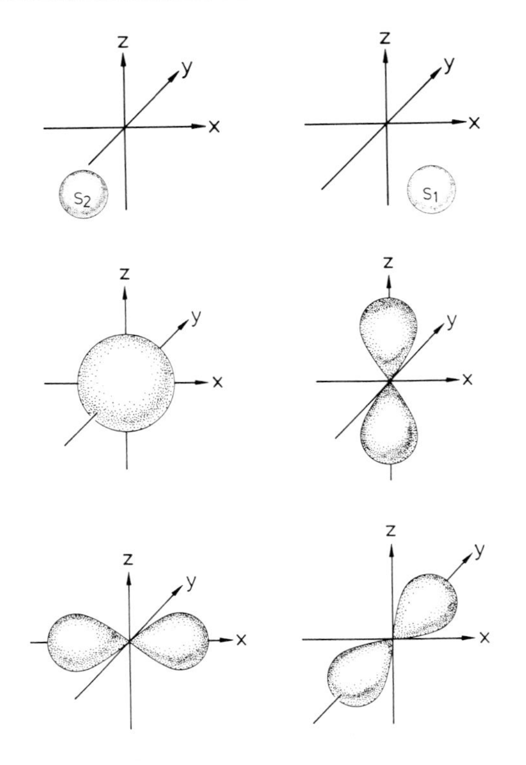

Fig. 6.16. Basis wavefunctions for H_2O (schematic drawing). *Upper row*: the $1s$ functions from the two hydrogen atoms 1 and 2; *middle* and *lower rows*: the $2s$ and $2p$ functions of the oxygen atom

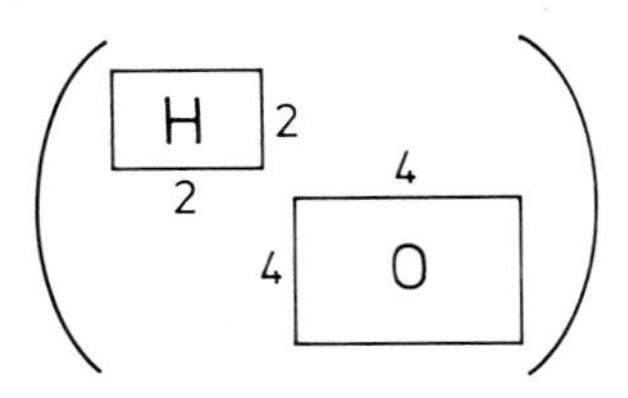

Fig. 6.17. Matrix representation for the basis set (6.16). The functions belonging to H and to O are each transformed among themselves

molecule, are localised near the protons and otherwise correspond to the s functions of hydrogen (compare Sect. 4.3). We can examine just how these functions transform under the symmetry operations. For example, the operation E transforms s_1 and s_2 into themselves. On reflection in the σ_v plane, in contrast, the two hydrogen atoms exchange places and thus the two wavefunctions s_1 and s_2 are transformed into one another. If we apply similar considerations to all the other symmetry operations, we readily obtain the relations

$$
\begin{aligned}
E\begin{pmatrix} s_1 \\ s_2 \end{pmatrix} &= \begin{pmatrix} 1 & 0 \\ 0 & 1 \end{pmatrix}\begin{pmatrix} s_1 \\ s_2 \end{pmatrix}, \quad & C_2\begin{pmatrix} s_1 \\ s_2 \end{pmatrix} &= \begin{pmatrix} 0 & 1 \\ 1 & 0 \end{pmatrix}\begin{pmatrix} s_1 \\ s_2 \end{pmatrix}, \\
\sigma_v\begin{pmatrix} s_1 \\ s_2 \end{pmatrix} &= \begin{pmatrix} 0 & 1 \\ 1 & 0 \end{pmatrix}\begin{pmatrix} s_1 \\ s_2 \end{pmatrix}, \quad & \sigma_v'\begin{pmatrix} s_1 \\ s_2 \end{pmatrix} &= \begin{pmatrix} 1 & 0 \\ 0 & 1 \end{pmatrix}\begin{pmatrix} s_1 \\ s_2 \end{pmatrix},
\end{aligned}
\tag{6.56}
$$

from which the matrices of the reducible representation can be read off. Taking the traces of these matrices, we obtain the characters, which are collected in Table 6.15.

As in the previous section, we decompose the reducible representation that occurs in (6.56) into irreducible representations; this can be done in two ways: one is to use the formula (6.47) with which we met previously,

$$
n_i = \frac{1}{h}\sum_R \chi(R)\chi_i(R) ; \tag{6.57}
$$

Table 6.15. The characters of the representation given in (6.56)

C_{2v}	E	C_2	σ_v	σ_v'
2H(1s)	2	0	0	2

and the other is by means of direct comparison with the character table. We shall leave both methods as an exercise to the reader, since the procedure was covered in detail in the previous section, and simply give the results in Table 6.16.

It thus becomes clear that the reducible representation in (6.56) decomposes into the irreducible representations A_1 and B_2.

We must now deal with the problem of how to transform the set of matrices belonging to a reducible representation explicitly into the set of matrices belonging to the irreducible representation. We recall that in Sect. 6.4 we found that a matrix could be transformed into block form by carrying out a similarity transformation. This, however, means simply transforming to a new basis. The transformation from the basis of the reducible representation to the basis of the irreducible representation is naturally a complicated problem in the general case; fortunately there exists a procedure for generating an irreducible representation from a given basis. For this purpose, a so-called projection operator P_i is used. Intuitively speaking, this operator projects the basis of the reducible representation onto a basis of the irreducible representation. The derivation of the following formula goes beyond the framework of this book; we therefore simply state it and show how it can be applied by giving an example. The formula is:

Table 6.16. The characters of A_1 and B_2 and their sums, which yield the characters of 2H(1s)

C_{2v}	E	C_2	σ_v	σ_v'
A_1	1	1	1	1
B_2	1	-1	-1	1
2H(1s) = $A_1 + B_2$	2	0	0	2

$$P_i = \frac{1}{h} \sum_R \chi_i(R^{-1}) \hat{R} \, . \tag{6.58}$$

Here, P_i is the projection operator, which projects the original basis, $\begin{pmatrix} s_1 \\ s_2 \end{pmatrix}$ in the present case, onto a new basis belonging to the irreducible representation denoted by the index i. How this "works" we shall see directly. The symbol h again denotes the order of the group, R are the group operations, χ_i is the character belonging to the i-th representation of the group operation R^{-1}, and $\hat{R}$ is the (in general reducible) representation matrix which corresponds to the group operation R. Let us first consider the irreducible representation A_1; the index i in (6.58) thus refers to "representation A_1". For R we insert the operations E, C_2, σ_v and σ_v', we use the characters given in Table 6.14, and we denote the matrices belonging to E, C_2, σ_v, and σ_v' as $\hat{E}, \hat{C}_2, \hat{\sigma}_v$, and $\hat{\sigma}_v'$, respectively; then we obtain

$$P_{A_1} = \tfrac{1}{4}(1 \cdot \hat{E} + 1 \cdot \hat{C}_2 + 1 \cdot \hat{\sigma}_v + 1 \cdot \hat{\sigma}_v']) \, . \tag{6.59}$$

If we now put the matrices given in (6.56) into this expression, we find

$$P_{A_1} = \frac{1}{4}\left[1 \cdot \begin{pmatrix} 1 & 0 \\ 0 & 1 \end{pmatrix} + 1 \cdot \begin{pmatrix} 0 & 1 \\ 1 & 0 \end{pmatrix} + 1 \cdot \begin{pmatrix} 0 & 1 \\ 1 & 0 \end{pmatrix} + 1 \cdot \begin{pmatrix} 1 & 0 \\ 0 & 1 \end{pmatrix}\right] , \tag{6.60}$$

i.e.

$$P_{A_1} = \frac{1}{2}\begin{pmatrix} 1 & 1 \\ 1 & 1 \end{pmatrix} . \tag{6.61}$$

In a similar manner, for the irreducible representation B_2 we obtain the result

$$P_{B_2} = \frac{1}{4}\left[\hat{E} - \hat{C}_2 - \hat{\sigma}_v + \hat{\sigma}_v'\right] = \frac{1}{2}\begin{pmatrix} 1 & -1 \\ -1 & 1 \end{pmatrix} . \tag{6.62}$$

What do the results (6.61) or (6.62) mean in terms of the basis? To answer this question, we apply P_{A_1} to the original basis $\begin{pmatrix} s_1 \\ s_2 \end{pmatrix}$; this yields the result:

$$P_{A_1}\begin{pmatrix} s_1 \\ s_2 \end{pmatrix} = \frac{1}{2}\begin{pmatrix} 1 & 1 \\ 1 & 1 \end{pmatrix}\begin{pmatrix} s_1 \\ s_2 \end{pmatrix} = \frac{1}{2}\begin{pmatrix} s_1 + s_2 \\ s_1 + s_2 \end{pmatrix} . \tag{6.63}$$

No matter which wavefunction we start with, i.e. with s_1 or with s_2, we always obtain the projection onto a certain linear combination, namely $s_1 + s_2$. If, on the other hand, we apply P_{B_2}, then the plus sign in (6.63) becomes a minus sign:

$$P_{B_2}\begin{pmatrix} s_1 \\ s_2 \end{pmatrix} = \frac{1}{2}\begin{pmatrix} 1 & -1 \\ -1 & 1 \end{pmatrix}\begin{pmatrix} s_1 \\ s_2 \end{pmatrix} = \frac{1}{2}\begin{pmatrix} s_1 - s_2 \\ -s_1 + s_2 \end{pmatrix} . \tag{6.64}$$

From this we can see that the basis wavefunctions for the irreducible representations A_1 and B_2 are given by

$$\begin{aligned} A_1 : \ \psi_1 &= \frac{1}{2}(s_1 + s_2) , \\ B_2 : \ \psi_2 &= \frac{1}{2}(s_1 - s_2) . \end{aligned} \tag{6.65}$$

One can, in addition, show that the projection operators belonging to A_2 or B_1 yield zero, i.e. (6.65) are in fact the basis functions for the irreducible representations belonging to the group C_{2v} which are generated by the basis functions s_1 and s_2. The result (6.65) should naturally not be at all new or surprising to us: recalling the hydrogen molecule-ion, for which quite similar symmetry considerations hold, we remember that there, too, we found these two wavefunctions, the symmetric and the antisymmetric function. However, there it was accomplished without using group theory, but rather by solving directly the equations for the coefficients.

We now turn to the somewhat more complicated case of the basis wavefunctions for the oxygen atom. Here, as we remember, the basis consists of the functions $2s, 2p_x, 2p_y$, and $2p_z$. We thus initially have a 4-dimensional reducible representation. Let us consider the effect of the symmetry operations individually; we again assume that the oxygen atom is located at the origin of a Cartesian coordinate system. Application of the identity operation E naturally yields the unit matrix. Considering the application of a rotation by 180° about the z-axis, we must take into account the fact that such a rotation changes the signs of the functions p_x and p_y, while the s function and the p_z function are left unchanged. A reflection through the σ_v plane changes the sign of the p_y function, but leaves all the other wavefunctions unchanged. Using these facts, we can immediately write down the matrices of the representation. We summarise them in the formulas (6.66):

$$\begin{aligned} E \begin{pmatrix} 2s \\ 2p_x \\ 2p_y \\ 2p_z \end{pmatrix} &= \begin{pmatrix} 1 & 0 & 0 & 0 \\ 0 & 1 & 0 & 0 \\ 0 & 0 & 1 & 0 \\ 0 & 0 & 0 & 1 \end{pmatrix} \begin{pmatrix} 2s \\ 2p_x \\ 2p_y \\ 2p_z \end{pmatrix} , \\ C_2 \begin{pmatrix} 2s \\ 2p_x \\ 2p_y \\ 2p_z \end{pmatrix} &= \begin{pmatrix} 1 & 0 & 0 & 0 \\ 0 & -1 & 0 & 0 \\ 0 & 0 & -1 & 0 \\ 0 & 0 & 0 & 1 \end{pmatrix} \begin{pmatrix} 2s \\ 2p_x \\ 2p_y \\ 2p_z \end{pmatrix} , \\ \sigma_v \begin{pmatrix} 2s \\ 2p_x \\ 2p_y \\ 2p_z \end{pmatrix} &= \begin{pmatrix} 1 & 0 & 0 & 0 \\ 0 & -1 & 0 & 0 \\ 0 & 0 & 1 & 0 \\ 0 & 0 & 0 & 1 \end{pmatrix} \begin{pmatrix} 2s \\ 2p_x \\ 2p_y \\ 2p_z \end{pmatrix} , \\ \sigma_v' \begin{pmatrix} 2s \\ 2p_x \\ 2p_y \\ 2p_z \end{pmatrix} &= \begin{pmatrix} 1 & 0 & 0 & 0 \\ 0 & 1 & 0 & 0 \\ 0 & 0 & -1 & 0 \\ 0 & 0 & 0 & 1 \end{pmatrix} \begin{pmatrix} 2s \\ 2p_x \\ 2p_y \\ 2p_z \end{pmatrix} . \end{aligned} \tag{6.66}$$

It is now an easy matter to set up the character table for the reducible representations given in (6.66) and, e.g. by trial and error, to find the irreducible representations contained in this reducible representation. Or we can use (6.57) as we did before, which can again be left as an exercise to the reader. The result is found to be the decomposition of the representation (6.66) into the representations $2A_1 + B_1 + B_2$. Application of the projection operator (6.58) allows us to find the basis which belongs to each irreducible representation. We thus obtain the following schematic result:

$$\begin{aligned} A_1 &: 2s, 2p_z \ , \\ A_2 &: - \ , \\ B_1 &: 2p_y \ , \\ B_2 &: 2p_x \ . \end{aligned} \tag{6.67}$$

Table 6.17. The basis functions for the irreducible representations

	O Orbitals	H Orbitals
A_1	$2s, 2p_z$	$\psi_1 = \frac{1}{2}(s_1 + s_2)$
A_2		
B_1	$2p_y$	
B_2	$2p_x$	$\psi_2 = \frac{1}{2}(s_1 - s_2)$

The empty space following A_2 indicates that there is no wavefunction which can be constructed from the original basis and which transforms according to the symmetry operations in the representation A_2.

Let us summarise our results concerning the new basis functions of the molecular orbitals using both the hydrogen and the oxygen wavefunctions; these are set out in Table 6.17.

From the original 6 atomic orbitals, a basis of 6 new molecular orbitals has been constructed.

What can we now expect from group theory, and what are its limitations? In Table 6.17, we have collected wavefunctions which have the same symmetry properties. Thus in the case of the irreducible representation B_1, only the wavefunction belonging to the state $2p_y$ has the corresponding symmetry properties (Fig. 6.18). In the case of the irreducible representation B_2, the two wavefunctions $2p_x$ and ψ_2 have, in contrast, the same symmetry behaviour (Fig. 6.19). The advantage is now found in the fact that in choosing the wavefunctions which are to be used as basis functions in formula (6.53), we need consider only those functions belonging to the same irreducible representation. For example, for B_1 the entire molecular orbital ψ

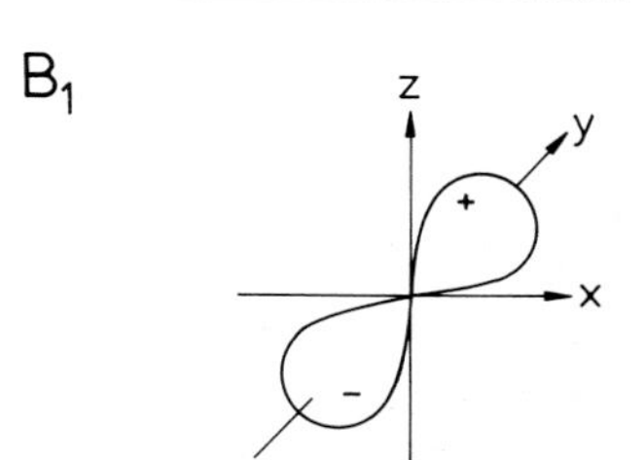

Fig. 6.18. The $2p_y$ function of oxygen, which belongs to the representation B_1

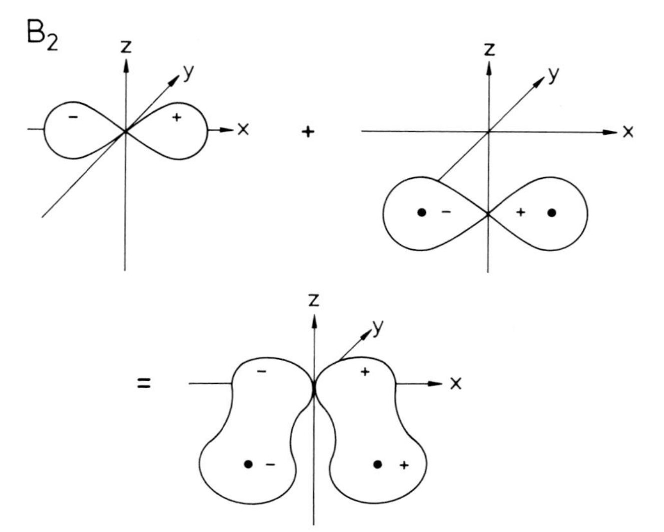

Fig. 6.19. The function $2p_x$ of the oxygen atom which belongs to the representation B_2 forms a bonding state together with the ψ_2 function of the hydrogens. (For the antibonding state cf. Fig. 6.21)

is reduced to the wavefunction which comes from the oxygen $2p_y$ state, i.e. $\psi = \phi_{2p_y}$. This is quite clearly a non-bonding orbital. For the representation B_2, we must however use a linear combination of the wavefunctions $2p_x$ and ψ_2, i.e. $\psi = c_1(2p_x) + c_2\psi_2$. If we insert this wavefunction into the extremal condition (6.54), we obtain two equations for the unknown coefficients c_1 and c_2. Setting the determinant of the coefficients equal to zero then yields two energy eigenvalues. Here, one state is bonding and the other is antibonding. For the irreducible representation A_2, we have no basis functions, while for A_1 there are three wavefunctions which are to be used (Fig. 6.20). The wavefunction for the molecular orbital then takes on the form $\psi = c_1(2s) + c_2(2p_x) + c_3\psi_1$. In this case, there are three wavefunctions with three energy eigenvalues which are obtained from the solution of the secular equation. The result is shown in Fig. 6.20 for the bonding state.

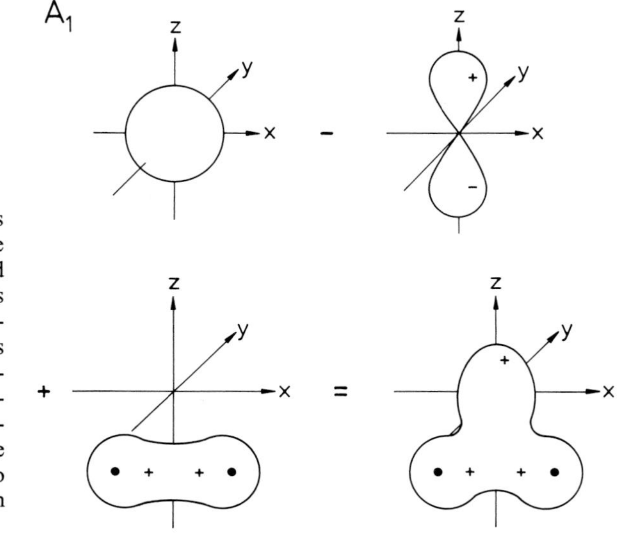

Fig. 6.20. The atomic functions of oxygen which belong to the representation A_1 (*above*) and those of the hydrogen atoms (*below left*) yield the wavefunction shown at the lower right as a bonding function. The function $2p_z$ does not play a major role here. (For the antibonding state cf. Fig. 6.21. The $2p_z$ wavefunction, which has no great influence, is not shown there)

A schematic overview of the wavefunctions obtained for H_2O is given in Fig. 6.21. A qualitative energy term diagram, as found from the solution of the secular determinant, is reproduced in Fig. 6.22. We begin with the lowest energy values: there are clearly two bonding orbitals (symmetry A_1, B_2 which are occupied by four electrons in total. Then (in the centre of the diagram) there are two nonbonding states (of symmetry A_1 and B_1) which are likewise occupied by four electrons; these come from the $2p_x$ and $2p_y$ orbitals of the oxygen atom. Finally, there are the antibonding states (symmetry B_2, A_1) which remain unoccupied. We leave it as an exercise to the reader to find the wavefunctions of the ammonia molecule in a similar manner.

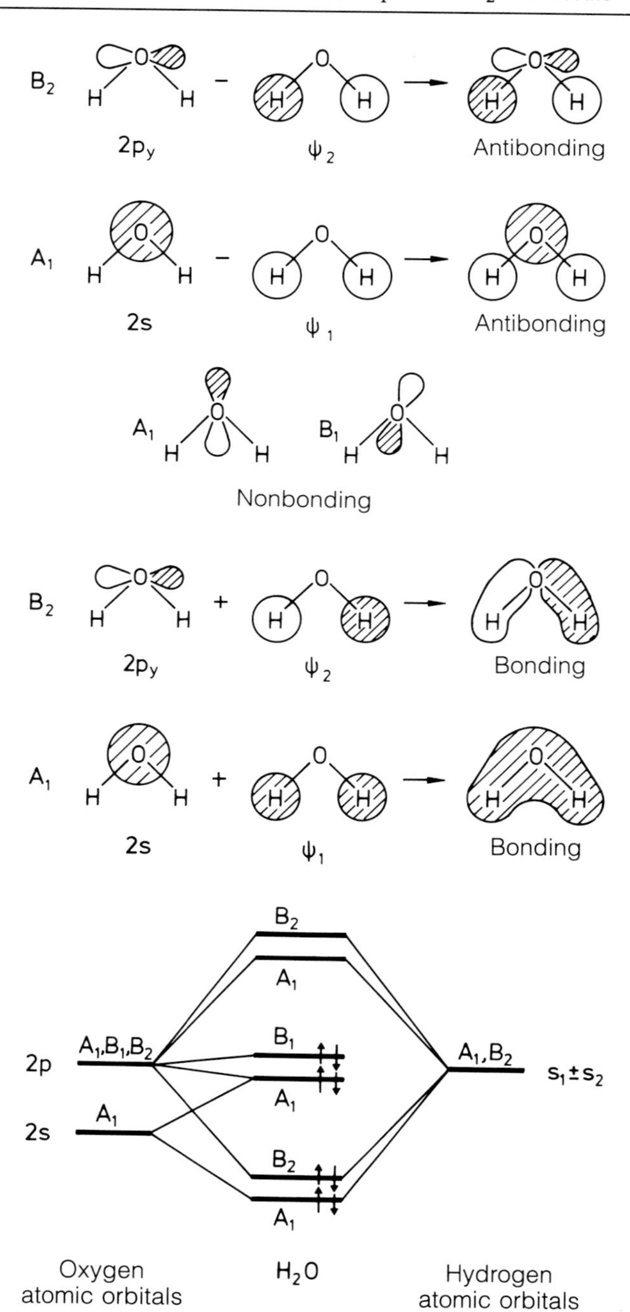

Fig. 6.21. An overview of the H_2O wavefunctions found in this section. Their arrangement here corresponds to the energy-level diagram Fig. 6.22

Fig. 6.22. The energy term scheme of H_2O

7. The Multi-Electron Problem in Molecular Physics and Quantum Chemistry

In this chapter, we shall meet up with some approaches to treating the multi- or many-electron problem in molecular physics and quantum chemistry. Among them are the Slater determinant approach and the Hartree-Fock equations to which it leads, which we will discuss for both closed and open electronic shells. An important concept is the correlation energy between electrons, and we will introduce several general methods for dealing with it.

7.1 Overview and Formulation of the Problem

7.1.1 The Hamiltonian and the Schrödinger Equation

In the following sections, we continue what was begun in Chaps. 4 and 5, where we already introduced some important methods using simple molecules as examples. Here, we deal with approaches to finding the electronic wavefunctions of molecules in general, including complex molecules. In the general case, N electrons with the coordinates $\boldsymbol{r}_j$, $j = 1, \ldots, N$ move in the Coulomb field of the M nuclei with coordinates $\boldsymbol{R}_K$, $K = 1, \ldots, M$ and nuclear charge numbers Z_K, and are also coupled to each other via the Coulomb interactions. The nuclei are taken to be fixed at their equilibrium positions $\boldsymbol{R}_K$, which they possess in the molecule under consideration. For an electron with the coordinate $\boldsymbol{r}_j$, we thus find an overall potential given by:

$$V(\boldsymbol{r}_j) = \sum_K V_K(\boldsymbol{r}_j) , \tag{7.1}$$

where the individual contributions consist of the Coulomb interaction energies between the electron j and the nucleus K:

$$V_K(\boldsymbol{r}_j) = -\frac{Z_K e^2}{4\pi\varepsilon_0 |\boldsymbol{R}_K - \boldsymbol{r}_j|} . \tag{7.2}$$

The Hamiltonian for the electron with index j then contains the operators for the kinetic energy and the potential energy, i.e. it is given by:

$$H(\boldsymbol{r}_j) \equiv H(j) = -\frac{\hbar^2}{2m_0}\nabla_j^2 + V(\boldsymbol{r}_j) . \tag{7.3}$$

(In a more exact treatment, the spin-orbit interaction would also have to be taken into account, but we shall neglect it here.) Between the electron with index j and an electron with index l there is in addition a Coulomb interaction, whose potential energy is given by:

$$W_{jl} = \frac{e^2}{4\pi\varepsilon_0|\boldsymbol{r}_j - \boldsymbol{r}_l|} . \tag{7.4}$$

The interaction energy of all the electrons may then be written as:

$$H_{\mathrm{int}} = \frac{1}{2}\sum_{j\neq l} \frac{e^2}{4\pi\varepsilon_0|\boldsymbol{r}_j - \boldsymbol{r}_l|} . \tag{7.5}$$

The factor 1/2 guarantees that the Coulomb interactions between each pair of electrons are not counted twice in the sum, since the indices j and l run over all electrons independently of one another, the only limitation being that an electron does not interact with itself, i.e. $j \neq l$.

After these preparatory definitions, we are ready to write down the Hamiltonian of the overall system; it has the form:

$$H = \sum_{j=1}^{N} H(j) + H_{\mathrm{int}} . \tag{7.6}$$

The Schrödinger equation is then

$$H\,\Psi(\boldsymbol{r}_1, \ldots, \boldsymbol{r}_N) = E\,\Psi(\boldsymbol{r}_1, \ldots, \boldsymbol{r}_N) , \tag{7.7}$$

where the wavefunction Ψ depends on all the electronic coordinates. Although the Hamiltonian H does not explicitly contain the electron spins, it is still important that the wavefunction Ψ also be a function of the spin coordinates, so that we can take the Pauli exclusion principle into account in a suitable manner, as we have already seen in Sect. 4.4. While it is in fact possible to solve the one-electron Schrödinger equation corresponding to the Hamiltonian (7.3) by using suitable approximations or numerical methods, the solution of the many-electron problem described by (7.6) and (7.7) presents considerable difficulties, since the electrons interact with each other. Even when there are only two electrons moving in a predetermined potential field (7.1), the problem cannot be solved exactly. We must therefore search for suitable approximate approaches; this process can be aided by applying our physical intuition.

7.1.2 Slater Determinants and Energy Expectation Values

One such approximate method can be found in the form of the Slater determinant, which we have already introduced in Sect. 4.4. Each individual electron is described by a wavefunction ψ; these are distinguished by their quantum numbers, denoted as q. In addition, an electron can be in either a spin-up state α or in a spin-down state β. The electron with index j thus can occupy states of the type

$$\psi_q(\boldsymbol{r}_j)\alpha(j) \tag{7.8}$$

or

$$\psi_q(\boldsymbol{r}_j)\beta(j) . \tag{7.9}$$

For the following considerations, it is expedient to introduce a uniform notation for the wavefunctions with spin up or with spin down. We call them s_m and adopt the convention:

$$s_{1/2} = \alpha ,$$
$$s_{-1/2} = \beta . \tag{7.10}$$

The index $m = 1/2$ clearly refers to spin up and the index $m = -1/2$ to spin down electrons. We can now combine (7.8) and (7.9) into the form

$$\chi_k(j) = \psi_q(\boldsymbol{r}_j) s_m(j) . \tag{7.11}$$

We have abbreviated the functions on the right-hand side of (7.11) as the wavefunction $\chi_k(j)$; here, k is a quantum number which includes the quantum numbers q and m:

$$k = (q, m) . \tag{7.12}$$

In order to avoid an overly complicated notation for our method, we let the index k take on the successive values $1, \ldots, N$. This scheme is quite sufficient to allow us to distinguish the different quantum states, and by a suitable renumbering it can be related to (7.12); we will not concern ourselves here with the details of this purely formal correspondence. Using the wavefunction χ_k, we can write the Slater determinant in a simple way:

$$\Psi = \frac{1}{\sqrt{N!}} \begin{vmatrix} \chi_1(1) & \ldots & \chi_N(1) \\ \chi_1(2) & \ldots & \chi_N(2) \\ & \vdots & \\ \chi_1(N) & \ldots & \chi_N(N) \end{vmatrix} . \tag{7.13}$$

As we have already seen in some examples in Sect. 4.4, the Slater determinant takes the Coulomb interaction of the electrons among themselves into account in a summary manner. We will now prove this in general. To this end, we first formulate the expectation value for the energy using the Hamiltonian (7.6) and the determinant (7.13):

$$\overline{E} = \left\langle \int \Psi^* H \Psi \, dV_1, \ldots, dV_N \right\rangle , \tag{7.14}$$

where the angular brackets refer to the spin functions. The evaluation of the integrals on the right-hand side of (7.14) is a tedious matter, and we relegate it to Appendix A1. Here, it suffices to give the final result:

$$\overline{E} = \sum_k H_{k,k} + \tfrac{1}{2} \sum_{k,k'} (V_{kk',kk'} - V_{kk',k'k}) . \tag{7.15}$$

In this expression, the symbol

$$H_{k,k} = \left\langle \int \chi_k^* \, H(\boldsymbol{r}) \, \chi_k dV \right\rangle \tag{7.16}$$

represents the expectation value of the Hamiltonian (7.3) for a single electron in the quantum state k. The angular brackets imply an expectation value with respect to the spin functions, as already noted, while the integral over dV refers to the spatial coordinates of the electrons. Because we have assumed product wavefunctions (7.11) and owing to (7.12), (7.16) can be simplified to

$$H_{qq} = \int \psi_q^*(\boldsymbol{r}) \, H(\boldsymbol{r}) \, \psi_q(\boldsymbol{r}) \, dV . \tag{7.16a}$$

We have already met the quantities $V_{kk',kk'}$ and $V_{kk',k'k}$ as special cases in previous sections.

$$V_{kk',kk'} = \left\langle \iint \chi_k^*(1)\chi_{k'}^*(2) \frac{e^2}{4\pi\varepsilon_0|\boldsymbol{r}_1 - \boldsymbol{r}_2|} \chi_k(1)\chi_{k'}(2)\, dV_1 dV_2 \right\rangle \tag{7.17}$$

represents the interaction of the charge density of electron (1) in state k with the charge density of electron (2) in state k'. This is a Coulomb interaction energy, which has an obvious classical interpretation. On the other hand, we also obtain the expression

$$V_{kk',k'k} = \left\langle \iint \chi_k^*(1)\chi_{k'}^*(2) \frac{e^2}{4\pi\varepsilon_0|\boldsymbol{r}_1 - \boldsymbol{r}_2|} \chi_{k'}(1)\chi_k(2)\, dV_1 dV_2 \right\rangle , \tag{7.18}$$

which, generalising our earlier results, can be termed the Coulomb exchange interaction energy.

Equation (7.15) is the most important result of this chapter. As we know from Sect. 4.4, a variational principle holds in quantum mechanics, and it states that the energy $\overline{E}$, which is calculated approximately in (7.15), is always greater than or at most equal to the exact energy. The attempt to minimise this energy $\overline{E}$ by making a suitable choice of the wavefunctions ψ_q leads to the so-called Hartree-Fock equations, which we shall present below for various important special cases. In solving these Hartree-Fock equations, we will arrive at the "self-consistent field" (SCF) method.

7.2 The Hartree-Fock Equation. The Self-Consistent Field (SCF) Method

Depending on how the individual electronic states are filled with electrons having parallel or antiparallel spins, (7.15) takes on various explicit forms. We will find different expressions for closed shells and open shells. Then, in Sects. 7.5 through 7.7, we explore the limits of the Hartree-Fock method described here, and try to show what approaches must be taken in order to improve the technique. As the reader will see, an extensive field remains to be explored, including the rational application of high-speed computers to the problem of calculating energy expectation values and wavefunctions.

As a first step, we attempt to simplify the expressions V in (7.17) and (7.18), recalling the assumption of product wavefunctions in (7.8) and (7.9). Inserting these into (7.17), we can split the right side into an integral over the spatial functions and a matrix element referring to the spin functions:

$$V_{kk',kk'} \equiv V_{qq',qq'} \langle s_m(1)s_m(1)\rangle \, \langle s_{m'}(2)s_{m'}(2)\rangle ,$$

where

$$V_{qq',qq'} = \int \psi_q^*(\boldsymbol{r}_1)\psi_{q'}^*(\boldsymbol{r}_2) \frac{e^2}{4\pi\varepsilon_0|\boldsymbol{r}_1 - \boldsymbol{r}_2|} \psi_q(\boldsymbol{r}_1)\psi_{q'}(\boldsymbol{r}_2)\, dV_1 dV_2 . \tag{7.19}$$

Since the spin functions are normalised, we find immediately that

$$\langle s_m s_m \rangle = 1 . \tag{7.20}$$

For the exchange interaction, we obtain:

$$V_{kk',k'k} = V_{qq',q'q} \langle s_m(1) s_{m'}(1) \rangle \langle s_m(2) s_{m'}(2) \rangle ,$$

where

$$V_{qq',q'q} = \int \psi_q^*(\boldsymbol{r}_1) \psi_{q'}^*(\boldsymbol{r}_2) \frac{e^2}{4\pi\varepsilon_0 |\boldsymbol{r}_1 - \boldsymbol{r}_2|} \psi_{q'}(\boldsymbol{r}_1) \psi_q(\boldsymbol{r}_2) \, dV_1 dV_2 . \tag{7.21}$$

But now we have

$$\langle s_m s_{m'} \rangle \neq 0 \text{ only when } m = m' . \tag{7.22}$$

The exchange interaction thus operates only between electrons having the same spins.

7.3 The Hartree-Fock Method for a Closed Shell

In this section, as we have already said, we shall investigate some special cases of the energy expression (7.15), and in the process introduce the Hartree-Fock method. We first consider the problem of so-called closed shells. In this case we are dealing with electronic levels characterised by quantum numbers q which are filled with pairs of electrons having their spins antiparallel. There are thus $N/2$ electrons with spin up and $N/2$ electrons with spin down. Let us take a closer look at the terms in (7.15) keeping this aspect in mind: the energy expression (7.16) now occurs twice with the same quantum numbers q, since it refers once to the spin up and once to the spin down electrons. Instead of the sum over all quantum numbers k, we can therefore replace the first sum in (7.15) by quantum numbers over q if we multiply the sum by a factor of 2. The Coulomb interaction (7.19) refers to both electronic spin directions, so the *double* sum $\sum_{kk'}$ in (7.15) requires a factor of 4. In the case of the exchange interaction (7.21), which enters (7.15) with a negative sign, the spin quantum numbers belonging to k and k' are the same, so that once for "spin up" and once for "spin down" results in a factor of just 2. It should thus be clear that for the case of a closed shell, expression (7.15) reduces to:

$$\overline{E} = 2 \sum_q H_{qq} + \sum_{qq'} (2V_{qq',qq'} - V_{qq',q'q}) . \tag{7.23}$$

Equation (7.23) can be used as the starting point of a variational calculation, for which we normalise it, taking into account the condition that the individual electronic wavefunctions be normalised. We need not explicitly apply the condition that they are mutually orthogonal, since it can be shown that the wavefunctions can always be chosen to be orthogonal within the determinant. This follows from the fact that columns or rows can be added together without changing the value of the determinant; using the Schmidt orthogonalisation scheme, it can be seen that the wavefunctions can always chosen to be mutually orthogonal as long as they were linearly independent to begin with. We thus require:

$$\langle \Psi^* | H_{\text{tot}} | \Psi \rangle = \text{Min.!} \tag{7.24}$$

and take as a supplementary condition:

$$\int \psi_q^* \psi_q dV = 1 , \quad q = 1, \ldots, N , \tag{7.25}$$

from which the normalisation of the overall wavefunction naturally follows:

$$\langle \Psi^* \mid \Psi \rangle = 1 \ . \tag{7.26}$$

Variation with respect to a wavefunction ψ_q means that we formally differentiate the right-hand side of the energy expression (7.23) with respect to ψ_q and drop the integration over the corresponding electronic coordinates. Applying this method, we immediately obtain the relation

$$\begin{aligned} H(1)\psi_q(1) &+ 2\sum_{q'}\int |\psi_{q'}(2)|^2 \frac{e^2}{4\pi\varepsilon_0 r_{12}}\, dV_2\, \psi_q(1) \\ &- \sum_{q'}\int \psi^*_{q'}(2)\psi_q(2)\frac{e^2}{4\pi\varepsilon_0 r_{12}}\, dV_2\, \psi_{q'}(1) = \varepsilon_q\psi_q(1)\ , \end{aligned} \tag{7.27}$$

where the ε_q are Lagrange multipliers which take into account the supplementary condition (7.25).

The resulting equation for ψ_q can be interpreted as a kind of Schrödinger equation. The first term in (7.27) represents the operators for the kinetic and potential energy of the wavefunction ψ_q in the field of the fixed atomic nuclei. The second term can be interpreted in a simple way if we remind ourselves that

$$e\,|\psi_{q'}(2)|^2 \tag{7.28}$$

is the charge density of electron (2) in the state q'. The sum then clearly represents the Coulomb interaction energy of electron (1) in the field of the charge densities (7.28). This term can be understood in terms of classical physics. Important and new, in contrast, is the third term in expression (7.27), which describes the Coulomb exchange interaction. Here, electron (1) is in wavefunction ψ_q and experiences the exchange density of electron (2); the latter is given by the expression

$$e\,\psi^*_{q'}(2)\psi_q(2)\ . \tag{7.29}$$

From the physical meaning of the terms on the right-hand side of (7.27) which we have just discussed, it follows that the parameter ε_q can be seen as the energy of an electron in the quantum state q. The set of equations (7.27) are distinguished from the usual Schrödinger equation in that they contain non-linear expressions in ψ_q, $\psi_{q'}$.

These equations (7.27) can be solved by an approach which is referred to as the "self-consistent field" method. The first step is to assume that the wavefunctions ψ_q are already known, at least approximately. In the next step, these assumed wavefunctions are inserted into the expressions for the charge density (7.28) and the exchange density (7.29), while the wavefunctions which follow H and which occur behind the integrals are taken as still to be determined. The set of equations (7.27), which has thus been linearised, is then solved for the ψ_q, and the resulting wavefunctions are reinserted into the charge and exchange densities (7.28) and (7.29), giving improved starting values for a new iteration. This procedure is continued until, at least in principle, the wavefunctions obtained are practically identical to those assumed in the previous step. The method thus leads to an internally consistent set of wavefunctions, as is implied by the name "self-consistent field" method.

7.4 The Unrestricted SCF Method for Open Shells

If closed shells are present, as was assumed in the previous section, then in the Hartree-Fock approach the individual electronic states are each occupied by two electrons having antiparallel spins. In the case of open shells, the electronic states which refer to the orbital motion of the electrons and which correspond to a pair of electrons with spin up and spin down may be *different* states. The Slater determinant then takes on the following form [using the notation of (5.65)]:

$$\Psi = |\psi_1\psi_2 \dots \psi_M \overline{\psi}_{M+1} \overline{\psi}_{M+2} \dots \overline{\psi}_{M+N}| , \tag{7.30}$$

where the functions $\psi_1 \dots \psi_M$ refer to electrons having spin up, and the functions $\overline{\psi}_{M+1} \dots \overline{\psi}_{M+N}$ to electrons with spin down. In the following, we shall also allow wavefunctions belonging to different spin directions to have the same dependence on the spatial coordinates. We therefore allow the case that some of the orbital quantum numbers of the group $M+1, \dots,$ $M+N$ are identical to some of those of the group $1, \dots, M$. Since the spins of these two groups are different, the determinant (7.30) does not vanish. The normalisation factor of the determinant is given by:

$$[(M+N)!]^{-1/2} . \tag{7.31}$$

By a proper choice of the ψ's, the expectation value of the energy of the molecular Hamiltonian is to be minimised. We assume that the wavefunctions in (7.30) are mutually orthogonal. This expectation value can then be obtained directly from (7.15) by a specialisation analogous to Sect. 7.2, so that we simply give the result here. It is

$$\begin{aligned} \overline{E} = \langle\Psi|H|\Psi\rangle = &\sum_{j=1}^{M+N} H_{jj} + \frac{1}{2}\sum_{i=1}^{M+N}\sum_{j=1}^{M+N} V_{ij,ij} \\ &- \frac{1}{2}\Bigg(\underbrace{\sum_{i=1}^{M}\sum_{j=1}^{M} V_{ij,ji}}_{\uparrow-\text{spins}} + \underbrace{\sum_{i=M+1}^{M+N}\sum_{j=M+1}^{M+N} V_{ij,ji}}_{\downarrow-\text{spins}}\Bigg) . \end{aligned} \tag{7.32}$$

Let us consider the different terms in (7.32). The H_{jj} in the first sum are defined by (7.16a). We remember that they refer to the energy expectation value for a single electron in the state j, where the energy consists of the kinetic energy of the electron and its potential energy in the field of the nuclei. In the following double sum, the quantities $V_{ij,ij}$ represent the Coulomb interaction energies between the charge densities of the electrons in states i and j. This interaction includes both electrons of like spin and those of opposite spin. The next two sums are expressions for the exchange interaction, which acts only between electrons having the same spin direction. When we vary the energy $\overline{E}$ in (7.32) by varying a wavefunction ψ_i or ψ_j, keeping the normalisation condition (7.25) in mind, we obtain the corresponding Hartree-Fock equations.

7.5 The Restricted SCF Method for Open Shells

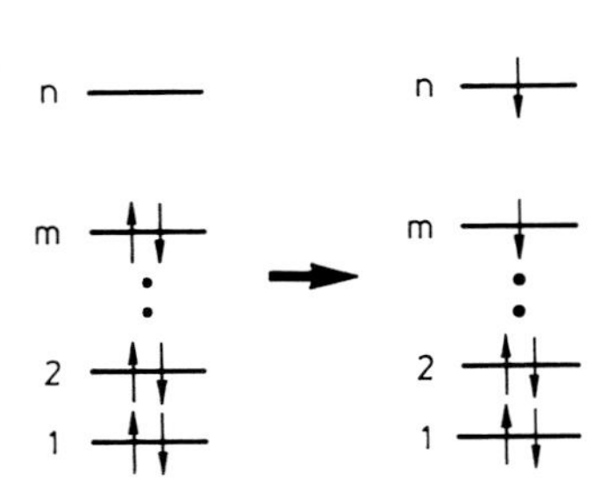

Fig. 7.1. The excitation of an electron from the state m into the state n, accompanied by a spin flip

In the preceding section, we met the so-called unrestricted open-shell SCF method. There, the wavefunctions for the many-electron problem were taken to have the form of Slater determinants, which are easy to deal with. However, this type of wavefunction is not necessarily an eigenfunction of the total spin. We will now introduce a formulation for the wavefunctions which is already an eigenfunction of the total spin operator. This approach is threfore called the "restricted open-shell method". We shall treat triplet wavefunctions; the approach is principally due to *Roothaan*. We assume that one electron from a filled shell, where the electronic states each contain one spin-up and one spin-down electron, is taken from a state m and put into the state n, and that its spin is flipped in this process (see Fig. 7.1). A wavefunction thus results in which the z-component of the total spin is equal to $-1 \cdot \hbar$. We write this wavefunction as ${}_{-1}^{3}\Psi_m^n$. The indices m and n indicate that the electron was excited from the state m into the state n. The number 3 at the upper left means that the wavefunction belongs to a triplet state, and the lower-left index -1 indicates that the z-component of the total spin has the quantum number $S_z = -1$. This wavefunction can be written as a determinant in the abbreviated form

$$
{}_{-1}^{3}\Psi_m^n = |\psi_1 \overline{\psi}_1 \dots \psi_g \overline{\psi}_g \overline{\psi}_m \overline{\psi}_n| \,, \tag{7.33}
$$

where the normalisation factor is still to be included. In order to go from this state to one where the z-component of the total spin is $S_z = 0$, we need only use the ladder operator for the z-component of the total spin; it is given by:

$$
S_+ = \sum_j [\sigma_x(j) + \mathrm{i}\sigma_y(j)] \,. \tag{7.34}
$$

In this equation, σ_x and σ_y are the usual Pauli spin matrices, and the arguments (j) enumerate the electrons which are acted upon by the spin operators. An elementary but tedious calculation then gives (leaving out the normalisation factor) the wavefunction

$$
{}_{0}^{3}\Psi_m^n = \{|\psi_1 \overline{\psi}_1 \dots \psi_g \overline{\psi}_g \psi_m \overline{\psi}_n| - |\psi_1 \overline{\psi}_1 \dots \psi_g \overline{\psi}_g \psi_n \overline{\psi}_m|\} \,, \tag{7.35}
$$

which belongs to the total spin $S = 1$. If we apply the raising operator for the z-component of the spin a second time, we obtain the wavefunction

$$
{}_{1}^{3}\Psi_m^n = |\psi_1 \overline{\psi}_1 \dots \psi_g \overline{\psi}_g \psi_m \psi_n| \,. \tag{7.36}
$$

As we already have seen in the unrestricted open-shell model, we can calculate the energy in a relatively simple manner. We obtain the following expression:

$$
\begin{aligned}
{}^{3}\overline{E} = & \underbrace{2\sum_{k=1}^{g} H_{kk} + \sum_{k=1}^{g}\sum_{l=1}^{g}(2V_{kl,kl} - V_{kl,lk})}_{\text{closed shell}} \\
& \underbrace{+H_{mm} + H_{nn} + V_{mn,mn} - V_{mn,nm}}_{\text{open shell}} \\
& \underbrace{+\sum_{k=1}^{g}(2V_{km,km} - V_{km,mk}) + \sum_{k=1}^{g}(2V_{kn,kn} - V_{kn,nk})}_{\text{closed} - \text{open shell interaction}}
\end{aligned} \tag{7.37}
$$

which reflects the interaction energies within the closed shell, the interaction energy between the two electrons in the now effectively open shell, and the interaction energy between the two shells. The quantities which occur in (7.37) are the same as in the preceding section. By variation of the energy $\overline{E}$ with respect to the individual wavefunctions, the Hartree-Fock equations can again be derived. As was shown by the examples in Sects. 7.2–7.4, the energy eigenvalues calculated using Slater determinants can be interpreted in a quite simple manner. This should, however, not obscure the fact that we are dealing here only with an approximation.

7.6 Correlation Energies

The Hartree-Fock method, which begins with the Slater determinants, is the most widely-used computational technique in atomic and molecular physics. It allows the exact calculation of the interaction effects between the electrons and the nuclei, and the approximate calculation of the overall interaction effects of the electrons among themselves. As one can readily see, the energy would be reduced even further by allowing the electrons to avoid each other spatially, not only in a global way by applying the Pauli exclusion principle, which requires the probability density for two electrons having the same spin at the same point in space to vanish. Electrons with antiparallel spins also have a Coulomb repulsion and will try to avoid each other in order to reduce the total energy. Compared to the Hartree-Fock energy, in which the Pauli principle has been taken into account via the Slater determinants, there remains an additional energy reduction which would occur in an exact calculation and which results from taking correlations into account, i.e. the tendency towards mutual avoidance by the electrons. The definition of the correlation energy is thus

Correlation energy = exact nonrelativistic energy − Hartree-Fock energy.

7.7 Koopman's Theorem

Once the electronic wavefunctions and the corresponding energies have been calculated for a molecule with a closed-shell configuration using the SCF method, the ionisation of the molecule can also be treated, at least approximately. This is done by applying Koopman's theorem, which might better be called Koopman's approximation. It states the following: ionisation, consisting of the removal of an electron from a molecule with closed shells, can be represented as the removal of an electron from a given self-consistent field orbital, leaving the other electrons unaffected. This is, in general, a good approximation, although it neglects the following effects:

1) the reorganisation energy of the electrons in the ion;
2) the difference between the correlation energy of the neutral molecule and that of the ion.

The second point is clear, since the correlation energy is generally neglected in the Hartree-Fock method. The first point is due to the fact that the charge distribution of the electrons gives rise to an effective potential for each particular electron. If an electron is then removed, this effective potential is naturally altered. Koopman's theorem thus states that, in general, the alteration is small.

7.8 Configuration Interactions

As we mentioned above, the Hartree-Fock method leaves an important effect out of consideration, by not taking into account the correlations between the electrons. For this reason, other methods have been developed which can treat electron correlations, at least partially. We begin with a single Slater determinant:

$$\Psi_{k_1,\dots,k_N} = \frac{1}{\sqrt{N!}} |\chi_{k_1}(\boldsymbol{r}_1)\chi_{k_2}(\boldsymbol{r}_2)\dots\chi_{k_N}(\boldsymbol{r}_N)| , \tag{7.38}$$

where we want to assume, in contrast to the Hartree-Fock method, that the wavefunctions χ_k are already known. The indices k naturally denote the quantum numbers of the individual electrons. For simplicity, we represent these quantum numbers by a single symbol, which however places no limitation on the method. Since the wavefunctions Ψ remain the same (except perhaps for a factor of -1) when we permute the indices k_j, we can assume that the quantum numbers k are already ordered in some particular fashion, e.g. in the sequence

$$k_1 < k_2 \dots < k_N . \tag{7.39}$$

If the wavefunctions χ_k of the individual electrons form a complete set in the mathematical sense, then the determinants (7.38) also form a complete set for each antisymmetric wavefunction Ψ of N electrons. This means that we can represent any arbitrary wavefunction Ψ, even in a many-electron problem, as a linear combination of determinants like (7.38). If we take as a trial function

$$\Psi = \sum_{k_1<k_2<\dots k_N} C_{k_1 k_2 \dots k_N} \Psi_{k_1,\dots,k_N} , \tag{7.40}$$

then the wavefunction we are seeking can be determined by finding the coefficients $C_{k_1,k_2\dots}$. In principle, the method for solving the many-electron problem is no different from that for a single-electron problem, where we can represent the wavefunction we are seeking as a linear combination of known wavefunctions; the only difference is that the combinations of indices become somewhat more complicated. We insert (7.40) into the Schrödinger equation:

$$H\Psi = E\Psi , \tag{7.41}$$

where H is the Hamiltonian for the kinetic energy of the electrons and their potential energies in the field of the nuclei and of the other electrons [cf. (7.6)]. We add primes to the indices k in (7.40), multiply the equation thus obtained by $\Psi^*_{k_1\dots k_N}$, integrate over all the electronic coordinates, and take the expectation value with respect to the spin variables. We thus obtain expressions of the type:

$$\left\langle \int \Psi^*_{k_1,\dots,k_N} H \Psi_{k'_1,\dots,k'_N} dV_1 \dots dV_N \right\rangle , \tag{7.42}$$

where the angular brackets represent the calculation of the expectation value with respect to the spin variables. The evaluation of (7.42) is given in Appendix A1 for the special case that the set of quantum numbers $k'_1, \dots, k'_N$ is the same as the set $k_1, \dots, k_N$. It is not difficult to generalise this result to the case (7.42), so that we simply give the final answer here:

$$\sum_{j=1}^{N}\sum_{k'_j} H_{k_jk'_j} C_{k_1\ldots k'_j\ldots k_N}$$

$$+\sum_{ij=1}^{N}\sum_{k'_ik''_j} V_{k_ik_jk'_ik''_j} C_{k_1\ldots k'_i\ldots k''_j\ldots k_N} = E\, C_{k_1\ldots k_N}\ . \tag{7.43}$$

The quantities $H_{kk'}$ and $V_{kk'k''k'''}$ are generalisations of those previously defined in (7.16–18), i.e.

$$H_{kk'} = \int \psi_k^*(\boldsymbol{r}) \left[\frac{-\hbar^2}{2m_0}\nabla^2 + V(\boldsymbol{r})\right] \psi_{k'}(\boldsymbol{r})\, dV\, \langle s_k|s_{k'}\rangle\ . \tag{7.44}$$

The angular brackets indicate the orthogonality relations between the spin wavefunctions $s_k, s_{k'}$, which denote spin up ($s_k \equiv \alpha$) or spin down ($s_k \equiv \beta$). $V_{kk',k''k'''}$ describes the Coulomb interaction energy between the electrons:

$$V_{kk',k''k'''} = \int \psi_k^*(\boldsymbol{r}_1)\psi_{k'}^*(\boldsymbol{r}_2)\frac{e^2}{4\pi\varepsilon_0 r_{12}} \cdot \psi_{k''}(\boldsymbol{r}_1)\psi_{k'''}(\boldsymbol{r}_2)\, dV_1 dV_2\, \langle s_k|s_{k''}\rangle\, \langle s_{k'}|s_{k'''}\rangle\ . \tag{7.45}$$

The sums on k'_j and k''_j run over all the quantum numbers in (7.43). In order to keep the notation in agreement with (7.40), we, however, must introduce the following convention: the coefficients C in (7.40) are defined only for quantum numbers which fulfill the condition (7.39). We stipulate that:

1) Coefficients occurring in (7.43) vanish if two or more of the quantum numbers indicated by their indices are identical.
2) If the rule (7.39) is broken, the indices of the coefficients will be reordered in such a way that it is restored to validity. Depending on whether the permutation is even or odd, the sign can change.

Equations (7.43) are a system of linear, homogeneous equations which can be solved numerically using modern techniques with a digital computer, as long as the number of coefficients has a limit and is not allowed to become infinite. This method is often combined with the LCAO approach. In that case, the matrix elements $H_{kk'}$ and $V_{kk',k''k'''}$ are evaluated by setting the electronic wavefunctions ψ_k equal to linear combinations of atomic orbitals ϕ_j with free coefficients. These coefficients can then be fixed in a first step, e.g. by solving the Hartree-Fock equations. The matrix elements $H_{kk'}$ and $V_{kk',k''k'''}$ can then be given as particular linear combinations of integrals over atomic orbitals ϕ_j. For the numerical solution of the many-electron problem using a supercomputer, the following steps are thus necessary:

1) Evaluation of integrals of the type

$$\int \phi_j^*\, H\, \phi_j\, dV\ ,$$

$$\int \phi_j^*(1)\phi_{j'}^*(2)\frac{e^2}{4\pi\varepsilon_0 r_{12}}\phi_{j''}(1)\phi_{j'''}(2)\, dV_1 dV_2\ . \tag{7.46}$$

The latter integrals are referred to as multiple-centre integrals.

2) Solution of the linear equations, i.e. calculation of the coefficients $C_{k_1,\ldots,k_N}$ and the corresponding energy eigenvalues.

7.9 The Second Quantisation*

The results of the preceding section can be formulated in a much more elegant fashion by using the so-called *second quantisation*. As we have already seen in I, the photon field can be quantised by establishing a correspondence between each light wave with a particular wavevector $\boldsymbol{k}$ (and a given polarisation direction) and a harmonic oscillator which describes the energy of the wave. The energy expression can be written in harmonic form and thus gives rise to a Hamiltonian H_L which can be expressed in terms of creation and annihilation operators b_k^+, b_k for the light quanta: $H_L = \sum_k \hbar\omega_k b_k^+ b_k$. Starting with classical waves, we can thus describe the creation and annihilation of light quanta, or photons. The operators b_k^+, b_k obey the following commutation relations:

$$b_k^+ b_{k'}^+ - b_{k'}^+ b_k^+ = 0 \, , \tag{7.47}$$

$$b_k b_{k'} - b_{k'} b_k = 0 \, , \tag{7.48}$$

$$b_k b_{k'}^+ - b_{k'}^+ b_k = \delta_{kk'} \, . \tag{7.49}$$

Now, however, we know that *electrons* also have a wave character, which is reflected in the Schrödinger equation. If we quantise this electron-wave field, we arrive at the particle character of the electrons. Just as in the quantisation of the photon field, where the creation and annihilation operators describe the creation or annihilation of light quanta, they here describe the creation or annihilation of electrons. Denoting the electronic state by its quantum numbers, e.g. k or j, we postulate the following commutation relations:

$$a_k^+ a_j^+ + a_j^+ a_k^+ = 0 \, , \tag{7.50}$$

$$a_k b_j + a_j a_k = 0 \, , \tag{7.51}$$

$$a_k^+ a_j + a_j a_k^+ = \delta_{jk} \, . \tag{7.52}$$

These differ from the relations for photons in that there is a (+)-sign in the middle, which is due to the fact that, in contrast to photons (bosons), two electrons (fermions) cannot be in the same quantum state. Equation (7.50) fulfills this requirement; if $j = k$, then it follows from (7.50) that

$$a_k^+ a_k^+ = 0 \, . \tag{7.53}$$

That is, if we try to create two electrons in the same state, this double creation, no matter which state we apply it to, always yields zero. The other commutation relations with the (+) signs then follow from self-consistency requirements which we cannot treat in detail here. The method is now applied as follows: the Schrödinger equation

$$H \Psi = E \Psi \tag{7.54}$$

has the following form in the second quantisation:

$$H = \sum_{ij} a_i^+ a_j H_{ij} + \tfrac{1}{2} \sum_{ijkl} a_i^+ a_j^+ a_k a_l V_{ijkl} \, , \tag{7.55}$$

where the matrix elements are given by (7.44) and (7.45). The expression (7.55) has the advantage that it holds for any number of electrons. If one is treating a particular problem in which a certain number N of electrons is present, then the Schrödinger equation can be solved, at least in principle, by constructing Ψ as a linear combination of all possible functions in which precisely N electrons occur. We denote the vacuum state as Φ_0; it is characterised by the relation

$$a_j \Phi_0 = 0 . \tag{7.56}$$

Then a state having N electrons with the quantum numbers $k_1 \ldots k_N$ can be built up by N-fold application of the creation operator to the state Φ_0:

$$\Psi_{k_1 \ldots k_N} = a_{k_1}^+ \ldots a_{k_N}^+ \Phi_0 . \tag{7.57}$$

The complete trial wavefunction is then given by a linear combination of the functions defined in (7.57),

$$\Psi = \sum C_{k_1 \ldots k_N} a_{k_1}^+ a_{k_2}^+ \ldots a_{k_N}^+ \Phi_0 . \tag{7.58}$$

The coefficients in this equation are still unknown quantities, which must be determined through, e.g. a minimisation of the expectation value of E. The trial function (7.58) may contain completely unrestricted sums over the individual quantum numbers k. If two quantum numbers are the same, the wavefunction (7.58) vanishes by construction, due to (7.53). Furthermore, if sets of quantum numbers are identical, then by reordering the operators a^+ they can be brought into a special form, e.g. in agreement with (7.39), whereby depending on whether the permutation is even or odd, the sign remains unchanged or is reversed. As can be seen by comparing the method of 2nd quantisation with that described in Sect. 7.8, the two are equivalent, but the 2nd quantisation is more elegant, because the resulting equations can be very simply found by substituting (7.58) into (7.54). Also, it is clear from the beginning which quantum numbers are to be used. In addition, the 2nd quantisation permits some novel approaches to the explicit solution of the problem.

7.10 Résumé of the Results of Chapters 4–7

In Chaps. 4–7, we have gained an overview of the methods available for determining the electronic wavefunctions in molecules and their energy eigenvalues. Chapter 4 was devoted in particular to the LCAO method, i.e. the construction of molecular orbitals for a single electron by taking linear combinations of atomic orbitals; as an illustration, we treated there the simple molecules H_2^+ and H_2. Furthermore, the hybridisation of the wavefunctions of carbon was introduced there. Chapter 5 presented a first insight into the way in which the calculation of the coefficients in the LCAO method can be simplified or eliminated by making use of the symmetry properties of the molecule; this was demonstrated for benzene and ethene. In Chap. 6, we then made a systematic survey of symmetries and symmetry operations as well as of the basic concepts and methods of the theory of group representations. These methods were then applied in detail to the wavefunctions of the H_2O molecule. As we saw, it is possible to reduce considerably the number of equations required for the LCAO approach by making use of symmetry. Finally, Chap. 7 introduced a series of methods for dealing with the many-electron problem. Simple approaches are based on the Slater determinant and the Hartree-Fock method which is associated with it. In order to take electron correlations into account, linear combinations of Slater determinants must be employed. An equivalent, but more elegant method is found in the 2nd quantisation, which we also treated briefly.

In the following Chaps. 8–10, we now turn to the experimental results obtained on small, simple molecules.

8. Overview of Molecular Spectroscopy Techniques

Spectroscopy using electromagnetic radiation in all wavelength regions, in the radiofrequency range, with microwaves, in the infra-red, in the visible spectral region and in the ultraviolet region, extending out to the spectral region of extremely short-wavelength gamma radiation, is the most important source of experimental information for molecular physics. In this and the following chapters, we shall deal with this topic. The experimental methods associated with these spectroscopies will be described in more detail where necessary. In this chapter, in Sects. 8.1 and 8.2, we first give the classification of the spectroscopic methods according to the spectral region studied, as required to obtain the desired information. This will serve as an introduction to the following Chaps. 9–14. In Sect. 8.3, we then indicate some additional methods, namely laser spectroscopy, photoelectron spectroscopy, and magnetic resonance, which are treated in the later Chaps. 15, 18, and 19.

8.1 Spectral Regions

Here, we first want to give a summary of the various spectral regions in the electromagnetic spectrum. See also Fig. 8.1.

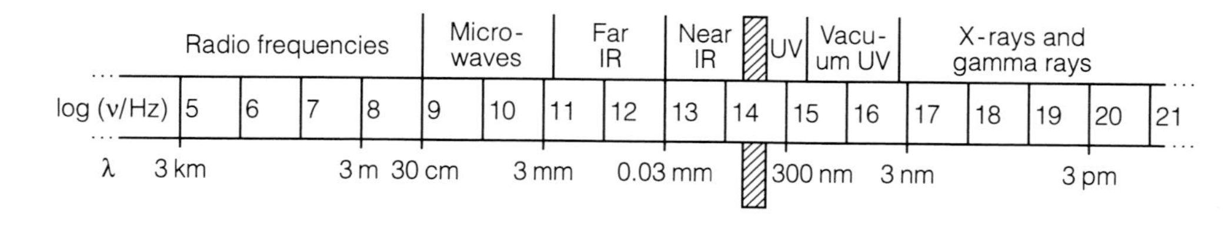

Fig. 8.1. The spectrum of electromagnetic radiation from the radiofrequency range up to gamma radiation, in units of frequency and of wavelength. The visible region is shaded

Beginning with the smallest energies, the spectral regions can be classified and characterised; we note that the boundaries between the regions are, however, not sharply defined. They were determined in the past by the different methods and instrumentation available for the production, transmission, and detection of the radiation, as well as by convention.

- In the region of *radiofrequencies*, i.e. in the range from a few kHz up to several 100 MHz, we find the nuclear resonance transitions.
- The term *microwaves* refers to electromagnetic waves in the range from about 1 to 100 GHz. This is the region of electron spin resonance spectroscopy, but also that of rotational spectroscopy, especially on small molecules in the gas phase. The upper end of this range already overlaps with the spectral region of the far infra-red.

- The *infra-red* spectral region extends from the upper part of the microwave range to the beginning of the visible region, at a wavelength near 800 nm. The long wavelength part, the *far infra-red* region ($\lambda = 0.1$–1 mm) is applicable to the excitation of rotational spectra, while the short wavelength end (the *near infra-red*, $\lambda = 10^{-3}$–10^{-1} mm) is the region where the characteristic vibrational spectra of molecules are observed: the so-called rotational-vibrational spectra.
- Electronic transitions of the valence electrons begin already in the infra-red; however, they lie mostly in the *visible* and the *UV* spectral regions. Here, the band spectra of molecules in the proper sense are observed, i.e. spectra consisting of electronic transitions with superimposed rotational and vibrational transitions.
- Beyond the short wavelength end of the ultraviolet region, and overlapping with it, is the X-*ray* region and then the region of γ-*radiation.* With radiation of such high quantum energies, transitions and states of the inner electrons, i.e. those in inner shells, can be investigated, especially by *photoelectron spectroscopy.*

In the different spectral regions, and also in the various scientific disciplines, a variety of units for measuring the frequencies and the wavelengths of the radiation are in conventional use, in part for practical reasons and in part for historical ones. Some important conversion formulas for units which measure energy are the following:

$$1\ \text{cm}^{-1} = 29.979\ \text{GHz} = 1.2398 \cdot 10^{-4}\ \text{eV} \tag{8.1}$$

$$1\ \frac{\text{kcal}}{\text{kmol}} = 0.349\ \text{cm}^{-1}\ . \tag{8.2}$$

Measuring energies in [cm^{-1}] or in [s^{-1}] is a widespread and convenient practice, but strictly speaking, it is incorrect. The unit $\overline{\nu}$, or wavenumber, is defined by the relation

$$\overline{\nu} = \frac{1}{\lambda} = \frac{\nu}{c} = \frac{\text{energy}}{hc} \quad [\text{cm}^{-1}]\ . \tag{8.3}$$

For the unit of frequency, we have

$$\nu = \frac{c}{\lambda} = \frac{\text{energy}}{h} \quad [\text{s}^{-1}]\ . \tag{8.4}$$

Energy may also be expressed in terms of $\hbar\omega$, that is by $(h/2\pi)2\pi\ \nu$.

8.2 An Overview of Optical Spectroscopy Methods

We can, to a good approximation, express the total excitation energy E of a molecule as the sum of the above-mentioned individual excitations, in particular as the sum of the partial excitations of the rotational, the vibrational, and the electronic levels. We thus have

$$E = E_{\text{el}} + E_{\text{vib}} + E_{\text{rot}}\ , \tag{8.5}$$

where el, vib, and rot refer to electronic, vibrational, and rotational excitations, respectively.

Figure 8.2 illustrates the vibrational and rotational levels in two different electronic excitation states of a molecule, I and II, and the possible transitions between them. According to this diagram, one can distinguish between three types of optical spectra, as follows:

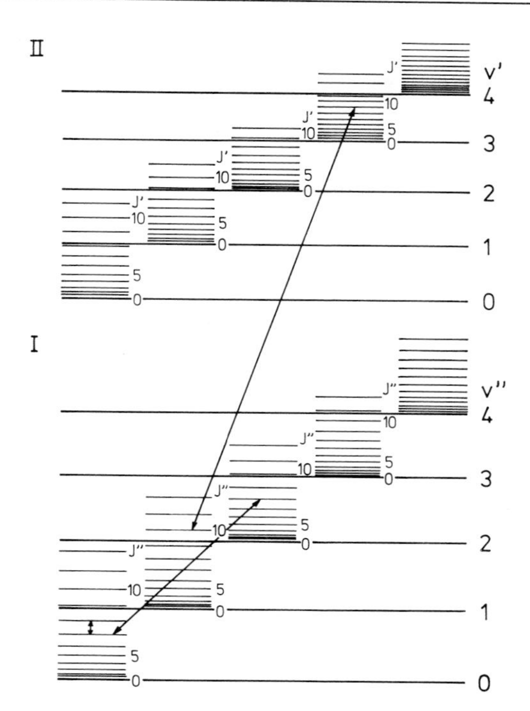

Fig. 8.2. Vibrational levels (quantum numbers v) and rotational levels (Quantum numbers J) of two electronic excitation states of a molecule, denoted by I and II. The three arrows refer (from *left* to *right*) to transitions in the rotational, the rotational-vibrational, and in the electronic spectra of the molecule

- *Rotational spectra* are transitions between the rotational levels of a given vibrational level in a particular electronic state. Only the rotational quantum number changes in these transitions; we denote it by J. These spectra lie in the region of microwaves or in the far infra-red. They are treated in the following Chap. 9. They consist typically of a large number of closely spaced, nearly equidistant spectral lines. Rotational spectra may also be observed by means of Raman spectroscopy; see Chap. 12.
- *Rotational-vibrational spectra* consist of transitions from the rotational levels of a particular vibrational state to the rotational levels of another vibrational state in the same electronic term. The electronic excitation state thus remains the same. The quantum numbers J and v change; v characterises the quantised vibrational levels. These spectra lie in the infra-red spectral region. Rotational-vibrational spectra are treated in Chap. 10. They consist of a number of "*bands*", i.e. groups of closely-spaced lines, the so-called band lines. These spectra can also be observed with Raman spectroscopy, as well as with infra-red spectroscopy.
- *Electronic spectra* consist of transitions between the rotational levels of the various vibrational levels of one electronic state and the rotational and vibrational levels of a different electronic state. This is termed a *band system*. It contains all the vibrational bands of the electronic transition being observed, each one with its rotational structure. In general, all three quantum numbers change in these transitions, i.e. J, v, and those which characterise the electronic state. The spectra lie in the near infra-red, the visible, or the ultraviolet regions. Electronic transitions in molecules are treated in Chap. 13. The band systems of all the allowed electronic transitions of a molecule together make up the *band spectrum* proper of the molecule.

In molecular spectroscopy, it is generally accepted practice when referring to transitions between two terms to list first the energetically higher-lying term, then the lower one. The direction of the transition, i.e. absorption or emission, can be denoted by an arrow between the two term symbols. If the various terms in a series are not numbered, then one frequently denotes the upper term with a prime, e.g. J' or v', and the lower term with a double prime, e.g. J'' or v''.

The spectral lines in molecular spectra, i.e. transitions between two terms, may be described in the following manner:

$$\begin{aligned}\bar{\nu}hc &= E'_{\rm el} - E''_{\rm el} + E'_{\rm vib} - E''_{\rm vib} + E'_{\rm rot} - E''_{\rm rot} \quad \text{[Joule]} \\ &= \Delta E_{\rm el} + \Delta E_{\rm vib} + \Delta E_{\rm rot} \,,\end{aligned} \tag{8.6}$$

where el, vib, and rot again refer to the electronic, vibrational, and rotational energies. In general, the following relation holds:

$$\Delta E_{\rm el} \gg \Delta E_{\rm vib} \gg \Delta E_{\rm rot} \,. \tag{8.7}$$

For rotational spectra, we have $\Delta E_{\rm el} = \Delta E_{\rm vib} = 0$; only the rotational term changes in the transitions, i.e.

$$\bar{\nu}hc = E'_{\rm rot} - E''_{\rm rot} \,. \tag{8.8}$$

Rotational-vibrational spectra correspond to transitions with $\Delta E_{\rm el} = 0$; the transitions take place between the terms of vibration and rotation. We then have:

$$\bar{\nu}hc = E'_{\rm vib} - E''_{\rm vib} + E'_{\rm rot} - E''_{\rm rot} \,. \tag{8.9}$$

A rotational-vibrational *band* is the total of all the band lines $\Delta E_{\rm rot}$ which belong to a particular term transition $\Delta E_{\rm vib}$. If the electronic energy also changes, then all three terms in (8.6) change in the transition and the *band system* of the corresponding electronic transition $\Delta E_{\rm el}$ is obtained. It contains all the vibrational *bands* ($\Delta E_{\rm vib}$) with their characteristic rotational structures. The terminology *band spectrum* of a molecule (in the wider sense) refers to the band systems of all the possible electronic transitions.

The positions of the three types of spectra within the electromagnetic spectrum are indicated for a small molecule in Fig. 8.3.

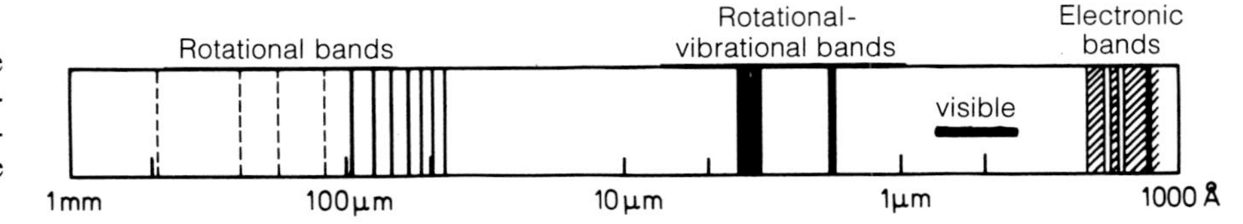

Fig. 8.3. An overview of the spectral positions of the absorption spectra of a small molecule. The numerical values are approximately correct for HCl

In molecular spectroscopy, it is usual to employ the following notation for the *terms E/hc* (measured in wavenumbers):

For rotational terms

$$\frac{E_{\rm rot}(J)}{hc} \equiv F(J) \,; \tag{8.10}$$

for vibrational terms

$$\frac{E_{\mathrm{vib}}(v)}{hc} \equiv G(J)\,; \tag{8.11}$$

and for electronic terms

$$\frac{E_{\mathrm{el}}(v)}{hc} \equiv T^{\mathrm{el}}\,. \tag{8.12}$$

The overall term of a molecule can thus be written as

$$E_{\mathrm{tot}}/hc \equiv T = T^{\mathrm{el}} + G(v) + F(v, J)\,. \tag{8.13}$$

Spectral lines can then be denoted by:

$$\overline{\nu} = \Delta T^{\mathrm{el}} + \Delta G + \Delta F \quad [\mathrm{cm}^{-1}]\,. \tag{8.14}$$

8.3 Other Experimental Methods

We should mention at this point that there are other methods of investigation in addition to rotational, vibrational, and electronic spectroscopies, which also give us insights into the structure and dynamics of molecules.

Laser spectroscopy permits the study of molecules with a spectral resolution which was completely unattainable in earlier times. It also makes possible the time resolution of molecular spectra down to the femtosecond range, and thus allows the study of the dynamics of molecular states and processes. Important additional information is gained from *photoelectron spectra*, in particular with respect to the analysis of the energy states of inner electrons. More about these two topics will be presented in Chap. 15.

Magnetic resonance of nuclei and electrons gives particularly detailed structural information, which cannot be obtained with other spectroscopic methods. These techniques will be treated in Chaps. 18 and 19.

9. Rotational Spectroscopy

The rotational energies of molecules are *quantised*: that is, they can be changed only through the absorption or emission of energy quanta. Rotational spectroscopy permits the measurement of these energy levels; from them one obtains information about the structure and bonding of the molecules. The essential concepts can be explained and understood using the simplest molecules as examples, i.e. the diatomic molecules. Sects. 9.1–3 are devoted to this task. The multiplicity of possible rotations in larger molecules can be only briefly touched upon in this book; we do this in Sect. 9.5.

9.1 Microwave Spectroscopy

The rotational spectra of molecules are observed almost exclusively as absorption spectra, because the spontaneous emission probability is very small as a result of the low transition energies; see also Chap. 16 and Sect. 5.2.3 in I. The rotational spectra lie in the microwave region of the electromagnetic spectrum, so that one requires a far infra-red (Fourier) spectrometer or a microwave spectrometer to observe them.

In a microwave spectrometer, the source of radiation is often a klystron. Since klystrons can, however, be tuned over only a narrow spectral region, tunable oscillators such as the backwards-wave generator (also called a carcinotron) are sometimes preferable. These are travelling-wave tubes in which the oscillation frequency (in the GHz range) can be tuned by 50% or more through variation of the electrical operating conditions. Another tunable generator is the so-called magnetron. Detection of the microwaves is usually performed using a microwave diode. Owing to the small absorption coefficients and to the necessity to work at low (sample) gas pressures and thus avoid pressure broadening of the spectral lines as much as possible, the longest possible absorption paths are employed (in the range of several metres). For quantum energies larger than ca. 10 cm^{-1}, infra-red Fourier spectrometers can be used to measure rotational spectra.

To improve the detection sensitivity and for a more exact frequency determination, one normally employs an effect-modulation technique. This means that the energy levels under investigation are modulated in such a way that the intensity of absorption and thus of the observed signals are also modulated. With this technique, the signal/noise ratio and thereby the precision of the measurement can be improved; this can be achieved in microwave spectroscopy by allowing an oscillating electric field to act on the sample molecules: The field creates a periodic oscillating Stark effect that modulates the signal. The modulation is carried out with field strengths of typically 100 Vcm^{-1} and at frequencies between 50 Hz and 100 Hz, and is referred to as Stark modulation. In the detection section of the spectrometer, only the modulated signals are amplified and detected. This allows background and noise components to be separated from the radiation which is to be measured. The resonance between the radiation and the level being studied is thus periodically switched on and off. In this manner, the frequency of the microwave radiation and thus of the rotational transitions can be determined with an accuracy of better than 10^{-6}.

Corresponding to the *selection rules* for the interaction of molecules with electromagnetic radiation, only molecules with *permanent electric dipole moments* permit the observation of rotational spectra. This selection rule for electric dipole radiation can be intuitively understood: a polar molecule which is rotating appears to have a time-dependent dipole moment to a stationary observer. The rotation of such molecules is therefore active with respect to optical absorption, meaning that the rotation leads to the absorption of electromagnetic radiation when the frequencies match. For homonuclear diatomic molecules such as H_2, N_2, or O_2, this does not apply, because they have no permanent dipole moments; they thus exhibit no rotational spectra. The same is true of all larger molecules without a permanent dipole moment, for example CCl_4 – unless the rotation leads to a distortion and thereby to a rotationally-induced dipole moment, or unless the molecule is at the same time subject to an asymmetric vibration and thus has an induced dipole moment which can be acted on by the oscillating electric field of the radiation.

9.2 Diatomic Molecules

9.2.1 The Spectrum of the Rigid Rotor (Dumbbell Model)

Figure 9.1 shows as an example of a typical rotational spectrum of a diatomic molecule the spectrum of HCl. Figure 9.2 shows a schematic illustration of the rotation spectrum of another linear symmetric top molecule with a smaller line spacing, together with the coresponding energy term scheme, which we shall now derive. The spectrum consists of a large number of nearly equidistant lines with a characteristic temperature-dependent intensity distribution. This spectrum can be understood as the spectrum of a rigid rotor, i.e. the spectrum of a system in which the two atoms are rigidly attached to one another. This so-called *Dumbell Model* is the simplest model for the rotation of a diatomic molecule. In classical mechanics, the rotational energy of such a rotor can be calculated according to the equation

$$E_{\mathrm{rot}} = \tfrac{1}{2}\Theta\omega^2 \qquad [\mathrm{Joule}]\,, \tag{9.1}$$

where Θ is the moment of inertia about an axis of rotation perpendicular to the line joining the two masses m_1, m_2, and ω is the angular velocity of the rotation; cf. Fig. 9.3.

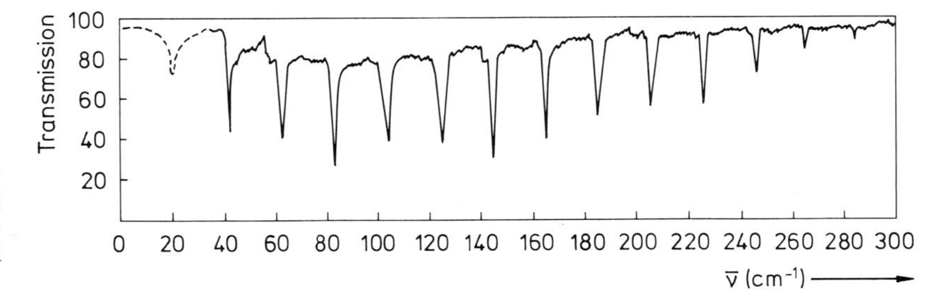

Fig. 9.1. The rotational spectrum of HCl in the gas phase; absorption spectrum. The minima in transmission correspond to maxima in the absorption

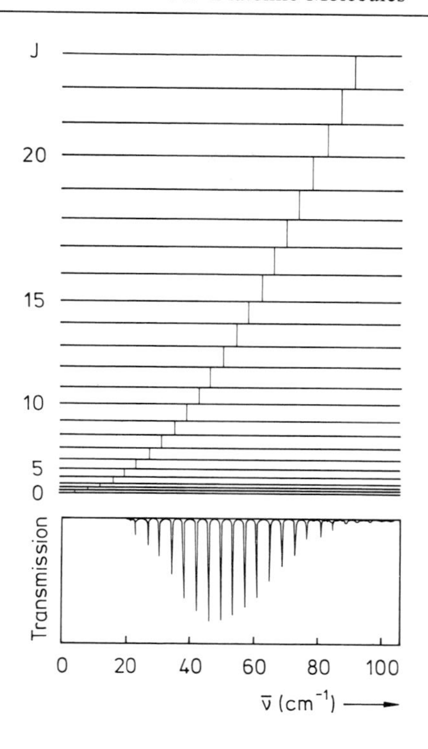

Fig. 9.2. The energy-level scheme for the rotation of a diatomic molecule (linear symmetric top) and its transmission spectrum. The energy is plotted in the *upper part* of the figure, increasing with increasing J; the *lower part* shows the transmission spectrum. The selection rule for optical transitions is $\Delta J = \pm 1$; the intensity distribution in the spectrum is explained in the text. The first few lines in the spectrum are so weak that they are not visible on the scale of this figure

The moment of inertia Θ of this dumbbell relative to its centre of gravity S is equal to

$$\Theta = m_1 R_1^2 + m_2 R_2^2 = m_r R^2 , \tag{9.2}$$

where R_1 and R_2 are the distances of the masses m_1 and m_2 from S and $R = R_1 + R_2$. The mass m_r is called the reduced mass and is given by:

$$m_r = \frac{m_1 m_2}{m_1 + m_2} . \tag{9.3}$$

The angular momentum (along an axis perpendicular to the molecular symmetry axis) is equal to

$$|\boldsymbol{L}| = \Theta \omega , \tag{9.4}$$

where $\boldsymbol{L}$ is the symbol for angular momentum and $|\boldsymbol{L}|$ or simply L stands for its magnitude.

We first make an estimate. Taking as a trial formula for the quantisation of the angular momentum:

$$|\boldsymbol{L}| = n\hbar \; (n = 0, 1, 2, \dots) \tag{9.5}$$

we obtain from (9.4) the smallest possible value of the rotational frequency $\omega = 2\pi\nu$:

$$\omega_{n=1} = \frac{L}{\Theta} = \frac{\hbar}{m_r R^2} . \tag{9.6}$$

If, for example, we insert the atomic masses of H and Cl, and their internuclear distance in the HCl molecule, known from gas-kinetic measurements to have the value $R = 1.28 \cdot 10^{-10}$ nm, we find

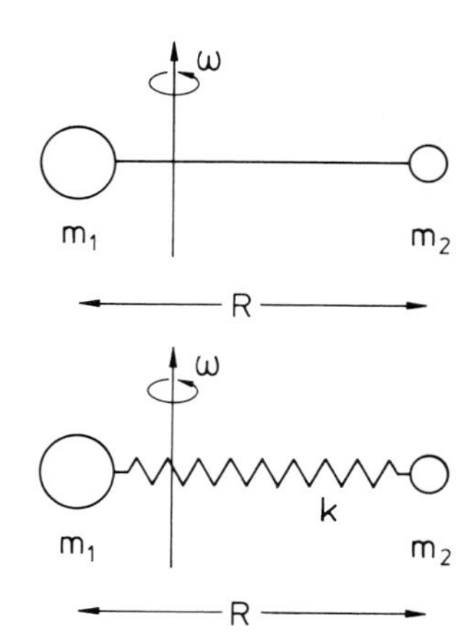

Fig. 9.3. The rotation of a diatomic molecule about its centre of gravity. In the case of a non-rigid rotor (*lower part of the figure*), the two atoms can oscillate relative to one another with the force constant k of the chemical bond

$$\nu_{n=1} = 6.28 \cdot 10^{11}\,\text{Hz}, \text{ or } \lambda = 0.47\,\text{mm} .$$

This rotational frequency, calculated in a semi-classical fashion, is very close to the smallest absorption frequency measured in the rotational spectrum of HCl, which has the value

$$\nu_{\text{min obs}} = 6.25 \cdot 10^{11}\,\text{Hz}, \text{ or } \lambda = 0.48\,\text{mm} .$$

This simple calculation indeed gives the order of magnitude of the frequency to a surprising degree of accuracy, but it is nevertheless too simple, if one wishes to understand the entire rotational spectrum. For the energy states of the rotor, we in fact find from (9.1) and (9.4):

$$E_{\text{rot}} = \frac{L^2}{2\Theta} . \tag{9.7}$$

With $L = n\hbar$, this becomes

$$E_{\text{rot}} = \frac{n^2\hbar^2}{2\Theta} . \tag{9.8}$$

This expression does not give a satisfactory result when compared with experiment, if it is assumed that the lines in the rotational spectrum are due to transitions between neighbouring quantum levels. Instead, the problem must be treated quantum mechanically from the beginning by solving the time-independent Schrödinger equation for the rotation. The orbital angular momentum $\boldsymbol{L}$ of a particle of mass m_r orbiting at a distance R from the origin can be calculated in just the same way as that of the electron in the hydrogen atom; we can therefore make use of the computation of the angular-momentum eigenfunctions for the H atom (compare I, Sect. 10.2, and Chap. 11 in this book). For a rigid rotor, we thus obtain the energy eigenvalues:

$$E_{\text{rot}} = \frac{\hbar^2}{2\Theta} n(n+1) , \tag{9.9}$$

i.e. instead of (9.5), we must introduce a different quantisation condition

$$|\boldsymbol{L}| = \hbar\sqrt{n(n+1)} . \tag{9.10}$$

In the case of rotation, it is usual to denote the quantum number by J instead of by n; thus for the rotational levels of the rigid rotor, we obtain the following expression in place of (9.8):

$$E_{\text{rot}} = \frac{\hbar^2}{2\Theta} J(J+1) \qquad [\text{Joule}] \qquad (J = 0, 1, 2, \ldots) . \tag{9.11}$$

Introducing term values $F(J)$, which in spectroscopy are usually given in the units cm^{-1}, we divide (9.11) by hc and obtain:

$$F(J) = \frac{E_{\text{rot}}}{hc} = B\,J(J+1) \qquad [\text{cm}^{-1}] \tag{9.12}$$

with the so-called rotational constant B,

$$B = \frac{h}{8\pi^2 c\Theta} \qquad [\text{cm}^{-1}] . \tag{9.13}$$

This constant is a characteristic value which can be extracted from the measured rotational spectrum. It is inversely porportional to the moment of inertia of the molecule; its determination therefore yields basic information about the structure of the molecule being investigated.

Each of the rotational eigenvalues (9.11) and (9.12) has its characteristic angular-momentum eigenfunction, whose squares give the probability that the angles ϑ and ϕ have values in the range $d\Omega = \sin\vartheta d\vartheta d\phi$. These are the same as those we met in I on solving the Schrödinger equation for the hydrogen atom (cf. Chap. 10 in I). Each eigenfunction with the quantum number J is associated with $2J+1$ functions having the "magnetic" quantum number $M = J, J-1, \ldots -J$; i.e. each state characterised by J is $(2J+1)$-fold degenerate, so long as no additional interaction is present which would lift the degeneracy.

The quantum number M is a measure of the components of angular momentum relative to a quantisation axis, which is defined for example by an applied electric field. In that case – see the Stark effect – the degeneracy with respect to M is lifted except with respect to the sign of M.

Summarising the results for the rigid rotor, we have

– a quantisation of angular momentum,

$$|\boldsymbol{L}| = \sqrt{J(J+1)}\,\hbar$$

with the quantised z-component $L_z = M\hbar$,

– energy eigenvalues $E_{\mathrm{rot}} = BhcJ(J+1)$

– with the rotational constant defined in (9.13),

$$B = \frac{h}{8\pi^2 c\,\Theta} \qquad [\mathrm{cm}^{-1}]\,.$$

The energy difference between two energy levels whose quantum numbers J differ by 1, that is the rotational quantum $E(J+1) - E(J)$, increases with increasing J. This means that the rotational energy increases with J for a constant internuclear distance, and we thus obtain a term scheme like the one shown in Figs. 9.2 and 9.4.

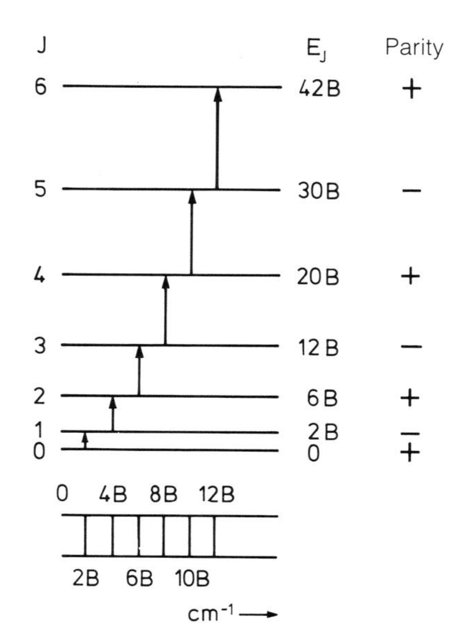

Fig. 9.4. Energy levels for the rotation of a rigid diatomic molecule, with parities as indicated. The selection rule $\Delta J = 1$ yields the spectrum consisting of equidistant lines, as shown

We now introduce the selection rules for optical transitions (electric dipole radiation), $\Delta J = \pm 1$ (and $\Delta M = 0, \pm 1$, with different polarisations referred to as σ and π transitions), and find the following expression for the quantum energy of the lines in the rotational spectrum corresponding to a transition between a level with the quantum number J and one with $J+1$; that is, from (9.12), we find that the term difference $F_{J+1} - F_J$ is given by the condition

$$h\nu = E_{J+1} - E_J \; . \tag{9.14}$$

For the wavenumbers of the rotational lines we then obtain

$$\overline{\nu}_{J\to J+1} = 2B(J+1) \qquad [\mathrm{cm}^{-1}] \, . \tag{9.15}$$

We thus calculate, as observed, a spectrum having equidistant lines with a spacing equal to $2B$, from which one can derive the rotational constant B; cf. Figs. 9.2 and 9.4. Since the moment of inertia of the molecule is in the denominator of B, heavier molecules have their spectra at longer wavelengths with a smaller energy spacing between the lines than lighter molecules with smaller moments of inertia.

Some examples for spectroscopically determined rotational constants are given in Table 9.1.

Table 9.1. Rotational constants of some diatomic molecules

$^1H_2$*	$2B = 121.6\ \mathrm{cm}^{-1}$
$^1H^{35}Cl$	20.79
$^{12}C^{16}O$	3.84
$^1H^{79}Br$	14.9
$^{39}K^{35}Cl$	0.257

* The rotational spectrum of H_2 is not directly observable; cf. Chap. 12

From the B values, as we have shown above, the internuclear distance R of the centres of gravity of the two atoms in the molecule can be determined. For $^1H^{35}Cl$, we obtain the equilibrium internuclear distance by using the numerical value for B given in Table 9.1:

$$R_e = \sqrt{\frac{\Theta}{m_r}} = 1.287 \cdot 10^{-10}\,\mathrm{m}$$

(the index e here stands for "equilibrium").

9.2.2 Intensities

The *intensities of the lines* (cf. Fig. 9.2) are given by the degree of degeneracy of the terms F_J for different values of J, by the thermal occupation probabilities of the rotational levels, and by the selection rules, taking the quantum-mechanical transition moments to be constant. As we mentioned above, each level with the quantum number J is $(2J+1)$-fold degenerate with respect to the magnetic quantum number M. The *degree of degeneracy* is thus $2J+1$. The statistical weight of the states corresponds to this value, as long as the degeneracy is not lifted. The *selection rules* follow from the symmetry of the wavefunctions on application of time-dependent perturbation theory; cf. Chap. 16 in I and Chap. 16 in this book.

The two important selection rules which we have already used above can be understood in an intuitive picture:

- only *polar molecules*, i.e. molecules with a permanent electric dipole moment, have a rotational spectrum which can be observed in optical spectroscopy;
- transitions with $\Delta J = \pm 1$ are optically allowed, i.e. transitions in which the angular momentum of the molecule changes by $\hbar$. This angular-momentum difference corresponds to the angular momentum of the optical quantum which is taken up on absorption or given off on emission, so that conservation of angular momentum is obeyed.

Finally, in order to understand completely the intensity distribution in the spectrum, we need to know which initial states for absorption are occupied at the absolute *temperature* T of the measurements. The thermal energy at room temperature corresponds to about 1/40 eV or 200 cm^{-1}; it is thus in general much larger than the rotational constant B which gives the spacing of the lowest rotational terms. In thermal equilibrium at room temperature, many rotational levels are therefore occupied. Quantitatively, the occupation probability N_J of a level with rotational quantum number J is given by:

$$\frac{N_J}{N_0} = \frac{g_J}{g_0}\,\mathrm{e}^{-(E_J-E_0)/kT} = \underbrace{(2J+1)}_{\text{Degeneracy}}\,\underbrace{\mathrm{e}^{-BhcJ(J+1)/kT}}_{\text{Thermal occupation}} \,. \tag{9.16}$$

In this expression, g_J and g_0 are the statistical weights of the states with the corresponding quantum numbers, and are equal to the degree of degeneracy $2J + 1$, with $g_0 = 1$. The intensity ratio of the lines in the absorption spectrum is proportional to the ratio of the occupation probabilities N_J and N_0. All together, we thus find from (9.16) an intensity profile like that shown in Figs. 9.1 and 9.2. For small values of J, the intensity of the lines increases with increasing J due to the increasing statistical weights; for larger values of J, the decrease of the exponential function in (9.16) dominates. Between these two extremes there is an intensity maximum. By differentiation of (9.16), we can readily show that the value of the quantum number at the maximum, J_{max}, is given by:

$$J_{\mathrm{max}} = \sqrt{\frac{kT}{2hcB}} - \frac{1}{2} \,, \tag{9.17}$$

where J_{max} is the integer which lies closest to the numerical value calculated from (9.17). The position of the most intense transition is given only approximately by (9.16), because the intensity distribution depends not only on the occupation probabilities alone, but also on the squares of the transition moments, which are calculated from the initial and final state wavefunctions and thus also depend on the quantum number J. A complete rotational spectrum, as shown in Fig. 9.1, cannot in general be registered with a single spectral apparatus, owing to the broad range of frequencies involved. Matching the spectral data of different devices with respect to intensities is not always a simple procedure. The intensity ratios of the absorption lines corresponding to different J values are therefore best determined from a rotational-vibrational spectrum; see Sect. 10.4.

9.2.3 The Non-Rigid Rotor

When rotational spectra are analyzed with a high precision, it becomes apparent that the absorption lines are not exactly equidistant; instead, their spacing becomes smaller and smaller as the quantum number J increases. To understand this, one has to assume that the internuclear distance of the atoms in the molecule changes with changing rotational quantum number J. It increases with increasing rotational energy, i.e. with increasing values of J, due to a centrifugal distortion of the molecule. The moment of inertia becomes larger as a result of this distortion. We thus must abandon the rigid rotor model in favor of the non-rigid rotor, in which the two nuclei are attached to one another by a bond with an elastic force constant k. This fact becomes particularly important for the analysis of rotational spectra in which molecular vibrations are also involved, so-called rotational-vibrational spectra. If the molecule is not only rotating, but also vibrating, the deviations from the rigid rotor model depend on the type and the frequency of the vibrations and are often much stronger than in the case of purely rotational motion.

But first we consider only the rotation of a diatomic molecule, that is the model of the *rotating non-rigid dumbbell.* For a quantitative description, one must assume that the rotor is not rigid, and that an elastic bond exists between the two atoms, having a force constant k (cf. Fig. 9.3). A rotation, or rather the centrifugal force that it generates, produces a stretching of the molecule. Classically, we can calculate the new internuclear distance R to be:

$$m_r R\omega^2 = k(R - R_e) , \tag{9.18}$$

where R_e denotes the equilibrium internuclear distance in the molecule at rest, and ω is the circular frequency of the rotation.

We thus have an equilibrium between the centrifugal force, which tends to stretch the molecule, and the elastic bonding force between the atoms. Qualitatively, it can be seen that a stretching increases the distance between the two masses m_1 and m_2 and thus the moment of inertia; this reduces B and the energy values E_J are lowered. The quantitative calculation follows from (9.18):

$$\Delta R \equiv R - R_e = \frac{m_r R\omega^2}{k} = \frac{m_r^2 R^4 \omega^2}{k m_r R^3} = \frac{(\Theta\omega)^2}{k m_r R^3} . \tag{9.19}$$

From this, we find

$$R - R_e = \frac{L^2}{k m_r R^3} \simeq \frac{L^2}{k m_r R_e^3} , \tag{9.20}$$

in which we have replaced R by R_e as a result of

$$R^3 = (R_e + \Delta R)^3 = R_e^3 \left(1 + \frac{\Delta R}{R_e} + \dots\right) , \quad \frac{\Delta R}{R_e} \ll 1 . \tag{9.21}$$

For the total energy,

$$E_{\text{rot}} = \frac{L^2}{2m_r R_e^2} - \frac{1}{2}k(R - R_e)^2$$

this model calculation based on classical physics then leads by a few simple steps to:

$$E_{\text{rot}} = \frac{L^2}{2m_r R_e^2} - \frac{L^4}{2k m_r^2 R_e^6} . \tag{9.22}$$

If we now make the transition from classical mechanics to quantum mechanics and replace L^2 by $J(J+1)\hbar^2$, as usual, we finally obtain for the rotational energy:

$$E_{\text{rot}} = \frac{\hbar^2}{2m_r R_e^2} J(J+1) - \frac{\hbar^4}{2k m_r^2 R_e^6} J^2(J+1)^2 \qquad \text{[Joule]} \tag{9.23}$$

and for the rotational terms:

$$F(J) = \frac{E_{\text{rot}}}{hc} = BJ(J+1) - DJ^2(J+1)^2 \qquad [\text{cm}^{-1}] , \tag{9.24}$$

where we have introduced a centrifugal stretching constant D defined by (9.23) in a way analogous to the definition of B. This constant D is much smaller than B; it follows from (9.23) that

$$D = \frac{\hbar^3}{4\pi k \Theta^2 R_e^2 c} \quad [\mathrm{cm}^{-1}] . \tag{9.25}$$

Insertion of numerical values into (9.25) and comparison with (9.13) yields approximately 10^{-3} to 10^{-4} for the relative magnitude D/B. The stretching term $DJ^2(J+1)^2$ in (9.23) is thus nearly negligible as long as J is small, but it may become important for large J. A measurement of D combined with (9.25) yields the force constant k of the bond and from it, the frequency

$$\nu = \frac{1}{2\pi}\sqrt{\frac{k}{m_r}} \quad [\mathrm{s}^{-1}] \quad \text{or} \quad \overline{\nu} = \frac{1}{2\pi c}\sqrt{\frac{k}{m_r}} \quad [\mathrm{cm}^{-1}] \tag{9.26}$$

of the valence oscillation along the direction of a line joining the two nuclei in the molecule, as we shall show in Sects. 10.2 and 10.3. There, we will also discuss more precise methods of studying these vibrations.

The term scheme of the non-rigid rotor can be found from that of the rigid rotor by shifting the terms as illustrated in Fig. 9.5. The spectrum is then slightly modified, as is also indicated schematically in the figure. The frequencies of the lines in the rotational spectrum of the non-rigid rotor may be found from (9.23) by applying the selection rule $\Delta J = \pm 1$ for a radiative transition:

$$\overline{\nu}_{J\rightarrow J+1} = F(J+1) - F(J) = 2B(J+1) - 4D(J+1)^3 \quad [\mathrm{cm}^{-1}] . \tag{9.27}$$

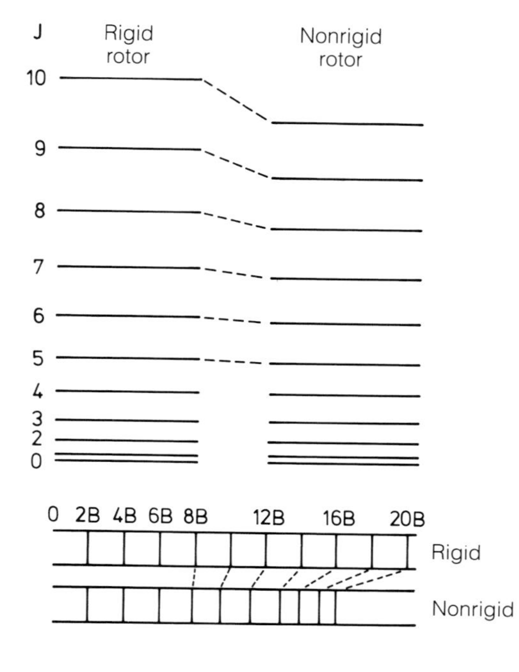

Fig. 9.5. The energy levels and the spectrum of a non-rigid rotor compared to a rigid rotor. A value $D = 10^{-3}\,B$ has been assumed for the stretching constant D. The levels of the rigid rotor, which are equidistant with a spacing of $2B$, are shifted towards lower energies in the non-rigid rotor, and the shift increases with increasing J. This effect is exaggerated in the figure

The selection rules remain unchanged, since the symmetries of the rotational states are not changed by the elastic force of the bond.

As an example we give the numerical values for the simple case of HCl. From Table 9.1, we find for this molecule, assuming that it is a rigid rotor, $2B = 20.79\ \mathrm{cm}^{-1}$. For the non-rigid rotor, the correction term is found to be $4D = 0.0016\ \mathrm{cm}^{-1}$; see Table 9.2.

Table 9.2. A comparison between experimental and calculated values for the positions of the rotational absorption lines of HCl, in cm^{-1} [from (9.24) and (9.27) with $2B = 20.79\ \mathrm{cm}^{-1}$ and $4D = 0.00016\ \mathrm{cm}^{-1}$]

$J \to J+1$	Experimental	Calculated for	
		rigid	non-rigid rotor
0 – 1	20.79	20.79	20.79
3 – 4	83.03	83.16	83.06
6 – 7	145.03	145.53	144.98
9 – 10	206.38	207.90	206.30

Table 9.2 compares the measured and calculated spectral line positions for the HCl molecule using the quoted values of the constants.

9.3 Isotope Effects

The extreme precision with which molecular moments of inertia can be determined with rotational spectroscopy by measuring B leads to an important application. From the line shifts, the masses of isotopes can be determined by investigating molecules containing different isotopes of the same elements. From the line intensities, the relative abundances of the isotopes can also be found. Since the moment of inertia is inversely proportional to the rotational constant B, molecules containing heavy isotopes have rotational lines corresponding to lower quantum energies and smaller line spacings. Naturally, the isotope effect is particularly large in the case of hydrogen. The rotational constant $2B$ of light hydrogen, H_2, is $121.52\ \mathrm{cm}^{-1}$; for heavy hydrogen, D_2 or ${}^2H_2$, it is found experimentally to be $2B = 60.86\ \mathrm{cm}^{-1}$, i.e. nearly exactly half as large due to the doubled mass and the resulting doubling of the moment of inertia. By the way, we can also see from this result that the internuclear distance in the H_2 molecule is hardly changed when the heavier isotopes are present. In other molecules, the differences are smaller. For example, for the ${}^{12}CO$ molecule, $2B$ is found to be $3.842\ \mathrm{cm}^{-1}$, and for the ${}^{13}CO$ molecule containing the heavy isotope of carbon, $2B$ is equal to $3.673\ \mathrm{cm}^{-1}$. Figure 9.6 shows as an example the differences in the rotational spectra of CO containing the isotopes ${}^{12}C$ and ${}^{13}C$.

In polyatomic molecules, the internuclear distances of the various atoms in the molecule can also be determined from the isotope effect. As an example, we consider here the linear molecule carbon oxysulphide, OCS. As we have already mentioned, a measurement of the rotational constant B of a linear molecule allows the determination of the moment of inertia perpendicular to the symmetry axis, but this quantity alone does not permit the calculation of both bond lengths from the central C atom to the O and S atoms. By carrying out measurements on molecules substituted with different isotopes, such as for example $CO^{32}S$ and $CO^{34}S$, one can, however, determine the individual CO and CS bond lengths from the

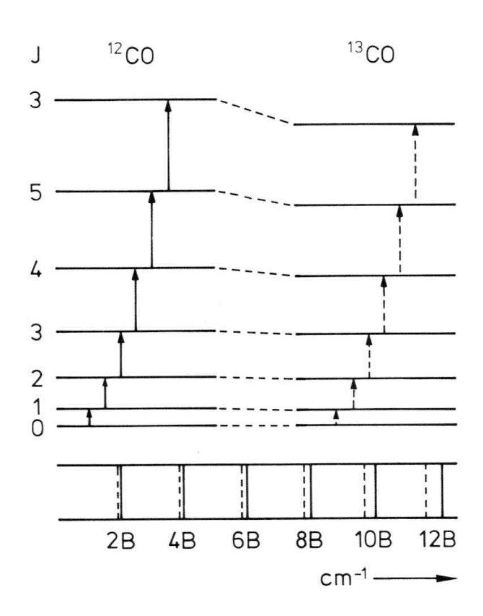

Fig. 9.6. The isotope effect on the rotational terms and the corresponding rotational spectrum of the CO molecule. The rotational constant B of the heavier molecule is smaller than that of ^{12}CO, and therefore the lines from ^{13}CO (*dashed*) are shifted to lower energies. The shift is exaggerated in this drawing

moments of inertia, assuming that the CS bond length does not change on changing the isotopic composition of the molecule. This is shown in the following.

We define the molecular centre of gravity by the equation

$$m_O R_O + m_C R_C = m_S R_S \ , \tag{9.28}$$

where R_O, R_C and R_S are the distances of the O, C, and S atoms from the centre of gravity (see Fig. 9.7); then the moment of inertia is found to be

$$\Theta = m_O R_O^2 + m_C R_C^2 + m_S R_S^2 \ . \tag{9.29}$$

In addition, for the distances we have

$$R_O = R_{CO} + R_C \qquad \text{and} \qquad R_S = R_{CS} - R_C \ , \tag{9.30}$$

where R_{CO} and R_{CS} are the internuclear distances of the O and S atoms from the central C atom (bond lengths).

Inserting (9.30) into (9.28) yields

$$M R_C = m_S R_{CS} - m_O R_{CO} \ , \tag{9.31}$$

where $M = m_O + m_C + m_S$ is the total mass.

Now we insert (9.30) into (9.29) and obtain

$$\begin{aligned} \Theta &= m_O (R_{CO} + R_C)^2 + m_C R_C^2 + m_S (R_{CS} - R_C)^2 \\ &= M R_C^2 + 2 R_C (m_O R_{CO} - m_S R_{CS}) + m_O R_{CO}^2 + m_S R_{CS}^2 \ . \end{aligned} \tag{9.32}$$

Using (9.31), we finally find the following expression for the moment of inertia:

$$\Theta = m_O R_{CO}^2 + m_S R_{CS}^2 - \frac{(m_O R_{CO} - m_S R_{CS})^2}{M} \ . \tag{9.33}$$

For molecules containing various isotopes, one finds different Θ values as a result of the differing masses.

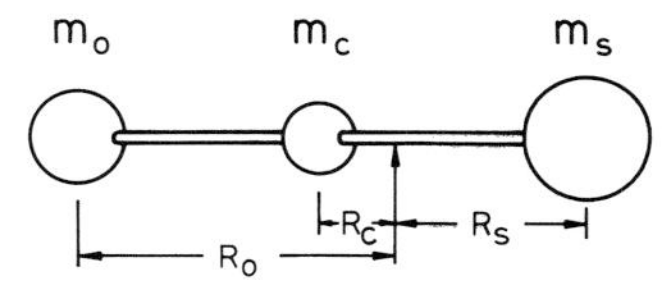

Fig. 9.7. The carbon oxysulphide molecule, OCS, indicating the definitions of the masses and the distances of the O, C, and S atoms from their common centre of gravity

Equation (9.33) connects a measurable quantity, the moment of inertia Θ, with two unknown quantities, the bond lengths R_{CO} and R_{CS}. If one wishes to determine these bond lengths, then it is necessary to measure the moments of inertia Θ of two molecules with different isotopic compositions; one then obtains two measured quantities Θ_1 and Θ_2 and two unknowns, the bond lengths which are to be determined. In this way, the bond lengths $R_{CO} = 1.16\,\text{Å}$ and $R_{CS} = 1.56\,\text{Å}$ were found by substituting sulphur isotopes of relative atomic masses 32 and 34 into the OCS molecule.

9.4 The Stark Effect

The modification of the quantum energy of spectral lines or the splitting of energy levels by a static electric field is known to us from atomic physics under the name *Stark effect*. In molecular physics, a static electric field leads to a lifting of the $(2J+1)$-fold degeneracy of the rotational levels, since different states belonging to the same J but with different values of the M quantum number have differing probability distributions for their charge densities relative to the molecular symmetry axes, and thus correspond to different polarisations by an electric field. For diatomic molecules, the energy shift can be written as:

$$\Delta E_J = \frac{p^2 E^2}{B} f(J, M^2) \,, \tag{9.34}$$

where the direction of the $\boldsymbol{E}$ field now gives the quantisation axis for the M quantisation.

In (9.34), p is the electric dipole moment of the molecule and E is the electric field strength; $f(J, M^2)$ is an abbreviation for an expression depending on the quantum numbers J and M, in which M enters only quadratically. One thus obtains a splitting into (J+1) sublevels, as shown in Fig. 9.8. As in atomic physics, the selection rule for optical transitions allows transitions with $\Delta M = 0$, so-called π transitions, as well as transitions with $\Delta M = \pm 1$, the σ transitions. Furthermore, the usual selection rule for electric dipole radiation holds, i.e. $\Delta J = \pm 1$. We will not concern ourselves here with the computation of the function $f(J, M^2)$. The splitting is very small; typical values of $\Delta\nu/\nu$ lie between 10^{-4} and 10^{-3} at an electric field strength of 10^3 V/cm.

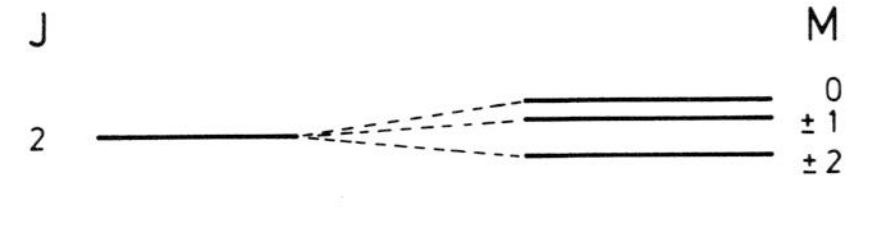

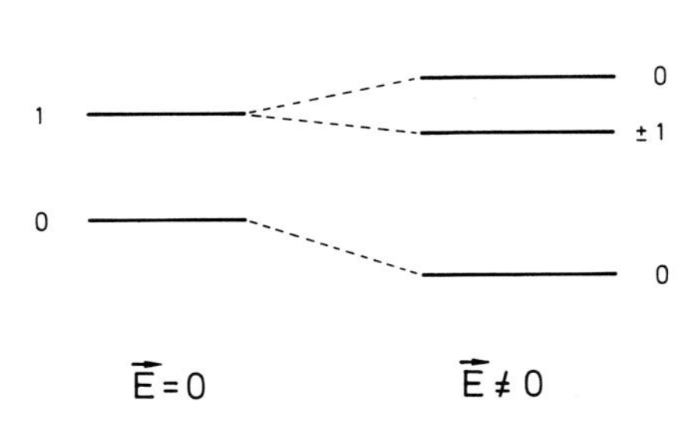

Fig. 9.8. The Stark splitting of the rotational terms $J = 0, 1, 2$, shown schematically. The Stark effect is the same for the positive and the negative signs of the quantum number M. The Stark shift is given by $\Delta E = -p^2 E^2/6hB$ for the state with $J = 0$

The Stark effect is important because it can be used relatively easily as an aid to the measurement of rotational spectra. One simply adds a central electrode to the microwave cavity in which the gas being studied absorbs microwave radiation, and uses it to apply the required static electric field. The energy terms to be measured can then be shifted or, if an alternating field is applied, they can be modulated at the frequency of the applied field. Some important applications of the Stark effect in molecular physics are:

- the determination of the quantum number J from the splitting pattern of individual rotational lines according to (9.34);
- the determination of molecular electric dipole moments $\boldsymbol{p}$ from the magnitude of the splitting or the term shifts in the applied field. This is an important method for measuring the dipole moments of molecules. It complements the usual method of measuring the dielectric constant ε to determine dipole moments, as discussed in Sect. 3.3;
- the Stark effect is essential for experimental rotational spectroscopy because it can be employed for effect modulation with a corresponding improvement in signal/noise ratio and precision in the measurement of rotational absorption spectra.

9.5 Polyatomic Molecules

In order to describe the rotation of a polyatomic molecule we require, as we know from classical mechanics, three *principal elements of the inertial tensor*, Θ_A, Θ_B and Θ_C in the general case, with respect to the three *principal axes* A, B and C. These are three mutually perpendicular directions about which the moment of inertia takes on maximal or minimal values. If the molecule has a symmetry axis, then it is one of the principal axes of the inertial tensor. Figure 9.9 shows some of the important types of small polyatomic molecules.

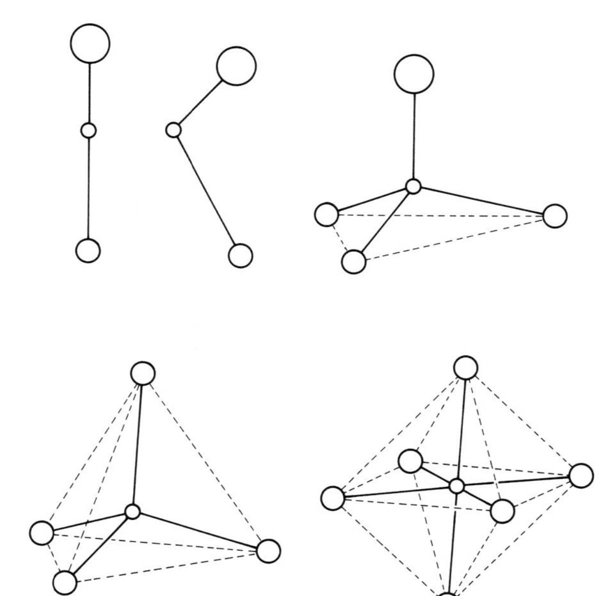

Fig. 9.9. Important molecular structural types of small polyatomic molecules. From *left* to *right* and from *top* to *bottom*: linear symmetric top, asymmetric top, symmetric top [distorted tetrahedron (example: CH_3Cl)], spherical top, tetrahedral (example: CCl_4), and octahedral (example: SF_6)

We denote the molecule-fixed coordinate system by x, y, z; then the kinetic energy of rotation of such a molecule is given by:

$$E_{\text{kin}} = \frac{L_x^2}{2\Theta_x} + \frac{L_y^2}{2\Theta_y} + \frac{L_z^2}{2\Theta_z} , \tag{9.35}$$

where L_x, L_y, and L_z are the components of angular momentum along the corresponding principal axes.

In order to calculate the rotational levels of a polyatomic molecule, the various axes and moments of inertia must be taken into account. In the general case, the so-called *asymmetric top molecule*, all of the principal elements of the inertial tensor are different from each other; an example is the H_2O molecule. The solution of the Schrödinger equation for such a molecule yields $(2J+1)$ different eigenvalues and eigenfunctions for each J value. There is, however, no general formula for such molecules, and each one must be analyzed individually. There is no preferred symmetry axis, and therefore none of the principal angular momentum components L_x, L_y or L_z is quantised. We shall treat this problem in more detail in Sect. 11.2. This general case of the asymmetric top will therefore not be discussed further at this point.

A simpler case is that of the *symmetric top molecule*. This term refers to a molecule in which two of the principal elements of the inertial tensor are the same, for symmetry reasons. Examples are NH_3, CH_3Cl, and C_6H_6. The solution of the Schrödinger equation in this case again yields a quantised total angular momentum according to

$$|\boldsymbol{L}| = \hbar\sqrt{J(J+1)} , \qquad J = 0, 1, 2 \ldots . \tag{9.36}$$

There is now a special symmetry direction within the molecule owing to the charge distribution. If we take the x-axis to be the direction whose moment of inertia is different from the other two, then x is this special direction, and a second quantisation condition holds for the component of the total angular momentum along the x-axis of the molecule:

$$|\boldsymbol{L}|_x = K \cdot \hbar . \tag{9.37}$$

The quantum number K introduced here can take on the values $0, \pm 1, \ldots, \pm J$. It refers to the molecular axis, while the quantum number M introduced earlier refers to an externally determined quantisation axis (e.g. by an applied field).

We thus now have a second quantisation condition for the angular momentum relative to the x-axis; cf. also Sect. 11.2. The energy of the rotational levels is then given by:

$$E_{\text{rot}} = BhcJ(J+1) + ChcK^2 \qquad \text{with}$$

$$B = \frac{h}{8\pi^2 c\Theta_x} \qquad \text{and} \qquad C = \frac{h}{8\pi^2 c}\left(\frac{1}{\Theta_x} - \frac{1}{\Theta_y}\right) . \tag{9.38}$$

The $(2J+1)$-fold degeneracy is thus lifted. However, for $K \neq 0$, the two-fold $\pm|K|$ degeneracy remains, since K enters (9.38) quadratically. This means that the rotational energy is the same for $+K$ and $-K$, since these two states differ only in the sense of their rotation. We can make further distinctions between

- cigar-shaped molecules with $\Theta_x < \Theta_y = \Theta_z$ (prolate spheroids). An example is CH_3Cl, where x is the direction of the axis of 3-fold symmetry between the Cl and the C atoms. In this case, $C < 0$ and the levels are shifted to smaller energies with increasing K;

– pincushion-shaped molecules, with $\Theta_x > \Theta_y = \Theta_z$ (oblate spheroids). An example is the benzene molecule; here, $C > 0$ and the levels are shifted to higher energies with increasing K values.

Alltogether, the spectrum of a symmetric top molecule is quite similar to that of a diatomic (or, more generally, of a linear) molecule, since the different K groups only contribute a parallel shift of the whole term scheme. As an example, Fig. 9.10 shows a portion of the rotational spectrum of CH_3F. The selection rules $\Delta J = \pm 1$ and $\Delta K = 0$ apply here.

For the intensities, the *degree of degeneracy* of the rotational levels is important. Corresponding to the quantum numbers J and K, and since K enters the energy expression quadratically, all the levels (except for $K = 0$) are doubly degenerate. In addition, the condition for the orientation of the angular momentum with respect to an externally determined quantisation axis, denoted by the quantum number M, must be considered; see Sect. 9.2.1. This quantum number is to be sure not required for the calculation of the rotational energy, for example using (9.38), but it is needed to characterise other properties of the states, e.g. their symmetries and degrees of degeneracy.

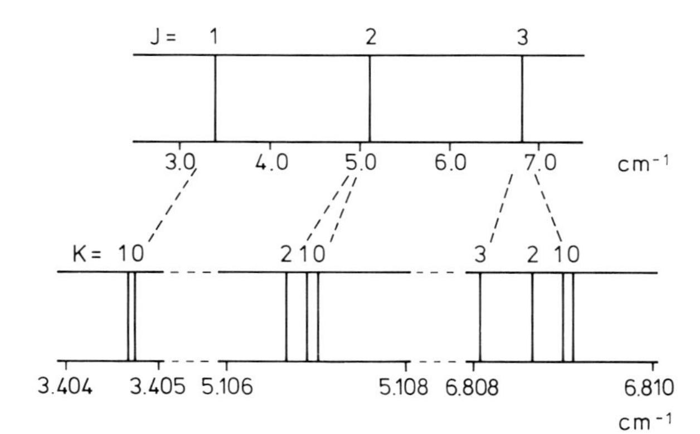

Fig. 9.10. A portion of the rotational spectrum of the symmetric top molecule CH_3F, shown schematically. The states characterised by the quantum number J (*upper part*) are split further according to the quantum number K (*lower part*). After Banwell

In the case of a linear symmetric top, with $K = 0$, each level has a $(2J + 1)$-fold degeneracy with respect to M (cf. Sect. 9.2.1). In the spherical top, in contrast, there is in addition to the $(2J+1)$-fold degeneracy with respect to an external quantisation axis a further $(2J + 1)$-fold degeneracy with respect to the orientation of the angular momentum relative to one of the molecular axes. Each level belonging to a particular J is thus $(2J + 1)^2$-fold degenerate. In molecules with still lower symmetry, the degeneracy is still more complicated. As we already discussed in Sect. 9.3, the M degeneracy can be lifted by an externally applied electric field (Stark effect), up to a two-fold degeneracy related to the sign of M.

It naturally holds equally well for polyatomic molecules that the approximation of a rigid molecular framework is only roughly applicable. In reality, distortions of the molecule due to rotations, and more especially to vibrations, must be taken into account. This can be done, as in the case of a dumbbell molecule, by introducing correction terms. In addition to the correction term which was already discussed for the rigid dumbbell rotor, there is a correction term proportional to K^2 for the symmetric top molecule.

Those molecules with tetrahedral symmetry occupy a special place among polyatomic molecules, e.g. CH_4 and CCl_4 (*spherical top*). In this case, all three moments of inertia with respect to the x-, y- and z-axes are equal for symmetry reasons, and the permanent electric dipole moment $\boldsymbol{p}$ is zero. These molecules therefore have no infra-red active rotational spectra. We mention here in advance that they are also inactive in Raman spectroscopy, since their polarisabilities are isotropic; more on this subject follows in Chap. 12.

At the conclusion of this section on molecular rotations, we should mention that a rotation about the cylinder axis of linear molecules (including polyatomic linear molecules), i.e. *streched linear symetric tops*, has not been taken into consideration so far. The reason is that the moment of inertia about this axis is practically equal to zero, owing to the distribution of mass in the molecule. The rotational constant B is therefore extremely large, larger than the binding energy of the molecule, and this rotation is practically unobservable spectroscopically. The quantum number K is zero.

The great precision with which one can obtain structural data even for somewhat larger molecules is illustrated in Fig. 9.11, using as an example the pyridine molecule. All of the internuclear distances and the bond angles in molecules which are not too large can be determined precisely using spectroscopic methods.

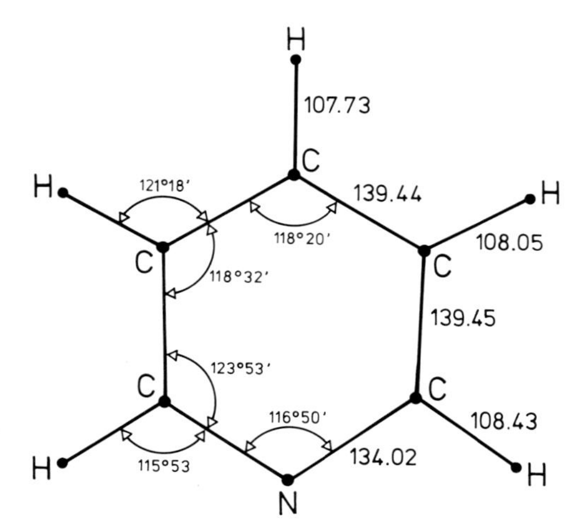

Fig. 9.11. Bond lengths in pm and angles in the planar molecule pyridine, derived from rotational spectra. After Labhart

10. Vibrational Spectroscopy

In contrast to atoms, molecules have internal degrees of freedom: their vibrational states can be excited, for example. The spectroscopy of these vibrations yields information on the structure and the bonding of the molecules. Here, as for molecular rotations, the fundamentals are best studied in diatomic molecules; this includes the coupling between vibrations and rotations (Sects. 10.1–10.4). Following an overview of the extensive field of vibrations of larger molecules in Sect. 10.5, we treat applications of molecular vibrations to radiation sources and lasers in Sects. 10.6–10.8.

10.1 Infra-red Spectroscopy

Within molecules, the atoms can undergo *vibrations* around their equilibrium positions, where they are located in the electronic ground state which we have considered up to now. These vibrations can appear in the optical spectra of the molecules; their frequencies lie in the infra-red spectral region. The measurement of spectra in the infra-red is at present carried out either with the aid of a grating spectralphotometer or, increasingly, using Fourier spectrometers. Light sources in the infra-red are thermal radiation sources such as the Nernst rod or the so-called Globar; the latter is a rod of SiC, which is heated to about 1500 K by means of an electric current. In the far infra-red region, gas plasma sources are superior; for example, the plasma in a mercury or xenon high-pressure lamp can be used.

For the detection of infra-red radiation, thermal detectors such as bolometers or the Golay cell, which is based on the heating of a volume of gas by absorbed IR radiation, may be employed. However, the most sensitive radiation receivers are special photoconductive detectors sensitised to the infra-red range, and photodiodes made of suitable semiconductor materials which can be tailored to the desired wavelength range, sensitivity, response time, and other parameters. Vibrational spectra are usually investigated in the form of absorption spectra; the transition probabilities for spontaneous emission from excited vibrational states are very small, so that vibrational spectroscopy in emission is hardly practicable. Another method for the study of vibrational spectra is Raman spectroscopy, which we shall treat in Chap. 12.

One can readily understand why molecular vibrations lie in the infra-red spectral region; this can be shown by a simple estimate for the HCl molecule. We assume that in this molecule, the H^+ and Cl^- ions are bound together at their equilibrium distance R_e by electrostatic attraction according to Coulomb's law. If we increase the bond length to R, we create a restoring force F_R which is given by:

$$F_R = -k(R - R_e) . \tag{10.1}$$

The index e stands here for "equilibrium".

The force constant k can be calculated in this model: assuming a pure Coulomb force, we find

$$k = \frac{dF}{dr} = \frac{2e^2}{4\pi\varepsilon_0 R_e^3} \; . \tag{10.2}$$

Inserting the measured equilibrium bond length $R_e = 1.28 \cdot 10^{-10}$ m, we find $k = 220\ \mathrm{Nm^{-1}}$. The eigenfrequency in this mass–and–spring model is given by

$$\omega = 2\pi\nu = \sqrt{\frac{k}{m_r}}\ \mathrm{s^{-1}} \; , \tag{10.3}$$

with m_r = reduced mass.

This is the classical oscillator frequency.
Inserting the numerical values, we find

$$\nu = \frac{\omega}{2\pi} = 5.85 \cdot 10^{13}\ \mathrm{Hz} \qquad \text{and} \qquad \lambda = 5.12\ \mu\mathrm{m} \; .$$

The measured values for HCl, $k = 516\ \mathrm{Nm^{-1}}$ and $\lambda = 3.5\ \mu$m, are of the same order of magnitude as those resulting from our greatly simplified model. We can conclude from this that the basic assumptions of the model are correct, but that we must refine it further.

At this point we give some typical numerical values for the force constants of different types of chemical bonds:

Covalent bonds, as in H_2:	$5 \cdot 10^2\ \mathrm{Nm^{-1}}$
Double bonds, as in O_2:	$12 \cdot 10^2\ \mathrm{Nm^{-1}}$
Triple bonds, as in N_2:	$20 \cdot 10^2\ \mathrm{Nm^{-1}}$
Ionic bonds, as in NaCl:	$1 \cdot 10^2\ \mathrm{Nm^{-1}}$.

10.2 Diatomic Molecules: Harmonic Approximation

We first consider again the vibrations of the simplest molecules, i.e. the diatomics. The vibrational spectrum of a diatomic molecule consists, when it is observed at low spectral resolution, of one line in the infra-red at the frequency ν, and a series of "harmonics" with strongly decreasing intensities at the frequencies $2\nu, 3\nu, 4\nu, \ldots$, as shown in Fig. 10.1, taking the CO molecule as an example. Here, ν is the frequency of the stretching vibration of the molecule. In this mode of vibration, the internuclear distance in the molecule changes periodically with the period of oscillation. If the resolution is sufficiently increased, it is seen that each of these lines has a characteristic substructure; they consist of a manifold of nearly equidistant lines. Figure 10.2 shows this structure, also for the case of CO. It is very similar to the rotational spectra treated in Chap. 9 and arises from the fact that the vibrating molecules also rotate, and that vibration and rotation are coupled. Such spectra are therefore called *rotational–vibrational spectra* or band spectra, since the lines occur in groups which form a "band". There are no vibrational spectra of free molecules without rotational structure. However, the structure does not appear when the spectral resolution is

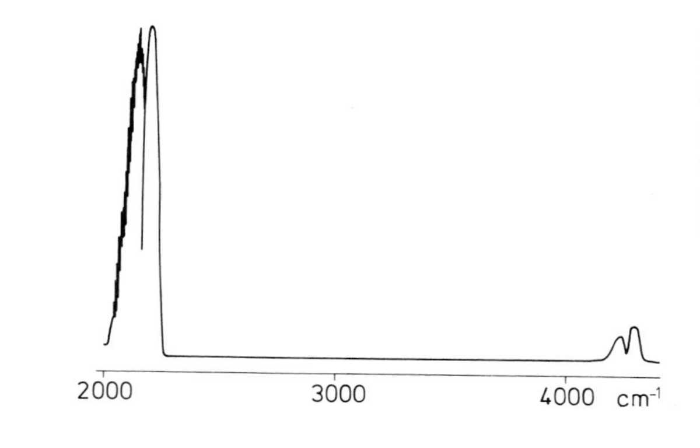

Fig. 10.1. The vibrational spectrum of CO in the gas phase. The fundamental vibration is at 2143 cm^{-1}, and the first harmonic is at 4260 cm^{-1}; measured with a poor spectral resolution. After Banwell

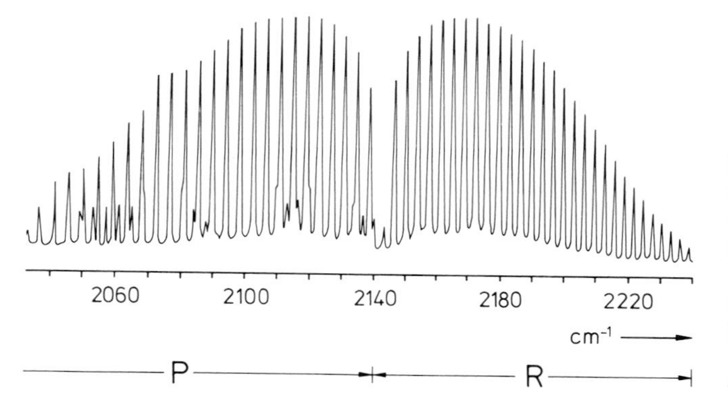

Fig. 10.2. The fundamental vibration of the CO molecule, measured at a high spectral resolution. Left and right of the centre at $\overline{\nu} = 2143.28$ cm^{-1} are the P and R branches. Evaluation according to (10.30–32) yields $\overline{\nu}_e = 2169.7$ cm^{-1}, $x_e = 0.0061$, $B_e = 1.924$ cm^{-1}, and $\alpha = 0.0091$ cm^{-1}

insufficient or when, as in the condensed phases, interactions with other molecules of the same or different types broaden the lines to such an extent that inhomogeneously broadened vibrational bands without resolved rotational structure result.

We at first leave rotational structure out of the discussion and consider only the vibrations. We calculate the energy levels of the vibrations of a diatomic molecule initially using the dumbbell model as introduced above, in terms of a harmonic oscillator with a force constant k along the line connecting the two nuclei in the molecule. We thus approximate the potential V of the bond as a parabolic potential with

$$V = \frac{k}{2}(R - R_e)^2 , \tag{10.4}$$

where R is the deviation from the equilibrium distance R_e. The quantum-mechanical calculation yields the following energy levels (see Sect. 9.4 in I):

$$E_{\text{vib}} = \hbar\omega(v + \tfrac{1}{2}) , \qquad v = 0, 1, 2, \dots \qquad [\text{Joule}] . \tag{10.5}$$

In this equation, ω is the classical oscillator frequency as in (10.3). The lowest energy (for $v = 0$) is the zero–point energy $(E_{\text{vib}})_0 = \hbar\omega/2$.

If we now use terms, measured in cm^{-1}, instead of the energy levels, we have to divide the levels E_{vib} in (10.5) by hc. In molecular spectroscopy, it is also usual to denote these vibrational terms by G_v and to write

$$G_v = \frac{E_{\text{vib}}}{hc} = \omega_e\left(v + \tfrac{1}{2}\right) \qquad [\text{cm}^{-1}] . \tag{10.6}$$

The vibrational constant introduced here, which is often used in molecular spectroscopy, is defined by

$$\omega_e = \frac{\hbar\omega}{hc} = \overline{\nu}_e \tag{10.7}$$

and is the wavenumber corresponding to the classical frequency, as calculated from (10.3).

In the following, we will not use this notation in terms of ω_e, in order to avoid confusion with the use of ω as a circular frequency; instead, we will use only the symbol $\overline{\nu}$ for the measured energies, when they are quoted in wavenumbers. The eigenfrequency of the harmonic oscillator, as in (10.5), will be denoted correspondingly by ν_e, and the wavenumber by $\overline{\nu}_e$.

A new quantum number v has also been introduced in (10.5); it measures the quantisation of the vibrations. With increasing quantum number $v = 2, 3, \ldots$, vibrational states with higher and higher energies are reached. For $v = 0$, we find from (10.5) the zero-point energy $(E_{vib})_0 = \hbar\omega/2$, which is not understandable in classical terms. Its existence results from the uncertainty relation for position and momentum (cf. I, Sect. 7.3). Even in the lowest vibrational level ($v = 0$), the vibrational energy is thus not equal to zero, but instead it has the value $\hbar\omega/2$. The vibration frequency $\omega = 2\pi\nu_e$ can again be calculated as

$$\omega = \sqrt{\frac{k}{m_r}} \quad s^{-1} . \tag{10.8}$$

Here, it is important to note that the vibration frequency depends on the *reduced mass* of the molecule. In molecules containing atoms of very different masses, $m_1 \gg m_2$, m_r is not very different from m_2. This can be understood intuitively, since in such a molecule, practically only the lighter mass m_2 is in motion, oscillating as if against a solid wall consisting of the greater mass m_1.

The energy levels in a parabolic potential according to (10.5), and the corresponding occupation probabilities $|\psi^2|$ of the oscillator, are shown in Fig. 10.3. From the figure it also becomes clear that for large vibrational quantum numbers, the occupation probabilities calculated from quantum mechanics become similar to those calculated classically. If we state in advance that the selection rule for optical transitions requires that the vibrational quantum number change by one unit, i.e. $\Delta v = \pm 1$, then we may expect a spectrum consisting of only a single line, owing to the fact that the energy levels are equidistant, with the quantum energy $E_{v+1} - E_v = h\nu_e$ or the wavenumber $\overline{\nu}_e$ (cm^{-1}).

As a general *selection rule* for the appearance of vibrational spectra, we find as for rotational spectra that the vibration of the molecule must be accompanied by an electric dipole moment, which changes in the corresponding transition. This is the selection rule for electric dipole radiation.

In the case that atoms of the same type oscillate relative to one another, for example in a homonuclear diatomic molecule such as H_2, N_2, or O_2, no dipole moment is present and there is no change in a dipole moment. In such molecules, vibrational or rotational–vibrational transitions are forbidden in the optical spectra. Their vibrational frequencies are therefore termed "optically inactive".

Nevertheless, these frequencies can be observed. On the one hand, in the discussion of the Raman effect in Chaps. 12 and 17, we shall see that they occur in Raman spectra owing to a change in the polarisability accompanying the vibrations. On the other hand, the frequencies can also be observed directly in the infra-red spectra – to be sure, with intensities

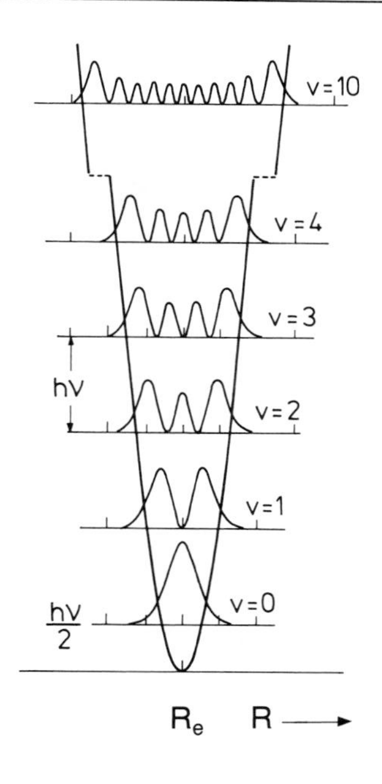

Fig. 10.3. A potential curve for the harmonic oscillator, with energy levels and occupation probabilities $|\psi_v(R-R_e)|^2$. After Hellwege

reduced by several orders of magnitude – because the dipole-free molecules usually have electric moments of higher orders. The path length in the absorbing gas must then be made accordingly long, since the corresponding transitions show considerably reduced transition probabilities.

10.3 Diatomic Molecules. The Anharmonic Oscillator

In reality, the potential curve of a diatomic molecule is not parabolic, as we assumed in the previous section. The true potential must be asymmetric with respect to the equilibrium distance R_e, as one can readily see. A reduction of the internuclear distance relative to R_e leads namely to an increase in the repulsion between the two atoms, since the attractive Coulomb potential is superposed with a repulsive potential of shorter range, which prevents the two atoms from penetrating each other and produces a stable equilibrium distance (see Fig. 1.2). The potential curve thus becomes steeper for $R < R_e$. On the other hand, an increase of the internuclear distance leads to a weakening of the chemical bond and finally to dissociation. In this range, i.e. for $R > R_e$, the potential curve becomes flatter. A more realistic potential curve than that of a harmonic oscillator is shown in Fig. 10.4, again using HCl as an example.

An often-used empirical approach which agrees well with experience is the so-called Morse potential:

$$V = D_e[1 - e^{-a(R-R_e)}]^2 . \tag{10.9}$$

Here, D_e is the dissociation energy and a is a quantity which is characteristic of the molecule under consideration:

$$a = (m_r/2D_e)^{1/2}\,\omega_e \qquad [\mathrm{cm}^{-1}] .$$

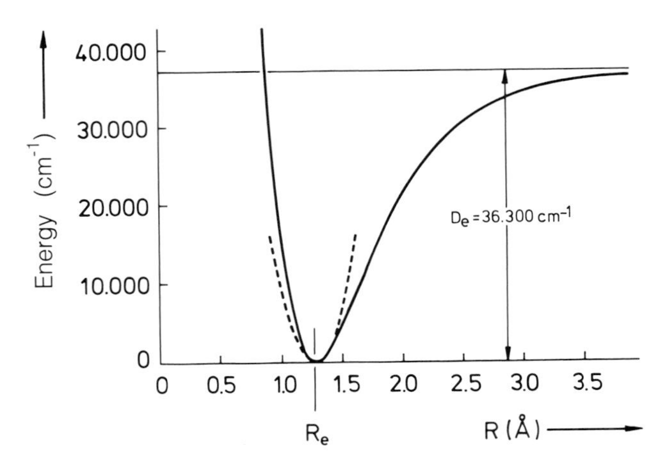

Fig. 10.4. The Morse potential curve for the HCl molecule. A harmonic potential is drawn in as a *dashed curve* for comparison. The dissociation energy from the potential minimum is called D_e

It depends on the reduced mass and the harmonic oscillator frequency.

The parameter a in the Morse potential thus contains the wavenumber corresponding to a harmonic oscillator, as well as the dissociation energy and the reduced mass, all quantities specific to the molecule.

In the neighbourhood of the minimum in the potential curve, the deviations of the Morse potential from a harmonic (parabolic) potential are in fact small, and the harmonic oscillator is a good approximation in this region. For $R = R_e$, $V = 0$, and for $R \to \infty$, V is equal to D_e. At small internuclear distances, $R \to 0$, the approximate potential of (10.9) is no longer valid.

At large deviations, $R > R_e$, the Schrödinger equation must be solved using the Morse potential for the potential energy, if one wishes to calculate the *anharmonic oscillator*. This is possible in closed form.

In this way, we arrive at the energy terms of the anharmonic oscillator; cf. Fig. 10.5. To a good approximation, they are given by:

$$E_v = \hbar\omega_e(v + \tfrac{1}{2}) - x_e\hbar\omega_e(v + \tfrac{1}{2})^2 \tag{10.10}$$

or

$$G_v = \overline{\nu}_e(v + \tfrac{1}{2}) - x_e\overline{\nu}_e(v + \tfrac{1}{2})^2 \,.$$

In fact, one often uses a generalisation of (10.10) for the evaluation of experimental data; it contains further terms with higher powers of $(v + \frac{1}{2})$, in particular the term $+y_e\hbar\omega_e(v + \frac{1}{2})^3$.

We note that here the symbol ω_e is used for the circular frequency $2\pi\nu_e$, and it should not be confused with the constant ω_e as frequently used in molecular spectroscopy; compare (10.6) and (10.7).

In (10.10), $\omega_e = 2\pi\nu_e$ is thus the value of the frequency of vibration, which we shall soon define more precisely, and x_e is the so-called anharmonicity constant, which is defined by the expression

$$x_e = \frac{\hbar\omega_e}{4D_e} \,. \tag{10.11}$$

The constant x_e is always positive and is usually of order 0.01.

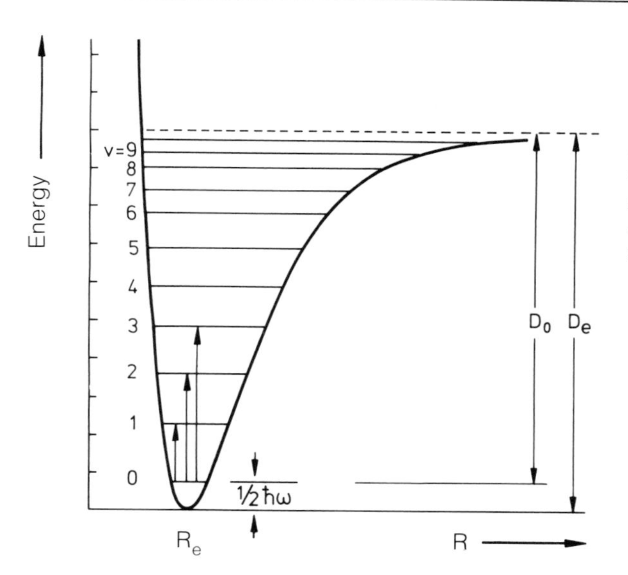

Fig. 10.5. The energy levels of an anharmonic oscillator. The three *arrows* correspond to the fundamental frequency and the first two harmonics in the vibrational spectrum. One can readily recognise the increase of the average internuclear distance with increasing quantum number v

Strictly speaking, still higher terms should be included in (10.10), as we mentioned above; they contain higher powers of $(v + \frac{1}{2})$. These are, however, very small corrections and will be neglected in the following.

The meaning of ω_e can be seen from a comparison of (10.10) with the terms of the harmonic oscillator, (10.5). We can rewrite (10.10) in the form:

$$E_v = \hbar\omega_e(v + \tfrac{1}{2})[1 - x_e(v + \tfrac{1}{2})] \qquad (10.12)$$

and see by comparing with (10.5) that we need to replace the vibration frequency ω in (10.5) by

$$\omega_v = \omega_e[1 - x_e(v + \tfrac{1}{2})] \qquad (10.13)$$

when we make the transition from the harmonic to the anharmonic oscillator. In the anharmonic oscillator, the vibration frequency as in (10.3) decreases with increasing quantum number v. In the hypothetical (because of the zero–point oscillation) case $E_v = 0$, i.e. $v = -1/2$, when the molecule would be in a state of no vibration and at rest, we would have

$$\omega = \omega_e \; . \qquad (10.14)$$

The vibration freqeuncy ω_e of the harmonic oscillator is thus a purely theoretical quantity, which is equal to the hypothetical vibration frequency of the anharmonic oscillator without zero-point oscillations. The index e means “equilibrium” here, too.

The highest vibration frequency in reality is that at $v = 0$; it is equal to:

$$\omega_{v=0} = \omega_e\left(1 - \frac{x_e}{2}\right) \; . \qquad (10.15)$$

Equation (10.10) thus describes the increasingly closer approach of the energy levels with increasing quantum number v, in agreement with the experimental evidence. The highest discrete bound level is at the energy D_e. Above D_e, there are only continuum states, and the molecule is dissociated. This region is called the dissociation-limit continuum.

The average internuclear distance of an anharmonic oscillator increases with increasing vibrational quantum number v, in contrast to the case of the harmonic oscillator, due to the asymmetric potential curve. This is clear from Figs. 10.5 and 10.6. This change in internuclear distance is also the cause of thermal expansion in solid materials: at higher temperatures the molecular oscillators are on the average in vibrational states with higher quantum numbers v, i.e. with larger intermolecular distances R.

As Fig. 10.5 illustrates schematically, it is necessary in quoting the dissociation energy to distinguish whether it is measured from the minimum of the potential curve or from the lowest term, with $v = 0$. We shall denote these two quantities by the symbols D_e and D_0. The values for the H_2 molecule can be read off from the experimental curve in Fig. 10.6.

We give a few numerical examples for clarification: in the $^1H^{35}Cl$ molecule, the wavenumber of the stretch vibration is found to be $\overline{\nu} = 2900$ cm^{-1} and $x_e = 0.0174$. Using (10.10), we calculate from this $D_e = 5.3$ eV. This quantity should be larger than the measured dissociation energy D_0 by an amount equal to the zero-point energy, here 0.2 eV; compare Fig. 10.5. The experimental value is $D_0 = 4.43$ eV. The agreement is thus not very good. The total number of discrete vibrational levels between the zero point energy and the energy value D_0 gives the largest possible quantum number v_{max}, with

$$\hbar\omega_e[(v_{max} + \tfrac{1}{2}) - x_e(v_{max} + \tfrac{1}{2})^2] = D_e \tag{10.16}$$

yielding $v_{max} = 22$, as compared to 14 if a harmonic oscillator were assumed, i.e. if $x_e = 0$.

Table 10.1 contains some further examples of measured values for diatomic molecules.

Table 10.1. Fundamental vibrational constants, force constants k, and dissociation energies D_0 of some diatomic molecules. After Engelke

Molecule	$\overline{\nu}$ [cm^{-1}] ($v = 0 \rightarrow v = 1$ trans.)	k [Nm^{-1}]	D_0 [kcal/mol]
H_2	4159.2	$5.2 \cdot 10^2$	104
D_2	2990.3	5.3	104
HF	3958.4	8.8	135
HCl	2885.6	4.8	103
HBr	2559.3	3.8	87
HI	2230.0	2.9	71
CO	2143.3	18.7	257
NO	1876.0	15.5	150
F_2	892.0	4.5	38
Cl_2	556.9	3.2	58
Br_2	321.0	2.4	46
I_2	231.4	1.7	36
O_2	1556.3	11.4	119
N_2	2330.7	22.6	227
Li_2	246.3	1.3	26
Na_2	157.8	1.7	18
NaCl	378.0	1.2	98
KCl	278.0	0.8	101

Since the experimentally determined dissociation energy of a molecule, D_0, measures the energy difference between the dissociation limit and the *zero-point energy* of the molecule, the dissociation energies of molecules with different isotopic compositions should differ

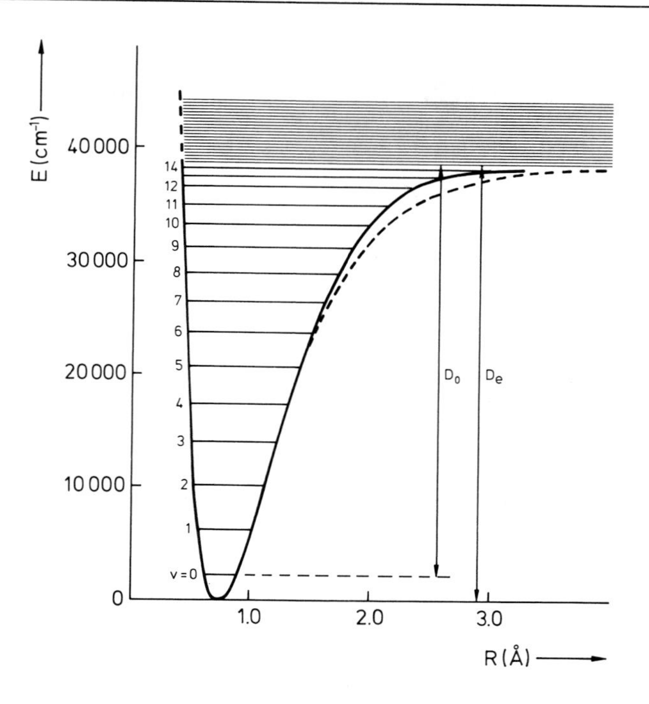

Fig. 10.6. The vibrational levels of the H_2 molecule and the potential curve which results from them. The *dashed curve* is the corresponding Morse potential. The continuum region above the dissociation energy is *shaded*. After Herzberg

by an amount equal to the difference in the zero-point energies, if – as is true to a good approximation – the energy of the chemical bonds depends only weakly or indetectably on the isotopic mass.

In this connection, the measured values for the hydrogen molecule are interesting. The numerical value for the dissociation energy of heavy hydrogen, $^2H_2$ or D_2, is 4.55 eV and is therefore 0.077 eV or 621 cm^{-1} larger than that for the light isotope $^1H_2$. This difference is close to the difference of the zero-point energies:

$$\tfrac{1}{2}\overline{\nu}_0(^1H_2) - \tfrac{1}{2}\overline{\nu}_0(^2H_2)\ ,$$

where $\overline{\nu}_0$ refers to the quantum energy of the valence vibration of the hydrogen molecule. The measured values for the lowest vibrational transition, i.e. for the transition $v = 0 \rightarrow v = 1$, are 4159 cm^{-1} for $^1H_2$ and 2990 cm^{-1} for $^2H_2$. The difference of the zero-point energies is thus equal to half the difference of the vibrational quantum energies, here 584 cm^{-1}, and is close to the value 621 cm^{-1} quoted above. This agreement can in fact be taken as experimental proof for the existence of zero-point oscillations if the assumption is made that the dissociation energy resulting from the potential curve, D_e, is the same for light and heavy hydrogen. In the case of heavy hydrogen, an amount of energy D_0 which is larger by the difference of the zero-point energies must be applied in order to reach the dissociation limit, if the lowest possible ground state of the molecules lies at an energy which is $(1/2)h\nu_0$ above the minimum of the potential curves.

A modern method for separating molecules with different isotopic compositions is based on this difference in dissociation energies of isotopically different molecules due to their different zero-point energies. The molecules to be separated are irradiated with intense light from a laser whose quantum energy has been chosen to be sufficiently high to cause dissociation of one type of molecule in the isotopic mixture, but not the other(s).

From the energy terms we can derive the *absorption spectrum* of an anharmonic oscillator by applying the selection rules. The selection rule $\Delta v = \pm 1$ for the harmonic oscillator must be modified somewhat in the case of the anharmonic oscillator; in addition to the singly-excited vibrations, harmonics can also be produced with a reduced transition probability. We have

$$\Delta v = \pm 1, \pm 2, \pm 3 \ldots , \tag{10.17}$$

where the relative intensities are roughly in the ratios $1 : x_e : x_e^2 : x_e^3 \ldots$.

Since x_e is a small number [cf. the values in Table 10.1, from which x_e may be calculated using (10.11)], the intensities decrease rapidly in the order shown. These are the "harmonics" which were mentioned earlier; compare also Fig. 10.1. The anharmonicity of the molecular vibrations is thus responsible for their occurrence.

The quantum energies of the transitions with $\Delta v = \pm 1$ are now no longer the same for all values of v, i.e. between all the vibrational terms in the potential curve; instead, they decrease with increasing v. In the harmonic approximation, the vibrational spectrum (without harmonics) contained only a single line $\overline{\nu}_e$, but with an anharmonic potential, we obtain a series of lines of decreasing intensity, in agreement with observations; it more or less converges for very large v.

The transitions from the ground state with $v = 0$ are by far the most important, since – as will be explained below – the higher vibrational levels are hardly occupied in thermal equilibrium and therefore play no significant role as initial states for absorption processes.

The energy of the most intense vibrational line from $v = 0$ to $v = 1$ is, from (10.10),

$$\Delta E = E_{v=1} - E_{v=0}$$

and for the wavenumber, we find by substitution

$$\overline{\nu}_{v \leftarrow 0} = \frac{\Delta E}{hc} = v\overline{\nu}_e[1 - x_e(v+1)] \tag{10.18}$$

and therefore

$$\overline{\nu}_{1 \leftarrow 0} = \frac{\Delta E}{hc} = \overline{\nu}_e(1 - 2x_e) . \tag{10.19}$$

The absorption transitions with $\Delta v = 2$ and $\Delta v = 3$, which we have called "harmonics", are correspondingly given by:

$$\overline{\nu}_{2 \leftarrow 0} = 2\overline{\nu}_e(1 - 3x_e)$$

and

$$\overline{\nu}_{3 \leftarrow 0} = 3\overline{\nu}_e(1 - 4x_e) .$$

A numerical example is given in Table 10.2.

Table 10.2. Vibrational transitions for $^1H^{35}Cl$, as described by (10.18) with $\overline{\nu}_v = v \cdot 2988.9\,[1 - 0.0174(v+1)]$ cm^{-1}

v	$\overline{\nu}$
$0 \rightarrow 1$	$\overline{\nu}_1 = 2885.9$ cm^{-1}
$0 \rightarrow 2$	$\overline{\nu}_2 = 5668.0$ cm^{-1}
$0 \rightarrow 3$	$\overline{\nu}_3 = 8347.0$ cm^{-1}
$0 \rightarrow 4$	$\overline{\nu}_4 = 10923.5$ cm^{-1}

Still higher harmonics have such small transition probabilities that they cannot be observed, in general.

The numerical value for the first vibrational transition, $\overline{\nu}_1 = \overline{\nu}_{1 \leftarrow 0}$, thus differs from the quantity $\overline{\nu}_e$ which we introduced above for the harmonic oscillator. For H_2, one finds for example $\overline{\nu}_1 = 4159.2$ cm^{-1}, and from it the calculated quantity $\overline{\nu}_e = 4395$ cm^{-1} with $x_e = 0.0168$.

In the following, we use numbers as indices on the frequency ν or the wavenumber $\overline{\nu}$ only to distinguish different vibrations of a molecule; this is necessary in polyatomic molecules, which have more than one type of vibration. Transitions between different quantum numbers v'' and v' of a vibration will be denoted by a parenthesis, e.g. $\overline{\nu}(v', v'')$. The symbol $\overline{\nu}_e$ will be used in the anharmonic oscillator (as already mentioned) for the calculated quantity obtained from the application of (10.19) to the observed vibrational transitions; it cannot be measured directly.

The occupation of the energy levels E_v having different vibrational quantum numbers v is, in thermal equilibrium, proportional to the Boltzmann factor $e^{-E_v/kT}$ and thus depends on the temperature. Since room temperature corresponds to 200 cm^{-1} as calculated from kT/hc, the occupation factor for HCl molecules, with a vibrational quantum energy of 2886 cm^{-1}, is very small at this temperature. Therefore, most HCl molecules at room temperature will be in the ground state, with $v = 0$. For this reason, the absorption spectrum consists for the most part only of the transition from $v = 0$ to $v = 1$. It is usual to denote this transition by $1 \leftarrow 0$, i.e. to write the higher level first. In order to observe the absorption transitions from levels with higher vibrational quantum numbers v, it is necessary either to raise the temperature of the molecules or to excite them into a higher quantum state directly by irradiation with light or by a chemical reaction. In this case, one can often also observe emission transitions between states having higher quantum numbers. However, thermal equilibrium is for the most part quickly reestablished via radiationless processes.

10.4 Rotational-Vibrational Spectra of Diatomic Molecules. The Rotating Oscillator and the Rotational Structure of the Bands

Vibrational spectra of molecules have, as we mentioned in Sect. 10.1, a clear-cut *rotational structure*, i.e. they consist of bands with many individual lines at a spacing of the order of a few cm^{-1}, when the spectrum from the gas phase is analyzed with a sufficient spectral resolution. This rotational structure is based on the fact that a rotational transition occurs at the same time as the vibrational transition. Now that we have studied the (hypothetical) non-rotating oscillator in Sect. 10.3 – it represents a relatively good approximation when the spectral resolution is not very high – we will take up the *rotating oscillator* in this section. It corresponds to the real behaviour of molecules in the gas phase. We will again explain the basic facts using a diatomic molecule as an example. A typical spectrum, that of the HBr molecule, is shown in Fig. 10.7.

The coupling of the vibrational and rotational motions in a molecule can be understood in terms of classical physics. If, however, we first ignore this coupling and consider the excitation of a diatomic molecule in the first approximation to be simply the sum of the excitation of a harmonic oscillator and of a rigid rotor, then in the simplest case we obtain the energy levels

$$\begin{aligned} E(v, J) &= E_{\mathrm{vib}}(v) + E_{\mathrm{rot}}(J) \\ &= \hbar\omega(v + \tfrac{1}{2}) + BhcJ(J+1) \end{aligned} \tag{10.20}$$

with the selection rules $\Delta v = \pm 1$ and $\Delta J = \pm 1$.

In the rotational-vibrational spectrum, transitions are naturally also allowed in which only the rotational quantum number changes, $\Delta v = 0$, $\Delta J = \pm 1$. These are the pure rotational transitions treated in Chap. 9. In contrast, in most cases (e.g. HBr, see Fig. 10.7) vibrational transitions without a change in the rotational quantum number, $\Delta v = \pm 1$, $\Delta J = 0$, are not allowed; that is, usually a change in the vibrational state must be accompanied by a change in the rotational state. We shall not derive the reason for this here.

However, this fact can be understood in an intuitive manner: a vibrational transition corresponds to a sudden change of the bond length. The classical analogy is an ice skater who changes his rotational velocity by extending or retracting his arms while performing a pirouette. One can imagine a change in the rotational state of a molecule during a vibrational transition in just this manner – the selection rule $\Delta J = 0$ is valid only when the angular momentum of the molecule is parallel to its cylinder axis.

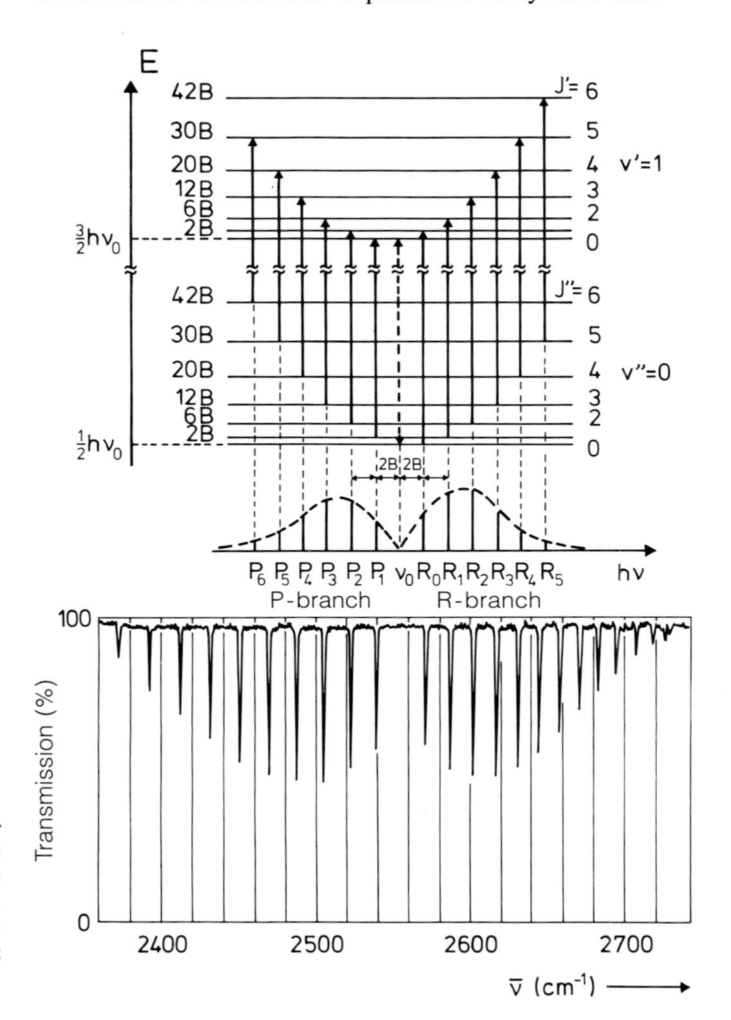

Fig. 10.7. A band in the rotational-vibrational spectrum of the HBr molecule, showing the term scheme and the transitions. The origin of the band is denoted by ν_0. A Q branch is not allowed here

These rotational and vibrational terms are illustrated schematically in Fig. 10.8 for a Morse potential. The corresponding transitions are shown in Fig. 10.7 for a portion of a typical rotational-vibrational spectrum. One observes different "branches" in the spectrum of a vibrational transition $(v + 1) \leftarrow v$, i.e. in a band. In the simplified case of a harmonic oscillator, these are

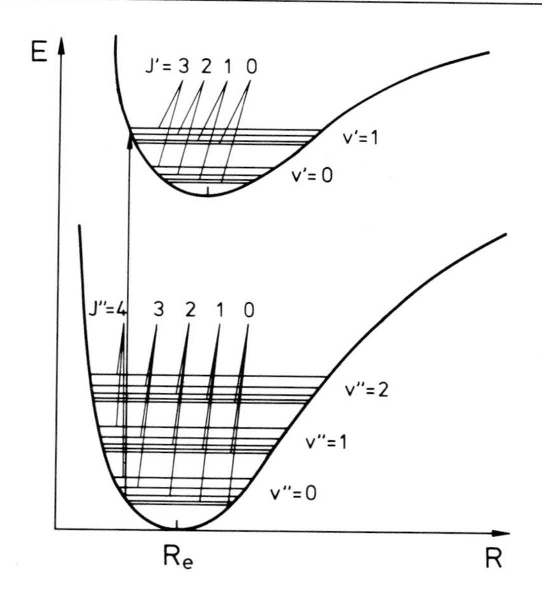

Fig. 10.8. Rotational-vibrational levels in the electronic ground state and in an excited electronic state. Only the lowest-lying rotational and vibrational terms are drawn. Transitions between the levels in the electronic ground state give rise to the rotational-vibrational spectrum. Transitions between the levels of different electronic states contribute to the electronic band spectrum; see Chap. 14

- the P branch, with $\Delta J = -1$. Taking $J'' = J' - 1$, we have $\overline{\nu} = \overline{\nu}_{1\leftarrow 0} - 2B(J' + 1)$, where $\overline{\nu}_{1\leftarrow 0}$ denotes the pure vibrational transition without rotation. Then the lines of the P branch have $\overline{\nu} < \overline{\nu}_{1\leftarrow 0}$, and the line spacings relative to $\overline{\nu}_{1\leftarrow 0}$ are $2B$, $4B$, ... as in the spectrum of a rigid rotor;
- the R branch, with $\Delta J = +1$. Taking $J' = J'' + 1$, we find $\overline{\nu} = \overline{\nu}_{1\leftarrow 0} + 2B(J'' + 1)$, i.e. $\overline{\nu} > \overline{\nu}_{1\leftarrow 0}$, and the line spacings relative to $\overline{\nu}_{1\leftarrow 0}$ are likewise $2B$, $4B$, ...;
- in some cases also a Q branch, with $\Delta J = 0$. If the rotational constant B is the same for both the vibrational levels which are involved in the transition, the Q branch (when it is allowed) consists of a single line at $\overline{\nu}_{1\leftarrow 0}$, the so-called band origin; otherwise, it contains a series of closely-spaced lines. In many cases, depending on symmetry, for example HBr (Fig. 10.7), the Q branch is not allowed.

The line spacings in the rotational-vibrational spectrum again yield the rotational constant B, as we have already seen in Chap. 9 for pure rotational spectra. This constant can thus be determined by infra-red absorption without resorting to microwave spectroscopy. The line intensities within the various branches are, in the first instance, determined by the occupation numbers of the rotational levels; cf. Sect. 9.3. Again, we remind the reader that the rotational quanta are usually very small compared to the thermal energy kT and therefore a Boltzmann distribution according to their degrees of degeneracy can be expected for the occupations of the rotational levels. On the other hand, in an absorption transition from the $v = 0$ to the $v = 1$ state, the upper vibrational level with its rotational sublevels is nearly unoccupied in thermal equilibrium owing to the large magnitude of the vibrational energy quanta. The thermal energy kT at room temperature corresponds to about 200 cm^{-1}, as mentioned above, while typical vibrational quanta are of the order of 1000 cm^{-1}. The Boltzmann factor $N_1/N = \mathrm{e}^{-\Delta E/kT}$ is thus much less than 1. The intensities in the absorption spectrum are then determined by the degeneracy factors, which increase with increasing J, and by the decreasing thermal occupation probabilities of the initial-state rotational levels of the $v = 0$ state with increasing J. The same is true of the transitions to $v = 2$, $v = 3$, etc. in the case of the anharmonic oscillator.

When the rotational-vibrational spectrum is measured at a sufficiently high spectral resolution, one finds that the lines within the branches are not exactly equidistant. The spacings become smaller with increasing distance from the origin at $\overline{\nu}_{1\leftarrow 0}$. This is due to the *coupling between vibrations and rotations*. The two motions are in fact not mutually independent; one cannot simply add the energies of the vibrational and rotational transitions, but instead must take the interaction of the two types of nuclear motion into account by introducing into the energy or term values mixed terms depending on both v and J.

As we have already seen, the vibrations of a molecule take place on a much faster time scale than its rotations. During a single rotation, a molecule vibrates several thousand times. The rotor therefore sees an internuclear distance $\langle R\rangle$ which is averaged over many vibrations. In the case of the anharmonic oscillator, the average internuclear distance $\langle R\rangle$ increases with increasing quantum number, v, i.e. with increasing vibrational excitation (see Sect. 10.3). The moment of inertia then also increases and the rotational constant B becomes smaller. In addition to the rotational stretching of the molecule which we have already treated in Sect. 10.3, there is thus a *vibrational stretching*.

This leads to the following relation for the time-averaged moment of inertia:

$$\langle\Theta(v+1)\rangle > \langle\Theta(v)\rangle > \Theta_e \,, \tag{10.21}$$

where Θ_e is the moment of inertia at the equilibrium bond length R_e. Correspondingly, the rotational constant B becomes dependent on the vibrational state v, so that we should write B_v to be more precise. Then B_v for $v > 0$ is smaller than the rotational constant B_0 for the ground state with $v = 0$.

This behaviour is described by the formula

$$B_v = B_e - \alpha(v + \tfrac{1}{2}) \quad (+\text{ terms of higher order}) \qquad [\mathrm{cm}^{-1}]\,. \tag{10.22}$$

Here, B_e means the rotational constant in the hypothetical state without vibrations, and α is a molecule-specific positive number, with $\alpha \ll B_e$. Due to the zero-point energy, from (10.22) we find for the quantity B_0 in the vibrational state with quantum number $v = 0$:

$$B_0 = B_e - \frac{\alpha}{2}\,. \tag{10.23}$$

In the same way, the stretching of the molecule by centrifugal force depends in the anharmonic oscillator on the vibrational quantum number v. The stretching constant D_e in the equilibrium state without vibrations, which was introduced in (9.24) and (9.25), thus becomes

$$D_v = D_e + \beta(v + \tfrac{1}{2})\,, \tag{10.24}$$

with a correction factor $\beta \ll D_e$. The rotational energy terms are thus changed by the vibrations. We note here that the factor β should not be confused with other quantities denoted by β, for example in Sects. 3.2 and 3.5.1.

Taking into account the anharmonicity, i.e. using a Morse potential, and considering (10.22) and (10.23), we now obtain an expression for the rotational energy which has been improved as compared to (10.21):

$$\begin{aligned} E_{v,J} &= \hbar\omega_e(v+\tfrac{1}{2}) - x_e\hbar\omega_e(v+\tfrac{1}{2})^2 + hcB_vJ(J+1) \\ &\quad - hcD_vJ^2(J+1)^2 \end{aligned} \tag{10.25}$$

and for the terms, measured in the unit cm^{-1},

$$T_{v,J} = G_v + F_{v,J}$$

$$= \overline{\nu}_e(v+\tfrac{1}{2}) - x_e\overline{\nu}_e(v+\tfrac{1}{2})^2 + B_v J(J+1) - D_v[J(J+1)]^2 \tag{10.26}$$

$$= \overline{\nu}_e(v+\tfrac{1}{2}) - x_e\overline{\nu}_e(v+\tfrac{1}{2})^2 + B_e J(J+1) - D_e[J(J+1)]^2$$
$$-\alpha J(J+1)(v+\tfrac{1}{2}) - \beta[J(J+1)]^2(v+\tfrac{1}{2})\,. \tag{10.27}$$

In these expressions, we have omitted correction terms of the form $(v+\frac{1}{2})^n$ with powers $n > 2$. A spectrum containing the "corrections" according to (10.26) and (10.27) and a term scheme which leaves them out for simplicity are shown in Fig. 10.7. When $\beta \ll \alpha$, the last term in (10.25), and correspondingly in (10.26) and (10.27), can be neglected in general.

The rotational-vibrational spectrum corresponds to transitions between the terms $E_{v,J}$ or $T_{v,J}$. The condition for it to be observable is again that the molecule be polar. Then for the observed transitions we have

$$\overline{\nu} = \frac{1}{hc}[E(v', J') - E(v'', J'')]\,, \quad \text{with the convention} \quad v' > v''\,. \tag{10.28}$$

Leaving out the stretching terms in (10.25), we find

$$\overline{\nu} = \overline{\nu}_e(v' - v'') - x_e\overline{\nu}_e[(v'+\tfrac{1}{2})^2 - (v''+\tfrac{1}{2})^2]$$
$$+B_{v'}J'(J'+1) - B_{v''}J''(J''+1)\,. \tag{10.29}$$

the selection rules for electric dipole radiation are given by:

$$\Delta J = \pm 1\,, \qquad \Delta v = 0, \pm 1, \pm 2 \ldots\,,$$

where for the harmonic oscillator, $\Delta v > 1$ is not allowed and $x_e = 0$. For $\Delta v = 0$, we obtain the pure rotational spectrum in a vibrational state with the quantum number $v' = v''$.

Owing to the selection rule $\Delta J = \pm 1$, there are two branches in the rotational-vibrational spectrum. The P branch refers, as above, to the series of transitions with $\Delta J = -1$, and the R branch to the series with $\Delta J = +1$.

The spectral lines in the P branch ($J' = J$, $J'' = J+1$) have the wavenumbers

$$\overline{\nu}_P = \overline{\nu}(v', v'') - 2B_{v''}(J+1) - (B_{v''} - B_{v'})J(J+1) \tag{10.30}$$

and those in the R branch ($J' = J+1$, $J'' = J$) are given by

$$\overline{\nu}_R = \overline{\nu}(v', v'') + 2B_{v''}(J+1) - (B_{v''} - B_{v'})(J+1)(J+2)\,. \tag{10.31}$$

The spectral lines of a band in the P branch therefore lie on the long-wavelength side of the pure vibrational line $\overline{\nu}(v', v'')$, the so-called zero line, which itself cannot be observed, while those in the R branch are on its short-wavelength side. The lines are no longer equidistant, due to the last terms in (10.30) and (10.31). The lines in the P branch move further apart with increasing J, while those in the R branch converge. This makes the structure of the spectrum in Fig. 10.7 understandable.

The zero line, $\overline{\nu}(v', v'')$, corresponds to the transition with $\Delta J = 0$, which is usually forbidden; it is the purely vibrational transition. It can thus not be observed directly, for the most part.

For this line (i.e. for the Q branch, when it is observable), we find

$$\overline{\nu}(v', v'') = (\overline{\nu}_e - x_e\overline{\nu}_e)(v' - v'') - x_e\overline{\nu}_e(v'^2 - v''^2)\,. \tag{10.32}$$

The first term in (10.32) yields the wavenumbers of the fundamental vibration and the harmonics $\Delta v > 1$, which occur because of anharmonicity, as multiples of the wavenumber $(\overline{\nu}_e - x_e\overline{\nu}_e)$. The second, much smaller term causes the harmonics to move closer together; compare Fig. 10.5. The wavenumbers of the purely vibrational lines have to be derived from the rotational-vibrational spectrum by applying (10.30) and (10.31).

Experimentally, one can then determine three quantities which are characteristic of the molecule under investigation: the pure vibration with the wavenumber $\overline{\nu}(v', v'')$, and the two rotational constants $B_{v'}$ and $B_{v''}$ (and from them, B_e and α). This is accomplished by measuring as many lines as possible in the spectrum and then finding the best fit to equations (10.30) and (10.31). As an example, we give the data derived from the spectrum of the CO molecule:

$$B_e = 1.924\,\mathrm{cm}^{-1}\,, \quad x_e = 0.0061\,, \quad \overline{\nu}_e = 2169.2\,\mathrm{cm}^{-1}\,, \quad \alpha = 0.0091\,\mathrm{cm}^{-1}\,;$$

see also Fig. 10.2.

As a further example of the analysis of the spectra, we give some experimental data for HCl. In this case, for the vibrational frequency with harmonics, the following values were measured:

$$\overline{\nu}(1 \leftarrow 0) = \overline{\nu}_e(1 - 2x_e) = 2885.9\,\mathrm{cm}^{-1}\,,$$

$$\overline{\nu}(2 \leftarrow 0) = 2\overline{\nu}_e(1 - 3x_e) = 5668.0\,\mathrm{cm}^{-1}\,,$$

$$\overline{\nu}(3 \leftarrow 0) = 3\overline{\nu}_e(1 - 4x_e) = 8347.0\,\mathrm{cm}^{-1}\,.$$

From these data, we calculate $x_e = 0.017$.

From the measured values

$$B_0 = B_e - \frac{\alpha}{2} = 10.440\,\mathrm{cm}^{-1} \quad \text{and}$$

$$B_1 = B_e - \frac{3\alpha}{2} = 10.137\,\mathrm{cm}^{-1}\,,$$

it follows that $B_e = 10.591\,\mathrm{cm}^{-1}$ and $\alpha = 0.303\,\mathrm{cm}^{-1}$. The eigenfrequency $\overline{\nu}_e$ is found to be 2989 cm^{-1}.

From B_0 and B_e, the internuclear distance R can furthermore be determined, as shown in Sect. 9.2; and from $\overline{\nu}_e$, the force constant and the vibrational frequency ν_e of the molecule can be derived. In this case, one obtains $R_0 = 1.2838 \cdot 10^{-8}$ cm for the internuclear distance in the $v = 0$ state, and the calculated quantity $R_e = 1.2746 \cdot 10^{-10}$ m in the hypothetical state without zero-point oscillation. The force constant k is found to be $4.8 \cdot 10^2$ Nm^{-1}, and the period of vibration $T_1 = \nu^{-1} = (c\overline{\nu}_e)^{-1} = 1.17 \cdot 10^{-14}$ s.

These are the measured data which infra-red spectroscopy gives us for the investigation of diatomic molecules. From the *force constants* and the *anharmonicities*, one can determine the shape of the *potential curve* and from it, can reach conclusions about *chemical bonding* in the molecule.

Strictly speaking, some additional influences on the energies of the levels and the transitions should be taken into account: the effect of the centrifugal stretching on the rotational constant B, its effect in turn on the vibrational potential, and the Coriolis coupling. These effects can, however, often be neglected at attainable spectral resolutions. We leave their treatment to the specialised literature.

10.5 The Vibrational Spectra of Polyatomic Molecules

Although diatomic molecules have only *one* vibrational degree of freedom and can oscillate only in the direction parallel to their bonding axes – hence the name valence or stretching vibration – molecules with more than two atoms have several vibrational degrees of freedom. In addition to stretching vibrations, they can undergo vibrations in which the bonding angles change: so-called *bending vibrations*. In order to describe their behaviour, we use the concept of *normal modes* of vibration, which we shall discuss in the following. However, we will need to make only minor extensions of our previous considerations in order to understand the vibrational spectra of polyatomic molecules.

The vibrations of a system of elastically coupled point masses can be described in terms of a superposition of the allowed normal modes of the system, as one learns from classical mechanics. The simplest case is that of two identical pendulums coupled by a spring. For this system, we find two normal modes with frequencies ν_1 and ν_2, the symmetric and the antisymmetric vibrations (Fig. 10.9), and these normal modes can be observed as spectral lines of the system by Fourier analysis of its motions. Spectroscopy does exactly this: the frequency analysis of a time-dependent behaviour.

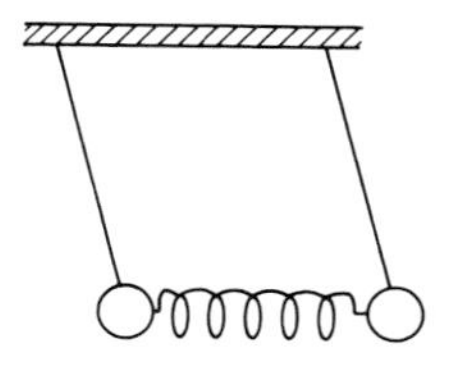

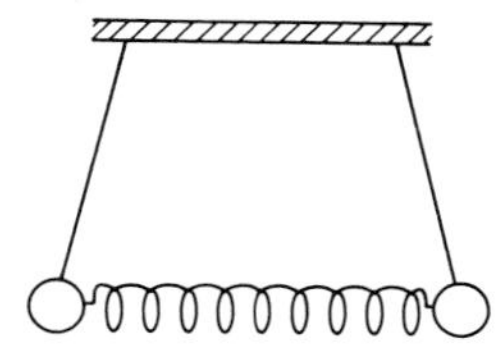

Fig. 10.9. The fundamental oscillations of two coupled pendulums: symmetric and antisymmetric oscillations (normal modes)

The normal modes are defined as motions in which all the point masses of the system move with the same frequency and with a fixed phase relation. The motion of the whole system is a pure harmonic oscillation. One normal mode can be excited without exciting any others, i.e. they can be completely decoupled from each other, as long as the amplitudes are kept small and nonlinearities are thus avoided.

The number f of the normal modes of a system is equal to the number of its *degrees of freedom*, which are not already occupied by other forms of motion. A system of N point masses initially has $3N$ degrees of freedom. If the masses are coupled together to form a molecule, then there are 3 degrees of freedom for the translational motion (motions of the centre of gravity of the whole molecule) and 3 degrees of freedom for rotations (only 2 in the case of a linear molecule, because the rotation about the cylinder axis does not contribute); we thus have for the internal motions of the molecule

$$f = 3N - 6 \tag{10.33}$$

as the number of degrees of freedom, or, for a linear molecule, $f = 3N - 5$.

In the case of a diatomic molecule, we find $f = 3 \cdot 2 - 5 = 1$; there is only one normal mode, namely the stretching vibration. For a linear triatomic molecule, we have $f = 9 - 5 = 4$. As an example, we consider the linear CO_2 molecule. Here, the vibrations can be described as the superpositions of the four normal modes sketched in Fig. 10.10, with the eigenfrequencies ν_1, ν_2, and ν_3. Just these vibrations are in fact observed. The vibrational patterns are shown in Fig. 10.10. One of the vibrations, the bending mode, is doubly degenerate and is therefore to be counted twice, since the bending can occur within the plane of the

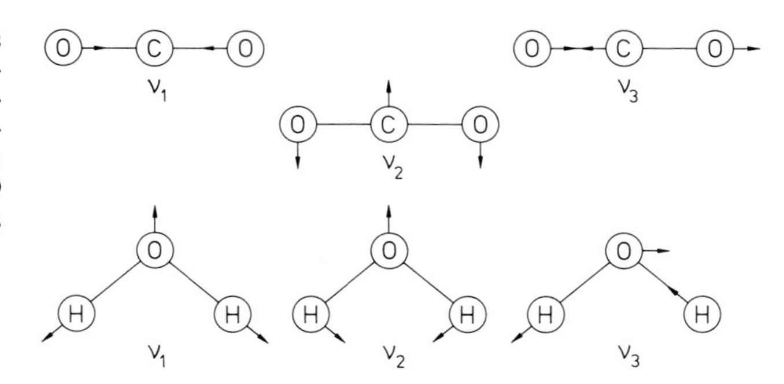

Fig. 10.10. The normal modes of the CO_2 and H_2O molecules. The following wavenumbers correspond to the vibrations: CO_2 $\overline{\nu}_1$: 1337 cm^{-1}; $\overline{\nu}_2$: 667 cm^{-1}; $\overline{\nu}_3$: 2349 cm^{-1}. H_2O $\overline{\nu}_1$: 3657 cm^{-1}; $\overline{\nu}_2$: 1595 cm^{-1}; $\overline{\nu}_3$: 3756 cm^{-1}

figure or perpendicular to it. The relative frequencies of these vibrations can be estimated: the highest frequency, corresponding to $\overline{\nu}_3 = 2349$ cm^{-1}, is that of the asymmetric stretching vibration, because it stretches the "springs" most strongly. The symmetric stretching vibration has a wavenumber of $\overline{\nu}_1 = 1337$ cm^{-1}, and the bending mode has $\overline{\nu}_2 = 667$ cm^{-1}. In general, the frequencies of stretching modes are higher than those of bending modes.

However, for the CO_2 molecule we can readily see that not all the vibrational modes can be observed in infra-red absorption, i.e. not all are infra-red active. For infra-red activity, a periodic change of the electric dipole moment is required, as we have seen. The symmetric CO_2 molecule has no electric dipole moment in its equilibrium state. When it oscillates in the symmetric stretching mode ν_1, its symmetry is maintained and no dipole moment is produced. In contrast, the asymmetric stretching mode ν_3 and the bending mode ν_2 are infra-red active. The dipole moment which is induced in the ν_2 mode is perpendicular to that induced in the ν_3 mode and thus also perpendicular to the cylinder axis of the molecule, so the corresponding rotational-vibrational bands are referred to as the perpendicular band (for ν_2) and the parallel band (for ν_3).

As an example of a nonlinear triatomic molecule, Fig. 10.10 also shows the normal modes of the water molecule, H_2O. Here, again, the frequency of the bending mode ν_2 is lower than those of the two other vibrational modes, in which the force constants are more strongly loaded. In the normal modes ν_1 and ν_2, the axis of two-fold symmetry through the centre of the molecule is maintained; both modes are thus referred to as symmetric, in contrast to the ν_3 mode. One can readily see by examining Fig. 10.10 that the dipole moment of the H_2O molecule changes periodically in all three normal mode of vibration; they are thus all infra-red active. Due to the presence of these molecules in the air, the H_2O and CO_2 lines are observed in every infra-red spectrum, unless the optical path of the infra-red spectrometer is evacuated.

An additional example of an experimentally observed spectrum is given by Fig. 10.11, which shows a portion of the infra-red spectrum of the HCN molecule. One can see the two rotational-vibrational bands belonging to the two normal mode vibrations ν_2 and ν_3, as well as a harmonic band at $2\nu_2$. The selection rules are again $\Delta v = \pm 1$ and $\Delta J = \pm 1$ for the stretching vibrations of linear molecules, but $\Delta J = \pm 1$ and $\Delta J = 0$ for the bending mode vibrations of linear molecules and for the vibrational bands of symmetric top molecules, such as CH_3I, NH_3, or C_6H_6.

One can also readily understand that no change in the rotational state of the molecule occurs along with the normal vibrational modes just mentioned. We thus observe allowed transitions in which only the vibrational quantum number v changes, i.e. the spectrum contains not only the P and R branches, but also the (narrow) Q branch. The Q branch in a

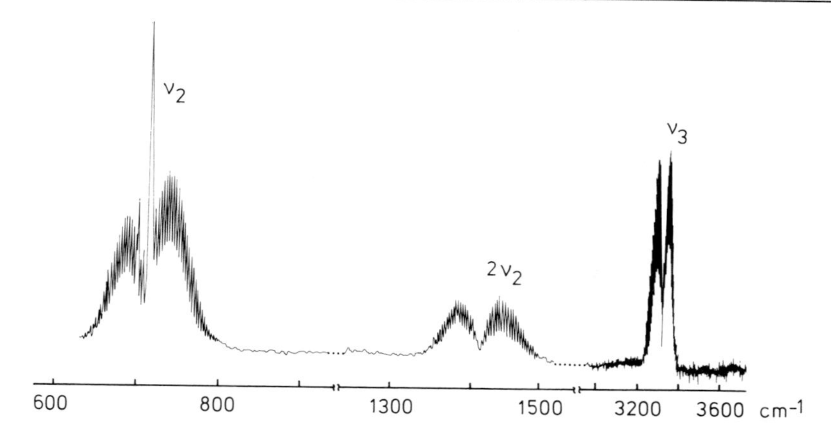

Fig. 10.11. A portion of the rotational-vibrational spectrum of the HCN molecule. The ν_2 vibration is a bending mode. Since it is a so-called perpendicular band, P, Q and R branches are allowed. For the $2\nu_2$ harmonic band and the parallel band ν_3 (stretching mode), only the P and R branches are allowed for reasons of symmetry. After Steinfeld

rotational-vibrational spectrum, as mentioned in Sect. 10.4, refers to all of the transitions between two vibrational states v' and v'' in which the rotational quantum number J remains unchanged. If the spacing of the rotational levels were the same in both vibrational states, this would be a single line. In fact, the rotational constants $B_{v'}$ and $B_{v''}$, and therefore the rotational level spacings, differ somewhat from each other; for this reason, the Q branch consists of a number of closely-spaced lines. Figure 10.12 shows the rotational-vibrational

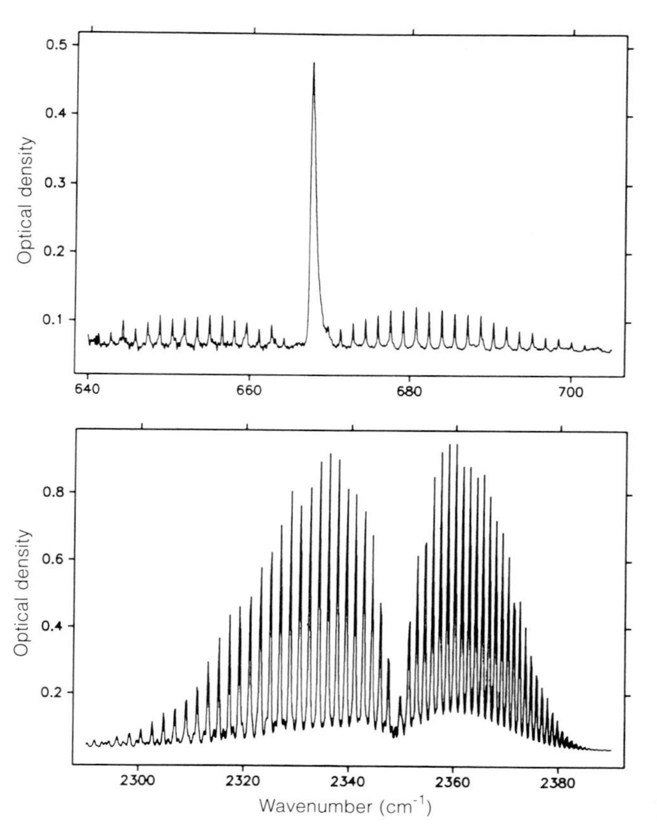

Fig. 10.12. A portion of the rotational-vibrational spectrum of CO_2. *Upper part*: The band of the bending mode, $\overline{\nu}_2$; *Lower part*: the band of the asymmetric stretching mode, $\overline{\nu}_3$ as examples for the differing selection rules: in the upper spectrum, there are P, Q and R branches; in the lower spectrum, the Q branch is forbidden. The quantum numbers J for the rotational levels are not given here. In Sect. 12.4, we explain that, due to the inversion symmetry of the CO_2 molecule and the nuclear spin $I = 0$ of the O atom, every second rotational level is suppressed. The line spacing is therefore $4B$ instead of the usual $2B$. If the inversion symmetry is destroyed by substituting one of the ^{16}O atoms by a heavier ^{18}O isotope (i.e. in the molecule ^{16}O–C–^{18}O), one observes that the spacing of the rotational lines is halved

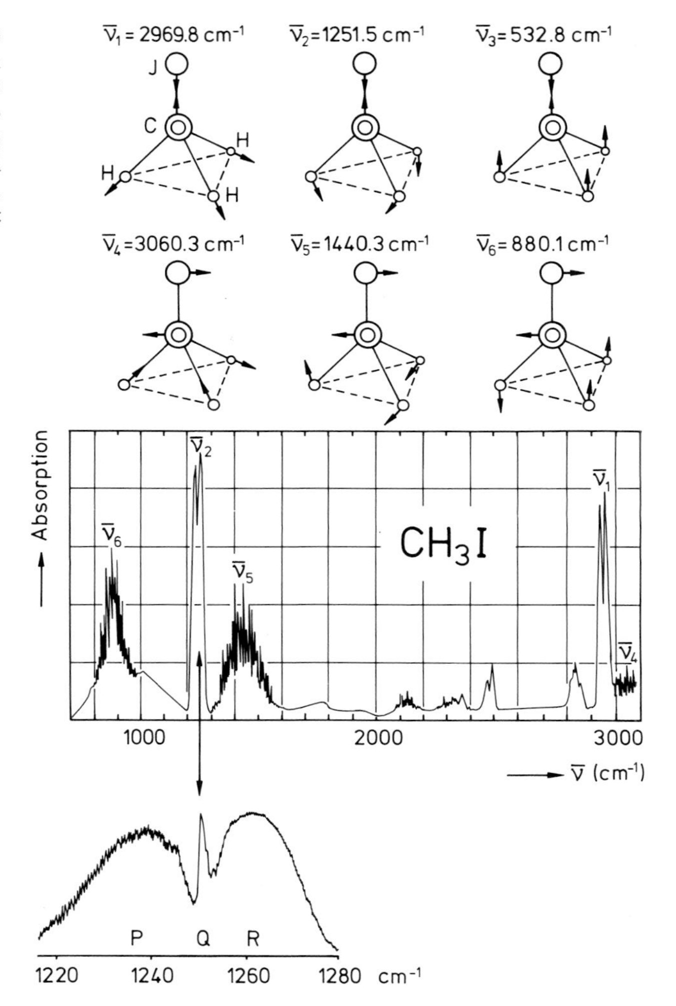

Fig. 10.13. A section of the rotational-vibrational spectrum of the CH_3I molecule. Besides the fundamental vibrations $\overline{\nu}_1$ to $\overline{\nu}_6$, the following combinations occur: at $1770\,\mathrm{cm}^{-1}$ ($2\overline{\nu}_6$ and $\overline{\nu}_2+\overline{\nu}_3$), at $2130\,\mathrm{cm}^{-1}$ ($\overline{\nu}_2+\overline{\nu}_6$), at $2320\,\mathrm{cm}^{-1}$ ($\overline{\nu}_5+\overline{\nu}_6$), and at $2480\,\mathrm{cm}^{-1}$ ($2\overline{\nu}_2$)

band of the bending mode ($\overline{\nu}_2 = 667\,\mathrm{cm}^{-1}$) of CO_2 as an example of a band with a Q branch, in contrast to the stretching mode ($\overline{\nu}_3 = 2349\,\mathrm{cm}^{-1}$), where the Q branch, with $\Delta J = 0$, is forbidden.

In the case of the symmetric top molecules, the quantum number K also becomes important; see Sects. 9.7 and 11.2. The selection rules, which we give here without derivation, are $\Delta K = 0$ for parallel bands and $\Delta K = \pm 1$ for perpendicular bands. These selection rules can also be understood intuitively: for vibrations parallel to the molecular axis, the projection of the angular momentum on this axis does not change during a vibration, i.e. $\Delta K = 0$.

Of course, every vibrational transition is surrounded by its accompanying rotational transitions, i.e. the whole band spectrum, as can be clearly seen in Figs. 10.11 and 10.12. Naturally, polyatomic molecules can also exhibit anharmonicity; accordingly, as in the diatomic molecules, one observes *harmonics* at 2ν, 3ν, etc. with strongly decreasing intensities. In addition, the deviation from purely harmonic behaviour leads to *combined vibrations*, such as $\nu_1 + \nu_2$, $\nu_1 - \nu_2$, or $2\nu_1 + \nu_2$. Some examples in the case of the CH_3I molecule are shown in Fig. 10.13. In molecules with several normal mode vibrations, it can happen that

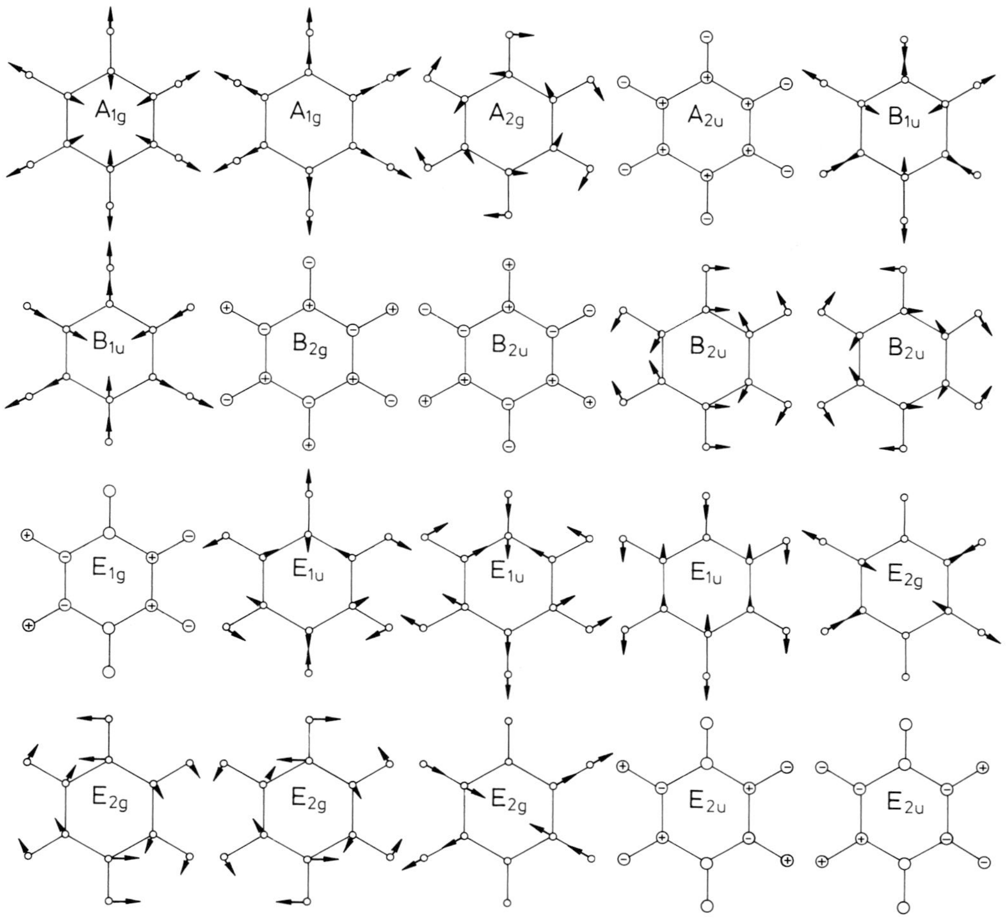

Fig. 10.14. The normal modes of the benzene molecule, C_6H_6. In the case of degenerate vibrational modes, only one component is shown. After Herzberg

one normal mode has nearly the same frequency as a harmonic or a combined vibration of other normal modes. Such a *Fermi resonance* can lead to an apparent strong increase in the intensity of the affected harmonic band or the band of the combined vibration.

Molecules containing more nuclei have larger numbers of normal mode vibrations. In order to resolve and classify them, one requires symmetry considerations, which we shall not treat further here; however, see Chaps. 5 and 6. Figure 10.14 shows the normal modes of the benzene molecule; they include vibrations which are not infra-red active. We shall return to this question and to the possibility of nevertheless observing them in Chap. 12, which deals with the Raman effect. If one does not take care to observe isolated molecules, i.e. at a high dilution in the gas phase, and to use a high spectral resolution, then only a single, unresolved line is observed for each vibrational transition instead of the rotational-vibrational bands with their well-defined structure; this is equally true in the case of polyatomic molecules as for diatomics. It is especially the case for molecules in the condensed phases.

10.6 Applications of Vibrational Spectroscopy

From a precise analysis of their vibrational spectra, one can obtain important data concerning the structure and bonding in molecules. The spectra can allow bonding angles and bond lengths, force constants and the potential curve for bonding to be calculated with high precision. For polyatomic molecules, infra-red spectroscopy is therefore an important method for *structural analysis*. In analytical chemistry, infra-red spectroscopy is furthermore a useful aid to the *identification of molecules or of molecular fragments*. The frequencies at which particular molecular subunits absorb in the infra-red are characteristic of those units. Even with unresolved rotational structure, i.e. in the condensed phases, the presence of particular molecular subunits in a sample can be determined by the detection of these *characteristic frequencies*. Table 10.3 gives typical numerical values for the quantum energies of some important vibrations.

Table 10.3. Wavenumber values of some typical subunit vibrations

C – H stretch	2850 – 3000 cm^{-1}
C – H bend	1350 – 1460 cm^{-1}
C – C stretch	700 – 1250 cm^{-1}
C = C stretch	1600 – 1700 cm^{-1}

When two groups that would have similar vibration frequencies if they were measured individually, are present in a molecule, then *resonance* between their vibrations can occur, with a resulting frequency shift similar to that seen in the Fermi resonances; cf. Sect. 10.5. A well-known example is the case of the carbonyl group C = O with $\overline{\nu} = 1715$ cm^{-1}, and the C = C double bond with $\overline{\nu} = 1650$ cm^{-1}. In the ketene radical, C = C = O, where these two frequencies should be observed, one instead finds the wavenumbers 2100 and 1100 cm^{-1}; the values are thus strongly shifted from those which would be observed for the isolated molecular subunits.

The characteristic frequencies are also influenced by the surroundings of the molecule, in particular by the *state of matter*. As a rule, one finds $\nu_{gas} > \nu_{liquid} > \nu_{solid}$, i.e. the frequencies are lowered by interactions with the molecular surroundings. The stretch mode vibration of HCl decreases by about 100 cm^{-1} on liquefaction and by an additional 20 cm^{-1} on solidification. These changes produced by the environment however depend strongly on the type of vibrations and on the molecule considered.

10.7 Infra-red Lasers

The basic principles of the laser, a light source with unusual and for many applications revolutionary properties, were already treated in I, Chap. 21. The laser has opened many new possibilities in experimental molecular physics. There are some important types of laser in which the laser-active medium consists of molecules; an example is the CO_2 laser. Although we have thus far spoken only of the absorption in the rotational-vibrational spectra, we shall now make use of the fact that the corresponding transitions can also be observed in emission, in particular by *stimulated emission*.

The vibrational spectra of the CO_2 molecule are employed in the CO_2 laser for the production of infra-red laser radiation. The laser tube contains a mixture of N_2 and CO_2 molecules as the laser-active medium. The vibrational transition of the N_2 molecule at 2360 cm^{-1}, with its associated rotational levels, is excited in a gas discharge by collisions with electrons and ions. As indicated in Fig. 10.15, the N_2 molecules can transfer their excitation energy in a radiationless manner to the CO_2 molecules through collisions. This process has a high yield, since the asymmetric stretching vibration $\overline{\nu}_3$ of CO_2 lies at 2349 cm^{-1}. There is thus a resonance between the rotational-vibrational levels of the nitrogen molecule and those of CO_2. In addition, this excited state of N_2 is metastable and thus long-lived, because a radiative transition to the ground state is forbidden.

Induced emission is now possible starting from the rotational-vibrational levels of CO_2 in the 2349 cm^{-1} region into the symmetric stretch vibration band $\overline{\nu}_1$ at 1390 cm^{-1}. All the transitions from the excited state at $\overline{\nu}_3$ into the rotational levels of the $\overline{\nu}_1$ state, taking the selection rule $\Delta J = \pm 1$ into consideration, are allowed and contribute to the laser process. There are thus many (about 100) discrete laser frequencies in a range of ca. 1000 cm^{-1}, the difference between $\overline{\nu}_1$ and $\overline{\nu}_3$. This corresponds to microwave radiation with a wavelength of about 10.6 μm. The CO_2 laser is particularly important in the field of materials processing, because it can be made to yield high energy densities in a relatively simple manner.

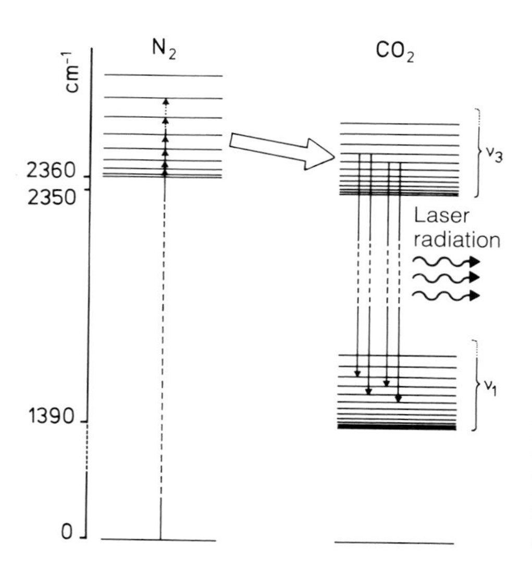

Fig. 10.15. The rotational and vibrational energy levels of the N_2 and CO_2 molecules which are employed in the CO_2 laser. A description is given in the text

10.8 Microwave Masers

The laser principle, i.e. the production of coherent radiation through stimulated emission, was first demonstrated in the microwave range by using the inversion vibration of the NH_3 molecule. In the year 1955, *Gordon, Zeiger,* and *Townes* reported the construction of the first ammonia maser. The word *maser* is an acronym for *M*icrowave *A*mplification by *S*timulated *E*mission of *R*adiation. When, a short time later, the principle was applied to visible light, the word *laser* was coined; it refers to *L*ight *A*mplification by *S*timulated *E*mission of *R*adiation.

The NH_3 molecule belongs to the class of symmetric top molecules. It has the form of a triangular pyramid with the three H atoms at the vertices of the base and the N atom at the apex. One of the normal mode vibrations of this molecule, with the frequency ν_2, is a motion in which the N atom oscillates towards the basal plane containing the H_3 group, periodically changing the height of the molecule. The value of ν_2 is $2.85 \cdot 10^{13}$ s^{-1}, corresponding to 950 cm^{-1}. The potential barrier for the passage of the N atom from one side of the molecule to the other through the plane of the H atoms is high, and in classical physics an excitation energy of $3h\nu_2 = 0.3$ eV would be required for the motion to occur. This mode is called the *inversion vibration* because it leads to an inversion of the molecule, like an umbrella reversed by the wind. Figure 10.16 shows the structure of the NH_3 molecule, the double potential for the inversion vibration, and the vibrational levels.

A similar doubling of the potential curve due to inversion is found in all XY_3 molecules with the structure of a non-planar top. The two configurations with X above or below the Y_3 plane cannot be converted into one another by any rotation of the molecule; but they are energetically in resonance. This leads to a splitting into pairs of levels, the inversion doubling. To be sure, it can be understood only in the framework of quantum mechanics.

Quantum-mecanically, the tunnel effect (cf. I, Sect. 23.3) can be invoked to explain how the N atom can pass through the barrier represented by the plane of the three H atoms even though it has an excitation energy less than 0.3 eV. It can thus oscillate continuously between the two sides of the H_3 plane. If the potential barrier were infinitely high, there would be two degenerate wavefunctions for the N atom on the one side and on the other side of the H-atom plane.

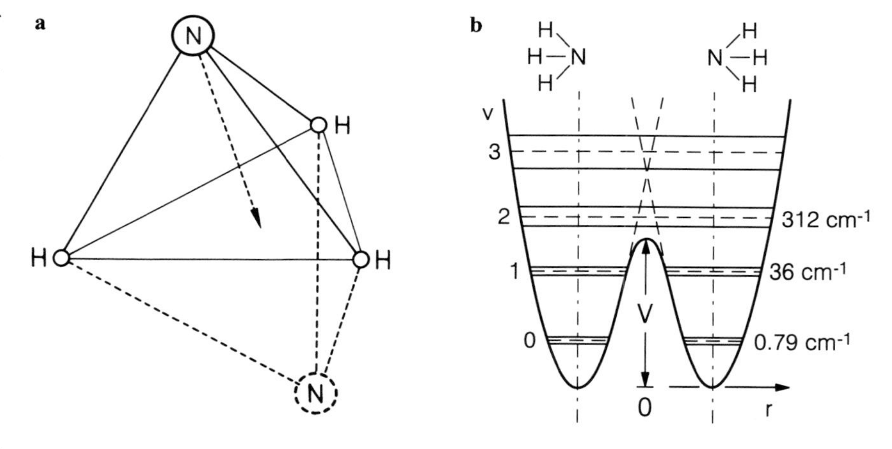

Fig. 10.16. (a) The structure of the NH_3 molecule. The N atom can occupy a position above or below the plane formed by the three H atoms. This leads to a splitting of the energy levels and a double potential curve whose barrier can be penetrated due to the tunnel effect in the inversion vibration. **(b)** The potential curve for an NH_3 molecule. The distance of the N atom from the H_3 plane is denoted by r; the scale is arbitrary. V is the height of the potential barrier. The *dashed horizontal lines* are the vibrational levels which would be found if the potential curve had only a single minimum. The inversion splitting increases strongly with increasing quantum number v. After Herzberg

This degeneracy is called the inversion degeneracy. When the height of the potential barrier is finite, the degeneracy is lifted, and pairs of symmetric and antisymmetric wavefunctions are formed from the previously isolated, degenerate wavefunctions. The vibrational levels split into pairs of levels; this is called inversion splitting. This process is quite analogous to the formation of symmetric and antisymmetric wavefunction pairs in the H_2^+ molecule-ion from the previously degenerate H_2 wavefunctions; compare I, Sect. 23.4. If a wavepacket is formed from these new wavefunctions at a time $t = 0$, it describes in the course of time the oscillation of the N atom between the two localised states, with an oscillation frequency $\omega = \Delta E/\hbar$, where ΔE is the splitting energy which is defined in Fig. 10.16. This frequency is called the inversion frequency. In the NH_3 molecule in its ground state, the frequency ν_i

(where i stands for inversion) has the value 23 870 MHz or $\bar{\nu}_i = 0.796\,\mathrm{cm}^{-1}$; it thus lies in the microwave range. Between the symmetric and the antisymmetric states, an electric dipole transition at this frequency is allowed: it is the maser transition.

In order to build a practical maser (see Fig. 10.17), one forms a beam of NH_3 molecules and passes it through an inhomogeneous electric quadrupole field. The electric field induces a dipole moment in the normally non-polar molecules and produces a quadratic Stark effect. The quadrupole field distinguishes molecules in the symmetric and the antisymmetric states of the inversion vibration. If the dimensions are properly chosen, only the molecules in the antisymmetric state can pass through the quadrupole field region and enter the resonator which is tuned to the frequency of 23 870 MHz.

In this resonator, a radiation field builds up, at first through spontaneous emission and then through stimulated emssion from the antisymmetric state into the symmetric state, which is initially not populated, due to the selection by the quadrupole field; the device acts as a self-exciting oscillator. This arrangement can also be used as a very narrow-band amplifier at the inversion frequency. The accuracy of the frequency of such a molecular amplifier, $\nu/\Delta\nu$ ($\Delta\nu$ is the bandwidth at the frequency ν) is very great, more than 10^{10}.

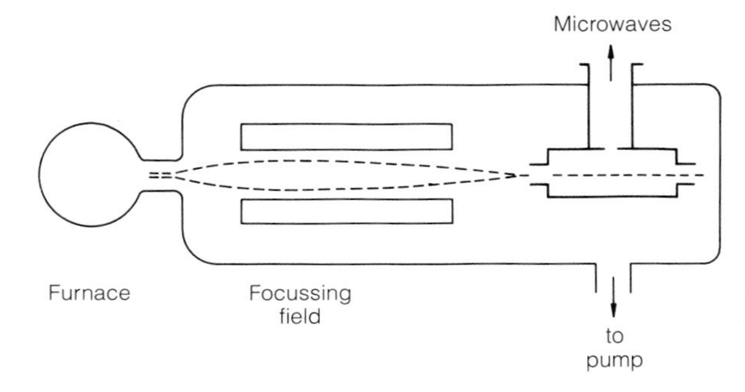

Fig. 10.17. An NH_3 maser. The molecules are selected by means of the Stark effect and focussed into a resonant cavity. A resonant radiation field is built up there; it can serve as an oscillator or as an amplifier

11. The Quantum-Mechanical Treatment of Rotational and Vibrational Spectra

Taking the diatomic molecule as an example, we introduce the Born-Oppenheimer approximation, which then permits the approximate separation of the molecular wavefunctions into electronic and nuclear parts. The rotational and vibrational motions are described by the nuclear wavefunction. We then consider the rotation of tri- and polyatomic molecules, giving a detailed discussion of both the symmetric and the asymmetric tops. In the treatment of molecular vibrations, the concept of normal coordinates plays an important role; here, again, symmetry considerations prove to be very useful, and we discuss them in the final sections of this chapter.

11.1 The Diatomic Molecule

11.1.1 The Born-Oppenheimer Approximation

In the preceding chapters, we dealt with the rotational and vibrational spectra of molecules and treated them in a "semi-classical" way. This means that we first investigated the motions of the atomic nuclei in the molecule using classical mechanics, and then applied empirically established quantisation rules. In this chapter, we justify that procedure by carrying out a strict quantum-mechanical calculation, based from the beginning on the Schrödinger equation. The various stages of approximation will be clarified in the process.

We begin with diatomic molecules; they will allow us to learn some important concepts which can later be applied to molecules with more than two atoms. As a first step, we show how the motions of the nuclei can be separated from those of the electrons. As we shall see, the electronic motions, which are quantum-mechanical in nature, mediate an interaction between the nuclei. The nuclear motions may in turn be separated to a good approximation into rotations and vibrations, and quantum mechanics also gives us the coupling terms between these motions.

The simplest examples of diatomic molecules are the hydrogen molecule-ion, H_2^+, and the hydrogen molecule, H_2. If we include different isotopes of hydrogen, the two atoms may have differing masses, m_1 and m_2. The atomic nuclei have coordinates denoted by $\boldsymbol{R}_1$ and $\boldsymbol{R}_2$. The following treatment could just as well be extended to molecules with several electrons, whose coordinates are given by $\boldsymbol{r}_1, \boldsymbol{r}_2, \ldots, \boldsymbol{r}_n$; however, in order to keep the procedure as clear as possible and the notation as simple as possible, we choose as an explicit example the hydrogen molecule-ion with a single electronic coordinate $\boldsymbol{r}$. The wavefunction Ψ then depends on the corresponding coordinates, i.e. $\boldsymbol{r}$, $\boldsymbol{R}_1$, and $\boldsymbol{R}_2$, and obeys a Schrödinger equation containing the following contributions: the kinetic energy of the electron and the two nuclei, the energy of the Coulomb interaction of the electron with the nuclei, and the Coulomb interaction energy between the nuclei. The Schrödinger equation is then given by

$$\left(-\frac{\hbar^2}{2m_0}\nabla^2 - \frac{\hbar^2}{2m_1}\nabla_1^2 - \frac{\hbar^2}{2m_2}\nabla_2^2 - \frac{e^2}{4\pi\varepsilon_0|\boldsymbol{r}-\boldsymbol{R}_1|} - \frac{e^2}{4\pi\varepsilon_0|\boldsymbol{r}-\boldsymbol{R}_2|}\right.$$
$$\left.+\frac{e^2}{4\pi\varepsilon_0|\boldsymbol{R}_1-\boldsymbol{R}_2|}\right)\Psi(\boldsymbol{r},\boldsymbol{R}_1,\boldsymbol{R}_2) = E\,\Psi(\boldsymbol{r},\boldsymbol{R}_1,\boldsymbol{R}_2)\ . \qquad (11.1)$$

The Laplace operator ∇ refers to the electron, while the two other Laplace operators refer to the two nuclei as denoted by their indices 1 and 2. The solution of this equation is a complicated many-body problem, which we can reduce considerably by applying some physical considerations. We use the fact that the masses of the nuclei are much greater than that of the electron: we can thus expect that the nuclei move much more slowly than the electron, or in other words, the electron can immediately follow the nuclear motions. This is the basic idea of the *Born-Oppenheimer approximation*.

In a first step, we find the wavefunction for the electronic motions while holding the nuclei fixed; the nuclear coordinates are then merely parameters in the electronic wavefunction. As we know, a direct interaction between the nuclei is caused by the electron(s), and is added to their Coulomb repulsion energy. In a second step, we then find the wavefunctions for the nuclear motions in the overall potential just described.

In order to carry out this program, we require a trial wavefunction Ψ which is expressed as a product, whose first factor represents the electronic motions with fixed nuclear coordinates, while the second factor takes the nuclear motions themselves into account. The trial function is thus given by:

$$\Psi(\boldsymbol{r},\boldsymbol{R}_1,\boldsymbol{R}_2) = \psi(\boldsymbol{r},\boldsymbol{R}_1,\boldsymbol{R}_2)\,\Phi(\boldsymbol{R}_1,\boldsymbol{R}_2)\ . \qquad (11.2)$$

If we insert this function into the Schrödinger equation (11.1), we must apply the product rule for differentiation, since the nuclear coordinates occur in both factors on the right-hand side of (11.2):

$$\frac{\partial^2}{\partial X_1^2}(\psi\Phi) = \psi\,\frac{\partial^2}{\partial X_1^2}\Phi + 2\frac{\partial\psi}{\partial X_1}\frac{\partial\Phi}{\partial X_1} + \Phi\,\frac{\partial^2}{\partial X_1^2}\psi\ . \qquad (11.3)$$

We then obtain in place of (11.1) the Schrödinger equation

$$\Phi\left(-\frac{\hbar^2}{2m_0}\nabla^2 - \frac{e^2}{4\pi\varepsilon_0|\boldsymbol{r}-\boldsymbol{R}_1|} - \frac{e^2}{4\pi\varepsilon_0|\boldsymbol{r}-\boldsymbol{R}_2|}\right)\psi$$
$$+\psi\left(-\frac{\hbar^2}{2m_1}\nabla_1^2 - \frac{\hbar^2}{2m_2}\nabla_2^2 + \frac{e^2}{4\pi\varepsilon_0|\boldsymbol{R}_1-\boldsymbol{R}_2|}\right)\Phi$$
$$-\frac{\hbar^2}{m_1}(\nabla_1\psi)\nabla_1\Phi - \frac{\hbar^2}{m_2}(\nabla_2\psi)\nabla_2\Phi - \frac{\hbar^2}{2m_1}\Phi\nabla_1^2\psi - \frac{\hbar^2}{2m_2}\Phi\nabla_2^2\psi = E\psi\Phi\ . \qquad (11.4)$$

Here, we have already collected the terms in a manner which will make them easier for us to interpret physically. The brackets in the first line clearly contain the Hamiltonian of the electron which is moving in the field of the *fixed* nuclear coordinates $\boldsymbol{R}_1$ and $\boldsymbol{R}_2$. We now choose the wavefunction ψ of the electron in such a way that it obeys the corresponding Schrödinger equation

$$\left(-\frac{\hbar^2}{2m_0}\nabla^2 - \frac{e^2}{4\pi\varepsilon_0|\boldsymbol{r}-\boldsymbol{R}_1|} - \frac{e^2}{4\pi\varepsilon_0|\boldsymbol{r}-\boldsymbol{R}_2|}\right)\psi = W\psi\ . \qquad (11.5)$$

As we know from the theory of the Schrödinger equation, it has in general a whole series of eigenvalues or energy values W, which we can distinguish by the values of the corresponding quantum numbers. Furthermore, the energy W also depends on the nuclear coordinates $\boldsymbol{R}_1$ and $\boldsymbol{R}_2$, which enter (11.5) as fixed parameters. We thus write

$$W = W(\boldsymbol{R}_1, \boldsymbol{R}_2) \ . \tag{11.6}$$

If we make a shift of the coordinates in (11.5):

$$\boldsymbol{r} \rightarrow \boldsymbol{r} + \boldsymbol{R}_1 \tag{11.7}$$

which is simply a shift of the coordinate origin, then we can see that the whole problem, including the determination of W, depends only on the difference of the nuclear coordinates

$$\boldsymbol{R}_1 - \boldsymbol{R}_2 \ . \tag{11.8}$$

We can thus take W to be a function of the nuclear coordinate difference

$$W = W(\boldsymbol{R}_1 - \boldsymbol{R}_2) \ . \tag{11.9}$$

As we shall show in more detail below, the last line on the left-hand side of (11.4) is a perturbation term, which is smaller than the first term by a factor of m_0/m_1 or m_0/m_2. We therefore initially neglect this term. The method we are applying here is called the Born-Oppenheimer approximation, as we mentioned above.

We now replace the left-hand side of (11.5) in the first line of (11.4) by its right-hand side, thus obtaining a Schrödinger equation which refers only to the nuclear coordinates:

$$\left[-\frac{\hbar^2}{2m_1}\nabla_1^2 - \frac{\hbar^2}{2m_2}\nabla_2^2 + \frac{e^2}{4\pi\varepsilon_0|\boldsymbol{R}_1 - \boldsymbol{R}_2|} + W(\boldsymbol{R}_1 - \boldsymbol{R}_2) \right] \Phi = E\Phi \ . \tag{11.10}$$

As we can see, a direct interaction energy between the nuclei comes about via the electronic energy which occurs in (11.5). This should not be surprising, since we already know from the theory of homopolar bonding (cf. Sects. 4.3 and 4.4) that a force between the nuclei is generated by the exchange of their electrons.

The Schrödinger equation (11.10) exhibits a formal analogy to the problem of the motion of an electron around a nucleus of finite mass. As we know from atomic physics, we can introduce new coordinates to simplify a problem of this type, namely the centre-of-mass coordinates:

$$\boldsymbol{R}_\mathrm{S} = (m_1\boldsymbol{R}_1 + m_2\boldsymbol{R}_2)/(m_1 + m_2) \tag{11.11}$$

and the relative coordinates:

$$\boldsymbol{R} = \boldsymbol{R}_1 - \boldsymbol{R}_2 \ . \tag{11.12}$$

In order to recalculate the Schrödinger equation for these new coordinates, we need also the mass at the centre of gravity,

$$m_\mathrm{S} = m_1 + m_2 \tag{11.13}$$

and the reduced mass m_r

$$m_\mathrm{r} = \frac{m_1 m_2}{m_1 + m_2} \ . \tag{11.14}$$

Using the corresponding transformation formulas, we can immediately write down the resulting Schrödinger equation for the wavefunction $\tilde{\Phi}(\boldsymbol{R}_S, \boldsymbol{R}) = \Phi(\boldsymbol{R}_1, \boldsymbol{R}_2)$:

$$\left(-\frac{\hbar^2}{2m_S}\nabla_S^2 - \frac{\hbar^2}{2m_r}\nabla_r^2 + V(\boldsymbol{R})\right)\tilde{\Phi}(\boldsymbol{R}_S, \boldsymbol{R}) = E\,\tilde{\Phi}(\boldsymbol{R}_S, \boldsymbol{R})\,, \tag{11.15}$$

where the operator for the potential energy is given explicitly by

$$V(\boldsymbol{R}) = \frac{e^2}{4\pi\varepsilon_0 R} + W(\boldsymbol{R})\,. \tag{11.16}$$

V is thus the sum of the energy of the electron in the field of the two nuclei and the Coulomb interaction energy of the nuclei with each other.

The separation of the coordinates into centre-of-mass and relative coordinates makes it possible for us to attempt a separation of the wavefunctions of the Schrödinger equation (11.15) into functions for the centre-of-mass and the relative motions:

$$\tilde{\Phi}(\boldsymbol{R}_S, \boldsymbol{R}) = \mathrm{e}^{\mathrm{i}\boldsymbol{K}\cdot\boldsymbol{R}_S}\chi(\boldsymbol{R})\,. \tag{11.17}$$

The energy eigenvalues of (11.15) then take on the form

$$E = \tilde{E} + \frac{\hbar^2 K^2}{2m_S}\,, \tag{11.18}$$

where $\tilde{E}$ results from the relative motion, while the second term represents the kinetic energy of the centre-of-mass motion.

We now turn to a discussion of the relative motion. As we know from the theory of electronic structure of diatomic molecules, the potential energy $V(\boldsymbol{R})$ depends only on the magnitude of the position vector; in other words, V is spherically symmetrical. In analogy to the treatment of the electronic motions in the hydrogen atom, we introduce polar coordinates at this point. The coordinates R, ϑ, and ϕ naturally refer here to the relative motions of the two nuclei, and we are only making use of a formal similarity to the hydrogen-atom problem. We transform the kinetic-energy operator to these polar coordinates and immediately carry out the separation of the wavefunctions into radial and angular parts, as is familiar from the hydrogen-atom problem:

$$\chi(\boldsymbol{R}) = f(R)F(\vartheta, \phi)\,. \tag{11.19}$$

We then obtain the Schrödinger equation

$$\left[-\frac{\hbar^2}{2m_r}\frac{1}{R^2}\frac{\partial}{\partial R}\left(R^2\frac{\partial}{\partial R}\right) + \frac{\boldsymbol{L}^2}{2m_r R^2} + V(\boldsymbol{R})\right] fF = \tilde{E}\, fF\,. \tag{11.20}$$

In this equation, $\boldsymbol{L}$ is the operator for the angular momentum. As we know from the hydrogen-atom problem, we can choose the angular functions F in such a way that they are at the same time eigenfunctions of the square of the angular momentum operator and of one of its components, e.g. the z-component:

$$\boldsymbol{L}^2 F_{J,M} = \hbar J(J+1)F_{J,M}\,, \tag{11.20a}$$

$$L_z F_{J,M} = \hbar M F_{J,M}\,, \tag{11.20b}$$

where we have used the standard notation for the angular momentum quantum numbers in molecules, J and M; they take on the roles of l and m in the hydrogen-atom problem. The quantisation rules here, as there, are given by:

$$J = 0, 1, 2, \ldots \tag{11.20c}$$

and

$$-J \leq M \leq J\,, \quad M \text{ an integer}\,. \tag{11.20d}$$

The angular dependence of F is illustrated for several cases in Fig. 4.1 in this book, as well as in I, Fig. 10.2. Due to (11.20a), we can replace $\boldsymbol{L}^2$ in (11.20) by $\hbar^2 J(J+1)$ and then divide F out of the resulting Schrödinger equation. We then obtain in place of (11.20) the Schrödinger equation

$$\left[-\frac{\hbar^2}{2m_\mathrm{r}}\frac{\partial^2}{\partial R^2} - \frac{\hbar^2}{m_\mathrm{r}}\frac{1}{R}\frac{\partial}{\partial R} + \frac{\hbar^2 J(J+1)}{2m_\mathrm{r}R^2} + V(\boldsymbol{R})\right] f = \tilde{E}\, f\,, \tag{11.21}$$

which refers only to the radial motion. Since the quantum number M does not occur in (11.21), but there are $2J+1$ different M values for each J according to (11.20d), and thus just as many different wavefunctions, the energy $\tilde{E}$ is $(2J+1)$-fold degenerate. In (11.21), we have rewritten slightly the kinetic-energy operator from (11.20).

The Schrödinger equation (11.21) is a second-order linear differential equation, which could be integrated using standard methods from the theory of differential equations. However, we are more interested here in elucidating the physical content of (11.21), in order to establish the connection to the results of Chaps. 9 and 10. To this end, we make two approximations, which are well justified in the present example. For one, we assume the expression

$$-\frac{\hbar^2}{m_\mathrm{r}}\frac{1}{R}\frac{\partial}{\partial R} \tag{11.22}$$

to be a small perturbation. Secondly, we search for a suitable explicit representation of the potential $V(R)$. The theory of molecular bonding indeed gives us such a representation; it is shown in Fig. 4.12. We of course are interested in stable bound states of the two nuclei, and so we need consider only values of the nuclear coordinates corresponding to internuclear distances in the neighbourhood of the energy minimum. We denote the resulting equilibrium internuclear distance by R_e. Expanding the potential function V in powers of the displacement $R - R_\mathrm{e}$, we obtain for the vicinity of the minimum the following expression:

$$V(R) \approx V(R_\mathrm{e}) + \tfrac{1}{2}k(R - R_\mathrm{e})^2\,. \tag{11.23}$$

We first consider the case of angular momentum $J = 0$. Neglecting (11.22) and using the approximation (11.23), we find that (11.21) becomes

$$\left[-\frac{\hbar^2}{2m_\mathrm{r}}\frac{\partial^2}{\partial R^2} + \tfrac{1}{2}k(R - R_\mathrm{e})^2\right] f = E_v\, f\,, \tag{11.24}$$

where we have set $\tilde{E} = E_v + V(R_\mathrm{e})$. Equation (11.24) is none other than the Schrödinger equation of a harmonic oscillator, which we already know well; in this case, the oscillator is centred around a point that is displaced from the coordinate origin by a distance R_e. The wavefunction of the ground state is thus given by

$$f(R) = N \exp[-\tfrac{1}{2}(R - R_e)^2/R_0^2] , \quad (11.25)$$

where N is a normalisation factor and R_0 is given by

$$R_0 = \sqrt{\frac{\hbar}{m_r \omega}} . \quad (11.26)$$

The energy values for the ground state and the excited states are then enumerated by the quantum number $v = 0, 1, \ldots$; the eigenvalues are

$$E_v = (v + \tfrac{1}{2})\hbar\omega , \quad (11.27)$$

with the frequency ω given by

$$\omega = \sqrt{\frac{k}{m_r}} . \quad (11.28)$$

We thus see that the two nuclei can carry out a harmonic oscillation along the axis joining their centres. Since the amplitude of this oscillation, as we know from experiment and can also calculate theoretically, is much smaller than the internuclear distance, it is certainly a good approximation to express the contribution from the rotational energy, proportional to $1/R^2$, through a replacement of R^2 by R_e^2. If we include states having $J \neq 0$, then the energy is given by:

$$\tilde{E} = V(R_e) + \left(v + \tfrac{1}{2}\right)\hbar\omega + \frac{\hbar^2 J(J+1)}{2m_r R_e^2} . \quad (11.29)$$

This energy expression is identical to that of (10.20), except for the first term in the sum, and the constants can be identified immediately. The first term comes from the binding energy of the nuclei at the equilibrium distance (and assuming infinite nuclear masses), and the second term from the vibrational energy, while the last term arises from the rotation. The rotational quantum number takes on the usual values, $J = 0, 1, \ldots$.

If we introduce the moment of inertia Θ, defined by

$$\Theta = m_r R_e^2 , \quad (11.30)$$

then (11.29) can also be written in the form

$$\tilde{E} = V(R_e) + \hbar\omega\left(v + \tfrac{1}{2}\right) + \frac{\hbar^2 J(J+1)}{2\Theta} . \quad (11.31)$$

The last expression in (11.31) is the quantum-mechanical analogue of the classical formula for the kinetic energy of a rotating dumbbell consisting of two masses $m_r/2$ at a distance of $2R_e$. We have thus reproduced the dynamics of a diatomic molecule, as they were already derived in the previous chapters as a model to explain the experimental data [cf. (10.20)]. Here, however, we have seen how a systematic quantum-mechanical treatment leads to the correct expressions for the wavefunctions and the energies, and which approximations were made in the process. The latter can be improved as a next step, e.g. by using a perturbation-theory approach. We then find the results which were introduced and discussed in Chaps. 9 and 10, and which for example take into account the anharmonicity. In the following section, we merely want to show that the terms which we neglected in this chapter are, in fact, small. The reader who wishes to hurry on can skip over these estimates of the relative importance of the neglected terms without missing information which is vital to understanding the rest of the book.

11.1.2 Justification of the Approximations

We begin with the expression (11.22). This operator is to be applied to a wavefunction, so that in order to estimate its importance, we must consider the whole expression, operator applied to wavefunction. We compare the resulting quantity arising from the term in (11.22) with the kinetic energy in the Hamiltonian (11.24). As we know from the theory of harmonic oscillators, the kinetic energy is of the same order of magnitude as the total energy, so that for the first term in (11.24), we can make the estimate:

$$-\frac{\hbar^2}{2m_r}\frac{\partial^2}{\partial R^2} f = \hbar\omega f \; . \tag{11.32}$$

For the corresponding expression resulting from (11.22), we obtain by using the explicit form in (11.25) an estimate:

$$-\frac{\hbar^2}{m_r}\frac{1}{R}\frac{\partial}{\partial R} f \approx \frac{\hbar^2}{m_r}\frac{1}{R_e}(R - R_e)\frac{m_r\omega}{\hbar} f \; , \tag{11.33}$$

which may be rewritten in the form

$$= \hbar\omega\frac{(R - R_e)}{R_e} f \; . \tag{11.34}$$

Here, the variable R has been weighted by the distribution function (11.25), according to which $R - R_e$ decreases as a quadratic exponential function over a distance R_0 as in (11.26). As can readily be seen from numerical examples, the relation

$$R - R_e \ll R_e \tag{11.35}$$

holds, so that we finally obtain, instead of (11.34), the estimate

$$(11.34) \ll \hbar\omega f \; . \tag{11.36}$$

Thus we have demonstrated that this term is considerably smaller than the term (11.32). A similar type of estimate also allows us to justify the replacement of R^2 by R_e^2 in the expression for the rotational energy. Let us now turn to the *justification of the Born-Oppenheimer approximation*. We consider the effect of the Laplace operator with respect to the nuclear coordinates $\boldsymbol{R}_1$ on the electronic wavefunctions:

$$\nabla_1^2 \psi \; . \tag{11.37}$$

We want to proceed as explicitly as possible and therefore use wavefunctions which are known from the theory of molecular bonding (cf. Sect. 4.3):

$$\psi = \phi_1(\boldsymbol{r} - \boldsymbol{R}_1) + \phi_2(\boldsymbol{r} - \boldsymbol{R}_2) \; , \tag{11.38}$$

where ϕ_1 and ϕ_2 are the electronic wavefunctions which are localised on atoms 1 and 2, respectively. Since ϕ_1 depends only on the relative coordinates $\boldsymbol{r} - \boldsymbol{R}_1$, we immediately obtain

$$\nabla_1^2 \psi = \nabla^2 \psi \; . \tag{11.39}$$

We now recall that on the average, the kinetic energy is of the same order as the total energy W:

$$-\frac{\hbar^2}{2m_0}\nabla^2\phi \approx W\phi \,. \tag{11.40}$$

However, since in (11.4) the perturbation term contained a factor $1/m_1$ and not a factor $1/m_0$, we multiply (11.40) by m_0/m_1 and obtain:

$$-\frac{\hbar^2}{2m_1}\nabla^2\phi \approx \frac{m_0}{m_1}W\phi \,. \tag{11.41}$$

From this we see that the energy which results from the perturbation term is a factor of m_0/m_1 smaller than the energy of the electron according to (11.40). If m_1 becomes sufficiently large, then the term (11.41) is arbitrarily small. Here, we still have to take account of the fact that W is independent of the nuclear masses m_1 and m_2. Since the Laplace operator ∇^2 means effectively that we multiply the wavefunctions by a factor 1/Length^2, we can formally take the square root of the left side of (11.41) to obtain an estimate for $\hbar\nabla\psi$:

$$\hbar\nabla\psi \approx \sqrt{m_0 W}\psi \,. \tag{11.42}$$

In precisely the same way, we obtain for the function of the nuclear coordinates an estimate of the order of magnitude:

$$\hbar\nabla_{R_1}\Phi \approx \sqrt{m_1 E_\mathrm{K}}\,\Phi \quad (E_\mathrm{K}: \text{ energy of the nuclear motions}) \,. \tag{11.43}$$

Using (11.42) and (11.43), we can multiply the left and the right sides by each other as follows:

$$\frac{\hbar^2}{2m_1}\nabla\psi\nabla_{R_1}\Phi \approx \sqrt{\frac{m_0}{m_1}}\sqrt{WE_\mathrm{K}}\,\psi\Phi \,. \tag{11.44}$$

We now recall the fact that the nuclei are oscillating and that the energy of the oscillations depends on the nuclear masses according to

$$E_\mathrm{K} \propto \hbar\omega \propto \frac{1}{\sqrt{m_1}} \,. \tag{11.45}$$

Furthermore, we recall the hydrogen atom, remembering that the electronic energy is proportional to the electron's mass:

$$W \propto m_0 \,. \tag{11.46}$$

If we now introduce (11.45) and (11.46) into (11.44), we can see that the left-hand side of the resulting expression is proportional to m_0/m_1, i.e. that this term can be neglected relative to the other terms which we considered in (11.4), presuming that the mass of the nucleus is sufficiently large; this is already the case to a good approximation even for the proton.

The estimates which we have just made may appear to the precise-minded reader to be somewhat superficial, since we have repeatedly estimated the magnitude of operators applied to wavefunctions. The estimates can, however, be made quite exact by calculating the corresponding expectation values.

11.2 The Rotation of Tri- and Polyatomic Molecules

11.2.1 The Expression for the Rotational Energy

In the previous section, we carried out a strict quantum-mechanical calculation for a diatomic molecule. In the process, we saw how assumptions which appeared reasonable in view of classical mechanics could be justified in detail. We also saw where approximations were to be made and how they could be put on a firm foundation. The insights we gained will now allow us to treat tri- and polyatomic molecules, letting ourselves be guided by the principles of classical mechanics without neglecting the correct quantum-mechanical procedures.

The starting point for our treatment is again the Schrödinger equation, which we write down for the example of a triatomic molecule; in the present chapter, we take only the nuclear motions into consideration. The potential energy then depends only on the relative coordinates, which we abbreviate as follows:

$$\boldsymbol{R}_{1,3} \equiv \boldsymbol{R}_3 - \boldsymbol{R}_1 , \; \boldsymbol{R}_{2,3} \equiv \boldsymbol{R}_3 - \boldsymbol{R}_2 , \; . \tag{11.47}$$

Here, $\boldsymbol{R}_j$, $j = 1, 2, 3$ is the coordinate of the nucleus having index j, as before. It is known from classical mechanics that conservation of the total linear momentum holds for such a system. This is also true in quantum mechanics, where only those quantities can be exactly measured simultaneously whose associated operators commute with one another. In our case, the momentum operator $\boldsymbol{P}$, given by

$$\boldsymbol{P} = \sum_j \boldsymbol{p}_j = \frac{\hbar}{\mathrm{i}} \sum_j \nabla_j , \tag{11.48}$$

commutes with the Hamiltonian of the Schrödinger equation:

$$\left[-\frac{\hbar^2}{2m_1} \nabla_1^2 - \frac{\hbar^2}{2m_2} \nabla_2^2 - \frac{\hbar^2}{2m_3} \nabla_3^2 + V(\boldsymbol{R}_{1,3}, \boldsymbol{R}_{2,3}) \right] \Phi = E \Phi . \tag{11.49}$$

One can readily convince oneself of this, since each differentiation of a coordinate commutes with another differentiation with respect to the same or another coordinate. $\boldsymbol{P}$ therefore certainly commutes with the Hamiltonian of the kinetic energy in (11.49). Furthermore, it is easy to see that $\boldsymbol{P}$ also commutes with the potential energy V, since the relation

$$\nabla_j V = -\nabla_k V , \quad j \neq k \tag{11.50}$$

holds, due to the dependence on the relative coordinates, and therefore the individual terms in the commutator of $\boldsymbol{P}$ with V cancel out. We can then split off the centre-of-mass motion with the wavefunction $\mathrm{e}^{\mathrm{i}\boldsymbol{K}\cdot\boldsymbol{R}}$ from the overall wavefunction Φ for the nuclear motion; here, $\hbar\boldsymbol{K}$ is the total linear momentum of the molecule.

Let us now consider the angular momentum. Here, we can easily verify that the angular momentum operator

$$L^2 = L_x^2 + L_y^2 + L_z^2 \tag{11.51}$$

commutes with the kinetic-energy operator in (11.49). Similarly, we can show that (11.51) commutes with the potential energy V, as long as V is invariant with respect to the common rotation of the relative coordinates $\boldsymbol{R}_{jk}$. We will assume this in the following. Let us now investigate the remaining degrees of freedom and take the triatomic molecule as an example.

Each nucleus carries three degrees of freedom, since we are assuming the nuclei to be point masses. The whole molecule then has 9 degrees of freedom, of which 3 refer to the motions of the centre of mass and 3 to the rotational motions. The remaining 3 degrees of freedom must then refer to the internal motions, which, analogously to the diatomic molecule, will be vibrations. We have already seen this in Sect. 10.5.

We can make these considerations intuitively clear in another way: the coordinates of the three nuclei define a plane which contains the centre of mass of the molecule. This plane can istelf rotate about two perpendicular axes, giving two degrees of freedom; in addition, there is the rotation of the overall nuclear coordinates within the plane. Thus, three oscillations in the plane also remain in this way of considering the molecule.

We turn now to the investigation of the rotational motions and the corresponding energies; this topic continues the discussion of Sect. 9.4. As we know from classical mechanics, a fundamental role in rotational motion is played by the *inertial tensor*:

$$\Theta = \begin{bmatrix} \Theta_{xx} & \Theta_{xy} & \Theta_{xz} \\ \Theta_{yx} & \Theta_{yy} & \Theta_{yz} \\ \Theta_{zx} & \Theta_{zy} & \Theta_{zz} \end{bmatrix} . \tag{11.52}$$

Its individual elements are given by the relations

$$\begin{aligned} \Theta_{xx} &= \sum_i m_i (Y_i^2 + Z_i^2) , \\ \Theta_{yy} &= \sum_i m_i (X_i^2 + Z_i^2) , \\ \Theta_{zz} &= \sum_i m_i (X_i^2 + Y_i^2) , \\ \Theta_{xy} &= \sum_i m_i X_i Y_i , \\ \Theta_{xz} &= \sum_i m_i X_i Z_i , \\ \Theta_{yz} &= \sum_i m_i Y_i Z_i . \end{aligned} \tag{11.53}$$

Since the nuclei are assumed to be point masses, the sum extends over the three nuclei with the point-like mass distributions m_i. Here, in analogy to the diatomic molecule, we assume that the nuclear coordinates lie at the potential minima and no oscillations are taking place, in the approximation we are presently considering (the vibrations will be investigated later). We choose the coordinate system for the triatomic molecule so that the plane of the molecule is the xy-plane and the z-axis is perpendicular to the molecular plane; then all the coordinates $Z_i = 0$. With the definitions (11.53), the inertial tensor then reduces to

$$\Theta = \begin{bmatrix} \Theta_{xx} & \Theta_{xy} & 0 \\ \Theta_{yx} & \Theta_{yy} & 0 \\ 0 & 0 & \Theta_{zz} \end{bmatrix} . \tag{11.54}$$

By rotating the coordinate system around the z-axis, we can cause the non-diagonal elements in (11.54) to vanish, so that we reduce the tensor to its principal-axis values:

$$\Theta = \begin{bmatrix} \Theta_{x} & 0 & 0 \\ 0 & \Theta_{y} & 0 \\ 0 & 0 & \Theta_{z} \end{bmatrix} . \tag{11.55}$$

(This form of the inertial tensor can always be obtained, even for polyatomic molecules.)

In classical mechanics, the angular momentum vector is related to the vector of the angular velocity of rotation by the equation:

$$\boldsymbol{L} = \Theta \boldsymbol{\omega} \ . \tag{11.56}$$

In this expression, $\boldsymbol{L}$ and $\boldsymbol{\omega}$ are vectors, while Θ is a tensor of rank 2, i.e. a matrix. The kinetic energy may then be written in the form:

$$E_{\text{rot}} = \tfrac{1}{2}\tilde{\boldsymbol{\omega}}\Theta\boldsymbol{\omega} \ , \tag{11.57}$$

where $\tilde{\boldsymbol{\omega}}$ is the transposed vector $\boldsymbol{\omega}$. In the case that Θ is diagonal, the following relations are valid:

$$\begin{aligned} L_x &= \Theta_x \omega_x \ , \\ L_y &= \Theta_y \omega_y \ , \\ L_z &= \Theta_z \omega_z \ . \end{aligned} \tag{11.58}$$

$$E_{\text{rot}} = \frac{1}{2}\frac{L_x^2}{\Theta_x} + \frac{1}{2}\frac{L_y^2}{\Theta_y} + \frac{1}{2}\frac{L_z^2}{\Theta_z} \ . \tag{11.59}$$

The total angular momentum $\boldsymbol{L}$ in classical mechanics is given by a vector sum of the angular momenta of the individual atoms:

$$\boldsymbol{L} = \sum_i \boldsymbol{L}_i \equiv \sum_i [\boldsymbol{R}_j \, , \boldsymbol{P}_j] \ . \tag{11.60}$$

In quantum mechanics, the momenta $\boldsymbol{P}_i$ are replaced in the well-known manner according to Jordan's rule by the gradient operators:

$$\boldsymbol{P}_k \rightarrow \frac{\hbar}{\mathrm{i}} \nabla_k \ . \tag{11.61}$$

For the angular momenta of each nucleus, the usual commutation relations hold:

$$L_{x,i} L_{y,i} - L_{y,i} L_{x,i} = \mathrm{i}\hbar L_{z,i} \tag{11.62}$$

and, in addition, the relations found from cyclic permutations of the indices i, j and k. For *different* nuclei, the angular momentum operators all mutually commute. The relations (11.62) can be applied equally well to the total angular momentum operator $\boldsymbol{L}$ (11.60), as one can readily calculate.

11.2.2 The Symmetric Top

Let us see what the quantisation of angular momentum means for the energy E_{rot} (11.59). We first consider the case that the moments of inertia obey the relation

$$\Theta_x = \Theta_y \ . \tag{11.63}$$

As a result of the equation

$$\boldsymbol{L}^2 = L_x^2 + L_y^2 + L_z^2 \ , \tag{11.64}$$

which is also valid for the operators, we can write (11.59) in the form

$$H_{\text{rot}} = \frac{1}{2}\frac{1}{\Theta_x}(\boldsymbol{L}^2 - L_z^2) + \frac{1}{2}\frac{1}{\Theta_z}L_z^2 , \tag{11.65}$$

where we have written H (for the Hamiltonian operator) instead of the energy E. Since the operators $\boldsymbol{L}^2$ and L_z commute with each other and with the Hamiltonian, we can choose the wavefunctions for the Schrödinger equation in such a way that they are simultaneously eigenfunctions of all three operators. In this way, we arrive again at (11.20a) and (11.20b); following the usual convention, we denote the quantum numbers belonging to L_z by k, with $-J \leq k \leq J$ and k an integer. Often, the notation $K = |k|$ is used. The quantity $\hbar k$ can be interpreted as the component of the total angular momentum along the z direction, which was chosen to coincide with the principal z-axis of the molecule. We now imagine that the operator for E_{rot} is applied to the corresponding wavefunction $F_{J,K}$; then we can replace the operators $\boldsymbol{L}^2$ and L_z^2 by their eigenvalues $\hbar^2[J(J+1)]$ and $\hbar^2 k^2$. Equation (11.65) thus becomes

$$E_{\text{rot}} = \frac{1}{2}\frac{\hbar^2}{\Theta_x}[J(J+1) - K^2] + \frac{1}{2}\hbar^2\frac{K^2}{\Theta_z} , \quad 0 \leq K \leq J , \tag{11.66}$$

and we have thus succeeded in calculating the rotational energy.

Since each value of K refers to the quantum numbers $k = K$ and $k = -K$, with the corresponding wavefunctions, the energy is doubly degenerate for $K \neq 0$ and only non-degenerate for $K = 0$. While the quantum number k relates to the projection of the angular momentum onto the molecule-fixed z axis, the quantum number M, that we found in (11.20b), refers to the projection onto the space-fixed axis (which we will now call the z'-axis). This leads to the question as to whether an M quantum number is also valid in the present case. We consider a rotation of the coordinate system between the space-fixed axes x', y', and z' and the molecule-fixed axes x, y, z. In particular, for the z'-component of the angular momentum, $L_{z'}$, we find:

$$L_{z'} = aL_x + bL_y + cL_z , \tag{11.66a}$$

where the constants a, b, and c depend on the rotation angles. Relation (11.66a) initially holds for the classical vectors, but it can be transferred directly to the quantum-mechanical calculation, so that it also holds for the angular momentum operators.

We now ask whether we can measure the square of the angular momentum (corresponding to the operator $\boldsymbol{L}^2$) and also its component along the z'-axis (operator $L_{z'}$) exactly and simultaneously. As we know from I, Sect. 10.2, $\boldsymbol{L}^2$ commutes not only with L_z, but also with L_x and L_y and thus also with the linear combination (11.66a). $\boldsymbol{L}^2$ and $L_{z'}$ thus can be simultaneously and exactly measured. The eigenvalue of the operator $\boldsymbol{L}^2$ is of course $\hbar^2 J(J+1)$, while we denote the eigenvalue of $L_{z'}$ again by $\hbar M$, in agreement with the previous notation. As we showed in I, it follows from the operator algebra for L_x, L_y, and L_z (or, equivalently, for $L_{x'}$, $L_{y'}$, and $L_{z'}$) and from the requirement of continuity of the wavefunctions, that $0 \leq J$ and J is an integer, while $-J \leq M \leq J$, with M also an integer. The quantum number M can be measured by applying an electric or a magnetic field to the molecules along the z' direction. Since however $L_{z'}$ does not commute with L_z^2 (except when $a = b = 0$), but the energy operator depends upon L_z^2, then in this case (of an applied field along the z'-axis), the energy (11.65) is no longer exactly measureable.

11.2.3 The Asymmetric Top

We now take up the case that the moments of inertia of the molecule along the principal axes are all different. The kinetic energy of the rotational motion can be written as

$$H_{\text{rot}} = A_x L_x^2 + A_y L_y^2 + A_z L_z^2 \,, \tag{11.67}$$

where we have used the abbreviation

$$A_j = \tfrac{1}{2}\Theta_j^{-1} \,, \quad j = x, y, z \,. \tag{11.68}$$

We introduce creation and annihilation operators for the z-component of the angular momentum, which we define as in I, Sect. 10.2, using the equation:

$$L_\pm = L_x \pm L_y \,. \tag{11.69}$$

The application of the operator L_+ to a wavefunction with angular momentum quantum number k transforms it into a wavefunction with the quantum number $k+1$, and the operator L_- reduces k by one unit in a corresponding manner. Equation (11.69) may be resolved into the operators for L_x and L_y as follows:

$$L_x = \tfrac{1}{2}(L_+ + L_-) \tag{11.70}$$

and

$$L_y = \tfrac{1}{2}(L_+ - L_-) \,. \tag{11.71}$$

If we insert L_x and L_y from these equations into (11.67), rearrange and introduce the abbreviations

$$\alpha = \tfrac{1}{2}(A_x - A_y) \,, \quad \beta = \tfrac{1}{2}(A_x + A_y) \,, \tag{11.72}$$

then we find the following expression for the Hamiltonian of the rotation:

$$H_{\text{rot}} = \alpha(L_+^2 + L_-^2) + \beta(L_+L_- + L_-L_+) + A_z L_z^2 \,. \tag{11.73}$$

We now make use of the property that (11.69) raises or lowers the quantum number k by one unit. Applying the corresponding operator twice to the angular-momentum eigenfunctions $F_{J,k}$, we obtain

$$L_+^2 F_{J,k} = \hbar^2 a_{J,k} F_{J,k+2} \tag{11.74}$$

and

$$L_-^2 F_{J,k} = \hbar^2 b_{J,k} F_{J,k-2} \,. \tag{11.75}$$

The coefficients a and b are given explicitly by

$$\begin{aligned} a_{J,k} &= \sqrt{J(J+1) - k(k+1)}\sqrt{J(J+1) - (k+1)(k+2)} \,, \\ b_{J,k} &= \sqrt{J(J+1) - k(k-1)}\sqrt{J(J+1) - (k-1)(k-2)} \,, \end{aligned} \tag{11.76}$$

as one can verify by calculation using the formulas (10.36) and (10.37) from I.

Furthermore, we of course have

$$L_z^2 F_{J,k} = \hbar^2 k^2 F_{J,k} \tag{11.77}$$

and, because of $\boldsymbol{L}^2 - L_z^2 = L_x^2 + L_y^2 = \frac{1}{2}(L_+L_- + L_-L_+)$, the eigenvalue equation:

$$\frac{1}{2}(L_+L_- + L_-L_+)F_{J,k} = \hbar^2[J(J+1) - k^2]F_{J,k} \tag{11.78}$$

holds. We now have the task of finding the eigenvalues and eigenfunctions of the operator (11.73), i.e. we must solve the Schrödinger equation

$$H_{\text{rot}} f(\vartheta, \phi) = \hbar^2 \lambda f(\vartheta, \phi) \,. \tag{11.79}$$

Here, we have expressed the eigenvalue on the right-hand side in the special form $\hbar^2\lambda$. Due to the appearance of the creation and annihilation operators in H_{rot} (11.73), we cannot simply assume that k is a good quantum number; instead, we expect linear combinations of the form

$$f = \sum_{k=-J}^{+J} c_J F_{J,k} \,. \tag{11.80}$$

In calculating these eigenfunctions, we can limit ourselves to functions $F_{J,k}$ with *fixed* values of J, since $\boldsymbol{L}^2$ and H_{rot} commute. Inserting (11.80) into (11.79) and using equations (11.74–78), we find the relation

$$\begin{aligned}\sum_{k=-J}^{+J} \big(&\alpha c_k a_{J,k} F_{J,k+2} + \alpha c_k b_{J,k} F_{J,k-2} \\ &+ \{2\beta[J(J+1) - k^2] + A_z k^2 - \lambda\} c_k F_{J,k}\big) = 0 \,,\end{aligned} \tag{11.81}$$

in which we have transferred the eigenvalue λ to the left-hand side and have divided out the common factor $\hbar^2$. The coefficients α, β, and A_z are contained in the Hamiltonian H_{rot}; $a_{J,k}$ and $b_{J,k}$ are defined by (11.76). In the following, we also use the abbreviations

$$g_{J,k} = \beta d_{J,k} + A_z k^2 \equiv 2\beta J(J+1) + (A_z - 2\beta)k^2 \,. \tag{11.82}$$

In the following equations, since J has a fixed value, we leave it out of the notation; i.e. we make the replacements:

$$\begin{aligned} a_{J,k} &\to a_k \,, \\ b_{J,k} &\to b_k \,, \\ g_{J,k} &\to g_k \,. \end{aligned} \tag{11.83}$$

Since the angular-momentum eigenfunctions F for different values of the quantum number k are linearly independent, a relation of the type (11.81) can be satisfied in general only if the coefficients of the same angular-momentum eigenfunctions vanish individually. Renumbering the indices, we obtain the equations

$$\alpha c_{k-2} a_{k-2} + \alpha c_{k+2} b_{k+2} + (g_k - \lambda) c_k = 0 \,. \tag{11.84}$$

Here, we have to remember that due to (11.76), the coefficients a_k and b_k can be equal to zero. In order to illustrate the meaning of (11.84), we consider some special cases, from which the reader can draw conclusions about the general method.

For the case that the angular-momentum quantum number $J = 0$, we find directly:

$$J = 0 \qquad (g_0 - \lambda)c_0 = 0 \,, \tag{11.85}$$

i.e. the eigenvalue is given simply by g_0, and because of (11.82) with $J = k = 0$, we have $g_0 = 0$. We therefore find that the rotational energy $E_{\text{rot}} = 0$.

For $J = 1$, there are two possibilities, depending on whether k is even or odd. If k is even, then the only possibility is that $k = 0$ and the corresponding equation is:

$$J = 1 \qquad (g_0 - \lambda)c_0 = 0 \,. \tag{11.86}$$

In this case, from (11.82), $g_0 \equiv g_{1,0} = 4\beta$, so that after reversing the replacements (11.72) and (11.68), we have

$$E_{\text{rot}} = \frac{\hbar^2}{2}\left(\frac{1}{\Theta_x} + \frac{1}{\Theta_y}\right) \,.$$

In contrast, if k is odd, we obtain two equations from (11.84), having the form:

$$\begin{aligned} (g_1 - \lambda)c_1 + 0 \quad &= 0 \,, \\ 0 + (g_{-1} - \lambda)c_{-1} &= 0 \,, \end{aligned} \tag{11.87}$$

where $g_1 \equiv g_{1,1}$ and $g_{-1} \equiv g_{1,-1}$ yield $g_1 = g_{-1} = 2\beta + A_z$. The eigenvalues which result from (11.87) are the same, $\lambda = 2\beta + A_z$; i.e. the energy $E_{\text{rot}} = \hbar^2\lambda$ is doubly degenerate and is equal to

$$E_{\text{rot}} = \frac{\hbar^2}{4}\left(\frac{1}{\Theta_x} + \frac{1}{\Theta_y}\right) + \frac{\hbar^2}{2}\frac{1}{\Theta_z} \,.$$

Now we examine the case of $J = 2$. The Eqs. (11.84) divide into one set for even k and one set for odd k. For even k we find the three coupled equations

$$\begin{aligned} J = 2 \,, \qquad &(g_2 - \lambda)c_2 + \alpha c_0 a_0 = 0 \,, \\ &\alpha b_2 c_2 + (g_0 - \lambda)c_0 + \alpha c_{-2} a_{-2} = 0 \,, \\ &\alpha c_0 b_0 + (g_{-2} - \lambda)c_0 = 0 \,, \end{aligned} \tag{11.88}$$

where the following relations for the coefficients hold:

$$g_2 \equiv g_{2,2} = 4\beta + 4A_z \,, \quad g_0 \equiv g_{2,0} = 12\beta \,, \quad g_{-2} \equiv g_{2,-2} = 4\beta + 4A_z \,,$$

$$a_0 \equiv a_{2,0} = 2\sqrt{6} \,, \quad a_{-2} \equiv a_{2,2} = 2\sqrt{6} \,,$$

$$b_0 \equiv b_{2,0} = 2\sqrt{6} \,, \quad b_2 \equiv b_{2,2} = 2\sqrt{6} \,.$$

These equations have a non-trivial solution only if the determinant

$$\begin{vmatrix} g_2 - \lambda & \alpha a_0 & 0 \\ \alpha b_2 & g_0 - \lambda & \alpha a_{-2} \\ 0 & \alpha b_0 & g_{-2} - \lambda \end{vmatrix} = 0 \tag{11.89}$$

vanishes, which, as always, yields a secular equation for the eigenvalues λ. With the known coefficients, this third-order equation for λ can be readily solved.

In the case that k is odd, Eqs. (11.84) split into two uncoupled equations:

$$(g_1 - \lambda)c_1 = 0 \tag{11.90}$$

and

$$(g_{-1} - \lambda)c_{-1} = 0 \,, \tag{11.91}$$

with

$$\begin{aligned} g_1 &\equiv g_{2,1} &= 10\beta + A_z \,, \\ g_{-1} &\equiv g_{2,-1} &= 10\beta + A_z \,, \end{aligned}$$

from which again the eigenvalues λ can be read off directly. Finally, for $J = 3$, the system of Eqs. (11.84) splits into equations for the case that k is even or odd. For the latter, we find a system of equations whose determinant of coefficients has the form:

$$\begin{vmatrix} x-\lambda & x & 0 & 0 \\ x & x-\lambda & x & 0 \\ 0 & x & x-\lambda & x \\ 0 & 0 & x & x-\lambda \end{vmatrix} = 0 \,. \tag{11.92}$$

The symbol x indicates here that non-zero coefficients occur at the corresponding positions in the determinant. From this form, the general rule for setting up the equations and the associated secular determinant can be seen. Non-zero terms occur only along the main diagonal and the neighbouring two subdiagonals. This, by the way, makes it possible to solve the system of equations in a step-by-step manner. The last equation represents a condition of self-consistency, which is equivalent to the vanishing of the determinant (11.92). In this way, it is possible to calculate the energy eigenvalues $\hbar^2\lambda$ and the corresponding eigenfunctions for any desired value of the quantum number J. We have thus completed our task of computing the rotational energies of the molecule.

In the next section, we turn to molecular vibrations.

11.3 The Vibrations of Tri- and Polyatomic Molecules

In the diatomic molecule, we saw that the nuclei can undergo *harmonic* oscillations about their equilibrium positions, at least for small amplitudes. The molecule thus behaves as if the two point masses were joined together by an elastic spring. We now examine a molecule with three or more atoms, and consider the vibrations of its nuclei. Our goal is to derive the Hamiltonian for these vibrations; this will be simpler if we first treat the vibrations using classical mechanics. We thus obtain equations for coupled oscillators, which can be decoupled by introducing the normal coordinates. As a result, we can represent the Hamilton function as the sum of individual Hamilton functions (one for each normal mode) and can easily carry out the quantisation.

As we know from experiments, and as can be calculated from the quantum-mechanical theory of the bonding forces in molecules, the nuclei maintain average distances from one another, and these distances are determined by minima in the potential energy curve. They can undergo *harmonic* oscillations about these equilibrium positions, as long as the displacements are not too great. The potential energy V of the M nuclei, which depends on their positions as given by the vectors $\boldsymbol{R}_j$, $j = 1, \ldots, M$, can be expanded in a series about the equilibrium positions $\boldsymbol{R}_{j,0}$, with the displacements denoted as $\boldsymbol{\xi}_j$:

$$\boldsymbol{R}_j = \boldsymbol{R}_{j,0} + \boldsymbol{\xi}_j \,, \quad j = 1, \ldots, M \,. \tag{11.93}$$

We thus obtain the following expansion, up to terms of second order in $\boldsymbol{\xi}_j$:

$$V(\boldsymbol{R}_1, \boldsymbol{R}_2, \dots, \boldsymbol{R}_M) = V(\boldsymbol{R}_{1,0}, \boldsymbol{R}_{2,0}, \dots, \boldsymbol{R}_{M,0}) + \sum_{j=1}^{M} (\nabla_j V)\, \boldsymbol{\xi}_j$$

$$+ \frac{1}{2} \sum_{j,j',l,l'} \frac{\partial^2 V}{\partial \xi_{jl} \partial \xi_{j'l'}} \xi_{jl} \xi_{j'l'} \,. \tag{11.94}$$

The derivatives $\nabla_j V$ and $\partial^2 V / \partial \xi_{jl} \partial \xi_{j'l'}$ are of course to be evaluated at the positions $\boldsymbol{R}_j = \boldsymbol{R}_{j,0}$. Since these are the equilibrium positions, we have

$$\nabla_j V = 0 \,, \tag{11.95}$$

so that the second term on the right-hand side of (11.94) vanishes. In the third term of (11.94), the indices l and l' denote the coordinates x, y, and z, i.e. for example, ξ_{jx} means the x-component of the displacement vector $\boldsymbol{\xi}_j$. The sum over j, j' runs over $j = 1, \dots, M$ and $j' = 1, \dots, M$. Since $V(\boldsymbol{R}_{1,0}, \dots, \boldsymbol{R}_{M,0})$ is a constant, independent of the $\boldsymbol{\xi}$'s, the remaining part of the potential energy of the nuclear vibrations which interests us is

$$W(\boldsymbol{\xi}_1, \dots, \boldsymbol{\xi}_M) = \frac{1}{2} \sum_{j,j',l,l'} \frac{\partial^2 V}{\partial \xi_{jl} \partial \xi_{j'l'}} \xi_{jl} \xi_{j'l'} \,. \tag{11.96}$$

In order to be able to write W in a simple form, we introduce new variables η, defined by:

$$\begin{aligned} \xi_{l,x} &\to \eta_{3(l-1)+1} \,, \\ \xi_{l,y} &\to \eta_{3(l-1)+2} \,, \\ \xi_{l,z} &\to \eta_{3(l-1)+3} \,, \end{aligned} \tag{11.97}$$

which clearly requires us to change the numbering of the indices so that the index k of η_k takes on the following values:

$$k = 1, \dots, 3M = \tilde{M} \,. \tag{11.98}$$

For W, we obtain in this way a potential expression of the form

$$W = \tfrac{1}{2} \sum_k \eta_k A_{kk'} \eta_{k'} \,, \tag{11.99}$$

with

$$A_{kk'} = \frac{\partial^2 V}{\partial \eta_k \partial \eta_{k'}} \,. \tag{11.100}$$

From this, it follows immediately that $A_{kk'}$ is a symmetric matrix. To find the Hamilton function, we still need the kinetic energy, which is given by

$$E_{\text{kin}} = \tfrac{1}{2} \sum_j m_j \dot{\boldsymbol{\xi}}_j^2 \,. \tag{11.101}$$

It is reasonable to use the new definitions (11.97) here, also; this can be done easily with the aid of the relation:

$$\tilde{m}_k = m_{[(k-1)/3+1]} \,. \tag{11.102}$$

The square brackets in (11.102) mean that one is to take the largest whole number corresponding to the quotient $(k-1)/3$:

$$\text{For } k=1\,,\quad \text{we find } \left[\frac{k-1}{3}\right] \equiv \left[\frac{0}{3}\right] = 0\,;$$

$$\text{for } k=2\,,\quad \text{we find } \left[\frac{1}{3}\right] = 0\,;$$

$$\text{for } k=3\,,\quad \text{we find } \left[\frac{2}{3}\right] = 0\,;\ \text{and only}$$

$$\text{for } k=4\,,\quad \text{do we find } \left[\frac{3}{3}\right] = 1\,;$$

$$\text{for } k=5\,,\quad \text{we find } \left[\frac{4}{3}\right] = 1\,,\ \text{etc.}\,.$$

Inserting the masses $\tilde{m}_k$ defined in (11.102), we can write the kinetic energy of the vibrations of the nuclei in a readily understandable way as

$$E_{\text{kin}} = \frac{1}{2}\sum_k \tilde{m}_k \dot{\eta}_k^2\,. \tag{11.103}$$

Using (11.99) and (11.103), we obtain as our first important result the Hamilton function:

$$H = E_{\text{kin}} + W = \frac{1}{2}\sum_k \tilde{m}_k \dot{\eta}_k^2 + W\,, \tag{11.104}$$

in which we should still express the velocities $\dot{\eta}_k$ in terms of the corresponding momenta, $p_k = \tilde{m}_k\dot{\eta}_k$, in keeping with the spirit of the Hamilton formalism. Since W is quadratic or bilinear in the displacements η_k, equation (11.104) represents the energy of coupled oscillators. Their equations of motion are:

$$\tilde{m}_k\ddot{\eta}_k = -\frac{\partial W}{\partial \eta_k} \tag{11.105}$$

or, when we insert W from (11.99),

$$\tilde{m}_k\ddot{\eta}_k = -\sum_{k'=1}^{3M} A_{kk'}\eta_{k'}\,, \tag{11.106}$$

where $A_{kk'}$ is given by (11.100).

The coefficients $A_{kk'}$ in (11.99) can be combined into a matrix A according to:

$$A = (A_{kk'})\,, \tag{11.107}$$

and likewise the masses $\tilde{m}$ in (11.103), which form a diagonal matrix $\tilde{m}$:

$$\tilde{m} = \begin{pmatrix} \tilde{m}_1 & & \\ & \tilde{m}_2 & \\ & & \ddots \end{pmatrix}\,. \tag{11.108}$$

W and E_{kin} then take on the simple forms

$$W = \frac{1}{2}\boldsymbol{\eta}^{\mathrm{T}} A\, \boldsymbol{\eta} \tag{11.109}$$

and

$$E_{\mathrm{kin}} = \frac{1}{2}\dot{\boldsymbol{\eta}}^{\mathrm{T}} \tilde{m}\, \dot{\boldsymbol{\eta}}\,. \tag{11.110}$$

Our next goal is to carry out a transformation of the vector $\boldsymbol{\eta}$ into another vector which will allow W to be expressed in a form analogous to (11.109), but with the matrix A in diagonal form. The first difficulty is that when we transform the $\boldsymbol{\eta}$ to new coordinates, we have no guarantee that E_{kin} will remain in the form (11.110) with $\tilde{m}$ as a diagonal matrix. In order to carry out a simultaneous diagonalisation of W and E_{kin}, we resort to a trick, by making the following transformation to a new vector $\boldsymbol{\zeta}$:

$$\tilde{m}^{1/2}\boldsymbol{\eta} = \boldsymbol{\zeta}\,. \tag{11.111}$$

The square root of the diagonal matrix $\tilde{m}$ is defined simply as a new diagonal matrix whose elements are the square roots of the diagonal elements of the original matrix:

$$\tilde{m}^{1/2} = \begin{bmatrix} \tilde{m}_1^{1/2} & & \\ & \tilde{m}_2^{1/2} & \\ & & \ddots \end{bmatrix}\,. \tag{11.112}$$

Inserting (11.111) in (11.109), we find

$$W = \frac{1}{2}\boldsymbol{\zeta}^{\mathrm{T}} \tilde{m}^{-1/2} A\, \tilde{m}^{-1/2} \boldsymbol{\zeta}\,. \tag{11.113}$$

In the following, we abbreviate the matrix product in (11.113) as

$$\tilde{m}^{-1/2} A\, \tilde{m}^{-1/2} = B\,. \tag{11.114}$$

As can easily be verified, B, like A, is a symmetric matrix. With these changes, (11.109) and (11.110) take on the form

$$W = \frac{1}{2}\boldsymbol{\zeta}^{\mathrm{T}} B\, \boldsymbol{\zeta} \tag{11.115}$$

and

$$E_{\mathrm{kin}} = \frac{1}{2}\dot{\boldsymbol{\zeta}}^2\,. \tag{11.116}$$

In all these considerations, we must take into account the fact that $\boldsymbol{\eta}$ or $\boldsymbol{\zeta}$ are simply time-dependent vectors, just as the variables $\boldsymbol{\xi}$ from which they are derived. It is easy to see that the equations of motion (11.107) for the η_k take on the following form as equations for the ζ_k:

$$\ddot{\zeta}_k = -\sum_{k'} B_{kk'}\, \zeta_{k'} \tag{11.117}$$

or, written as a matrix equation,

$$\ddot{\boldsymbol{\zeta}} = -B\, \boldsymbol{\zeta}\,. \tag{11.118}$$

To solve this differential equation, we use the trial function

$$\boldsymbol{\zeta}(t) = \mathrm{e}^{\mathrm{i}\omega t} \boldsymbol{\zeta}(0) + \mathrm{c.c.}\,. \tag{11.119}$$

The equations represented by (11.118) are linear, so it is sufficient to substitute the first part of (11.119) into (11.118), to determine ω and $\zeta(0)$, and only then to include the complex conjugate part of the solution from (11.119), so that ζ will be real. We thus insert $\zeta(t) = e^{i\omega t}\zeta(0)$ into (11.118), carry out the differentiation with respect to time, and divide out the common factor $e^{i\omega t}$. We are then left with the linear algebraic equations

$$\omega^2 \zeta(0) = \boldsymbol{B}\,\zeta(0) \ . \tag{11.120}$$

They have a non-trivial solution only if the determinant

$$\det(\boldsymbol{B} - \omega^2\,\mathbf{1}) = 0 \tag{11.121}$$

vanishes. Here, $\mathbf{1}$ is the unit matrix. A thorough mathematical analysis shows that the frequencies ω are, in fact, real. We of course might have expected this from the beginning, since for our system, energy conservation must hold and a complex quantity ω would mean that the system were damped or would show exponential amplitude growth, which would certainly be in contradiction to the energy conservation theorem.

Equations of the form (11.120) lead, as is well known, to a whole set of frequencies, the so-called eigenfrequencies, which in general are different from one another. Each eigenfrequency corresponds to a particular eigenvector $\zeta(0)$, and thus, according to (11.111), to a particular vector $\boldsymbol{\eta}(0)$; they are uniquely determined up to a constant factor (as long as degeneracies do not occur). The vibrations which are thus described are the *normal modes*. If we decompose the time-dependent vectors ζ or $\boldsymbol{\eta}$ into their components, then it follows from (11.119) that all the components oscillate with the same frequency; this is a property typical of the normal modes. For the determination of the eigenvectors and their symmetry behaviour, group theory, which we introduced in Chap. 6, again proves to be a fundamental aid. We shall discuss this point in more detail in Sect. 11.4.

We now distinguish the frequencies and eigenvectors by means of the index k, so that instead of (11.119), we can more precisely write

$$\zeta_k(t) = e^{i\omega_k t}\zeta_k(0) + \text{c.c.} \ . \tag{11.122}$$

As a more detailed analysis shows, three of the eigenfrequencies vanish, owing to the fact that the equations allow solutions which correspond to a uniform translation of all the nuclei at the same time. In what follows, we ignore these frequencies and their eigenfunctions, since they were already included in the treatment of the centre-of-mass motion.

As we already noted, A and therefore B are symmetric matrices. The eigenfunctions of symmetric matrices have the property that they obey the orthogonality relations

$$\zeta_j^{\mathrm{T}} \cdot \zeta_k = \delta_{jk} \begin{cases} 0 & \text{for } j \neq k \\ 1 & \text{for } j = k \end{cases} \ . \tag{11.123}$$

We expand the general vector $\zeta(t)$ as a superposition of the eigenvectors $\zeta(0)$:

$$\zeta(t) = \sum_j \beta_j(t)\,\zeta_j(0) \ . \tag{11.124}$$

Since the eigenvalues ω of (11.120) occur in pairs with positive and negative signs, we must assume that the time-dependent function β_j has the general form

$$\beta_j = a_j e^{i\omega_j t} + b_j e^{-i\omega_j t} \ . \tag{11.125}$$

We now recalculate the expression for the kinetic energy E_{kin} in terms of the variables β_j by using the expansion (11.124); we find

$$E_{\text{kin}} = \tfrac{1}{2}\dot{\zeta}^2 = \tfrac{1}{2}\dot{\zeta}^{\mathrm{T}} \cdot \dot{\zeta} = \tfrac{1}{2}\sum_j \dot{\beta}_j \zeta_j^{\mathrm{T}}(0) \sum_k \dot{\beta}_k \zeta_k(0) , \tag{11.126}$$

or, if we make use of the orthogonality relations (11.123),

$$\tfrac{1}{2}\sum_j \dot{\beta}_j^2 . \tag{11.127}$$

We proceed in a similar way with W, by inserting (11.124) into (11.115) and thus obtaining

$$W = \tfrac{1}{2}\sum_j \beta_j(t)\zeta_j^{\mathrm{T}}(0)\, B \sum_k \beta_k(t)\zeta_k(0) . \tag{11.128}$$

We take into account the fact that the relations (11.120) are fulfilled for each value of the index k; i.e.

$$\boldsymbol{B}\zeta_k(0) = \omega_k^2 \zeta_k(0) , \tag{11.129}$$

and finally apply the orthogonality relations (11.123). We can then express W in the simple form:

$$W = \tfrac{1}{2}\sum_k \omega_k^2 \beta_k^2 . \tag{11.130}$$

Let us examine the expression for the total energy which we obtain as the sum of E_{kin} and W, i.e. $H_{\text{vib}} = E_{\text{kin}} + W$; using (11.127) and (11.130), we find the result

$$H_{\text{vib}} = \sum_k \tfrac{1}{2}(\dot{\beta}_k^2 + \omega_k^2 \beta_k^2) . \tag{11.131}$$

This expression has a very simple interpretation: we can see that *the total energy can be written as the sum of independent oscillators*. This fact becomes particularly clear if we identify β_k with the spatial coordinate Q_k of an oscillator and $\dot{\beta}_k$ with its momentum, P_k, setting the mass m formally equal to 1:

$$\left.\begin{array}{l} \beta_k \rightarrow Q_k \\ \dot{\beta}_k \rightarrow P_k \\ m = 1 \end{array}\right\} . \tag{11.132}$$

Let us summarise: we have seen that in the case of elastic bonds between the nuclei of a molecule, the kinetic and potential energies can be written with the aid of a coordinate transformation in such a way that they appear as the sum of the energies of uncoupled harmonic oscillations.

We can now complete the final step. As we know, the energy of a harmonic oscillator is readily quantised. The Schrödinger equation is then:

$$H_{\text{vib}}\, \Phi = E_{\text{vib}}\, \Phi ,$$

where H_{vib} is given by

$$H_{\text{vib}} = \sum_k \tfrac{1}{2}(P_k^2 + \omega_k^2 Q_k^2) .$$

It is frequently more elegant to introduce the creation and annihilation operators for vibrational excitations in place of the variables Q_k and P_k:

$$B_k^+ = \tfrac{1}{2}(Q_k - \mathrm{i}P_k) \,,$$
$$B_k = \tfrac{1}{2}(Q_k + \mathrm{i}P_k) \,.$$

Then $H_{\rm vib}$ can be written as

$$H_{\rm vib} = \sum_k \hbar\omega_k (B_k^+ B_k + \tfrac{1}{2}) \,,$$

where $\frac{1}{2}\sum_k \hbar\omega_k$ represents the zero-point oscillation energy.

A few concluding remarks are appropriate here. Quite obviously, we have not treated tri- and polyatomic molecules in such a consistent way as the diatomic molecules, where we started with the Schrödinger equation and showed that a suitable coordinate transformation could bring it into a form where the quantisation of the translational, the rotational, and the vibrational motions became clearly evident. We could have done the same thing in the case of triatomic or polyatomic molecules, but showing the individual steps of the coordinate transformation would have then been extremely lengthy and tedious. We therefore chose a simpler approach, in which we started from classical mechanics, which also yields the correct result. Naturally, in a more precise treatment, we would have to consider additional effects; this is true for the diatomic molecules as well. For example, we would have to take into account anharmonicities in the potential energy curve when the displacements from the equilibrium nuclear positions become larger, etc.

11.4 Symmetry and Normal Coordinates

In Chap. 5, we have seen with the aid of some straightforward examples how the calculation of electronic wavefunctions can be simplified by making use of molecular symmetries. In Chap. 6, we presented the basic mathematical tools needed to utilise symmetry properties, giving a detailed treatment of the fundamentals of the theory of group representations. Symmetry considerations are also useful in dealing with molecular vibrations: here, they help to determine the normal coordinates in an easy way. We shall demonstrate this using a simple example, namely a diatomic molecule (cf. Fig. 11.1). In the major part of this section, knowledge of Chaps. 5 and 6 is unnecessary, so that the reader who has not yet studied those chapters can still profitably read the present one.

Our starting point is the classical Hamilton function for the nuclear vibrations,

$$H_{\rm vib} = E_{\rm kin} + V \,; \tag{11.133}$$

but we note that all our considerations could just as well be applied to a Hamiltonian *operator*. However, it is more intuitively clear to deal initially with the classical expression. The kinetic energy is given by

$$E_{\rm kin} = \frac{m_1}{2}\dot{\xi}_1^2 + \frac{m_2}{2}\dot{\xi}_2^2 \,, \tag{11.134}$$

and the potential energy may be written in the form

$$V(\boldsymbol{R}_1, \boldsymbol{R}_2) = \frac{K}{2}(\boldsymbol{R}_2 - \boldsymbol{R}_1 - \boldsymbol{a})^2 . \qquad (11.135)$$

We consider here a one-dimensional problem as shown in Fig. 11.1, so that the vectors all lie along the x-axis. As in Sect. 11.3, we introduce the coordinates of the zero-point of the vibrations, $\boldsymbol{R}_{j,0}$, so that we can give the potential energy (11.135) in the form:

$$V = \frac{K}{2}(\boldsymbol{R}_{2,0} - \boldsymbol{R}_{1,0} + \boldsymbol{\xi}_2 - \boldsymbol{\xi}_1 - \boldsymbol{a})^2 = \frac{K}{2}(\xi_2 - \xi_1)^2 . \qquad (11.136)$$

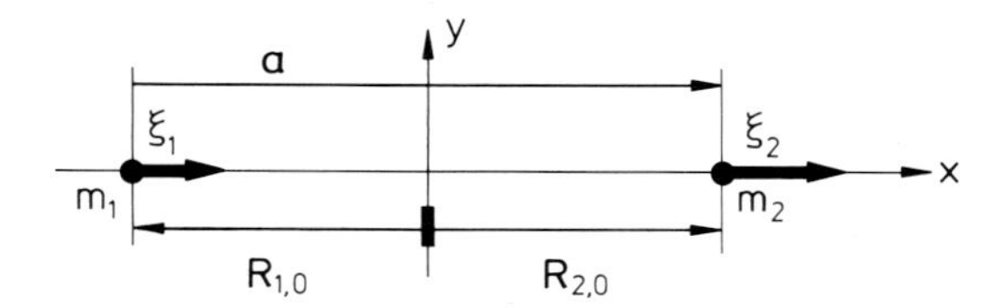

Fig. 11.1. The one-dimensional motion of two point masses with $m_1 = m_2$. Their equilibrium positions are separated by a distance vector **a** and are located symetrically with respect to the origin of the coordinate axes, i.e. $\boldsymbol{R}_{2,0} = -\boldsymbol{R}_{1,0}$. The displacements from the equilibrium positions are in the x-direction and are denoted by ξ_1 and ξ_2

We now examine symmetry operations T which leave the Hamilton function (11.133) invariant, i.e. we consider those transformations which transform H onto itself:

$$TH = H . \qquad (11.137)$$

The symmetry operation which we have in mind here is a reflection along the y-axis, which changes the sign of the x-component of every vector and simultaneously exchanges the indices 1 and 2. The reflection thus transforms both the equilibrium positions $\boldsymbol{R}_0$ and the displacements ξ, according to:

$$\begin{aligned} &\boldsymbol{R}_{1,0} \to \boldsymbol{R}_{2,0} , \quad \xi_{1,x} \to -\xi_{2,x} , \\ &\boldsymbol{R}_{2,0} \to \boldsymbol{R}_{1,0} , \quad \xi_{2,x} \to -\xi_{1,x} . \end{aligned} \qquad (11.138)$$

Let us first consider the kinetic energy (11.134). Since the transformation exchanges the indices on ξ_1 and ξ_2, we obtain an expression for the energy in which the masses m_1 and m_2 have been permuted; it is thus no longer of the form (11.134). We obtain the same form only in the case that $m_1 = m_2$. We thus recognise a first condition for the existence of molecular symmetry, i.e. that the mass points which are carried into one another by the symmetry operation must contain the same masses. If we insert the result of the transformation (11.138) into (11.136), we can immediately see that the potential energy V has remained unchanged.

We now investigate just how the equations of motion behave under these reflections. The equations of motion which belong to the energy (11.133) are:

$$m_1\ddot{\xi}_1 = K(\xi_2 - \xi_1) , \qquad (11.139)$$

$$m_1\ddot{\xi}_2 = -K(\xi_2 - \xi_1) . \qquad (11.140)$$

If we now carry out the transformation (11.138), then (11.139) is transformed into

$$-m_1\ddot{\xi}_2 = K(-\xi_1 + \xi_2) \qquad (11.141)$$

and (11.140) into

$$-m_1\ddot{\xi}_1 = -K(-\xi_1 + \xi_2) . \qquad (11.142)$$

Clearly, however, we could write (1.141) in the form:

$$m_1\ddot{\xi}_2 = -K(\xi_2 - \xi_1) \tag{11.143}$$

and (11.142) in the form:

$$m_1\ddot{\xi}_1 = K(\xi_2 - \xi_1) \ . \tag{11.144}$$

Aside from the order in which the equations occur, these expressions are identical with the original (11.139) and (11.140). The equations of motion are, as our example demonstrates, likewise invariant under this symmetry operation.

The variables ξ_1 and ξ_2 form the basis of the representation of the symmetry group. It contains only two elements, namely the identity operation E and the reflection σ. Let us see how these operations affect the basis vectors ξ_1 and ξ_2. For E, we find

$$E\begin{pmatrix}\xi_{1,x}\\ \xi_{2,x}\end{pmatrix} = \begin{pmatrix}\xi_{1,x}\\ \xi_{2,x}\end{pmatrix} = \begin{pmatrix}1 & 0\\ 0 & 1\end{pmatrix}\begin{pmatrix}\xi_{1,x}\\ \xi_{2,x}\end{pmatrix} \ , \quad \chi = 2 \ . \tag{11.145a}$$

Considering the transformations (11.138), we obtain

$$\sigma\begin{pmatrix}\xi_{1,x}\\ \xi_{2,x}\end{pmatrix} = \begin{pmatrix}-\xi_{2,x}\\ -\xi_{1,x}\end{pmatrix} = \begin{pmatrix}0 & -1\\ -1 & 0\end{pmatrix}\begin{pmatrix}\xi_{1,x}\\ \xi_{2,x}\end{pmatrix} \ , \quad \chi = 0 \ , \tag{11.145b}$$

where we have inserted matrices into the third term which would have the same effect as the corresponding operations E or σ on $\begin{pmatrix}\xi_{1,x}\\ \xi_{2,x}\end{pmatrix}$. The sums of the diagonal elements, i.e. the traces of the matrices, are the characters of the group, which we have indicated as the last entry in (11.145a) and (11.145b) for E and σ.

According to group theory, there are two irreducible representations for the group represented by E and σ; their characters are listed in the second and third rows of Table 11.1.

Table 11.1. The group characters; the upper row is for the reducible representation (11.145), and the middle and last rows for the two irreducible representations

	E	σ
Γ	2	0
Γ_1	1	1
Γ_2	1	−1

From the table, we can see immediately that the first row can be represented as the sum of the entries in the second and third rows. From this we know (cf. Chap. 6) that the reducible representation given in (11.145) can be decomposed into the two irreducible representations which we have called Γ_1 and Γ_2. The representations belonging to Γ_1 or Γ_2 can be constructed immediately by using (6.58), but, in the present case, they can also be seen directly. As we have already learned in Chap. 5 in the discussion of parity, functions may be classified in terms of their even or odd parity, i.e. whether they remain unchanged or change their signs under a reflection. If we set

$$\xi_{2,x} = \xi_{1,x} \ , \tag{11.146}$$

then we can readily calculate the following equation:

$$\sigma\begin{pmatrix}\xi_{1,x}\\ \xi_{1,x}\end{pmatrix} = \begin{pmatrix}-\xi_{1,x}\\ -\xi_{1,x}\end{pmatrix} = (-1)\begin{pmatrix}\xi_{1,x}\\ \xi_{1,x}\end{pmatrix} \ . \tag{11.147}$$

That is, this basis vector is transformed into itself with a negative sign by the reflection; it thus has odd parity. In the case

$$\xi_{2,x} = -\xi_{1,x} \ , \tag{11.148}$$

in contrast, the calculation yields

$$\sigma \begin{pmatrix} \xi_{1,x} \\ -\xi_{1,x} \end{pmatrix} = \begin{pmatrix} \xi_{1,x} \\ -\xi_{1,x} \end{pmatrix} = (+1) \begin{pmatrix} \xi_{1,x} \\ -\xi_{1,x} \end{pmatrix} ; \tag{11.149}$$

that is, the basis vector in this equation has even parity. Using (6.58), as we have already done in Sect. 6.8, we obtain the column vectors on the left-hand sides of (11.147) and (11.149) automatically; these are already the normal coordinates. In order to convince ourselves of this fact, we use the relation (11.146) associated with (11.147), or the corresponding relation (11.148) associated with (11.149). In the former case, we obtain on inserting (11.146) into (11.139) the following result:

$$m_1 \ddot{\xi}_1 = 0 , \tag{11.150}$$

with the solution

$$\xi_2 = \xi_1 = a + bt . \tag{11.151}$$

Here, we are dealing with a translational motion of both masses, i.e. not an oscillation at all. On the other hand, if we insert (11.148) into (11.139), we find

$$m_1 \ddot{\xi}_1 = -2K \xi_1 , \tag{11.152}$$

which is the equation of motion of a harmonic oscillator. The solution may be found immediately by using an exponential function as a trial solution, $\xi_1(t) = \xi_1(0)\mathrm{e}^{\mathrm{i}\omega t}$. For the frequency ω, we obtain the relation:

$$m_1 \omega^2 = 2K , \quad \text{that is} \quad \omega = \sqrt{\frac{2K}{m_1}} . \tag{11.153}$$

The solutions ξ_1, ξ_2 can then be written directly in the form

$$\xi_1 = a\mathrm{e}^{\mathrm{i}\omega t} + a^* \mathrm{e}^{-\mathrm{i}\omega t} \tag{11.154}$$

and

$$\xi_2 = -\xi_1 = -a\mathrm{e}^{\mathrm{i}\omega t} - a^* \mathrm{e}^{-\mathrm{i}\omega t} , \tag{11.155}$$

where we have again used relation (11.146). The solutions (11.154) and (11.155) can be rewritten in the form of a vector equation as follows:

$$\begin{pmatrix} \xi_1 \\ \xi_2 \end{pmatrix} = \begin{pmatrix} a \\ -a \end{pmatrix} \mathrm{e}^{\mathrm{i}\omega t} + \begin{pmatrix} a^* \\ -a^* \end{pmatrix} \mathrm{e}^{-\mathrm{i}\omega t} . \tag{11.156}$$

From this last equation, we can see that

$$\begin{pmatrix} a \\ -a \end{pmatrix} \mathrm{e}^{\mathrm{i}\omega t} \tag{11.157}$$

is the normal coordinate we were seeking; it becomes a real normal coordinate on addition of its complex conjugate, as indicated in (11.156).

From this simple example it is clear that the use of symmetries and the transformation properties of the normal coordinates reduces the computational effort required. Instead of solving the coupled Eqs. (11.139) and (11.140) directly, we had only to solve much simpler decoupled ones, (11.150) and (11.152).

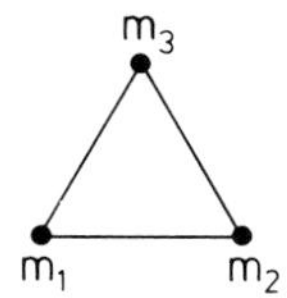

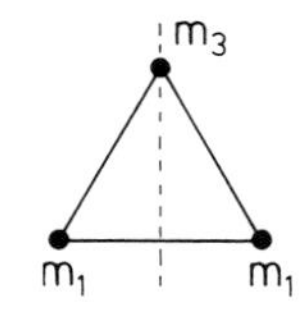

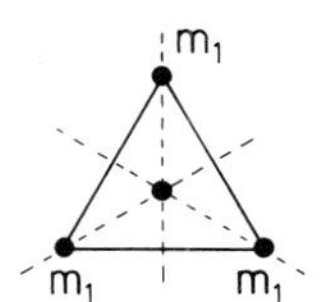

Fig. 11.2. *Left*: Three nuclei of a molecule in the plane of the page and equidistant from one another. There is no symmetry here, since the masses m_1, m_2 and m_3 are all different. *Centre*: Since here $m_1 = m_2$, but $m_1 = m_2 \neq m_3$, only the symmetry operation of reflection through the dashed vertical line is possible. *Right*: If all the masses are equal, then the symmetry operations of the group C_{3v} are possible. In each case we have assumed that the nuclear interactions also allow the symmetry operations indicated

Let us now see how the procedure works in the general case by referring to Fig. 11.2. In the case shown at the left of the figure, where the masses are all different, even the kinetic energy shows no symmetry at all, so that here, we cannot make use of symmetry considerations. In the case shown in the centre of Fig. 11.2, we find mirror symmetry with respect to a vertical axis passing through mass m_3, as long as the potentials are also invariant under the reflection. This is by no means obvious, since the potential depends on the electronic wavefunctions, and their distribution could, for example, be asymmetric. However, if we assume that the potential is also symmetric, then we can make use of the mirror symmetry of the molecule to find the normal coordinates. In the case of the right side of Fig. 11.2, the three identical point masses can be transformed into one another by a variety of symmetry operations, as discussed in detail in Sects. 6.2 and 6.3. If we choose the coordinates of the equilibrium positions as given in Fig. 11.3, then the effect of the symmetry operations can be described quite easily. Let us first consider the equilibrium positions themselves. On reflection through the vertical line passing through 3, the vectors $\boldsymbol{R}_{1,0}$ and $\boldsymbol{R}_{2,0}$ are converted into one another, while the third vector $\boldsymbol{R}_{3,0}$ is transformed into itself. The displacements $\boldsymbol{\xi}_1$ and $\boldsymbol{\xi}_2$ exchange their indices, with the x-coordinate changing sign while the y-coordinate remains unchanged. The index of $\boldsymbol{\xi}_3$ remains unchanged, while the x-coordinate changes its sign and the y-coordinate is unchanged. A rotation of the molecule by 60° about an axis perpendicular to the page causes a cyclic permutation of the indices of the equilibrium position vectors $\boldsymbol{R}_{j,0}$ and has the same effect on the displacements $\boldsymbol{\xi}_j$, with each vector $\boldsymbol{\xi}_j$ also being rotated by 60°. With the aid of a model in which each pair of masses is joined by a spring, and all the springs have the same force constants, one can readily see that the potential energy is also invariant with respect to these symmetry operations. We leave this as an exercise to the reader.

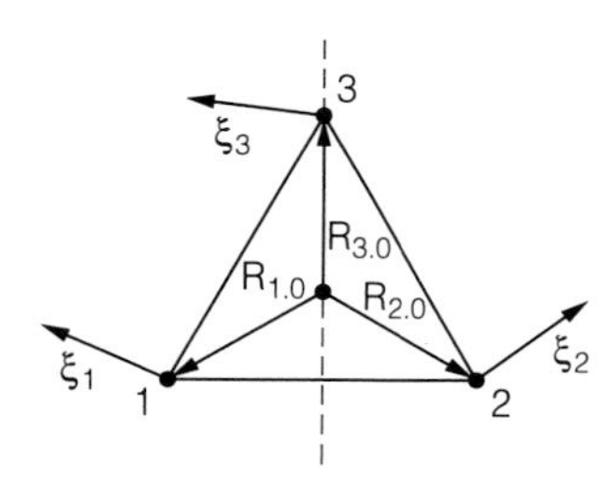

Fig. 11.3. A triatomic molecule having C_{3v} symmetry, showing the equilibrium positions $\boldsymbol{R}_{j,0}$ and the displacement vectors ξ_j for the nuclei which are indexed by $j = 1, 2, 3$

The basis of the reducible representation is determined by a vector made up of the two components of $\boldsymbol{\xi}_1$, the two components of $\boldsymbol{\xi}_2$, and the two components of $\boldsymbol{\xi}_3$. It is thus a six-dimensional vector, which we can write using the variables η introduced in Sect. 11.3 as

$$\boldsymbol{\eta} = \begin{pmatrix} \eta_1 \\ \eta_2 \\ \vdots \\ \eta_6 \end{pmatrix} .$$

By subjecting this vector to the various transformations of the group C_{3v}, i.e. the reflections and rotations given in Figs. 6.2 and 6.3: $E, C_3, C_3^2, \sigma_v, \sigma_{v'}$, and $\sigma_{v''}$, the matrices of the representation can be found in a similar manner to the relations (11.145). There, the matrices corresponding to the various operations were given in the third term, and the characters could be read off as their traces. Using the character table, Table 6.8, it is then possible to carry out the decomposition of the reducible representation into its irreducible representations, as we did here according to Table 11.1 and in (11.147) and (11.149). Now we must use relation (6.58). Analogously to (11.147) and (11.149), we then find fixed relationships between the individual components of the vectors of the representations. This then means that we have obtained the *normal coordinates* directly. We leave it to the reader as an exercise to derive these normal coordinates for the case shown in Fig. 11.4.

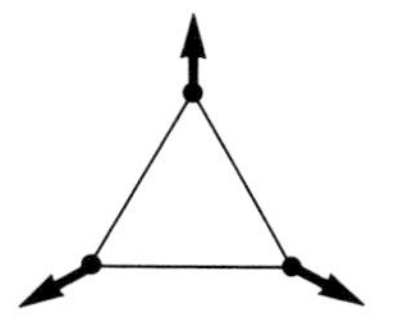
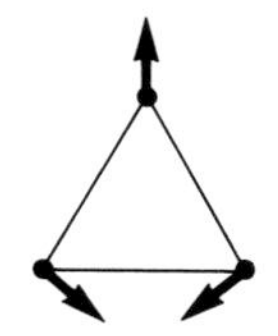
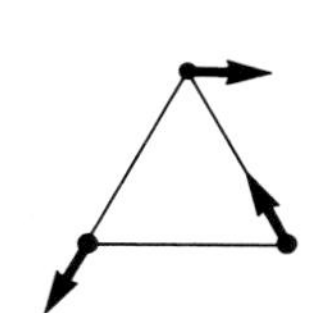

Fig. 11.4. The normal modes of the molecule shown in Fig. 11.3. In addition, there are three "improper" normal modes, namely two translations and one rotation (in the plane)

11.5 Summary

In this chapter, we have treated the motions of the atomic nuclei in molecules quantum-mechanically. We have seen, using the example of a diatomic molecule, how the motions can be split up into translations, rotations, and vibrations. The translational motion is not subject to a quantisation condition (aside from the "particle-in-a-box" quantisation given by the dimensions of the container, which gives practically continuous energy levels when a macroscopic container is considered). In contrast, we were able to derive the quantised energy levels for rotations and vibrations, and could make clear which approximations were introduced in the process. We then investigated the rotational motions of polyatomic molecules, including those of the asymmetric top, i.e. a molecule in which the moments of inertia along all three principal axes are different. Finally, we studied molecular vibrations, dividing the problem into two parts: the classical problem of determining the normal coordinates, and the problem of the quantisation. We described the derivation of the normal coordinates for a diatomic molecule, which can oscillate only along the x-axis, and – in less detail – for a triatomic molecule within the molecular plane. We saw in the latter case how useful group theory can be.

Having determined the normal coordinates classically, one can then construct the Hamiltonian in a very simple manner, as a sum of Hamiltonian operators for individual harmonic oscillators, so that the quantum-mechanical problem can be solved immediately. The wavefunction is found to be the product of the individual oscillator wavefunctions, each one referring to a particular normal mode. In determining the normal coordinates, we found that some of the degrees of freedom can automatically be dropped, since they are due not to vibrations, but instead to translational and rotational motions. We mention this fact for completeness, since we did not demonstrate it in the general case.

12. Raman Spectra

Along with infra-red spectroscopy and microwave spectroscopy, a further important method of investigating the rotational and vibrational spectra of molecules is *Raman spectroscopy*. It is based on the inelastic scattering of light from molecules, which is known as the Raman effect. The rotational and vibrational frequencies of the molecules which scatter the light are present in the scattered spectrum as difference frequencies relative to the elastically-scattered primary light. In this chapter, we shall explain how this scattered-light spectrum arises and what information it contains (Sects. 12.1–12.3). In the final section, 12.4, we then discuss the statistics of nuclear spin states and their influence on the rotational structure of the spectrum. This last section refers not only to the Raman effect, but also to rotational spectra in general, and can thus be regarded as a complement to Chap. 9.

12.1 The Raman Effect

As we have already seen, light can be emitted or absorbed by molecules, if the resonance condition $\Delta E = h\nu$ is fulfilled. In addition, however, as we know from classical physics, light of any wavelength can be scattered. Elastic or Rayleigh scattering, as described by classical physics, is explained in terms of the force acting on the electronic shells of the molecule due to the $\boldsymbol{E}$ vector of the light. This force induces an electric dipole moment $\boldsymbol{p}_{\text{ind}} = \alpha \boldsymbol{E}$, which oscillates at the frequency of the light and, acting as a Hertzian dipole, emits on its own part a light wave of the same frequency. This scattered radiation is coherent with the radiation field which induces it.

In the year 1928, *Raman* observed frequency-shifted lines in the spectrum of scattered light. The frequency shift relative to the primary light corresponded to the vibrational and rotational frequencies of the scattering molecules. This process, which had been predicted theoretically by *Smekal* in 1925, is called the Raman effect and forms the basis of Raman spectroscopy. The Raman-scattered light, in contrast to Rayleigh scattering, is not coherent with the primary light. Frequency shifts to smaller energies (Stokes lines) as well as shifts to higher energies (anti-Stokes lines) are observed. The frequency shift is independent of the frequency of the primary light and is a unique property of the scattering molecules. The entire Raman spectrum of a molecule is shown schematically in Fig. 12.1. It is the structure of this spectrum which we want to elucidate.

To observe the Raman spectrum, a high spectral resolution is required, owing to the smallness of the frequency shifts. The order of magnitude of these shifts is 1 cm^{-1} in the rotational Raman effect. Using primary light in the visible range, with a wavenumber $\overline{\nu}_{\text{p}}$ (p stands for primary) of the order of 20 000 cm^{-1}, we also need to supress the unshifted and many times more intense elastically scattered light in order to observe these small shifts;

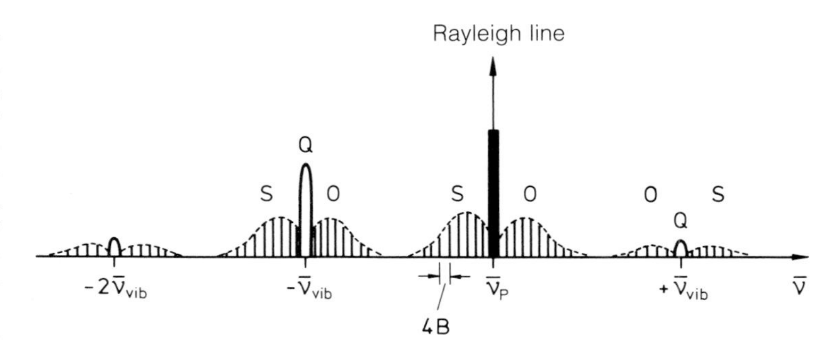

Fig. 12.1. The total Raman spectrum of a diatomic molecule, represented schematically. The Rayleigh line at the frequency $\overline{\nu}_p$ of the primary light is surrounded directly by the rotational Raman lines. Spaced at frequency shifts corresponding to the molecular vibrations, $\overline{\nu}_{vib}$, are the rotational-vibrational Raman lines, the Q, S, and O branches. The corresponding anti-Stokes lines at $\overline{\nu}_p + \overline{\nu}_{vib}$ are much weaker and usually cannot be observed. The same is true of the harmonics at $2\overline{\nu}_{vib}$

for this reason, a double or even a triple monochromator is often used for the detected beam. Furthermore, the smallest possible linewidth of the primary light is also necessary; otherwise, it is impossible to separate the weak Raman-scattered lines from the Rayleigh line, which is orders of magnitude more intense. For this reason, laser light is often used as the primary light source in Raman measurements. As is well known, lasers can be used to produce very intense monochromatic beams of light, with linewidths much smaller than the expected Raman shifts. Figure 12.2 gives a schematic drawing of a typical experimental set-up.

12.2 Vibrational Raman Spectra

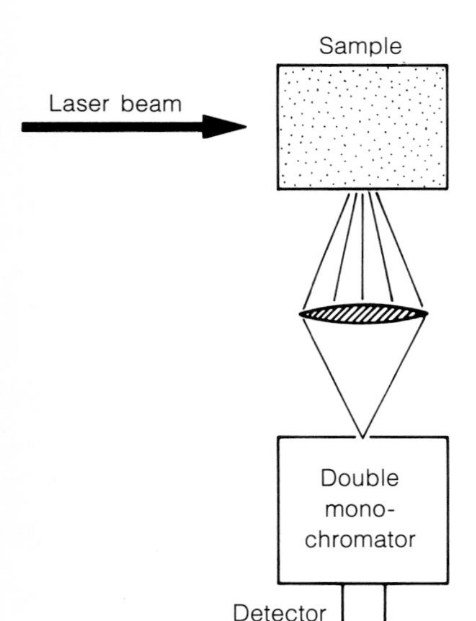

Fig. 12.2. An experimental set-up for observing the Raman effect. Detection is carried out preferentially perpendicular to the direction of the incoming primary light beam, in order to keep the intensity of the primary light in the detector as small as possible

We at first leave the molecular rotations out of consideration and assume that the molecule exhibits only vibrations. In the case of free molecules, we must naturally later revise this assumption.

The *classical* explanation of the *vibrational Raman effect* begins with the explanation of Rayleigh scattering. In this theory, it is assumed that the scattering molecule initially is not vibrating or rotating. When primary light of frequency ν_p (also denoted by ν_0) and an electric field strength $E = E_0 \cos(2\pi\nu_p t)$ strikes the molecule, a dipole moment is induced in its electronic shells, which oscillates with the same frequency ν_p as the E vector of the primary light. We then have

$$p(t) = \alpha E_0 \cos(2\pi\nu_p t) \; . \tag{12.1}$$

If, on the other hand, the molecule is already vibrating at one of its characteristic vibrational frequencies, then the oscillations of the induced moment are amplitude modulated at the frequency ν_{vib} of the molecular vibration, assuming that the polarisability α of the molecule changes as a function of the internuclear distance R of the vibrating atomic nuclei. The polarisability can be expanded as a series in powers of the internuclear distance R:

$$\alpha(R) = \alpha(R_0) + \frac{d\alpha}{dR}(R - R_0) + \text{higher order terms} \; . \tag{12.2}$$

Due to the molecular vibrations, R is time-dependent. It obeys the equation:

$$R = R_0 = q \cos(2\pi\nu_{vib}t) \; . \tag{12.3}$$

Combining this equation with (12.2), we find

$$p(t) = \alpha E = \left[\alpha(R_0) + \frac{d\alpha}{dR} q \cos(2\pi \nu_{vib} t)\right] E_0 \cos(2\pi \nu_p t) , \tag{12.4}$$

or, rewriting using well-known trigonometric identities,

$$p(t) = \alpha(R_0) E_0 \cos(2\pi \nu_p t) + \frac{d\alpha}{dR} E_0 q \{\cos[2\pi(\nu_p + \nu_{vib})t] + \cos[2\pi(\nu_p - \nu_{vib})t]\} . \tag{12.5}$$

In this way, sidebands are produced in the scattered light spectrum, having the frequencies $\nu_p \pm \nu_{vib}$.

This is the 1st order vibrational Raman effect. With decreasing intensity, one can also observe Raman lines with $\nu_p \pm 2\nu_{vib}$, $\nu_p \pm 3\nu_{vib}$, etc. due to the ever-present anharmonicity, i.e. the terms of higher order in the series expansion of $\alpha(R)$ in (12.2). These are called the Raman effect of second, third, ... order.

A vibration is thus Raman-active if $d\alpha/dR \neq 0$, that is, the polarisability α of the molecule must change as a function of the internuclear distance R during a vibration. This is always the case for diatomic molecules. For this reason, homonuclear and thus nonpolar molecules such as H_2 or N_2 are Raman-active. Their rotational and vibrational spectra can be measured using the Raman effect, although they are not accessible to microwave or infra-red spectroscopies because the transitions are forbidden by symmetry.

In the case of polyatomic molecules with centres of inversion, infra-red and Raman spectroscopies complement each other when one wishes to observe the molecular vibrations. In such molecules, the infra-red active normal modes are Raman-inactive, and the infra-red forbidden normal modes are Raman-allowed. This can be illustrated by the example of the CO_2 molecule: the symmetric stretching vibration ν_1 (cf. Fig. 10.10) is infra-red inactive, since the centres of positive and negative charge in the molecule coincide during the vibration. This motion is, however, Raman-active, since the polarisability changes periodically as a result of the stretching vibration. The asymmetric stretching vibration ν_2, in contrast, is infra-red active, since here, an electric dipole moment is present: however, it is Raman-inactive, because the changes in the polarisability due to the shortening and lengthening of the two C–O bonds in the molecule just compensate each other.

The classical theory of the Raman effect which we have described here explains many of the observations well, but it fails when the intensities are considered. In the classical picture, the same intensities would be expected for the lines which are shifted to lower energies and those shifted to higher energies, i.e. the Stokes and anti-Stokes lines. In fact, the Stokes lines are much more intense.

This is understandable in the light of the *quantum-mechanical* treatment of the Raman effect, which is given in Sect. 17.2. In this theory, the Raman Effect is treated as inelastic photon scattering, which in the case of the Stokes lines begins with a level having a small vibrational quantum number v (in particular $v = 0$) and ends with a level having a higher quantum number v' (for example $v' = 1$), while the reverse process occurs in the anti-Stokes effect: the scattering begins in an excited vibrational level and ends in a level with a smaller v, e.g. in the ground state; see Fig. 12.3. So, for the Stokes lines without rotation, the wavenumber of the scattered light is given by:

$$\overline{\nu} = \overline{\nu}_p - \overline{\nu}_{vib} \quad (\text{or } 2\overline{\nu}_{vib} \text{ etc.}) \tag{12.6}$$

and for anti-Stokes lines:

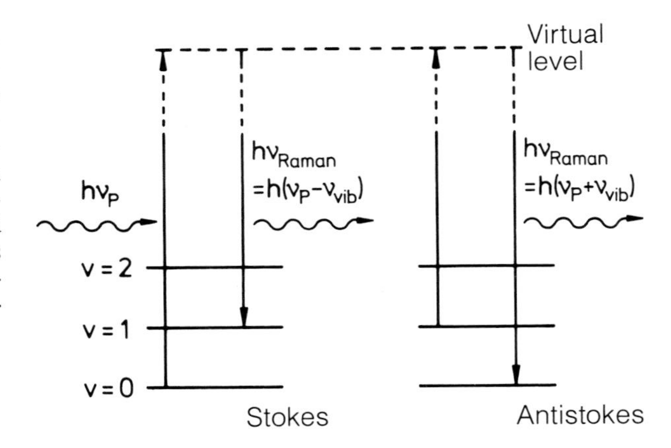

Fig. 12.3. A schematic representation of vibrational Raman scattering, to illustrate the Stokes and anti-Stokes scattering. Primary light of wavenumber $\overline{\nu}_p$ connects a real excitation state of the molecule with a virtual state. The Raman-scattered light has lost energy (Stokes lines) or gained energy (anti-Stokes lines) relative to the primary light

$$\overline{\nu} = \overline{\nu}_p + \overline{\nu}_{vib} \quad (\text{or } 2\overline{\nu}_{vib} \text{ etc.}) .$$

The difference $\overline{\nu}_p - \overline{\nu}$ is called the Raman shift. As we already pointed out, $\overline{\nu}_p$ is the wavenumber of the primary light which excites the transition and $\overline{\nu}_{vib}$ is the wavenumber of the molecular vibration.

In the case of Stokes Raman scattering, the molecule takes on energy from the photon; in anti-Stokes scattering, it gives up energy to the photon. The intensity ratios between Stokes and anti-Stokes lines are thus given by the occupation probabilities n of the initial states, and these can be calculated from the Boltzmann factor in thermal equilibrium. They are, in any case, different for Stokes and anti-Stokes transitions. The intensity of the anti-Stokes lines of course must decrease with decreasing temperature, since this process presumes that the molecule is initially in an excited vibration state, and the number of such molecules decreases when the temperature is lowered.

For the intensities, we then find

$$\frac{I_{\text{anti-Stokes}}}{I_{\text{Stokes}}} = \frac{n(v=1)}{n(v=0)} = e^{-h\nu_{vib}/kT} . \tag{12.7}$$

If we set $\nu_{vib} = 1000 \text{ cm}^{-1}$ and $T = 300$ K in this expression, we obtain a numerical value of e^{-5}, i.e. 0.7 ‰ for the relative intensity.

Quantum-mechanically, the selection rules are $\Delta v = \pm 1$ (and ± 2, ± 3, with much smaller probabilities, since here the nonlinear contributions to the polarisability are responsible for the transitions). This subject will be treated in more detail in Sect. 17.2.

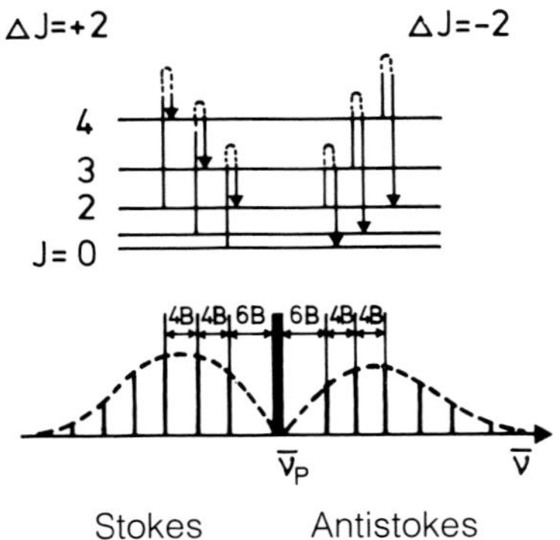

Fig. 12.4. A diagram explaining the occurrence of rotational Raman spectra. To the left and right of the Rayleigh line $\overline{\nu}_p$ are the Stokes and the anti-Stokes lines of the Raman spectrum

12.3 Rotational Raman Spectra

Now we turn to the *rotational Raman effect* (Figs. 12.4, 5). Here, again, one observes a series of scattering lines on both sides of the Rayleigh (i.e. the primary) line, but now with spacings corresponding to the rotational quanta. Here, also, many aspects of the observed effect can be understood classically. The polarisability of a nonspherical molecule is, as we have already discussed in Sect. 3.2, anisotropic and must be treated as a tensor with the principal polarisabilities $\alpha_{\parallel}$ and $\alpha_{\perp}$, where parallel and perpendicular refer to the long and short axes of the polarisability tensor, i.e. usually the body axis and an axis perpendicular to it. The rotation of a molecule therefore also leads to a periodic modulation of the dipole moment induced by the $\boldsymbol{E}$ field of the primary light and thus to a modulation of the frequency of the scattered radiation. This frequency modulation, however, occurs at $2\nu_{\mathrm{rot}}$, since the same polarisability as at the beginning of the rotation recurs after a rotation through 180° owing to the tensor symmetry. The additional lines accompanying the primary light thus occur at spacings corresponding to twice the rotational frequency.

This can be made clear by a simple demonstration experiment, which is shown in Fig. 12.6. The light from a lamp operated at a frequency of 50 Hz is reflected by two white balls in a dumbbell model which can rotate about an axis perpendicular to the line joining the balls and represents a linear dumbbell-molecule. The reflected light then contains, in addition to the unshifted component modulated at 50 Hz, two sidebands at the shifted frequencies $(50 \pm 2\nu_{\mathrm{rot}})$. This could be demonstrated by a simple frequency meter.

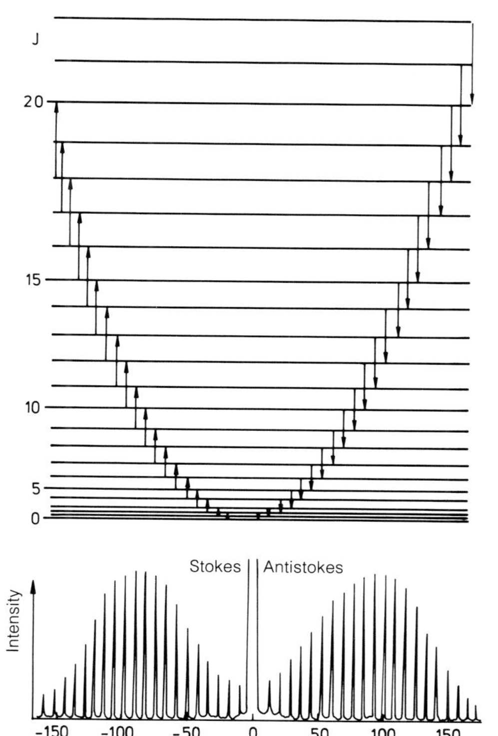

◀ **Fig. 12.5**

Fig. 12.5. The complete rotational Raman spectrum of a diatomic molecule consists of many nearly equidistant lines with a spacing of $4B$. Term scheme and intensity distribution are explained in the text. The selection rule is $\Delta J = \pm 2$. The Raman shift $\overline{\nu} - \overline{\nu}_{\mathrm{p}}$ is negative for Stokes lines and positive for anti-Stokes lines

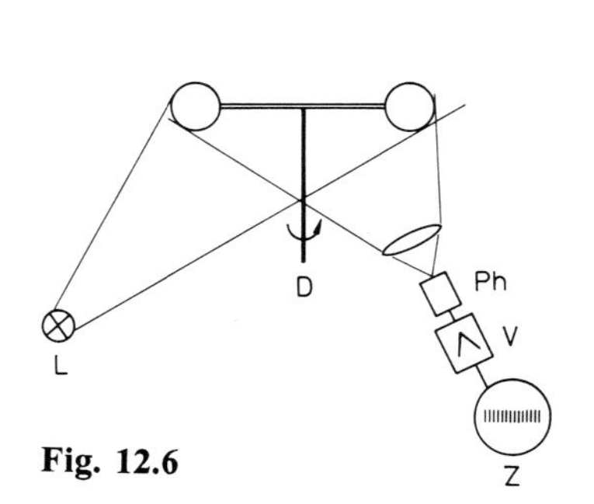

Fig. 12.6

Fig. 12.6. A demonstration experiment for the rotational Raman effect, after Auer. The light of a lamp L is reflected by two spheres which are rotating around the axis D. A frequency meter Z indicates the rotational frequency as an additional frequency to that of the modulation of the light (50 Hz). More details are given in the text

This classical model gives an explanation of the rotational Raman effect which is at least qualitatively satisfactory. It is determined by the difference $\alpha_{\parallel} - \alpha_{\perp}$, i.e. the anisotropy in the polarisability of the molecule, and occurs only when $\alpha_{\parallel} - \alpha_{\perp}$ is nonzero. This condition is fulfilled for all diatomic molecules, even for nonpolar molecules such as H_2, N_2, as well as for CO_2. For symmetric tetrahedral molecules such as CH_3 or CCl_4, in contrast, one finds $\alpha_{\parallel} = \alpha_{\perp}$, and there is thus no rotational Raman effect. The classical explanation for the occurrence of the *doubled* rotational frequency in the spacing of the Raman lines is reproduced in quantum mechanics in terms of modified selection rules which correspond to a two-photon process.

The quantum-mechanical treatment of the rotational Raman effect as inelastic photon scattering accompanied by the uptake or release of rotational quanta leads to the selection rule $\Delta J = \pm 2$ in the case of the linear rotor. For the rigid rotor with energy levels $E_J = BJ(J+1)$, the shift of the rotational Raman lines relative to the primary light is given by

$$\overline{\nu}_{\text{rot}} = \pm B[(J+2)(J+3) - J(J+1)] = \pm B[4J+6] \qquad (\text{cm}^{-1}) , \tag{12.8}$$

where the sign is relative to the primary frequency of the exciting light.

For the wavenumber of the Raman-scattered light, we find

$$\overline{\nu} = \overline{\nu}_{\text{p}} \pm \overline{\nu}_{\text{rot}} .$$

In a transition with $\Delta J = +2$, the molecule is raised to a higher rotational state by the scattering process. The wavenumber of the scattered light is therefore smaller than that of the primary light, $\overline{\nu}_{\text{p}}$. The Stokes lines of the spectrum thus appear on the low-frequency side of the primary light. For the anti-Stokes lines, with $\Delta J = -2$, the reverse is true.

The rotational Raman spectrum of a linear molecule thus has the structure which is shown in Fig. 12.4. The first Raman line, with $J = 0$, is located at a distance $6B$ from $\overline{\nu}_{\text{p}}$, the primary line, and then the other lines follow at a constant spacing of $4B$. The intensity distribution in a rotational Raman spectrum is given, as in the rotational spectra treated earlier, by the thermal populations and the multiplicities of the J terms. It is demonstrated in Fig. 12.5. In addition, this figure makes clear that due to the smallness of the rotational quanta, the number of rotational lines in the spectrum may be quite large. The difference in thermal populations between neighbouring levels is, as a result of the smallness of the rotational quanta, very little in comparison to kT; thus the difference in intensity between the Stokes and the anti-Stokes lines is also small.

The selection rules depend on the symmetry of the molecule. For the symmetric top, e.g. CH_3Cl, we have $\Delta J = 0, \pm 1, \pm 2$, and $\Delta K = 0$. Its rotational Raman spectrum has more lines than that of the linear rotor; we shall however not go into this further here.

We are now in a position to understand the *rotational structure in the vibrational Raman spectrum*. Each vibrational Raman line is accompanied by rotational lines, which can be understood in the same way as the rotational structure in a rotational-vibrational spectrum, but taking into account the different selection rules for Raman transitions. Figure 12.7 shows, as an example of a measured spectrum, a section of the Raman spectrum of $^{16}O_2$, namely the vibrational line at $\overline{\nu}_{\text{e}} = 1556\ \text{cm}^{-1}$, and its accompanying rotational lines. The structure of this Raman spectrum corresponds to that of the infra-red vibrational spectrum, but now the selection rules are $\Delta v = \pm 1\ (\pm 2, \ldots)$, $\Delta J = 0, \pm 2$. The comparison gives the differences between a rotational-vibrational spectrum and a rotational-vibrational Raman spectrum; this is illustrated schematically in Fig. 12.8.

Let us now compare the rotational-vibrational spectrum (without the Raman Effect) to the corresponding Raman spectrum, referring to Fig. 12.8.

In the *rotational-vibrational spectrum* of a diatomic molecule, for $\Delta v = 1$ there are

the R branch, $\overline{\nu}_{\text{vib}} + \overline{\nu}_{\text{rot}}$, $\Delta J = 1$, $\overline{\nu}_{\text{rot}} = 2B(J+1)$
the P branch, $\overline{\nu}_{\text{vib}} - \overline{\nu}_{\text{rot}}$, $\Delta J = -1$, $\overline{\nu}_{\text{rot}} = 2BJ$

and possibly

the Q branch, $\overline{\nu}_{\text{vib}}$, $\Delta J = 0$.

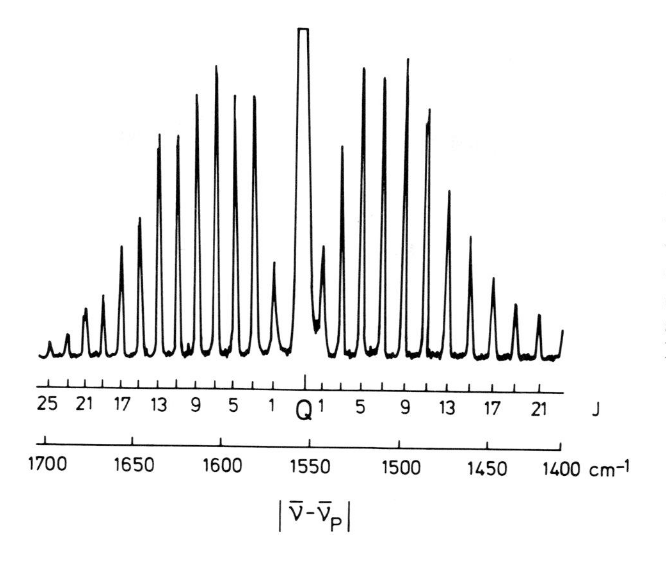

Fig. 12.7. A section of the rotational-vibrational Raman spectrum of oxygen, $^{16}O_2$. This is a vibrational line (Stokes line) with its accompanying rotational lines. In the centre, at the energy of the vibrational wavenumber $\overline{\nu}_e = 1556\ \mathrm{cm}^{-1}$, we see the Q branch ($\Delta J = 0$) as a broad line. For $^{16}O_2$ ($I = 0$), the lines with even J are missing; cf. Sect. 12.4. After Hellwege

By contrast, in the *Raman spectrum*, for Stokes lines with $\Delta v = +1$, we have

$$\begin{aligned} &\text{the } S \text{ branch,} \quad \overline{\nu}_p - (\overline{\nu}_{vib} + \overline{\nu}_{rot}), \quad \Delta J = 2 \\ &\text{the } Q \text{ branch,} \quad \overline{\nu}_p - \overline{\nu}_{vib}, \quad \Delta J = 0 \\ &\text{the } O \text{ branch,} \quad \overline{\nu}_p - (\overline{\nu}_{vib} - \overline{\nu}_{rot}), \quad \Delta J = -2\,. \end{aligned} \tag{12.9}$$

Here, $\overline{\nu}_{rot}$ is an abbreviation for the rotational energy, $B(4J+6)$; see (12.8).

Vibrational spectrum

$\Delta J = +1$ (R branch) $\bar{\nu}_{vib} + \bar{\nu}_{rot}$

$\Delta J = -1$ (P branch) $\bar{\nu}_{vib} - \bar{\nu}_{rot}$

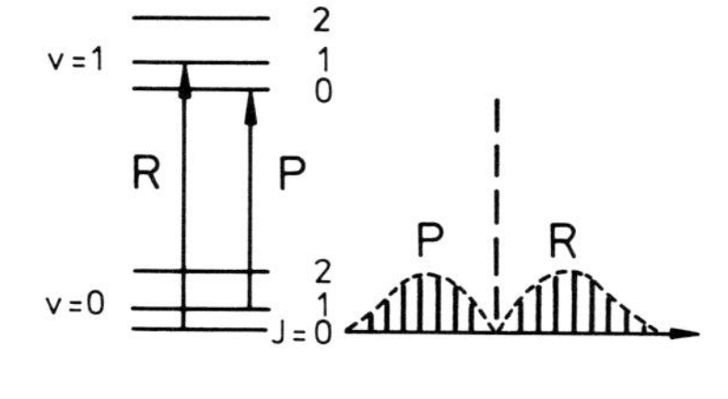

Vibrational raman spectrum

$\Delta J = 2$ (S branch) $\bar{\nu}_P - (\bar{\nu}_{vib} + \bar{\nu}_{rot})$

$\Delta J = 0$ (Q branch) $\bar{\nu}_P - \bar{\nu}_{vib}$

$\Delta J = -2$ (O branch) $\bar{\nu}_P - (\bar{\nu}_{vib} - \bar{\nu}_{rot})$

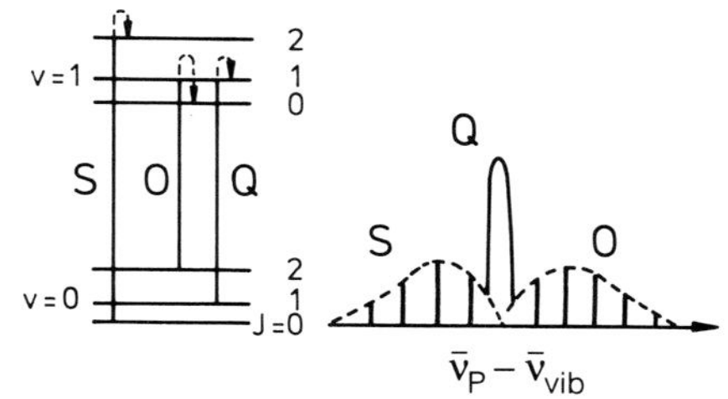

Fig. 12.8. A vibrational Raman spectrum (*right*) compared to a rotational-vibrational spectrum (*left*), schematically. Only a section of each spectrum is shown. The selection rules and therefore the rotational structure of the two spectra are different. Here, we show only Stokes lines in the Raman spectrum, i.e. lines having a lower frequency than that of the primary light. The Raman shift is $\overline{\nu}_p - \overline{\nu}_{vib}$, where $\overline{\nu}_p$ refers to the primary light and $\overline{\nu}_{vib}$ to a vibrational wavenumber

The anharmonicity and the coupling have been left out of these considerations; cf. Chap. 9. Both effects lead to small shifts in the spectra, as we have already seen in Sects. 9.2.3 and 9.5; these can usually not be determined so accurately in the Raman spectrum. For large molecules, the rotational fine structure of the Raman spectrum is barely resolvable.

The whole Raman spectrum of a molecule was already indicated schematically in Fig. 12.1, and now becomes clear. Immediately adjacent to the Rayleigh line, the line from the primary light, we see the rotational lines. At a spacing corresponding to the vibrational line,

$\bar{\nu}_{vib}$, one observes the Stokes region of the rotational-vibrational lines, and with reduced intensity on the other side, the anti-Stokes lines. This structure repeats with decreasing intensity as "harmonics" in the range $\pm 2\bar{\nu}_{vib}$.

Raman spectroscopy is thus a second method for measuring molecular rotational and vibrational quanta. What, then, are its advantages, or at least the differences, relative to infra-red or microwave spectroscopies?

- By properly choosing the primary light, one can shift the investigation of rotational and vibrational spectra from the microwave and infra-red spectral ranges to *more conveniently accessible spectral regions*, namely into the range of visible light.
- There are vibrations and rotations of molecules which are visible in the Raman spectrum, but not in the infra-red or microwave spectra. For example, diatomic homonuclear molecules such as H_2, N_2, or O_2 can be investigated *only with Raman spectroscopy*, since their rotations and vibrations are infra-red inactive.
- From the polarisation behaviour of Raman spectra, one can obtain information about the *polarisability tensor* of the molecules. A depolarisation of the Raman spectrum as compared to the primary light allows the motions of the molecules in the surrounding medium during the scattering process, particularly in liquids, to be studied; the molecules change not only their positions, but also their orientations in the course of these motions. When the molecule moves during the scattering process, the polarisation diagram for the scattered radiation can deviate from that calculated for a motionless molecule.
- In the Raman effect, in contrast to the usual single-photon spectroscopies, the *parity* of the state is conserved. The reason for this is the fact that two photons take part in the overall process. In single-photon dipole processes, the parity changes; in two photon processes, it changes twice, i.e. it returns to its initial value.
- The intensity of the Raman effect is largely independent of the frequency of the primary light, as long as the quantum energy of the light is sufficiently far removed from that of an electronic transition. The light quantum of excitation of the Raman effect, $h\nu_p$, ends in a so-called virtual level; cf. Fig. 12.3. If the quantum energy of the primary light approaches that of an electronic transition, i.e. a real excitation level, then the Raman scattering intensity increases. This amplification of the Raman spectrum is called the *resonant Raman effect*.

  Using the resonant Raman effect, particular parts of large, complex molecules can be investigated specifically, by intentionally using primary light whose frequency is near to that of a real excitation in the molecular subgroup. In that case, the portion of the Raman spectrum originating from the particular subgroup and containing its rotational and vibrational structure is strongly enhanced relative to the Raman spectra of other molecular subgroups.

12.4 The Influence of Nuclear Spins on the Rotational Structure

In the rotational spectrum and in the rotational-vibrational spectrum of homonuclear diatomic molecules such as H_2, N_2, and O_2, and quite generally in the spectra of molecules with a centre of inversion symmetry, such as CO_2, characteristic intensity differences are observed in the lines originating from levels with an even rotational quantum number J as compared to those originating from levels with an odd value of J. Examples are shown in Figs. 12.7

and 12.11. Since such spectra are often (but not exclusively – see for example the rotational-vibrational spectrum of CO_2 in Fig. 10.12) investigated as Raman spectra, we will consider this interesting and important aspect of molecular spectra at this point. The effect is due to the influence of the *nuclear spins* on the spectra.

The nuclear spins and the magnetic moments of the nuclei of course interact with the electronic shell of a molecule, but the influence of this interaction on the electronic spectra is comparatively small. The resulting hyperfine structure of the molecular terms and the spectral lines is based on the same interaction mechanism as in atoms. It can be investigated by using high-resolution spectroscopy, but we will not deal with this here; the subject is treated in more detail in I, Chap. 20.

The observed intensity variations in the Raman spectra are due, however, to a different phenomenon of fundamental importance: the influence of the nuclear spins on the *statistics*, that is the relative probability with which particular molecular states occur. In molecules with two identical nuclei, the observed intensity distribution in the spectrum is a result of the influence of the nuclear spins on the symmetry of the overall wavefunction of the molecular state. It is due to the Pauli exclusion principle, according to which the overall wavefunction of fermions, i.e. particles with half-integral spins, must be antisymmetric with respect to exchange of the particles. In the case of bosons, with integral spins, the wavefunction must be symmetric under particle exchange. Here, we are dealing with nuclear fermions (e.g. ^1H, with $I = 1/2$) and nuclear bosons (e.g. ^{16}O, with $I = 0$).

Let us consider the H_2 molecule in order to explain this effect; see Fig. 12.9. The two protons in the molecule are fermions with spin 1/2. The spins of the two protons may be parallel; then the molecule has a total nuclear spin quantum number of $I = 1$. The spin wavefunction is symmetric with respect to particle exchange, as the particles are indeed identical when their spins are parallel. This kind of hydrogen is called ortho hydrogen, o-H_2. The two nuclear spins could, however, also be antiparallel, and the total spin quantum number would then be 0. In this case, the spin wavefunction is antisymmetric with respect to

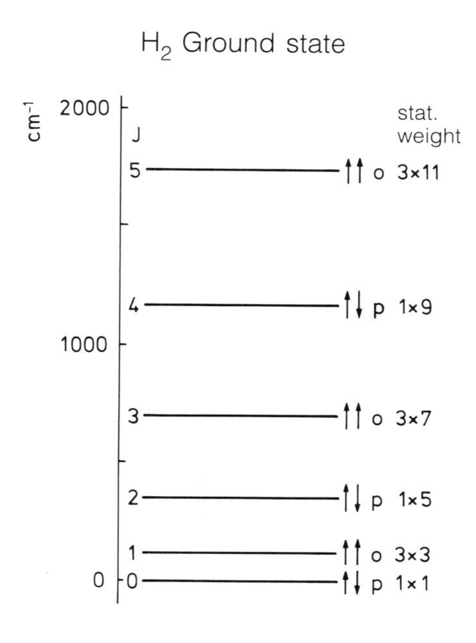

Fig. 12.9. The rotational levels of the H_2 molecule in the ground state, with the spin orientations of the nuclei (ortho- and para-H_2) and the statistical weights of the configurations. Ortho hydrogen has rotational levels with odd values of J, while para hydrogen has levels of even J

exchange of the nuclei, and this configuration is termed para hydrogen, p-H_2. The statistical weight of the two configurations is 3:1, as shown in Table 12.1.

Table 12.1. o- and p-hydrogen

	I	M_I	Wavefunction	Character
o-H_2	1	1	$\uparrow\uparrow$	
		0	$\frac{1}{\sqrt{2}}(\uparrow\downarrow + \downarrow\uparrow)$	triplet
		−1	$\downarrow\downarrow$	
p-H_2	0	0	$\frac{1}{\sqrt{2}}(\uparrow\downarrow - \downarrow\uparrow)$	singlet

The overall wavefunction of the molecule is the product of the spatial functions (including rotation) and the spin functions. Exchange of the nuclei means in the case of a dumbbell molecule simply a reversal of the dumbbell, equivalent to an inversion in space. Under this operation, the rotational eigenfunctions for $J = 1, 3, 5, \ldots$ change their signs (cf. Sect. 11.1); they have negative parity and are antisymmetric with respect to exchange. The rotational functions with $J = 0, 2, 4, \ldots$ remain unchanged; they have positive parity and are symmetric.

The overall parity is the product of the parities of the functions contributing to the total system. For particles with half-integral spin, it must be negative. Then o-H_2, i.e. hydrogen molecules with $I = 1$ and thus positive parity of the spin function, must have rotational states with negative parity, i.e. $J = 1, 3, 5, \ldots$ with the statistical weight 3, if the remaining spatial function has positive parity, as is in fact the case for the ground state of hydrogen. As we shall see in Sects. 13.3 and 13.4, this is true of the state denoted by $^1\Sigma_g^+$, the ground state of the hydrogen molecule. Para hydrogen, with $I = 0$ and negative parity of the spin function, must have rotational functions with $J = 0, 2, 4, \ldots$, so that the overall product gives a negative parity for the total wavefunction.

Between these two types of hydrogen, which by the way can be separated from each other macroscopically, transitions are rather strictly forbidden. Only transitions within the term system with even J and within that with odd J are possible, if the nuclei are completely uncoupled. The weak coupling between the nuclear spins and the electronic shells does, however, make transitions between the two systems possible, with a very small transition probability.

At the lowest temperatures, only p-H_2 is stable; o-H_2, due to its $J = 1$, i.e. because a rotational quantum is excited, is metastable. The spontaneous conversion of o-H_2 into p-H_2 by flipping of a nuclear spin occurs very slowly, over a time of years. This process can be accelerated by addition of paramagnetic materials or other catalysts, so that pure p-H_2 can be prepared at low temperatures. It remains in the p-H_2 state for some time even after warming and evaporation to H_2 gas. The case of deuterium or heavy hydrogen, $^2H_2$ or D_2, is just the reverse: the nuclear spin of ^2H is $I = 1$, the nucleus is a boson, and at low temperatures, ortho-D_2 is stable and para-D_2 is metastable.

Normally, thermal equilibrium is established between the two H_2 modifications. Hydrogen is a mixture of p-H_2 and o-H_2 in the ratio 1:3. This has the following consequences for the rotational spectrum (see Fig. 12.10):

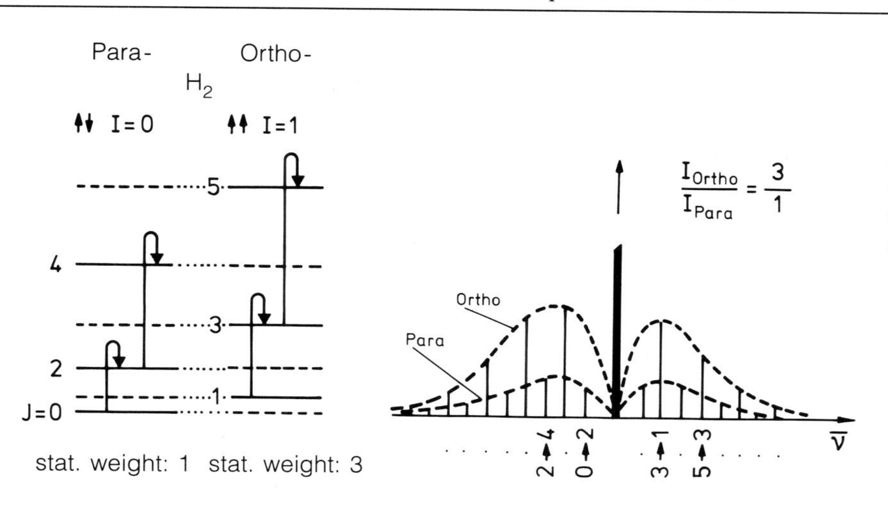

Fig. 12.10. The rotational Raman spectrum of the H_2 molecule. The overall spectrum is a superposition of the spectra of ortho and para hydrogen, with the intensity ratio 3:1. The direct line at the centre of the spectrum is the Rayleigh line

- In the rotational spectrum, there can be no transitions with $\Delta J = \pm 1$, and therefore no allowed transitions at all. They would in any case be infra-red inactive due to the lack of a dipole moment in H_2.
- Rotational Raman lines with $\Delta J = \pm 2$ are, in contrast, allowed. They belong alternately to o- and p-H_2. For this reason, alternating intensities are observed in the Raman spectrum, as can be seen in Fig. 12.10.

The observed alternating intensities in the spectra of other homonuclear molecules can be understood in an analogous manner. For $^{16}O_2$, with $I = 0$, all the levels with even J quantum number are lacking, for example. Here, the electronic wavefunction in the ground state has negative parity (term symbol $^3\Sigma_g^-$). In order to make the overall wavefunction symmetric (^{16}O is a boson) with respect to exchange, the rotational wavefunctions must also have negative parity. Thus, the rotational lines with even J are missing in the spectrum; compare Fig. 12.7. For ^{14}N, with $I = 1$, all the lines in the Raman rotational branches are observed in the spectrum of the N_2 molecule, but with alternating intensities in the ratio 1:2, which results from the possible spin configurations in the molecule. These are parallel, with $I_{tot} = 2$, and antiparallel, with $I_{tot} = 0$; see Fig. 12.11. For N_2 molecules with two different isotopes, i.e. $^{14}N^{15}N$, this alternating intensity is lacking in the Raman spectrum.

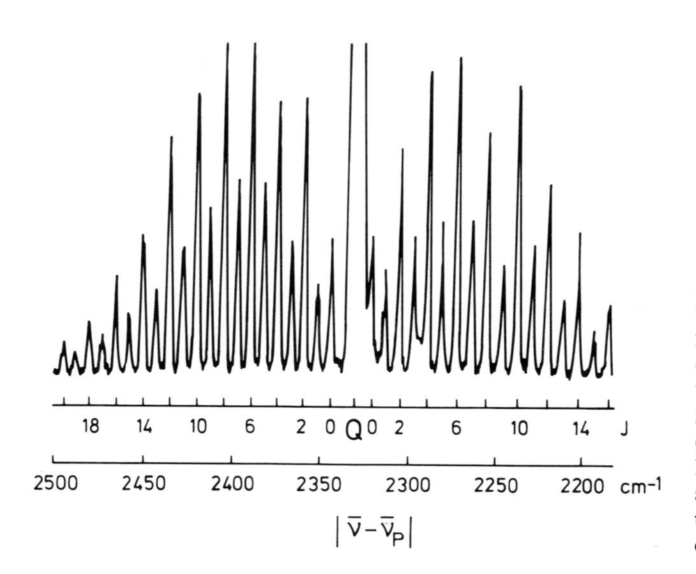

Fig. 12.11. The rotational-vibrational Raman spectrum of the nitrogen molecule, $^{14}N_2$. In the centre, at the position of the vibrational wavenumber $\bar{\nu}_e$ = 2330 cm^{-1}, the Q branch ($\Delta J = 0$) appears as a broad line. In $^{14}N_2$, with $I = 1$, an alternating intensity of the rotational lines with a ratio 1:2 is observed. After Hellwege

In general, the following rules hold, which we give here without a detailed derivation:

The ratio of the statistical weights of antisymmetric and symmetric states of the nuclear spins in a diatomic molecule with two identical nuclei having the nuclear spin quantum number I is

$$\frac{g_a}{g_s} = \frac{I}{I+1} . \tag{12.10}$$

This is then also the intensity ratio of alternate lines in the rotational spectrum; for H_2, with $I = 1/2$, we find 3:1, for N_2, with $I = 1$, we find 2:1, and for O_2, with $I = 0$, one of the two components is completely lacking, i.e. 1:0.

The molecule $C^{16}O_2$ is particularly illustrative; it is linear and therefore has a centre of inversion symmetry. Here, only rotational levels with an even quantum number J are allowed, since the electronic ground state has positive parity ($^1\Sigma_g^+$). We pointed this fact out already with respect to the rotational-vibrational spectrum of the CO_2 molecule, shown in Fig. 10.12, without being able to explain it at that point. On the other hand, if one investigates molecules having two different oxygen isotopes, e.g. $^{16}OC^{18}O$, then the symmetry is lowered, since the two O nuclei are now different. For this molecule, *all* the rotational terms are observed in the spectrum, which thus contains twice as many lines.

The influence of the nuclear spins on the statistics of the possible molecular states which we have described cannot be explained within classical physics and presents an impressive demonstration of the correctness of quantum-mechanical concepts, in particular the Pauli principle. Simply observing the intensity pattern in the rotational spectrum of a molecule permits the determination of the nuclear spin quantum number I from (12.10).

13. Electronic States

Having presented in Chaps. 9 and 10 a treatment of the rotational and rotational-vibrational spectra of molecules, which lie in the microwave and the infra-red spectral regions, respectively, we turn in this and in the following chapter to electronic transitions in molecules, and thus to band spectra in the general sense. These spectra are observed in the near infra-red, in the visible, and in the ultraviolet spectral ranges, and consist of a very large number of lines with a structure which is often difficult to analyze. When observed with a spectrometer having a limited resolution, they appear as structureless, band-like spectra; hence the name 'band spectrum'. In this chapter, we present the fundamentals needed to understand the molecular quantum numbers. Following another detailed treatment of diatomic molecules in Sect. 13.3, we give a preview of the subject for larger molecules in Sect. 13.4.

13.1 The Structure of Band Spectra

A band spectrum, as we indicated in Chap. 8, exhibits a threefold structure: it contains a number of often clearly separated groups of bands, the so-called *band systems*; each band system consists of a number of *bands*; and each band is made up of a large number of *band lines*, which occur in an ordered fashion.

This threefold structure of the spectrum corresponds to the three contributions to the total energy of a molecule, that is the electronic excitations, the vibrational excitations, and the rotational excitations of the molecule.

The electronic excitation determines the position of a band system in the spectrum. All the bands of a particular band system belong to the same electronic excitation. The position of a band within the band system is given by the change in the vibrational energy between the initial and the final states. The change in the electronic quantum number and the change in the vibrational quantum number are thus constant for all the lines of a band. The band lines are distinguished from each other by the differences of the rotational quantum numbers.

The complete spectrum of a molecule thus consists of the the rotational- vibrational and the electronic transitions. The terms for the possible excited states of the molecule can be written in the form

$$T = T^{\mathrm{el}} + G_v + F_{v,J} \; . \tag{13.1}$$

For the wavenumbers of the lines in the band spectrum, the corresponding formula is:

$$\begin{aligned} \bar{\nu} &= \Delta T^{\mathrm{el}} + \Delta G + \Delta F \\ &= T'^{\mathrm{el}} - T''^{\mathrm{el}} + G'_{v'} - G''_{v''} + F'_{v',J'} - F''_{v'',J''} \; . \end{aligned} \tag{13.2}$$

Electronic transitions occur between the various possible electronic excited states of one or more electrons and the ground state. We require an analysis of the electronic transitions if we wish to understand the electronic structure, bonding, excited states, behaviour towards chemical reaction, dissociation energies, and other physical properties of molecules. The ground state and the excited states can be characterised by eigenfunctions and quantum numbers, which must be derived and defined for molecules on the basis of the well-tested methods which have previously been applied to atoms.

13.2 Types of Bonding

Owing to the great variety of molecules and the marked differences in their structures, it is useful to distinguish among three limiting cases of the chemical bonding and thus of the type of molecule:

- *Ionic molecules* such as the alkali halides NaCl or LiF are strongly polar; i.e. an electron is transferred more or less completely from one bonding partner to the other. These molecules, in terms of their electronic excitations, are basically similar to the atoms or ions of which they are formed. In solution, they dissociate readily into their component ions, e.g. NaCl into $Na^+ + Cl^-$.
- *Van der Waals molecules* such as Hg_2 or Cd_2 are formed preferentially through bonding of neutral atoms, which contain only filled electronic shells and therefore can form only weak bonds with other atoms by means of induced dipole moments. The bonding is provided by so-called Van der Waals forces, which fall off rapidly with distance. The atomic states are maintained still more strongly than in the case of ionic molecules. These molecules dissociate into neutral atoms.
- The most important and physically most interesting class are the *atomic molecules* such as H_2, N_2, O_2, or AlH, as examples of diatomic molecules. Their bonding is due to formation of homopolar or covalent bonds, which comes about when electrons of the bonding partners in unfilled atomic orbitals form new orbitals, belonging in common to both atoms, i.e. molecular orbitals. These molecules can dissociate from their ground states into neutral atoms. This group contains a vast number of molecules known from organic chemistry.

13.3 Electronic States of Diatomic Molecules

We first consider the last-named type of molecules and attempt to establish a connection between atomic and molecular orbitals and to understand it. Here, as in the theory of chemical bonding which we treated in Chaps. 4 through 7, it is instructive to begin with the hydrogen molecule-ion, H_2^+, or with the hydrogen molecule, H_2. As we have seen in the latter case (Sect. 4.3), we can construct the electronic wavefunctions for all of the electrons in the molecule to a good approximation out of symmetrised or antisymmetrised products of one-electron wavefunctions. We therefore begin with such one-electron wavefunctions. On the other hand, atoms such as N, O, etc. bring with them a number of electrons in filled or partly-filled shells into the molecule. We must thus consider how these various atomic orbitals change or combine as linear combinations when we allow two such atoms to approach each other and form bonds.

We shall take the following approach:

- Molecular orbitals will be built up using linear combinations of atomic orbitals;
- The available electrons will be filled into the molecular orbitals in such a way that each orbital contains a maximum of two electrons (Pauli exclusion principle);
- In filling energetically degenerate orbitals, first each molecular orbital will be singly occupied, before a double occupation is allowed. The spins of the electrons will be preferentially all parallel (Hund's rule).

Now, in order to understand the molecular orbitals, we leave the vibrations and rotations out of consideration, as in Chaps. 4–7, and consider a fixed nuclear framework. In those chapters, we dealt primarily with the construction of molecular orbitals from atomic wavefunctions; here, instead, we want to introduce mainly the definition and derivation of the important quantum numbers. As an example of a simple molecule, we consider a homonuclear, diatomic molecule AB made up of the two atoms A and B according to the reaction $A + B \rightarrow AB$. In each of the two atoms, the electrons are associated with certain atomic orbitals A, B. Beginning with the separated atoms at a distance $R = \infty$, we create the molecule at its equilibrium bond length R_e by gradually bringing the atoms together. In this configuration, the electrons experience not only the Coulomb field of their original nucleus, but also that of the other nucleus; it has rotational symmetry with respect to the internuclear axis. If we now in a thought experiment further reduce the internuclear distance until the two nuclei merge, i.e. $AB \rightarrow (AB)$, then the electronic states of the molecule must be transformed into those of an atom with atomic number equal to the sum of the atomic numbers of A and B. By a mutual approach of the atoms, one thus arrives at the *combined-nucleus atom* with the molecular orbitals (AB).

The simplest example of a "molecule" is the H_2^+ ion, consisting of two protons and one electron. The associated combined-nucleus atom would be the He^+ ion, while the separation to $R = \infty$ would give an H atom and an H^+ ion. The potential curves for the H_2 and H_2^+ molecules are shown in Fig. 13.1.

The energy order of the electronic terms of normal, strongly bound molecules with small internuclear distances can be understood by referring to the combined-nucleus atom. A small separation of the nuclei, e.g. going from He^+ to H_2^+, gives rise to a strong electric field along the internuclear axis. This field affects the electronic terms in the manner known from the Stark effect in atoms. For more weakly bound molecules with larger internuclear distances, a better approximation is to start from the separated atoms. These two limiting cases can serve as an orientation and aid to understanding the term diagrams of molecules and the formation of molecular orbitals from atomic orbitals.

We start with the idea that we can form the molecule AB by allowing the atoms A and B to approach each other, simultaneously transforming the atomic orbitals of A and B into molecular orbitals AB. Then the following changes occur in the atoms (cf. Chap. 4):

- The central symmetry of the Coulomb potential is removed. An additional electric field along the molecular axis (z-direction) acts on the electrons.
- The electrons are associated with both atoms at the same time.
- The electronic terms which were originally degenerate are split.

This has the following consequences: first of all for the quantum numbers which characterise the states; at an infinite internuclear distance, the atomic electrons can be described

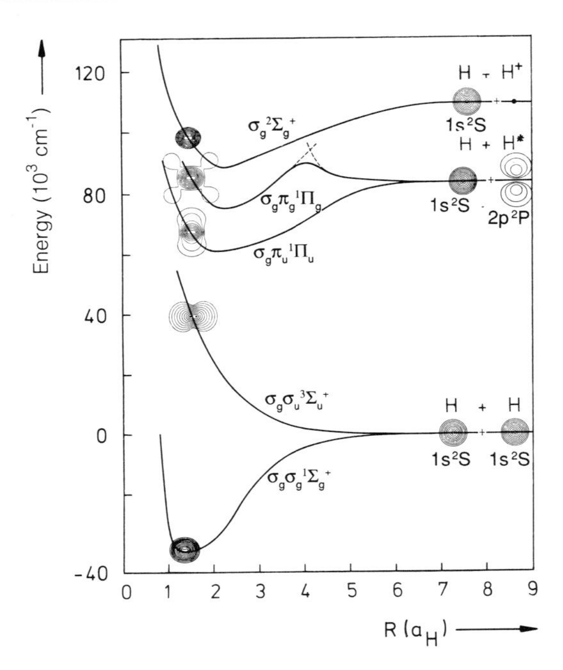

Fig. 13.1. The potential curves and electron density distributions for the ground and excited states of H_2 and H_2^+. These results were calculated using the Hund-Mulliken molecular orbital approximation method. The unit of the internuclear distance R here is the Bohr radius, $a_H = 0.529 \cdot 10^{-8}$ cm. After Hellwege

by eigenfunctions with the 4 quantum numbers n (principal quantum number), l (angular momentum), m_l (magnetic quantum number), and m_s (spin projection quantum number); but on close approach of the nuclei, whereby the Coulomb field loses its central symmetry, the angular momentum quantum number l is no longer a good quantum number, while the magnetic quantum number m_l, referred to the internuclear axis as quantisation axis, retains its validity. The orbital angular momentum $\boldsymbol{l}$ of the electrons precesses about the internuclear axis z with a quantised z-component. For this component, we find

$$l_z = m_l \hbar \, , \quad \text{with} \quad m_l = l, \; l-1, \ldots, -l \, . \tag{13.3}$$

The energy of these states in the axially symmetric field of the nuclei is, in contrast to its behaviour in a magnetic field, just the same for orientations of l_z along the $+z$-direction as along the $-z$ direction, since the effect of an electric field on a precessing electron is independent of the sense of the precessional motion. This degeneracy can, to be sure, be lifted by a perturbation, such as rotation of the molecule. Otherwise, the energy depends only on even powers of m.

For this reason, a new quantisation condition is introduced in molecular physics, characterised by the quantity λ:

$$\lambda = |m_l| = l, \; l-1, \ldots, 0 \, . \tag{13.4}$$

Although the orbital angular-momentum quantum number is no longer valid, it still has a meaning. For one thing, it is still approximately valid for the inner, well shielded electronic shells; but even where the interaction is stronger, it will tell us the origin of a particular atomic orbital which contributes to a molecular orbital. For this reason, the quantum number l is retained in molecular physics. We will now give a more detailed discussion of the notation for the electronic states with their quantum numbers. Orbital wavefunctions of electronic states with $\lambda = 0, 1, 2, \ldots$ are called σ, π, δ ... orbitals, analogously to the notation s, p, d

for electronic states in atoms with $l = 0, 1, 2 \ldots$. Molecular orbitals with $\lambda \neq 0$ are two-fold degenerate, corresponding to the quantum numbers $\pm m$. This degeneracy can be lifted by an additional perturbation such as a rotation of the whole molecule. As for the atomic electrons, the relations $l \leq n - 1$ and $\lambda \leq l$ apply. The various angular momentum states are thus denoted as in the following scheme:

m_l:	0	± 1	± 2	± 3
λ:	0	1	2	3
Symbol:	σ	π	δ	ϕ

Apart from the σ states, all the angular momentum states are doubly degenerate; furthermore, each of these states has two possible orientations of the spin relative to the quantisation axis, $m_s = \pm\frac{1}{2}$. Thus, σ states can hold 2 electrons and all other states can hold 4.

For an additional characterisation of orbitals with the same λ, the quantum numbers from the originally separated atoms are used. They are written after the symbol for λ, and the original term is denoted by an index, for example $\sigma 1s_A$, $\sigma 2p_B$, which expresses the fact that the corresponding electrons have the quantum number $\lambda = 0$ and were originally present, i.e. before the formation of the molecule, on atoms A and B as $1s$ or $2p$ electrons, respectively. We thus denote the molecular electronic states by the symbols $\lambda\, n\, l$.

The orbitals having the same value of λ can also be distinguished by putting the original quantum numbers n and l before the symbol for the molecular orbital, which is indicated by lower-case Greek letters , e.g. $1s\sigma$, $2s\sigma$, $2p\sigma$, or $2p\pi$. This notation is particularly common when the description begins with the combined-nucleus atom (AB). The quantum numbers n and l are then those referring to (AB). For example, a $3d\pi$ electron of a molecule is an electron with the quantum numbers $n = 3, l = 2$, and $\lambda = 1$. If the real molecule is approached from both limiting cases, infinite internuclear distance or zero internuclear distance, then the quantum numbers n and l which lead to a particular electronic function of the molecule are in many cases different for the originally-separated atoms and for the combined-nucleus atom. In contrast, the quantum number λ is always a "good" quantum number over the whole range from very small to very large internuclear distances. While the angular momentum precesses about the internuclear axis, which serves as quantisation axis, its projection on this axis remains constant.

Finally, one also indicates the symmetry of the orbitals. Orbital wavefunctions are termed "gerade" (from the German for "even"), index g, or "ungerade" (odd, index u) depending on whether they are symmetric or antisymmetric with respect to a centre of symmetry, i.e. to inversion. Thus, σ_g denotes an even, or inversion-symmetric, function and σ_u an odd function with $\lambda = 0$. This is particularly important in the case of homonuclear molecules.

If molecular orbitals are formulated by taking a linear combination of the atomic $1s$ orbitals from atom A and atom B, then one finds symmetric and antisymmetric combinations, i.e.:

$$\begin{aligned} \sigma_g 1s &= \frac{1}{\sqrt{2}}(\sigma 1s_A + \sigma 1s_B) \;, \\ \sigma_u 1s &= \frac{1}{\sqrt{2}}(\sigma 1s_A - \sigma 1s_B) \;. \end{aligned} \tag{13.5}$$

The function σ_u corresponds to an antibonding state, as was shown in Chap. 4. Antibonding states are also denoted by an asterisk, i.e. σ_u^*.

In (13.5), the factor $1/\sqrt{2}$ is a normalisation factor, as already introduced in Chap. 4; here we neglect the overlap integral S. For the hydrogen molecule, these two electron

Fig. 13.2. An energy level diagram of the molecular orbitals of the H_2 molecule. The linear combinations of the two s electrons $1s_A$ and $1s_B$ yield a bonding orbital σ and an antibonding orbital σ^*

Fig. 13.3. *Upper part*: In the H_2 molecule, the two electrons can both occupy the lowest, bonding orbital. *Lower part*: In the He_2 molecule, two electrons are in a bonding orbital and two in an antibonding orbital. The state is not bound, and a stable He_2 molecule does not exist in its ground state

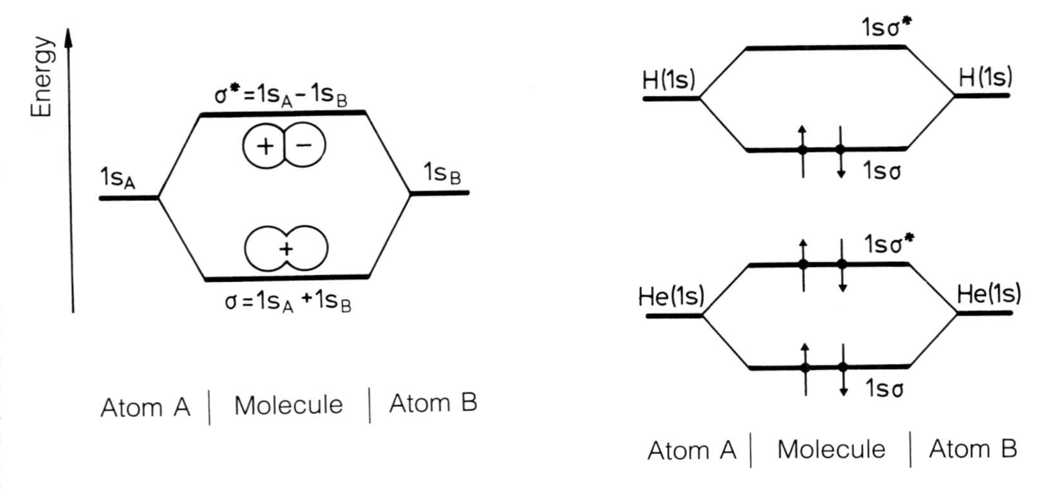

configurations are represented in Fig. 13.2. Each of these molecular orbitals can be occupied by at most two electrons, which differ in their spin projection quantum numbers $m_s = \pm 1/2$. The two electrons in the H_2 molecule can both occupy the bonding orbital σ, so that the molecule has a stable ground state. This is not the case for the molecule He_2, as shown in Fig. 13.3. Of the 4 electrons, two have to occupy the nonbonding orbital σ^*, so that the electron configuration is $1s\sigma^2\,1s\sigma^{*2}$. The ground state is thus unstable. However, if one of the electrons is excited from the antibonding orbital $1\sigma^*$ into a bonding state 2σ, then the bonding contribution predominates. The He_2 molecule, which is not stable in its ground state, thus has a bound excited state. A molecule of this type is termed an *excimer* (for excited dimer). Its electron configuration is then given by $1s\sigma^2\,1s\sigma^*\,2s\sigma$.

If we consider diatomic molecules with more electrons, then all the possible orbitals must be filled with electrons in the order of increasing energy, taking the Pauli principle into account. Each pair of electrons in the molecule, according to this principle, must differ in at least one of the four quantum numbers ($n, l, \lambda,$ and m_s). All the electrons with the same $n, l,$ and λ are combined into an electronic shell. Filled electronic shells have no spin or orbital angular momentum. Table 13.1 gives an overview of the possible electronic states and shells for $n = 1$ and 2.

Table 13.1. Possible electronic states in molecules. The atomic orbitals (AO) are combined into molecular orbitals (MO). Their spatial extent is indicated schematically, along with their multiplicities and spins

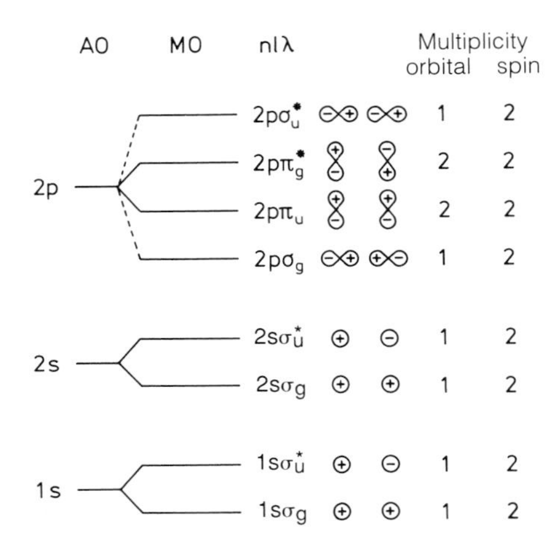

AO	MO	nlλ		Multiplicity orbital	Multiplicity spin
2p		$2p\sigma_u^*$		1	2
		$2p\pi_g^*$		2	2
		$2p\pi_u$		2	2
		$2p\sigma_g$		1	2
2s		$2s\sigma_u^*$	⊕ ⊖	1	2
		$2s\sigma_g$	⊕ ⊕	1	2
1s		$1s\sigma_u^*$	⊕ ⊖	1	2
		$1s\sigma_g$	⊕ ⊕	1	2

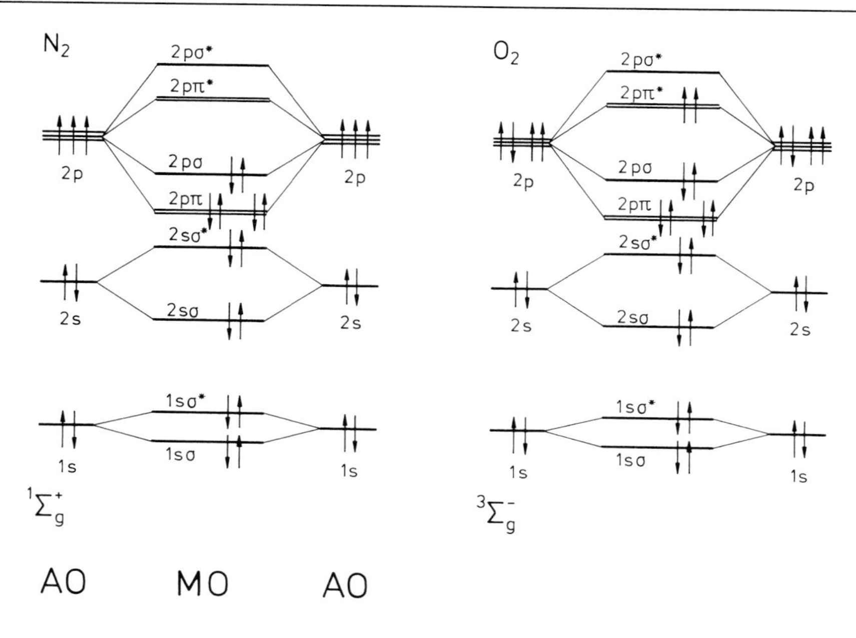

Fig. 13.4. *Left-hand part*: The molecular orbitals of N_2, showing their origins from atomic orbitals and their electron occupations in the ground state, following the molecular orbital scheme. The energy ordering of the molecular orbitals derived from the $2p$ atomic orbitals can be found from experiment and from detailed calculations including the electron-electron interactions. In a simple first-order picture, one would expect that the splittings $2p\sigma/2p\sigma^*$ and $2p\pi/2p\pi^*$ would be located symmetrically with respect to the original position, i.e. with respect to the atomic orbitals, as in Table 13.1. *Right-hand part*: The same scheme for the O_2 molecule

Figure 13.4 shows a further example, the lowest possible electron configuration for the nitrogen molecule N_2, as an MO (molecular orbital) diagram. The allowed electronic states are successively filled with the 14 available electrons. However, it should be noted that the order of the four highest orbitals in Fig. 13.4 does not agree with that given in Table 13.1. This is because the table contains the term ordering obtained from the separated-atom model, but in the N_2 molecule, a slightly different order is found for the molecular orbitals derived from the $2p$ atomic orbitals. This comes about because of the influence of the remaining (inner) electrons, which we have not yet considered. More about this subject is given in the correlation diagram, Fig. 13.5.

Two electrons each occupy the orbitals $1s\sigma$ and $1s\sigma^*$ and the orbitals $2s\sigma$ and $2s\sigma^*$ in Fig. 13.4. Of the remaining six electrons, four occupy the $2p\pi$ orbital, and the last two enter the $2p\sigma$ orbital. The six uppermost electrons in the energy-level diagram give rise to the bonding: they occupy 3 bonding orbitals. This corresponds to a triple bond, symbolised in the chemical literature by N≡N.

If we continue on from the N_2 molecule to the O_2 molecule (also shown in Fig. 13.4), we have two more electrons to place in orbitals. For them, the $2p\pi^*$ orbital is available. Its nonbonding character reduces the strength of the bonding in oxygen compared to N_2, giving a double bond, O=O.

Furthermore, according to Hund's rules, we expect that the spins of the two $2p\pi$ electrons will be parallel, since the $2p\pi$ orbital is only half filled with two electrons. This agrees with observations, which show the O_2 molecule to be paramagnetic, with the spin quantum number $S = 1$.

Alltogether, the electron configuration of the N_2 molecule can be given as $\sigma_g 1s^2 \sigma_u^* 1s^2 \sigma_g 2s^2 \sigma_u^* 2s^2 \sigma_g 2p^2 \pi_u 2p^4$, arising from the atomic electron configuration $1s^2 2s^2 2p^3$ of the N atom. In the O_2 molecule, we have an additional $\pi_g^* 2p^2$ molecular orbital.

Finally, Fig. 13.5 shows a so-called *correlation diagram*, indicating the relationship between orbitals at a large internuclear distance – where the ordering is obtained directly from the atomic orbitals of the separated atoms – and orbitals at a small internuclear distance, where it is determined by the combined-nucleus atom. Each orbital on the left side of the diagram becomes an orbital on the right side; the ordering of orbitals of different types, e.g. of π and σ orbitals, can change as a function of the intenuclear distance, with the electrons in a particular orbital having a bonding or an antibonding character. The internuclear distance increases from left to right in the diagram. We will not go into the details of this diagram, which indicates the energy ordering of the molecular orbitals. Light molecules such as H_2 lie to the left in the diagram, near to the combined-nucleus limit, while heavy molecules, such as P_2, lie to the right, near the separated-atom limit. Nitrogen is in between.

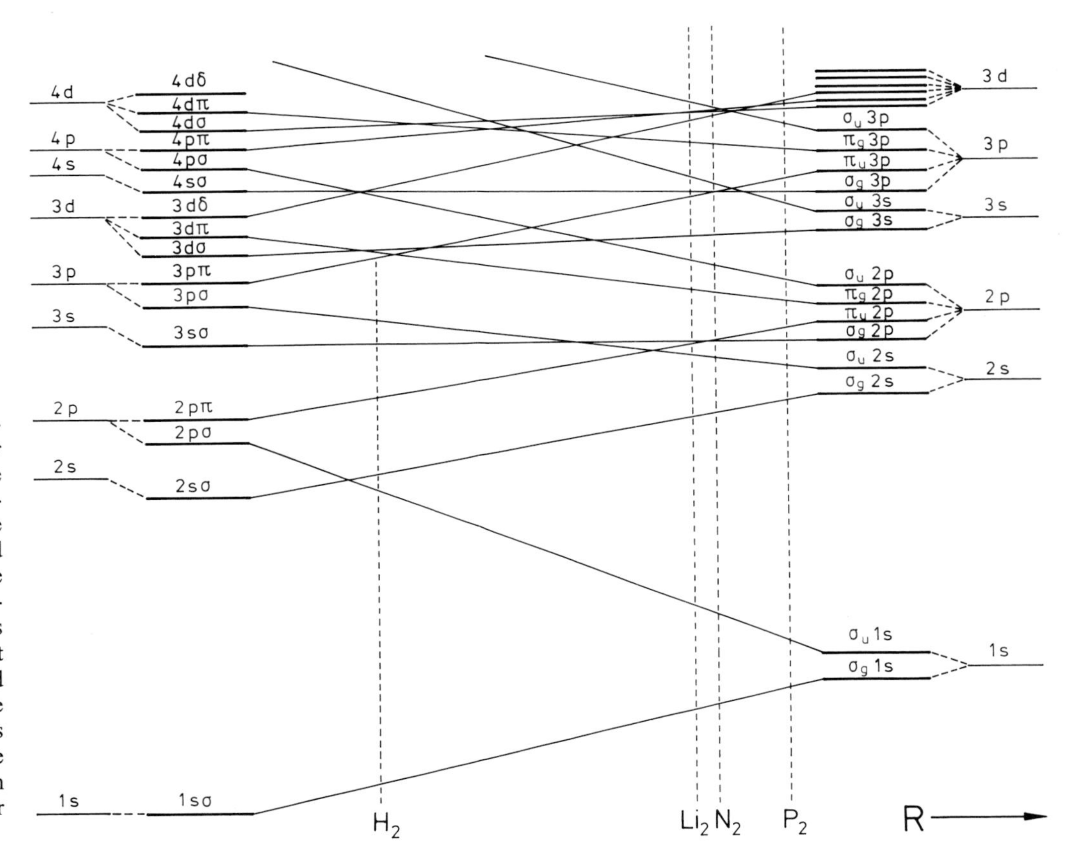

Fig. 13.5. Ordering of the molecular orbitals in a homonuclear diatomic system. At the left, the term symbols in the combined-nucleus atom are shown; on the right are those for the separated atoms, and in the centre, for the molecule. The connection between the left and the right sides of the diagram is in reality not linear; the lines were assumed here for simplicity, but the true variation of the term energies with R must be calculated. The positions of some molecules in the diagram are indicated. After Herzberg

13.4 Many-Electron States and Total Electronic States of Diatomic Molecules

For a molecule with several outer electrons, the mutual coupling of the electrons must be taken into account, in order to characterise the overall or *total electronic state*. As in an atom, the angular momenta of the inner electrons in closed shells add vectorially to give zero; a few outer electrons remain outside the closed shells. Their angular momentum coupling is referred to the molecular internuclear axis as quantisation axis. The coupling of the one-electron angular momenta $\boldsymbol{l}_i$, with the index i referring to the i-th electron, to a total orbital angular momentum $\boldsymbol{L}$ having the magnitude $\sqrt{L(L+1)}\,\hbar$ is taken into account by introducing a new quantum number Λ, which measures the component of $\boldsymbol{L}$ along the internuclear axis (i.e. m_L). It obeys the relation $\boldsymbol{L}_z = \Lambda\hbar$; Λ, in contrast to L, is a good quantum number even in the non-centrally symmetric potential of the molecule.

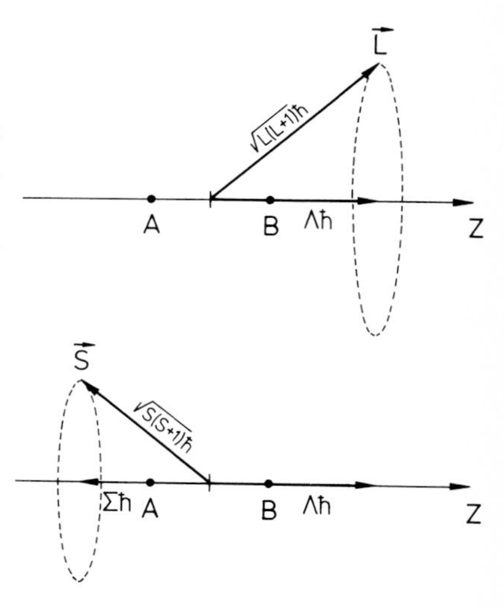

Fig. 13.6. Notation for defining the molecular angular momentum quantum numbers. *Upper part*: The orbital angular momentum $\Lambda\hbar$ as the component of $\boldsymbol{L}$ along the internuclear axis of the molecule AB. *Lower part*: The spin $\Sigma\hbar$ as the component of $\boldsymbol{S}$ along the magnetic field produced by $\Lambda\hbar$

We must keep in mind that the coupling of the $\boldsymbol{l}_i$ to one another to a resultant $\boldsymbol{L}$ in molecules is usually weaker than the coupling of each individual electron to the axially-symmetric field of the nuclei. The vectors $\boldsymbol{l}_i$ of the outermost electrons in unfilled shells therefore each precess alone around the internuclear axis with a quantised component $\pm\lambda$, where here, $\lambda = m_l$ [and not $|m_l|$ as in (13.4)]. For the resultant total orbital angular momentum along the internuclear axis, the quantisation condition of the axial or z-component is given by

$$\boldsymbol{L}_z = \pm\Lambda\hbar_i\,, \qquad \text{with} \qquad \Lambda = |\Sigma\lambda_i|\,; \tag{13.6}$$

see Fig. 13.6.

In this equation, the sum is algebraic, since all the components λ_i of the individual angular momenta lie along the internuclear axis. Due to the strength of the axially symmetric electric field, the coupling behaviour of the orbital angular momenta in the electronic shell of a molecule has a certain similarity to the Paschen-Back effect for atoms in a strong magnetic field. In general, one obtains several different overall states for a given electron configuration. These states are denoted by the values

$$\Lambda = 0, 1, 2, \ldots \tag{13.7}$$

and the corresponding symbols Σ, Π, Δ, ..., with a notation analogous to that for one-electron states, σ, π, δ ....

A Σ term, with $\Lambda = 0$, is simple; all the other terms are doubly degenerate, since each value of Λ corresponds to two opposite senses of precession of the electrons, according to (13.6). Here, again, states of different parity, even and odd states Σ_g, Σ_u, Π_g, Π_u, are possible.

Finally, we still have to consider the electron spins, i.e. the fourth quantum number m_s. The spins are hardly influenced by the electric field along the internuclear axis. Instead, they couple vectorially to a total spin $\boldsymbol{S}$ with the quantum number $S = \Sigma m_{s_i}$. Only the projection of the total spin on the quantisation axis, denoted by Σ, is of importance. The electronic motion produces a magnetic field which for $\Lambda > 0$ is parallel to the axis, and the vector $\boldsymbol{S}$ precesses around it, with only its component $\Sigma\hbar$ along the axis being quantised. This component can take on all integral values between $+S$ and $-S$; we thus have

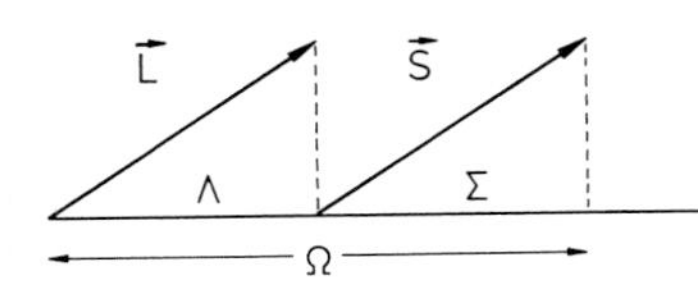

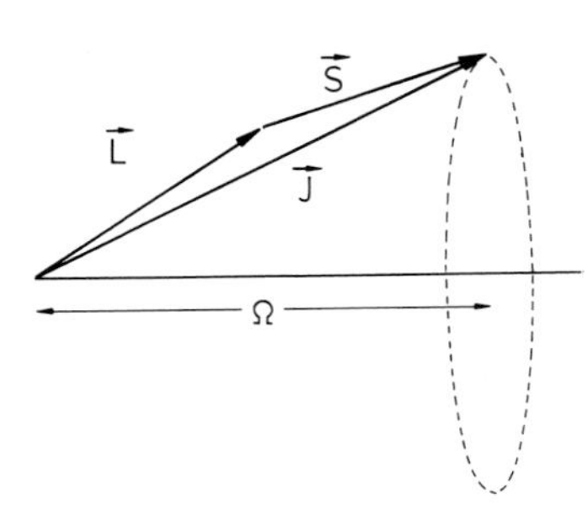

Fig. 13.7. The orbital angular momentum and the total spin yield the quantum number Ω as the sum of Λ and Σ, i.e. as the sum of the quantum numbers and therefore of the projections on the internuclear axis. The electronic angular momentum vector $\boldsymbol{J}$ is shown in the lower part of the figure. It should in fact be denoted by $\boldsymbol{J}'$ in order to distinguish it from $\boldsymbol{J}$ in Fig. 13.8

$$S_z = \Sigma\hbar \qquad \text{with} \qquad \Sigma = S, S-1, \ldots -S \tag{13.8}$$

(cf. Fig. 13.6).

Depending on the number of electrons, Σ is a whole integer or a half-integer.

The quantum number Σ is not to be confused with the term symbol Σ (see above), nor with a summation sign.

The associated quantum number S determines the multiplicity $(2S+1)$, which characterises each state of a given Λ. For two electrons, S can take on the values $S=1$ or $S=0$, i.e. the multiplicities 1 and 3 are possible, corresponding to singlet and triplet states.

As a result of the magnetic spin-orbit coupling between $\boldsymbol{L}$ and $\boldsymbol{S}$, each term belonging to a particular value of Λ splits into a muliplet of $2S+1$ terms. These are distinguished by their values of the quantum number of the resultant electronic angular momentum of the electronic shell along the internuclear axis, $\Omega = |\Lambda + \Sigma|$; see Fig. 13.7. Due to the strong coupling of the orbital and spin angular momenta to the axis, the axial component Ω of the total electronic angular momentum is in general more significant than the angular momentum itself. However, Λ and Σ can be oriented in the same direction as the internuclear axis or opposite to it; thus, for e.g. $\Lambda = 1$, $\Sigma = 1$, Ω can take on the values 2 or 0. The magnitude of the splitting can be calculated in terms of the spin-orbit interaction energy $W_{LS} = \boldsymbol{A L S}$. We shall not show the calculation here, however. This type of coupling of the one-electron angular momenta is not the only possible one; it corresponds to the coupling which we have referred to in atoms as Russell-Saunders coupling. Depending on the strength of the interaction between the various spin and orbital angular momenta, other coupling cases are possible.

It is important to realise that Ω is not a measure of the total angular momentum of the molecule, as is $\boldsymbol{J} = \boldsymbol{L} + \boldsymbol{S}$ in the case of an atom, but instead only of the electronic part. The total angular momentum also includes the important contribution from the molecular rotation, i.e. the dumbbell angular momentum $\boldsymbol{N}$. We shall return to this point later; see also Fig. 13.8.

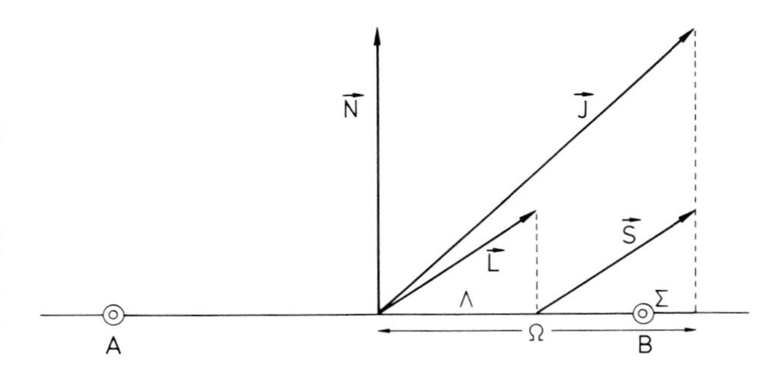

Fig. 13.8. The coupling of angular momenta in molecules. Orbital angular momenta and spins couple to the electronic angular momentum with component Ω along the internuclear axis. The dumbbell angular momentum of the molecular rotation is denoted by N. The vector sum of N and Ω gives the total angular momentum J. This is Hund's coupling case A

The quantum numbers of a molecular wavefunction are written in analogy to the atomic case as ${}^{2S+1}\Lambda_{\Omega}$; that is, one writes the multiplicity at the upper left and the quantum number Ω of the resultant angular momentum along the internuclear axis to the lower right of the Λ-term symbol. Frequently, Ω is left off. In addition, there are the symmetry symbols u and g (see above). In the case of homonuclear molecules, the eigenfunctions are even or odd depending on whether the molecule has an even or odd number of uneven electronic orbitals. We give as an example the state characterised by the configuration symbol $(2p\pi)(3s\sigma)(3d\pi)\,{}^4\Delta_{3/2}$; this means that there are three valence electrons with

$$n = 2, \quad l = 1, \quad \lambda = 1$$
$$n = 3, \quad l = 0, \quad \lambda = 0$$
$$n = 3, \quad l = 2, \quad \lambda = 1 .$$

For the orbital angular momentum we find

$$\Lambda = 1 + 1 = 2 , \qquad \text{therefore a } \Delta \text{ state} .$$

For the resultant spin, we have

$$S = \tfrac{1}{2} + \tfrac{1}{2} + \tfrac{1}{2} = \tfrac{3}{2} , \qquad \text{thus} \qquad 2S + 1 = 4 .$$

The spin and the orbital angular momentum are antiparallel, giving

$$\Omega = 2 - \tfrac{3}{2} = \tfrac{1}{2} .$$

Due to the multiplicity of 4, the configurations

$$^4\Delta_{7/2}, \ ^4\Delta_{5/2}, \text{ and } ^4\Delta_{3/2}$$

are also possible.

Furthermore, the term symbol takes into account whether a molecular function is symmetric or antisymmetric with respect to reflection in a plane through the internuclear axis. In this sense, Σ^+ and Σ^- mean symmetric or antisymmetric, respectively. Finally, the symbols g and u denote the parity, i.e. whether the wavefunction retains or changes its sign on inversion through a centre of symmetry of the molecule.

As a second example, we consider a system of two electrons, one of them in a σ_g orbital, the other in a π_u orbital. We obtain $\Lambda = 1$, i.e. we must have a Π state. Since there are two spins present, a triplet state with $S = 1$ or a singlet state with $S = 0$ are possible. One of the electrons has the index 'g' and the other the index 'u', so the overall wavefunction must be odd; we thus obtain the states $^3\Pi_u$ and $^1\Pi_u$ as possibilities for the configuration $\sigma_g\pi_u$. In a similar manner, one finds that, for example, a configuration π^2, i.e. two π electrons, can have the quantum numbers $\Lambda = 2$ and 0 and the overall states $^1\Sigma^+, {}^1\Sigma^-, {}^1\Delta, {}^3\Sigma^+, {}^3\Sigma^-$, and $^3\Delta$. Taking the Pauli principle into account, in the configuration $(2p\pi_u^-)^2$ for example, we find only the three possibilities $^3\Sigma_g^-$, $^1\Delta_g$, and $^1\Sigma_g^+$. This is demonstrated using the example of the electron configuration of the O_2 molecule in Table 13.2 and in Fig. 13.9. For the H_2 molecule, Fig. 13.10 shows some of the possible excited electronic configurations and the observed (simplified) term diagram of the lowest excited states. A complete term diagram for the singlet and the triplet systems of H_2 is given in Fig. 13.11.

In order to characterise a molecular term completely, the quantum symbols of the individual electrons are written before the symbol for the overall term.

The ground state of the H_2 molecule, which contains two electrons with the molecular quantum numbers $1s\sigma$, can be characterised by the symbol $(\sigma_g 1s)^2\,{}^1\Sigma_g^+$. Other electron configurations of the ground states of homonuclear diatomic molecules are shown in Table 13.3. It is noticeable that many, but not all of the ground states of molecules are singlet states, with $S = 0$. In Table 13.3, B_2 and O_2 offer exceptions to this rule, with $S = 1$ giving paramagnetic ground states.

When equivalent electrons are present in the molecule, i.e. electrons with the same quantum numbers n, l, and m or λ, then, as mentioned, we require the Pauli principle in order to decide which electronic states are allowed. To this end, we imagine the individual

Table 13.2. Possible combinations and quantum numbers for the electron configuration $(\pi 2p)^2$, corresponding to the ground state of O_2

λ_1	λ_2	s_1	s_2	Λ	Σ	State
1	1	+	+ −	Pauli	forbidden	–
1	1	+	−	2	0	$^1\Delta$
1	1	−	+			
−1	−1	+	−	−2	0	
−1	−1	−	+			
1	−1	+	+	0	1	$^3\Sigma^-$
−1	1	+	+			
1	−1	−	−	0	−1	
−1	1	−	−			
1	−1	+	−	0	0	$^1\Sigma^+$
1	−1	−	+	0	0	

electrons to be uncoupled, as we did in the atomic case, and assume a small internuclear distance R for the molecule, i.e. a strong electric field along the quantisation axis. Then each pair of electrons in the molecule must differ in at least one of the four quantum numbers $n, l, m_l = \pm\lambda$ and $m_s = \pm\frac{1}{2}$.

In this way, we arrive at the electronic shells of the molecule, i.e. at groups of electrons having the same n, l, and $\lambda = |m_l|$. An initial overview of this process is given by Table 13.1. For example, it becomes clear in this way that a given σ orbital can contain at most two electrons having antiparallel spins; in a π, δ, or higher orbital, there can however be four electrons with $m_l = \pm\lambda$ and $m_s = \pm\frac{1}{2}$.

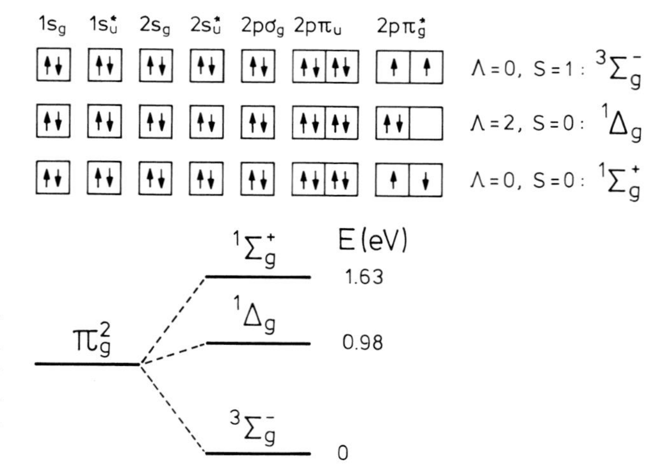

Fig. 13.9. Allowed configurations of the two outer electrons in the O_2 molecule, as an example of the configuration π_g^2, and the energies of the corresponding terms

An electron configuration $(1s\sigma_g)^2$ thus gives only the molecular state $^1\Sigma_g^+$, but the configuration $(2p\pi_u)^2$ gives three states, $^3\Sigma_u^-$, $^1\Delta_g$, and $^1\Sigma_g^+$. The indices g and u refer to a homonuclear molecule. Now, in order to denote the electronic states which actually occur in such a molecule, we have to consider how to distribute the available electrons among the allowed orbitals. By filling all the electrons into the orbitals in the order of increasing energy and taking the Pauli exclusion principle into account, we have defined the ground state of the molecule. For the H_2 molecule, for example, it is the $^1\Sigma_g^+$ state, in which both of the electrons occupy $1s\sigma$ orbitals. The ground state of the O_2 molecule, in agreement with Hund's rules for the successive filling of electronic shells, which we already know from atomic physics, is paramagnetic, $^3\Sigma_g^+$. See also Fig. 13.10.

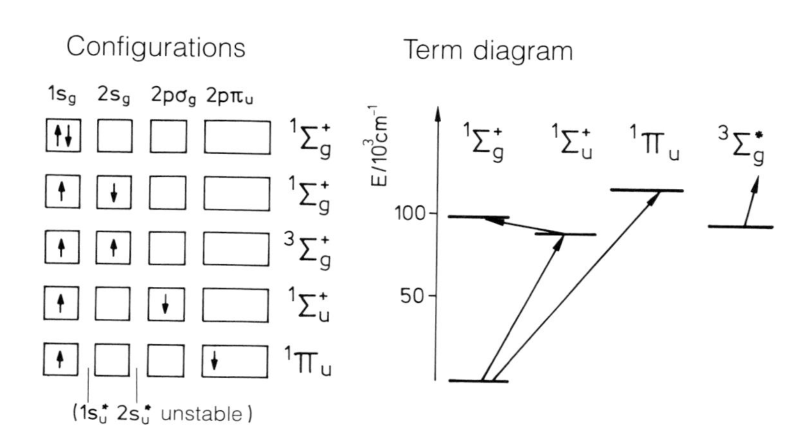

Fig. 13.10. The ground state and the lowest excited states of the H_2 molecule. The electron configurations and a simplified term diagram are shown

In order to obtain the energy ordering of the molecular orbitals, we start from the atomic orbitals in the two limiting cases of the combined-nucleus atom and the separated atoms, and make the connection to the molecular orbitals. This is indicated in Fig. 13.5 for diatomic, homonuclear molecules. With continuously increasing internuclear distance, the orbitals at the left side of Fig. 13.5 must be transformed into those on the right side. In drawing the (qualitative) correlation lines connecting these two limiting cases, one must keep in mind that orbitals of the same symmetry belong together. Lines between states of the same symmetry may not cross each other, as is shown by a quantum-mechanical calculation (lifting of the degeneracy!). In Fig. 13.5, the approximate positions of some molecules are indicated on the diagram. Hydrogen, H_2, lies near the combined-nucleus-atom limit; N_2, in contrast, is much nearer to the separated-atom limit. We are now in a much better position to understand the electron configurations of some homonuclear molecules which are given in Table 13.3.

Of course, the electron configurations described here determine not only the ground states of the molecules considered, but also their possible excited states. Figure 13.11 shows a term diagram derived in this way for the H_2 molecule, with singlet and triplet term systems. It also contains the notation for the terms and for the electron configurations.

Since the arrangement of the electrons also has a decisive influence on the bonding of the atoms in the molecule, it is understandable that a different potential curve is associated with each different arrangement of the electrons, i.e. with each excited state. As a rule, the outermost electron, the valence electron, is responsible for the bonding; if it is excited, the bond is generally weakened. For this reason, the equilibrium internuclear distance R_e of the

Table 13.3. Electron configurations of diatomic homonuclear molecules (the molecules in parentheses are not stable)

Molecule	Configuration								Ground state
	$\sigma_g 1s$	$\sigma_u^* 1s$	$\sigma_g 2s$	$\sigma_u^* 2s$	$\pi_u 2p$	$\sigma_g 2p$	$\pi_g^* 2p$	$\sigma_u^* 2p$	
H_2^+	↑								$^2\Sigma_g$
H_2	↑↓								$^1\Sigma_g$
He_2^+	↑↓	↑							$^2\Sigma_u$
(He_2)	↑↓	↑↓							$^1\Sigma_g$
Li_2	↑↓	↑↓	↑↓						$^1\Sigma_g$
(Be_2)	↑↓	↑↓	↑↓	↑↓					$^1\Sigma_g$
B_2	↑↓	↑↓	↑↓	↑↓	↑↑				$^3\Sigma_g$
C_2	↑↓	↑↓	↑↓	↑↓	↑↑↓↓				$^1\Sigma_g$
N_2	↑↓	↑↓	↑↓	↑↓	↑↑↓↓	↑↓			$^1\Sigma_g$
O_2	↑↓	↑↓	↑↓	↑↓	↑↑↓↓	↑↓	↑↑		$^3\Sigma_g$
F_2	↑↓	↑↓	↑↓	↑↓	↑↑↓↓	↑↓	↑↑↓↓		$^1\Sigma_g$
(Ne_2)	↑↓	↑↓	↑↓	↑↓	↑↑↓↓	↑↓	↑↑↓↓	↑↓	$^1\Sigma_g$

excited states is usually larger and the dissociation energy D is usually smaller than in the ground state. Figure 13.12 shows an example of this in the potential curves for the excited hydrogen molecule. These curves are calculated from the measured values of the dissociation energies and of the vibrational quanta.

We give here without further proof the selection rules for electric dipole transitions: they are $\Delta\Lambda = 0, \pm 1$; $\Delta S = 0$, i.e. intersystem crossings between states of different multiplicities

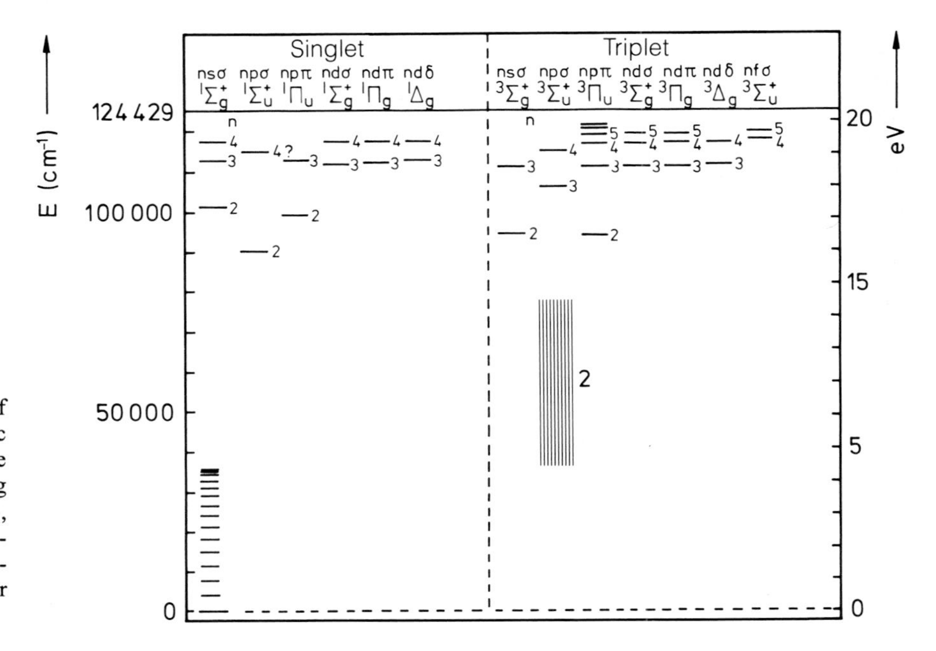

Fig. 13.11. Electronic terms of the H_2 molecule. The vibronic terms are indicated only for the ground state. The lowest-lying triplet state, $np_\sigma\, ^3\Sigma_u^+$ $(n = 2)$, is unstable; we have thus indicated only the dissociation continuum in the diagram. After Herzberg

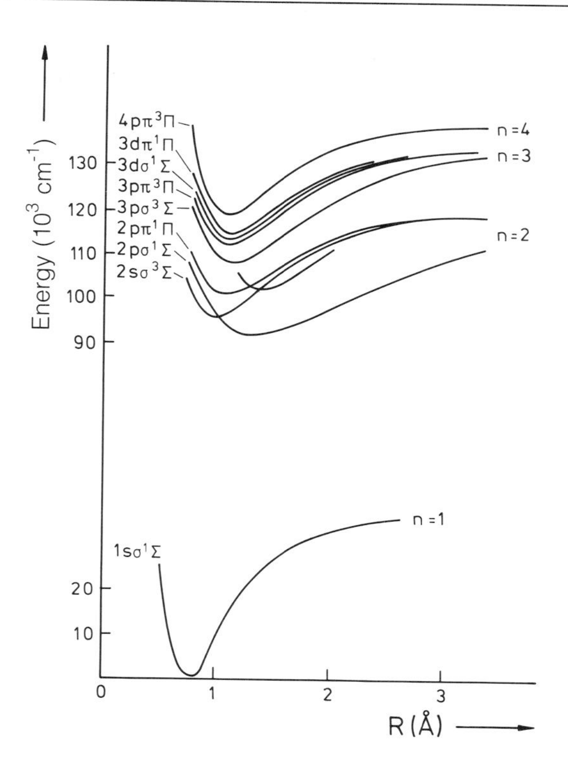

Fig. 13.12. Calculated potential curves for various excitation states of the H_2 molecule. After Finkelnburg

are forbidden. This rule is relaxed somewhat due to spin-orbit coupling, as is the case also in heavy atoms. An example of such intersystem bands is provided by the oxygen bands observed in the atmosphere, corresponding to the transition $^3\Sigma \rightarrow {}^1\Sigma$ of the O_2 molecule; see also Fig. 13.9. Furthermore, only transitions with a change in parity, u $\rightarrow$ g or g $\rightarrow$ u, are allowed.

The electronic states of larger molecules are classified according to the same principles as those of the diatomic molecules we have discussed here. We shall not treat this topic in detail.

As we mentioned above, in carrying out the coupling of angular momenta according to the scheme

$$\Lambda + S \rightarrow \Omega ,$$

where the vector $\boldsymbol{L}$ precesses about the internuclear axis and $\boldsymbol{S}$ about the magnetic field defined by Λ, we have still not taken the angular momentum of the molecular rotation into account. The total angular momentum of a molecule in fact is determined by the interaction of $\boldsymbol{N}$ with the electronic angular momentum $\boldsymbol{\Omega}$. We thus have $\boldsymbol{\Omega} + \boldsymbol{N} = \boldsymbol{J}$, where $\boldsymbol{J}$ denotes the total (spatially fixed) angular momentum of the molecule; see Fig. 13.8. Due to the fact that the rotation of the molecule as a whole produces an additional magnetic field, the possibilities for coupling of the various angular momentum vectors are numerous. The competition between the magnetic fields which give rise to the coupling between the vectors of the molecular rotation, the orbital angular momentum, and the electronic spin can give rise to a variety of different coupling cases, in a similar manner as we have seen in atomic physics; the Zeeman and the Paschen-Back effects describe the limiting cases.

These different coupling cases which can occur in the interaction between the electronic motions and the molecular rotation were explained and classified by *Hund*. They differ in the level of significance of the molecular axis as quantisation axis. Here, we have limited ourselves to Hund's Case A, out of the five cases A, B, C, D and E distinguished by Hund when the molecular axis dominates as quantisation axis (see Fig. 13.8). The remaining cases and their influence on the molecular terms and spectra, for example the molecular analogue of jj-coupling in atomic physics, will not be treated here, since that would go beyond the framework of an introductory text. Their precise treatment permits a complete analysis of the quantum numbers Λ, Σ, and Ω which characterise the electron configurations of the two electronic levels representing the initial and final states of a band system. It thus allows the elucidation of all the statements which can be made concerning a molecule on the basis of spectroscopic investigations.

14. The Electronic Spectra of Molecules

Electronic transitions in molecules lead to spectra which are extraordinarily rich in lines, because not only does the electronic state change during the transition, but so also do the vibrational and rotational states. The analysis of the resulting band spectra can be extremely tedious and complex. In Sects. 14.1 through 14.4, we discuss the most important concepts for the understanding and the analysis of these spectra. Section 14.6 again contains some remarks on the extension to larger molecules.

14.1 Vibrational Structure of the Band Systems of Small Molecules; The Franck-Condon Principle

In the preceding chapter, we considered the electron configurations of molecules as though they were stationary and rigid. We now proceed to include the vibrations of the atomic nuclei in our model. We recall the Born-Oppenheimer approximation discussed in Sect. 11.1, which allows us to separate the electronic and nuclear motions in thought, and to a considerable extent in practice, as well.

In this section, we will concern ourselves with the electronic band spectra of molecules, i.e. with electronic transitions which are accompanied by changes in their vibrational states. This defines in a certain sense the overall structure of the spectra. The finer structure can be described only by including also the effects of molecular rotations on the spectrum, which we shall do in the following section.

For each electronic state, we expect to find a specific potential curve giving the potential energy of the nuclear configuration as a function of the internuclear distance coordinate, as we have already seen in Chap. 10. In addition to the energies E_0 and E_1 for the ground and excited states of the electron, the most important quantities which we need to know and understand are the equilibrium internuclear distance R_e and the dissociation energy D in the various electronic states; cf. Figs. 13.1 and 13.12. The equilibrium distances R_e are usually larger in the excited states than in the ground state, since the bonding is normally weakened by excitation of an electron. The reverse case can occur, however, e.g. when the excitation lifts an electron from an antibonding orbital into a bonding orbital. Some additional examples of potential curves for excited states are given in Fig. 14.1, where we have chosen the O_2 molecule as an example. The potential curves shown there are those associated with the states given in the term diagram of Fig. 13.9. Each potential curve contains vibrational states, as we have already discovered in Chaps. 10 and 11. The frequency and thus the quantum energy of a particular normal mode of the molecule in general changes from one electronic state to another. This is particularly easy to understand in the case of a diatomic molecule: the frequency of the stretching vibration is determined by the binding force between the two

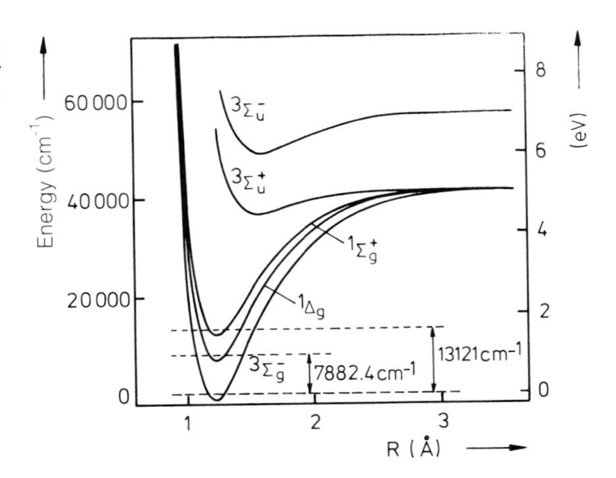

Fig. 14.1. Potential curves for the lowest electronic states of O_2; compare Fig. 13.9. After Herzberg

nuclei, and that force changes on electronic excitation. The vibrational quantum number v is denoted as v'' in the lower state and as v' in the upper state when referring to transitions.

Let us first consider the *absorption* of radiation. Electronic transitions from the ground state of the molecule, which has an electronic energy E''_{el} and the vibrational quantum number $v'' = 0$, into an excited electronic state generally lead to not only one absorption line, but instead to a large number of lines: a band or band system. The band structure is based upon the fact that, as explained in Fig. 14.2, the electronic transitions are accompained by changes in the vibrational quantum number; this means that the transitions are of the type $E''_{\text{el}}(v'') \rightarrow E'_{\text{el}}(v')$. Furthermore, a rotational structure is also superposed onto these vibrational transitions; we shall say more about this in the following section. The intensity of such electronic-vibronic transitions is determined by the transition matrix elements (and by the electronic selection rules), which we will treat in more detail later, especially in Sect. 16.4. We will see there that the purely electronic part of these matrix elements is independent of the vibrational quantum number to a first approximation. For the vibronic part of the transition matrix elements connecting vibrational states with the quantum numbers v' and

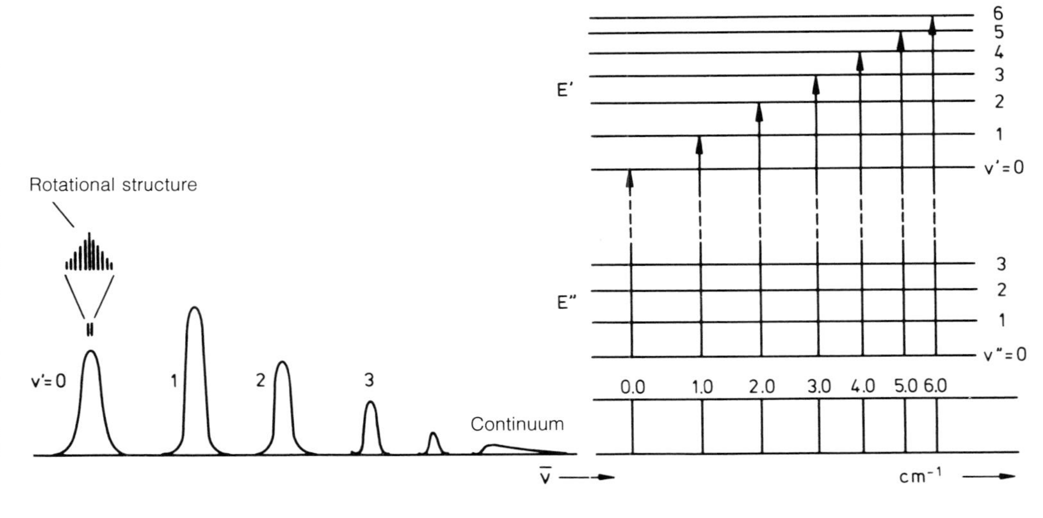

Fig. 14.2. *Left part of figure*: A schematic representation of a band system: transitions originate in the ground state ($v'' = 0$) and lead with differing intensities to various vibrational levels having the quantum numbers $v' = 0, 1, 2, \ldots$. The rotational structure is superposed onto these transitions. *Right part*: The vibrational structure in absorption from the electronic ground state E'' and from the level $v'' = 0$ into an excited electronic state E' with the vibrational levels $v' = 0, 1, \ldots 6$, as it may be observed in the spectrum (schematic)

v'', there are no strict selection rules; there are only rules which result from the important and intuitively evident *Franck-Condon principle*. This principle makes statements about the probabilities of the individual vibronic transitions, and therefore about the line intensities in the band spectrum, as well as about the general structure of the spectrum.

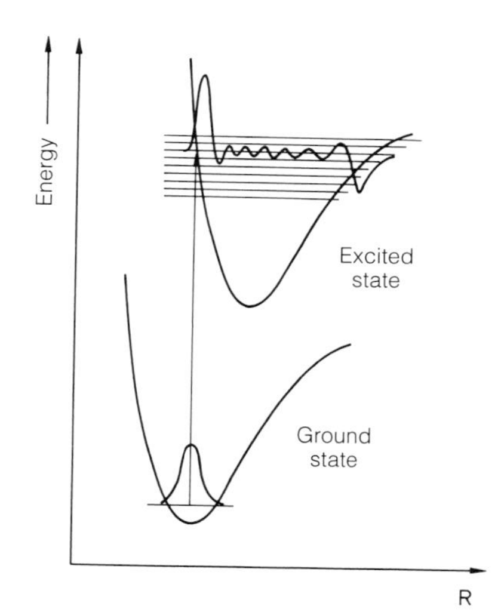

Fig. 14.3. An explanation of the Franck-Condon principle. Absorption from the vibrational ground state is strongest for that particular transition into a vibrational level of the excited electronic state whose wavefunction maximum lies directly over the maximum of the ground state in the potential diagram. Transitions to other vibrational terms are also possible, but with reduced probabilities. The total energy and two potential curves are plotted as functions of the nuclear coordinates (internuclear distance R). The zero points of the two curves are shifted due to the electronic excitation. Vibrations occur within each of the two potential curves. Here, the wavefunctions for $v'' = 0$ and $v' = 6$ are shown as an example

We will explain the Franck-Condon principle by referring to Fig. 14.3, initially leaving off the rotational terms and considering only the vibrations. It takes into account in an intuitively clear way the fact that the electronic motions are rapid in comparison to the nuclear motions: during an electronic transition, the position and velocity of the nuclear coordinates will thus not change noticeably. Electronic transitions then occur mostly vertically in the diagram shown in Fig. 14.3; that is, they maintain the internuclear distance R, and their highest probability is between those parts of the vibrational functions for which the amplitudes, and thus the occupation probabilities of the nuclei at that position, are greatest.

In a classical picture, the nuclei spend the most time at the turnaround points of the vibration, i.e. the points where the vibrational levels intersect the potential curves. (This is true in a quantum-mechanical picture, also, with the exception of the lowest vibrational level, $v = 0$: in this level, the occupation probability is highest in the centre of the level, midway between its intersections with the potential curve.) The transitions thus take place with the highest probability from or to these intersection points (or to the centre of the level $v = 0$). Owing to the finite width of the probability distribution, however, there is not a single sharp transition with a well-defined vibrational excitation energy, but rather a whole series of vibrational transitions of differing transition probabilities to neighbouring levels. The quantum-mechanical formulation (Sect. 16.4) states that the transition probability is determined by the *Franck-Condon integral*:

$$\int \chi_{v'}(\boldsymbol{R})\chi_{v''}(\boldsymbol{R})\, dV_{\text{nucl}} \,, \tag{14.1}$$

i.e. the overlap integral of the nuclear vibrational function $\chi_{v''}$ associated with the electronic ground state E'', and the function $\chi_{v'}$ belonging to the excited electronic state E', at the same internuclear distance $\boldsymbol{R}$, integrated over the entire volume of the molecule (nuclear coordinates); cf. Sect. 16.4.

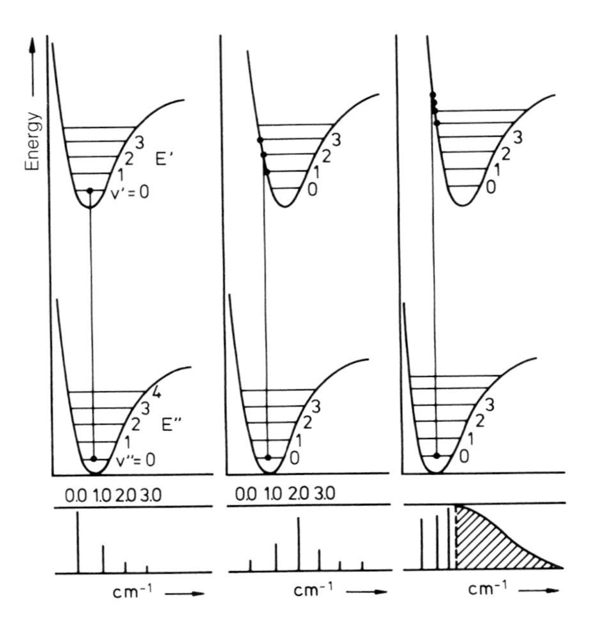

Fig. 14.4. The Franck-Condon principle. Depending on whether the internuclear distance in a molecule stays the same (*left*), increases somewhat (*centre*), or increases strongly (*right*) on electronic excitation, different intensity patterns are found for the various vibronic bands associated with an electronic transition. In each case, only one transition is indicated, beginning in the region of the highest occupation probability for the initial state, $v'' = 0$. In the spectra, the maximum intensity will be observed for the 0,0 transition (*left*), for a higher vibrational state (*centre*), and near to the dissociation limit (*right*)

In the special case that the internuclear distance does not change on electronic excitation, $R''_e = R'_e$, then the two potential curves have their minima directly above one another. The vertical transition from $v'' = 0$ to $v' = 0$ is then strongest and all the other vibronic transitions are weaker, assuming that the molecule is initially in its electronic ground state with $v'' = 0$. Compare Fig. 14.4, left-hand part.

However, usually the equilibrium distance is larger in the excited state, i.e. $R''_e < R'_e$, and the excitation tends to weaken the bonding. If only the $v'' = 0$ level, the zero-point vibrational level, is thermally occupied in the electronic ground state, then one will observe transitions from $v'' = 0$ into several v' levels of the excited electronic state, depending on the values of the Franck-Condon integrals. The energy differences between the initial and final states, that is the quantum energies of the vibronic lines in the band spectrum, are then given by

$$\begin{aligned}\Delta E = h\nu = E'_{\rm el} + h\nu'_e[(v' + \tfrac{1}{2}) - x'_e(v' + \tfrac{1}{2})^2] \\ - \{E''_{\rm el} + h\nu''_e[(v'' + \tfrac{1}{2}) - x''_e(v'' + \tfrac{1}{2})^2]\}\,. \qquad (14.2)\end{aligned}$$

In this expression, $E''_{\rm el}$ and $E'_{\rm el}$ are the electronic energies in the ground state and the excited electronic state, respectively, and $h\nu''_e$ and $h\nu'_e$ are the vibrational quanta in the ground and excited states, extrapolated to the equilibrium internuclear distance as described in Sect. 10.3. Since we have as yet not taken the rotation of the molecules into account, (14.2) is valid only for the zero lines of the bands. It can also be used to calculate the band edges (see Sect. 14.2).

In summary, the different intensity distributions within the band spectra shown in Fig. 14.4 can be understood. If the equilibrium internuclear distance R_e in the two electronic states is the same, then the bands with $v' \to v'' = 0 \to 0,\ 1 \to 1,\ 2 \to 2$ are the most intense. If the excited electronic state has a somewhat larger bond length, ($R'_e > R''_e$), then the vertical transition from v'' into e.g. the upper vibrational level $v' = 2$ can be observed, and it will be the strongest line in the absorption spectrum (Fig. 14.4, centre). Transitions to larger or smaller values of v' are less probable, and one finds the intensity distribution shown in the figure. If the change in bond length on electronic excitation is very large, as in Fig. 14.4 (right-hand part), then vertical transitions will lead to levels with still larger values of v', and even into the region of the dissociation continuum.

When absorption of radiation is accompanied by an increase in the bonding strength of the molecule, then the band spectrum has a different intensity distribution. The line with the smallest quantum energy no longer has the highest intensity; the distribution is similar to that shown in the centre of Fig. 14.4. In this case, one observes in addition to the line or band corresponding to $v'' = 0 \rightarrow v' = 0$ (0,0 line) also lines with $v' = 1$, $v' = 2$, etc., which are however more intense than the 0,0 line.

This intensity distribution in the band systems, which can be understood in terms of the Franck-Condon principle, is clarified with the help of a two-dimensional diagram, in which the transitions between two vibrational terms are plotted on a coordinate system v', v''. In these so-called *edge diagrams*, Fig. 14.5, the quantum number v'' of the vibrations in the lower electronic state is used as the abcissa and the quantum number v' of the vibration in the upper electronic state as ordinate. The intensity of the bands corresponding to transitions $v'' \rightarrow v'$ or $v' \rightarrow v''$ is then plotted in the resulting coordinate system. If the potential curves in the upper and lower electronic states are directly above one another (Fig. 14.4, left-hand part), the most intense bands lie on a diagonal in the edge diagram; otherwise (Fig. 14.4, centre and right-hand part), they lie on a parabola of variable curvature. In Fig. 14.5, the left branch of the parabola corresponds to absorption, and the right branch to emission. Instead of the band intensities of the observed transitions, one can also plot the quantum energies of the bands in an edge diagram; in this way, the term differences of the vibrational terms can be displayed and determined in a clearcut manner.

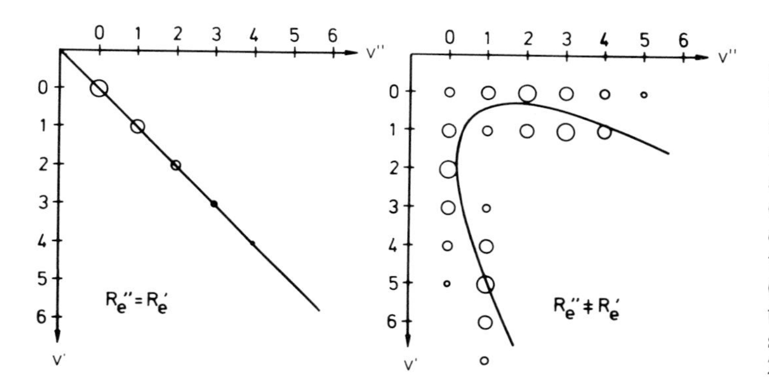

Fig. 14.5. Edge diagrams of band spectra for the case of no change in the internuclear distance on excitation (*diagonal line*) and for the case of a change on excitation (*parabolic curves*). The upper branch of the parabola refers to emission (v' = const., $v'' = 2, 3, 4 \ldots$); the lower branch refers to absorption (v'' = const., $v' = 2, 3, 4 \ldots$)

It follows from the Franck-Condon principle that the band system has very different shapes depending on how the internuclear distance changes on electronic excitation. The two extreme cases are particularly important: either no change of the internuclear distance on excitation, or an extremely large change. These two limits, called the *group spectrum* and the *series spectrum*, are explained in Fig. 14.6.

In the case $R_e' = R_e''$, the group spectrum shown in the left part of Fig. 14.6 is obtained. It consists of the transitions $\Delta v = 0$ (i.e. the diagonal line in the edge diagram) and the subdiagonals, with $\Delta v = \pm 1$ (at a lower intensity). The spectrum thus contains a few groups of closely-spaced bands; such a spectrum can be seen for example in the band spectra of CN and of C_2 (observable in a carbon arc). In the case that $R_e' > R_e''$, one sees in absorption, in

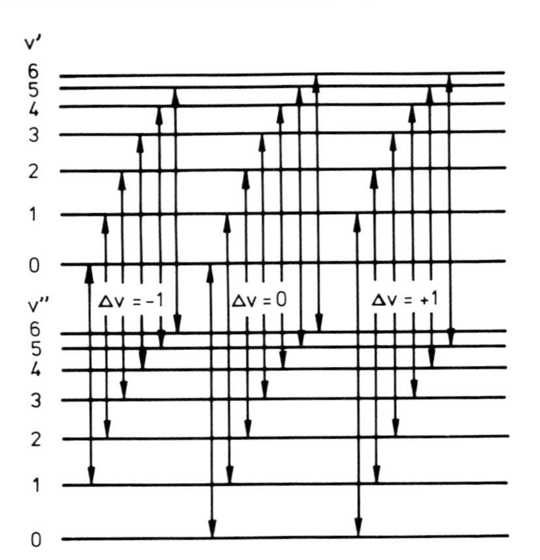

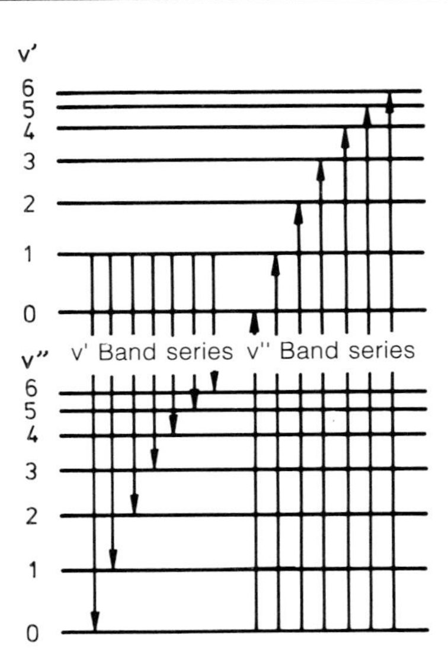

Fig. 14.6. The most intense vibrational bands in a "group spectrum" (*left part*; internuclear distance in both electronic states stays roughly the same) and in a "series spectrum" (*right part*, corresponding to a large change in the internuclear distance). Compare also Figs. 14.4 and 14.5

contrast, a series of bands having the same lower levels (cf. Fig. 14.4). In emission, there is a shifted series of bands having the same upper levels. Such a series spectrum, which can be observed for example from the I_2 molecule, is shown on the right in Fig. 14.6. The spacings of the bands give directly the vibrational quanta in the different electronic states.

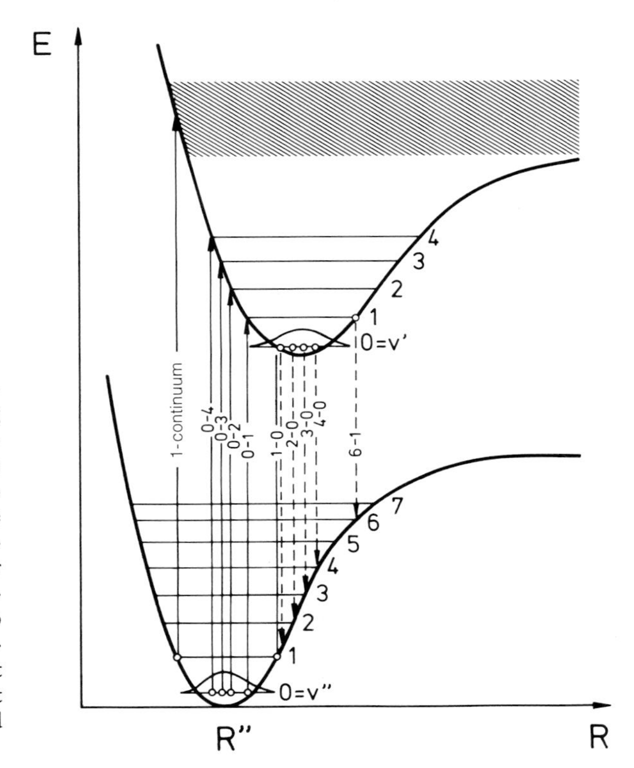

Fig. 14.7. Absorption and fluorescence (*dashed*) according to the Franck-Condon principle. When the potential curves of the ground and excited electronic states are shifted relative to one another, then the absorption transitions occur at higher quantum energies than the emission transitions. This is due to relaxation of the vibrational energy in the excited state, so that emission occurs from the lowest vibrational level of the excited electronic state

In this way, the intensity distribution within the vibronic components of the spectrum, i.e. within a band system, gives information on the change of the equilibrium internuclear distance R_e between the molecular ground and excited states. This is readily understood with the help of the Franck-Condon principle.

In *emission*, one in principle observes the same band spectra as in absorption. Many investigations of molecules are indeed carried out in emission, since it is readily observed in a gas discharge. Here, again, the Franck-Condon principle applies and one observes "vertical" transitions from the upper potential curve to that of the ground state, as shown in Fig. 14.7. However, the initial state for emission is not necessarily identical with the endpoint of an absorption process. The coupling of the radiation field to the excited state reached through absorption is often relatively weak, so that collisions with other molecules between the absorption and the emission processes permit relaxation of the excited molecules, accompanied by an adjustment to a new equilibrium internuclear distance.

How rapidly such a relaxation process of the excited state and its thermalisation take place depends sensitively on the probability of interactions of the molecules with their environment, i.e. on the pressure in the case of a gas sample. A molecule isolated in outer space has very few possibilities to give up its excess vibrational energy to its environment, because the probability for the emission of vibrational quanta as radiation having the same frequency as the vibrations is small, and collisions with other molecules, which would allow direct exchange of energy, are also rare. For a molecule in a condensed phase, on the other hand, the thermalisation process, that is the establishment of an equilibrium value of the vibrational energy corresponding to kT (T = absolute temperature), takes place on a time scale of picoseconds. The emission spectrum of a molecule thus depends not only on the temperature, but also on its possibilities to give up vibrational quanta with $v' \neq 0$ before the emission occurs – it then begins preferentially at the level $v' = 0$. This is indicated schematically in Fig. 14.8; it is clear from this figure that a sort of mirror-image relation between the absorption spectrum and the emission spectrum exists, with the 0,0 ($v' = v'' = 0$) transition at its midpoint.

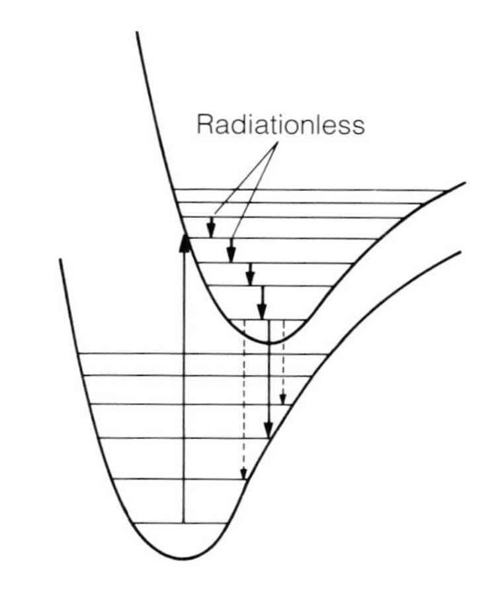

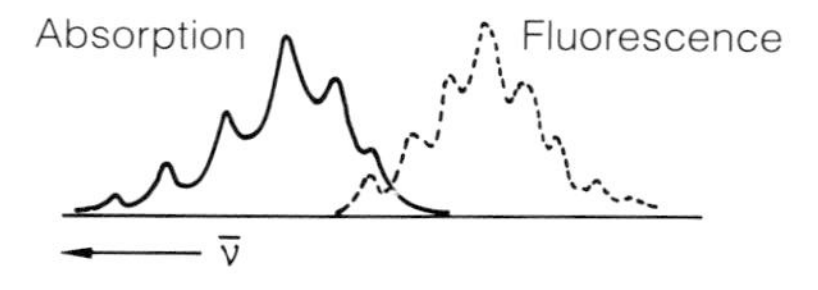

Fig. 14.8. Absorption and fluorescence in a molecule. In the absorption spectrum, the vibrational structure of the excited electronic state can be observed; in fluorescence, that of the ground state is seen. The fluorescence spectrum is shifted towards lower energies relative to the absorption spectrum. The vibrationless 0,0 transition ($v' = v'' = 0$) can be common to both spectra if the shift is small, and the remaining parts of the spectra are then mirror images. The linewidth shown here is typical of a spectrum taken in the condensed phase (in solution). Rotational structure is not resolved

14.2 The Rotational Structure of Electronic Band Spectra in Small Molecules; Overview and Selection Rules

The most noteworthy characteristic of molecular band spectra in comparison to atomic line spectra is the enormous number of lines they contain, when observed at a sufficiently high resolution. One requires not only a high spectral resolution of the apparatus, but also the elimination of Doppler and pressure broadening of the lines (cf. I, Sect. 16.2), as far as possible. Then a quite distinct "fine structure", consisting of a very large number of lines, is observed for the electronic-vibronic transitions described in the previous section. This is due in the main to rotational transitions. The spectrum of the iodine molecule, for example, exhibits more than 20 000 lines in the visible spectral region! A small fraction of these lines has already been explained as due to vibrational structure. In Fig. 14.9 (left-hand part), this is shown once again. However, the very large number of lines can be understood only by considering that each of the vibrational levels which we have treated thus far also contains rotational structure (Fig. 14.9, right-hand part). Without a sufficiently high spectral resolution of this structure, only "band edges" are seen, with a shading effect, i.e. a continuous decrease of the emission or absorption intensity towards one side of the band.

We shall at first limit ourselves to the simplest case of a diatomic dumbbell molecule. For the wavenumber of a transition between two terms,

$$T = T^{\mathrm{el}} + G_v + F_{v,J}, \tag{14.3}$$

we have

$$\begin{aligned} \bar{\nu} &= T' - T'' = T'^{\mathrm{el}} - T''^{\mathrm{el}} + G'_{v'} - G''_{v''} + F'_{v',J'} - F''_{v'',J''} \\ &= \Delta T^{\mathrm{el}} + \Delta G + \Delta F \; . \end{aligned} \tag{14.4}$$

The spectrum which one obtains on changing the electronic energy, ΔT^{el}, contains all the vibrational bands, ΔG_v, with their associated rotational structure, ΔF. The band systems are to be found mainly in the visible and ultraviolet spectral regions and, with all of the electronic transitions taken together, they make up *the band spectrum* of the molecule.

For electric dipole transitions, the most important selection rules are the following:

$$\Delta J = J' - J'' = 0, \pm 1 \qquad (\text{except for } J' = J'' = 0) \tag{14.5}$$

and

$$\Delta\Lambda = \pm 1 \qquad \text{and} \qquad \Delta\Sigma = 0 \tag{14.6}$$

(insofar as the spin-orbit coupling is not too strong). Furthermore, the parity selection rule, which requires that the initial and final states for the transition have different parities, applies here.

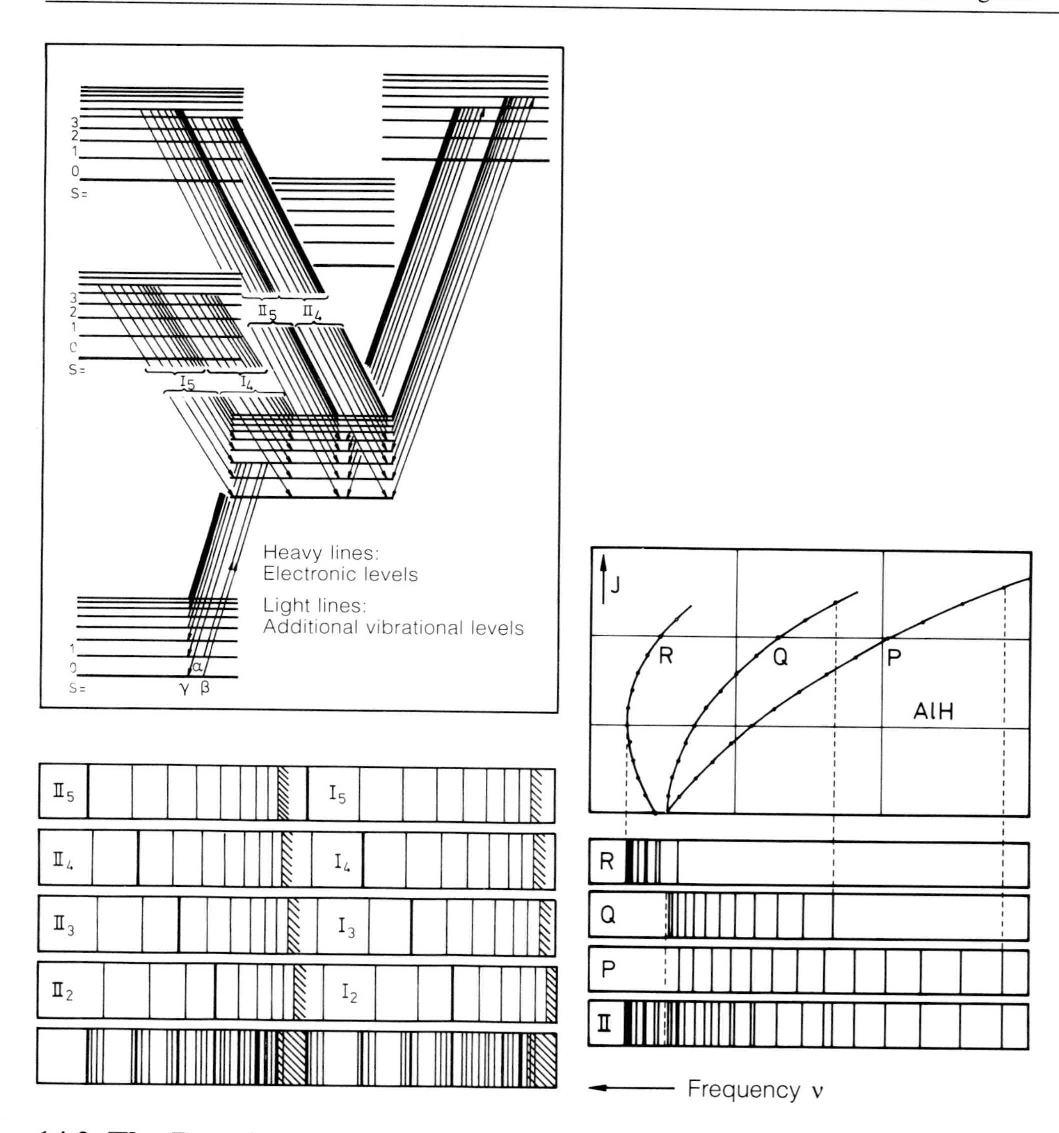

Fig. 14.9. *Left part*: An explanation of the large number of lines in a band spectrum. Each transition arrow in the left part of the figure refers not to an individual spectral line, but to the edge or zero line of a whole band; compare the right part of the figure. Below, some band systems from the term diagram are plotted schematically; each line here refers to the whole band. The observed spectrum is the superposition of band systems, which can also be combined into series of band systems. After R.W. Pohl. The vibrational quantum numbers are denoted by s in this figure, and the symbols for the subspectra are historical. The *right-hand part* of the figure shows on an enlarged frequency scale a band of the molecule AlH which is shaded towards its long-wavelength side. Its edge line is at $\lambda = 435$ nm, and the rotational structure is resolved. The observed spectrum (II) is the superposition of three branches (P, Q, R). These are plotted above as a Fortrat diagram on the right-hand side of the figure

14.3 The Rotational Structure of the Band Spectra of Small Molecules; Fortrat Diagrams

The typical structure of a band spectrum, including rotational transitions, is shown in Fig. 14.10. It can be decomposed into three "branches", which we have already introduced in Sect. 10.4 and called the P, Q, and R branches. An empirical description of the lines of a band spectrum was given more than 100 years ago in terms of the *Deslandres formula* (1885):

$$\bar{\nu} = A \pm 2Bm + Cm^2 \, . \tag{14.7}$$

In this formula, the index $m = \frac{1}{2}, \frac{3}{2}, \frac{5}{2}, \ldots$ is incremented beginning at a gap in the series of lines which is called a zero point; the two signs of the second term correspond to the R branch (+) and the P branch (−). For $B = 0$, the Q branch is obtained.

The explanation of the rotational structure of the band spectra should be clear in the light of what we have already said, and will be discussed by referring to Fig. 14.10. Each vibrational level with the quantum number v' or v'' has its associated rotational levels,

$$E_{J''} = B''hcJ''(J''+1)$$

and

$$E_{J'} = B'hcJ'(J'+1) \,,$$

if we consider only the limit of the rigid rotor. We must, however, take into account the fact that the rotational constant B'' in the lower potential curve is different from the constant B' in the upper curve. The lines in the band spectrum are due to transitions between potential curves, which correspond to the electronic energies E'' and E', and associated levels with the quantum numbers v'', J'' in the lower and v', J' in the upper electronic states (potential curves).

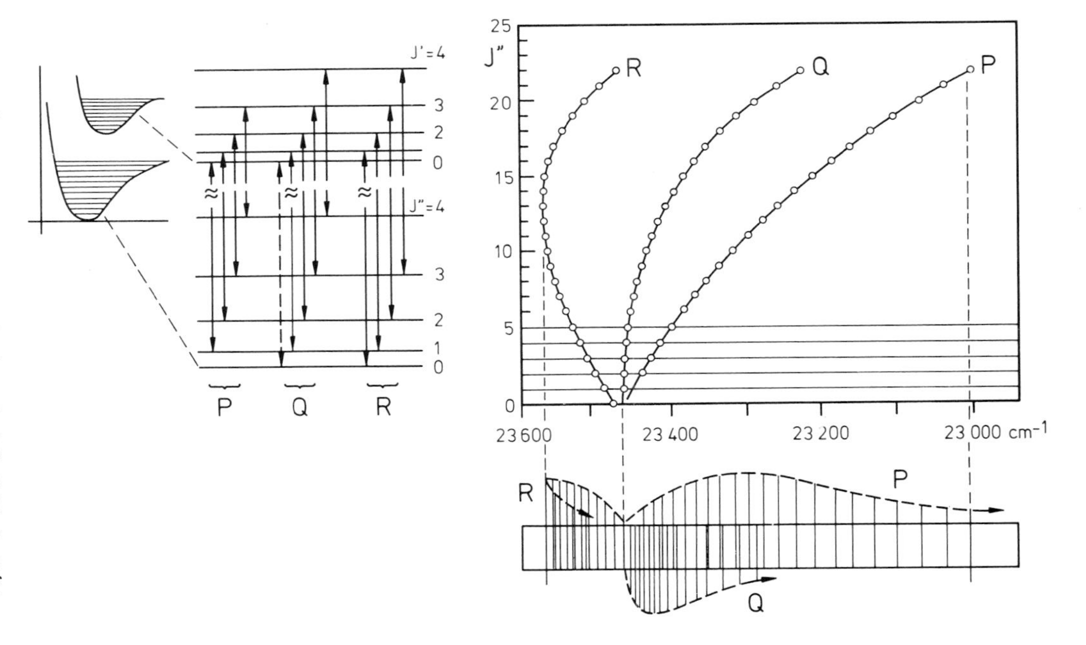

Fig. 14.10. The explanation of the rotational structure of bands. In the left half of the figure, the relation of the band branches P, Q, and R to the term diagram is shown schematically. For the analysis of the spectra, the wavenumbers of the lines are plotted in a Fortrat diagram against an index, which here is the quantum number J''. The observed total spectrum can thus be decomposed into its P, Q, and R branches. From the line spacings, one can determine the rotational constants B' and B''. The right half of the figure shows a spectrum of AlH, as in Fig. 14.9. Note the direction of the frequency axis

The overall energy of a molecular state is thus given by

$$E_{\text{tot}} = E_{\text{el}} + E_{\text{vib}} + BhcJ(J+1) \,, \tag{14.8}$$

where E_{el} refers to the electronic energy and E_{vib} to the vibronic energy. For a transition between two states we then have, in simplified form,

$$\Delta E_{\text{tot}} = \Delta(E_{\text{el}} + E_{\text{vib}}) + \Delta[BhcJ(J+1)] \tag{14.9}$$

and for the observable spectral lines,

$$\overline{\nu} = \overline{\nu}_{v',v''} + \Delta[BJ(J+1)]\,, \tag{14.10}$$

where $\overline{\nu}_{v',v''}$ refers to the electronic-vibronic transition without rotation, $J'=J''=0$, between the states (E'', v'') and (E', v'). This 0,0 transition (with $J'=J''=0$) is thus the reference point for the rotational structure. It is also called the zero line of a band; cf. Sect. 14.1. The selection rules for J depend on the type of electronic transition.

If both the electronic states which participate in the transition, i.e. the upper and the lower state, have no angular momentum relative to the molecular axis and are thus $^1\Sigma$ states, then the following selection rule holds:

$$\Delta J = \pm 1\,;$$

otherwise, the selection rule is

$$\Delta J = 0 \quad \text{or} \quad \pm 1\,, \quad \text{but not} \quad J'' = 0 \quad \text{to} \quad J' = 0\,.$$

From (14.10), we obtain

$$\overline{\nu} = \overline{\nu}_{v',v''} + B'J'(J'+1) - B''J''(J''+1)\,. \tag{14.11}$$

In contrast to the rotational-vibrational spectra which we treated earlier (Sects. 10.4 and 10.5), the rotational constants B' and B'' belong not only to different vibrational levels in the same electronic state, where they are usually not too different in magnitude, but instead to two different electronic states. Due to the fact that on electronic excitation of a molecule, the internuclear distance R and thus the moment of inertia can change by a large amount, B' and B'' will now often have quite different values. The quadratic terms in (14.11) are thus not just a small correction, as they were in the case of rotational-vibrational spectra, but instead can predominate over the linear terms for larger values of J.

We now denote the transitions with $\Delta J = -1$ as the P branch, those with $\Delta J = +1$ as the R branch, and those with $\Delta J = 0$ as the Q branch, as in Chap. 10; then for the spectral lines of these three branches, we obtain the following expressions:

P branch, $\Delta J = -1$, $J'' = J' + 1$

$$\overline{\nu}_P = \overline{\nu}_{v',v''} - (B'+B'')(J'+1) + (B'-B'')(J'+1)^2$$
$$\text{with } J' = 0, 1, 2 \ldots \tag{14.12}$$

R branch, $\Delta J = +1$, $J' = J'' + 1$

$$\overline{\nu}_R = \overline{\nu}_{v',v''} + (B'+B'')(J''+1) + (B'-B'')(J''+1)^2$$
$$\text{with } J'' = 0, 1, 2 \ldots \tag{14.13}$$

Q branch, $\Delta J = 0$, $J' = J''$

$$\overline{\nu}_Q = \overline{\nu}_{v',v''} + (B'-B'')\,J'' + (B'-B'')\,J''^2$$
$$\text{with } J'' = 1, 2, 3 \ldots\,. \tag{14.14}$$

It is important to note that no line is allowed at the band origin, $\overline{\nu}_{v',v''}$.

The expressions (14.12–14) for the allowed rotational transitions have the same form as the empirical relation (14.7) and demonstrate the physical significance of the coefficients A

$-C$ introduced there. We obtain the old Deslandres formula by identifying the index m with J.

A graphical representation of (14.12–14) as a $\overline{\nu}/J$ plot yields parabolas, which are called *Fortrat parabolas* (Fig. 14.11), since $\overline{\nu}$ is a quadratic function of J. These Fortrat diagrams present a clear overview of a band spectrum, because they show the spectral lines belonging to different spectral branches and different rotational quantum numbers, which in the measured spectrum are often mixed together, in a spatially ordered and separated form. Conversely, the observed spectrum of the bands is obtained from the Fortrat diagram by projecting the points in the diagram onto the $\overline{\nu}$ axis; cf. Figs. 14.9 and 14.10.

The band origin $\overline{\nu}_{v',v''}$, also called the zero line, is missing in all three branches. If $B' < B''$, meaning that $R_e' > R_e''$ and thus $\Theta' > \Theta''$, then the lines of the P branch lie on the low-energy side of the band origin, and their spacing increases with increasing J, while the R branch lies on the other side, with rapidly decreasing line spacing. In this case, as can be seen in Fig. 14.11, the order of the lines can even reverse for large J values and bend back towards lower energies. Bands with this characteristic are termed red-shaded. The lines of the Q branch likewise lie on the low-energy side of the origin, and the line spacing also increases with increasing J. In the less common case that $B' > B''$, i.e. $R_e' < R_e''$ and $\Theta' < \Theta''$, meaning that the electronic excitation increases the strength of the bonding, the shape of the Fortrat parabolas is reversed, and the bands are violet-shaded. The overall spectrum of a band is a superposition of the three (or two) branches.

In the special case that the internuclear distance remains unchanged on electronic excitation, and therefore $B' = B''$, one can see from (14.14) that the Fortrat parabolas degenerate to straight lines. The Q branch then consists of a single line, and the P and R branches are a series of equally spaced lines with the spacing $2B$, as in the pure rotational-vibrational spectra. There are also transitions with $\Delta J = \pm 2$, which however have much reduced intensities. They give rise in the spectrum to O and S branches.

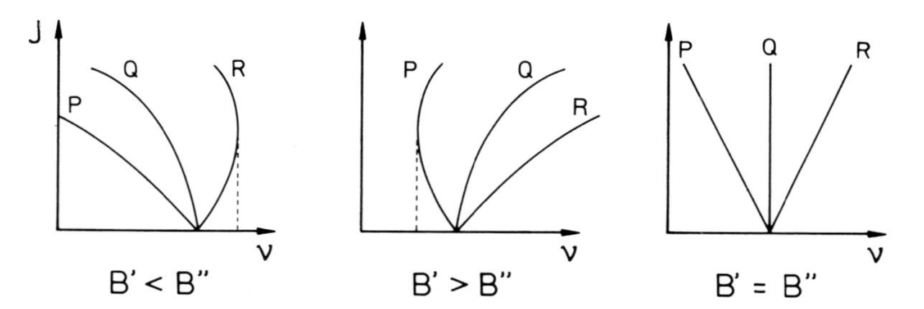

Fig. 14.11. Plots of the band lines as Fortrat diagrams (cf. Fig. 14.10) yield characteristic curves, depending on whether the rotational constant in the upper state is smaller, larger, or the same as in the lower state

Thus the "shading" of a band spectrum allows direct conclusions to be drawn as to whether the rotational constant of the molecule increases or decreases on electronic excitation, i.e. in the dumbbell model, whether the bond length decreases or increases on excitation.

The complete analysis of the numerous lines in a band spectrum is a tedious process, both in terms of their measurement and in terms of data reduction, and would in many cases be a nearly impossible task without the aid of computers. However, as we have already shown, even a rough analysis gives information about changes in the form of the molecule on electronic excitation. The rotational structure of even a *single* band is sufficient for the spectroscopic identification of the molecule; it is thus a kind of molecular fingerprint. In this way, for example in astrophysical observations the CO molecule could be detected in the atmosphere of the planet Venus by identification of its absorption bands.

When molecules become more complex, that is when they contain more than two atoms and are no longer linear, a complete analysis of their band spectra including the entire rotational structure becomes much more complicated. The spectra then contain many branches which interpenetrate and overlap one another, making an analysis of all the lines nearly impossible.

If, however, a quantitative analysis of the rotational structure of all the bands of a band system is possible, then it yields numerical values for a number of quantities characteristic of the molecule investigated:

- The rotational constants B' and B'' for all the vibrational levels v' and v'', and from them the moments of inertia for molecular rotation at the equilibrium internuclear distances R_e in the two electronic states participating in the transition, as well as the centrifugal stretching constant α (cf. Sect. 10.4), and finally the internuclear distances R'_e and R''_e of the two states;
- the vibrational wavenumbers of the molecular vibrations in the two electronic states including their anharmonicities, i.e. also the calculated quantities $\overline{\nu}'_e$ and $\overline{\nu}''_e$ as well as x'_e and x''_e;
- the spacing $\overline{\nu}(v' = 0, v'' = 0)$ between the lowest levels of the two potential curves, and from it the calculated energy spacing without vibrations;
- and finally, when the angular momentum coupling is analyzed (which we have not discussed in detail), the quantum numbers Ω' and Ω''.

When transitions to and between a number of different electronic excited states can be investigated, then the corresponding potential curves of the molecule, and thus its excitations, can be studied.

Furthermore, electronic band spectra with their complete rotational and vibrational structure can of course also be measured for symmetrical molecules such as H_2, N_2, and O_2, for which pure rotational or rotational-vibrational spectra are unobservable, or observable only with very low intensities, due to the lack of an electric dipole moment. The rotational and vibrational structure which accompanies allowed electronic transitions thus becomes observable.

Finally, we mention without giving further details that the intensity distribution of the lines in a band spectrum can be used for a spectroscopic determination of the temperature of the gas being investigated. The occupation of the rotational and vibrational levels in thermal equilibrium is determined by the temperature, and is in turn responsible for the relative intensity of the transitions in the spectrum.

14.4 Dissociation and Predissociation

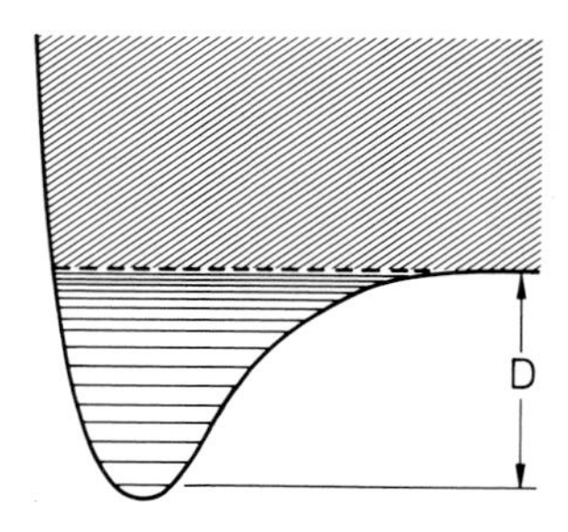

Fig. 14.12. The potential curve of a small molecule. The region of discrete vibrational levels merges into a region of continuous energy levels above the dissociation energy D

As can be seen in Fig. 14.12, the dissociation limit of a molecule merges into a continuum of states, and the potential curve becomes horizontal for large values of R. If a molecule is so strongly excited that its total energy lies in the region of the dissociation continuum of one of the electronic states, then it can dissociate. An excitation of this kind can take place in thermal collisions, at a sufficiently high temperature; however, here we wish to consider the possibility of dissociation and of the determination of the dissociation energy by means of the absorption of radiation.

The direct photodissociation of a molecule, as in the reaction $AB + h\nu \rightarrow A + B$, by *rotational-vibrational excitation* alone, without electronic excitation, is nearly always impossible. As a single-quantum process, it is forbidden by the selection rules for vibrational transitions, which allow only transitions with small changes in the vibrational quantum number v, as well as by the Franck-Condon principle (cf. Fig. 14.13). As a result of the availability of infra-red lasers with very high photon fluences (for example the CO_2 laser), photodissociation as a multiple-quantum process, with absorption of a number of vibrational quanta $h\nu_{\mathrm{vib}}$ according to the reaction scheme $AB + nh\nu_{\mathrm{vib}} \rightarrow A + B$, or $A^* + B^*$, has become feasible and is in fact used in molecular physics for the production of molecular fragments. The asterisk on A^* and B^* means that these fragments are in electronic excited states. We must, to be sure, keep in mind that such a multiple-quantum process is not to be understood as a cascade of successive multiple absorptions of a vibrational quantum $h\nu_{\mathrm{vib}}$; instead, it takes place via virtual states. (See Sect. 17.2 for a discussion of virtual states.) As we have seen, the spacing of the vibrational levels, i.e. the magnitude of the vibrational quanta, becomes smaller and smaller as the dissociation limit is approached along the potential curve (Sect. 10.3). The dissociation of a molecule AB according to the reaction scheme

$$AB + nh\nu_{\mathrm{vib}} = A + B^* \ (+E_{\mathrm{kin}}) \qquad (14.15)$$

is therefore not possible as a simple cascade process.

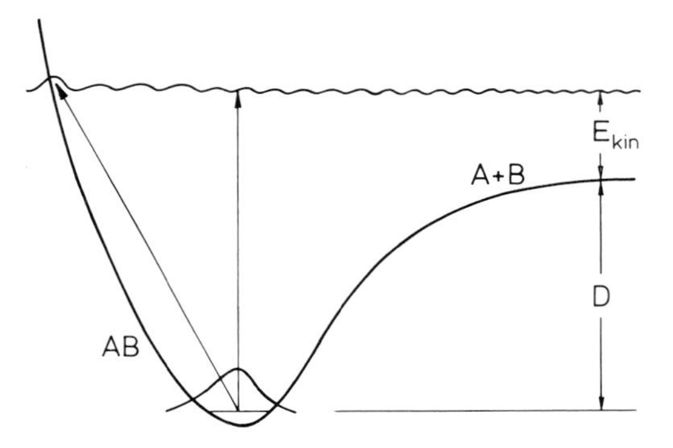

Fig. 14.13. The dissociation of a molecule by absorption of radiation without simultaneous electronic excitation is very improbable according to the Franck-Condon principle; it is in fact nearly impossible. The slanted transition arrow in the figure is forbidden because of the accompanying large change in the internuclear distance; the vertical transition is forbidden because it would require a sudden change in the velocities of the vibrating nuclei. In addition, it would involve a large Δv

Absorption of radiation can, however, readily lead to dissociation if it is accompanied by an *electronic excitation* into a higher-lying electronic state, and an additional strong vibrational excitation results, assuming that the two participating potential curves are suitably shifted relative to one another. According to the Franck-Condon principle, transitions from the vibrational ground state of the lower electronic state, i.e. $v'' = 0$, can be observed for a converging band series $v' = 1, 2, \ldots$ of the excited electronic state up to the point of convergence K. For the corresponding wavenumber of the point of convergence, $\overline{\nu}_K$, it follows from Fig. 14.14 that

$$hc\overline{\nu}_K = D'' + E_{\mathrm{At}} \, . \qquad (14.16)$$

Here, D'' refers to the dissociation energy in the lower electronic state, and E_{At} to the excitation energy of the atomic fragments A and B resulting from the dissociation. If this energy is in fragment B alone, i.e. B^*, and if it emits its excitation energy as a quantum of radiation, then the determination of E_{At} is straightforward; in other cases, it can often be problematic. Depending on which of the molecular or atomic excitation energies in Fig. 14.14 are known, the observation of the convergence point K can lead to the quantities D'' or D', that is the dissociation energy in the ground state or in the excited state. It must be remembered that the calculated dissociation energy, relative to the minimum of the potential curve, is to be distinguished from the real energy D which begins at the vibrational ground level $v'' = 0$; cf. Sect. 10.3. In any case, the dissociation energies of numerous small diatomic molecules such as H_2, O_2, and I_2 have been determined in this way. The molecular excitation energy E_{Mol} in Fig. 14.15 can be found directly from the transition $v'' = 0$ to $v' = 0$ in the band system. From it, the dissociation energy in the excited state,

$$D' = hc\overline{\nu}_K - E_{\text{Mol}} , \tag{14.17}$$

can be determined. In the case of the I_2 molecule, the spectrum of which contains a very large number of lines in the visible spectral region, the dissociation energy can be readily determined, and its measurement is even suitable as an experiment for a teaching laboratory. For the point of convergence to the continuum limit, one finds by extrapolating the spacing of neighbouring band edges to high values of the quantum number v', i.e. to small vibrational quanta, the value $K = 2.48$ eV. With $E_{\text{At}} = 0.94$ eV, it follows that I_2 has a dissociation energy $D'' = 1.54$ eV.

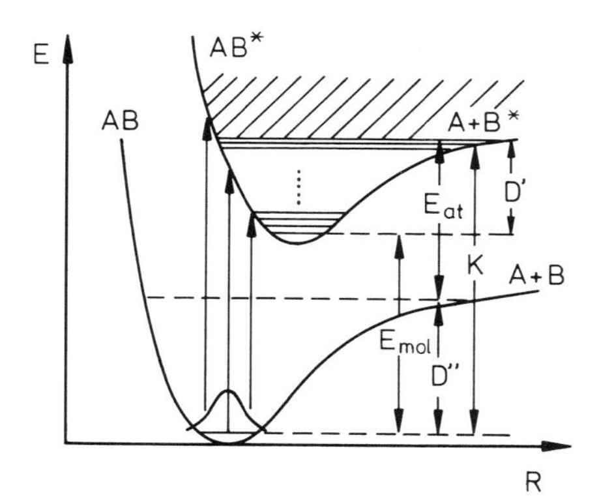

Fig. 14.14. The determination of the dissociation energy of a small molecule in its ground state (D'') and in an excited electronic state (D'). The excitation energy of the excited atom resulting from the dissociation is denoted by E_{At}, and that of the molecule by E_{Mol}, while K is the experimentally determined convergence point of the bands

If the point of convergence cannot be determined directly or precisely, then the spacing of neighbouring band edges, $\Delta\overline{\nu}$, is plotted as a function of v' for the transitions from $v'' = 0$ and extrapolated to $\Delta\overline{\nu} = 0$. From the formula (14.2) for the band edge,

$$\overline{\nu} = C_0 + C_1 v' + C_2 v'^2 , \tag{14.18}$$

it follows for the differences between the states v and $v + 1$:

$$\Delta\overline{\nu} = (C_1 - C_2) - 2C_2 v' , \tag{14.19}$$

i.e. a linear decrease of $\Delta\overline{\nu}$ with increasing v. The extrapolation to $\Delta\overline{\nu} = 0$ yields the quantum number v_k for the vibrational term corresponding to the band edge, and from it, using the edge formula,

$$\frac{K}{hc} = C_0 + C_1 v_k - C_2 v_k^2 \,. \tag{14.20}$$

If the excitation takes place with a quantity of energy $E > K$, then the excess energy can appear as kinetic energy of the fragments according to the relation

$$E_{\mathrm{kin}} = E - K \qquad \text{or} \qquad AB + E \to A + B^* + E_{\mathrm{kin}} \,. \tag{14.21}$$

Occasionally, one observes band series which are smeared out some distance below the dissociation limit, i.e. in a region which lies below the convergence point for the vibrational quanta, and which show no rotational structure. It can be demonstrated by chemical means that dissociation products result from the absorption of radiation by these smeared-out bands. This possibility of a dissociation induced by radiation which has a longer wavelength than that corresponding to the dissociation continuum is known as *predissociation*. This term refers to a radiationless transition of the molecule from a discrete rotational-vibrational level (v'', J'') of a bound electronic state into the dissociation continuum of another electronic state.

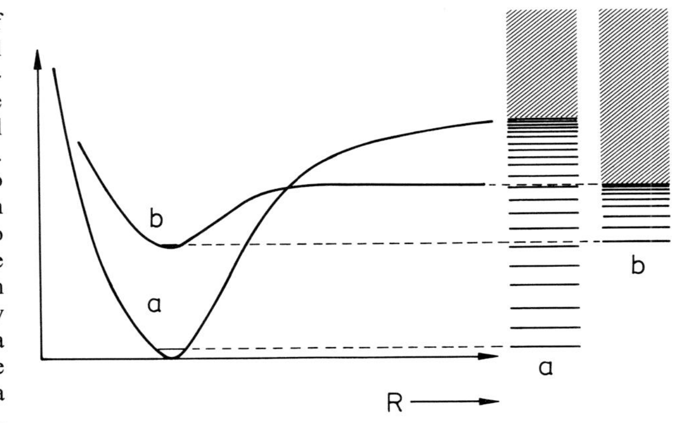

Fig. 14.15. The explanation of predissociation: two potential curves a and b of a small molecule, which cross each other, are illustrated with their vibrational terms and dissociation continua. At the crossing point of the two curves, a radiationless transition of the molecule from state a to the electronically excited state b can take place; this state can then dissociate with less energy than state a. This leads to a shortening of the lifetimes of the vibronic levels in a and thus to a line broadening in the spectrum. If, for example, light absorption from a lower-lying state a' (not shown) excites the molecule into state a, one would expect a series of vibronic bands up to the dissociation continuum of a, assuming that the potential curves are suitably shifted relative to one another. At the crossing point with the potential curve b, a dissociation can result from the excitation of a' to a, even when the excitation energy is smaller than the dissociation energy of potential curve a. This is the phenomenon of predissociation

Predissociation is explained in Fig. 14.15. The molecule being investigated has the potential curve a in the figure. By excitation with radiation, one can populate discrete levels below the dissociation limit and continuous states above it. If the molecule in an excited state has a potential curve like b, i.e. if it can dissociate from the state b with less energy than from state a, then there is a possibility that when it is in an excited level of a in the neighbourhood of the crossing point of the two potential curves, it can undergo a *radiationless* transition into the state b and dissociate with less energy than would be necessary from state a. The probability of such a transition is especially large at the crossing point, since the Franck-Condon principle applies to such radiationless processes and the two molecular states are the same both in terms of their energies and in terms of their nuclear coordinates at the crossing point. The bands belonging to state a become diffuse owing to the uncertainty relation for energy and lifetime, because the possibility of a transition from a to b shortens the lifetimes of the rotational-vibrational levels in a. This, in turn, leads to a broadening of the lines, so that the various rotational lines of a given band overlap those of the neighbouring bands.

At the crossing point of the two potential curves in Fig. 14.15, a transition can take place in a time which is typically of the order of 10^{-10} to 10^{-13} s. The lifetimes of the vibronic levels without the possibility of a radiationless deactivation is of the order of 10^{-8} s. The ratio of these times gives a measure of the shortening of the lifetimes of the levels and of the corresponding line broadening in the spectrum. An additional possibility for predissociation occurs in those cases where the crossing electronic state is a nonbonding state, which has no discrete rotational and vibrational levels, but instead only a dissociation continuum. An example is shown in Fig. 14.16.

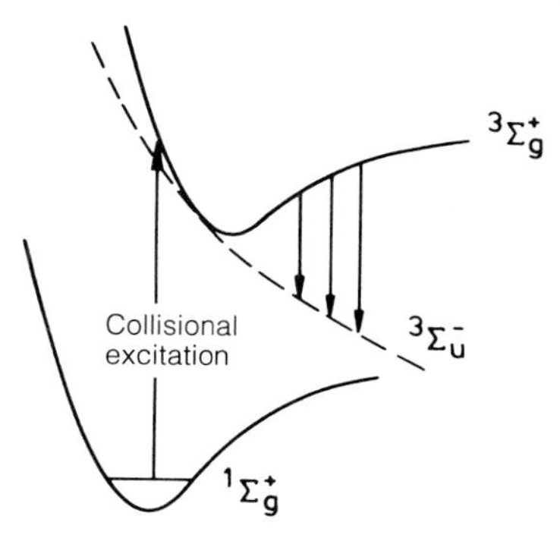

Fig. 14.16. The origin of the continuum in the spectrum of a hydrogen lamp. Collisions in a gas discharge produce excitations into the triplet system. The lowest state of this system is non-bonding; compare the term diagram in Fig. 13.9. The emission spectrum is therefore continuous, since there are no vibrational terms in a non-bonding state

14.5 Applications of Band Spectra

Band spectra have a number of *applications*. First of all, they are of course an indispensable aid to the investigation and elucidation of molecular structure and chemical bonding. Their analysis yields important information about the shape and position of the potential curves of molecules in their ground and excited states. Furthermore, the molecular vibrations and rotations can be detected and analyzed using spectroscopic methods in the visible or UV regions instead of microwave or infra-red spectroscopies, as we have seen. Vibration and rotation can also be investigated in emission; this is in general practically impossible in the microwave or infra-red spectral regions because of the small transition probabilities for spontaneous emission.

Molecular band spectra are also used in *light sources*. In the ultraviolet, the *hydrogen lamp* is frequently used as a source of continuous radiation. In this application, hydrogen molecules are excited by collisions in a gas discharge from their ground state $^1\Sigma_g^+$ into various excited states, including some whose population by absorption of radiation from the ground state is forbidden. One of these is the second-lowest triplet state, $^3\Sigma_g^+$. It has no allowed radiative transition to the ground state, but a transition to the $^3\Sigma_u^-$ state is allowed; this is a nonbonding state. Its potential curve is thus not parabolic, with a minimum corresponding to the bond length, but rather a curve which decreases continuously with increasing R (see Fig. 14.16). Transitions into this state yield a continuum, as shown in the figure.

There are several applications of the electronic band spectra of molecules for the generation of light in lasers. In the N_2 *laser*, transitions in the band spectrum of N_2 molecules are used to produce laser light. In *excimer lasers*, one makes use of the fact mentioned in Sect. 13.3 that there are molecules which have an excited electronic state which is bound, although the ground state is nonbonding, i.e. the ground state has no vibrational levels. Transitions from such a metastable excited dimeric state into the ground state then do not occur at discrete frequencies, but instead are continuous. Such molecules are called excimers (excited dimers), when they consist of dimers of two molecules $(MM)^*$ or of two atoms $(AA)^*$. Examples of the latter are the noble-gas fluorides used in excimer lasers. One thus obtains a "continuous" emission over a certain wavelength range, and the possibility of constructing a laser with tunable-wavelength emission.

Excimers are in addition interesting as the active medium in lasers because the emitting state, a dimeric state, is different from the absorbing monomeric state. The emission is therefore shifted to lower quantum energies, and the population inversion which is required for laser action is more easily obtained. Among organic molecules, excimers were first observed with pyrene in solution. Figure 14.17 shows the decrease of the blue monomer emission and the increase of the green excimer fluorescence with increasing concentration of pyrene molecules in a solution, due to the formation of molecular complexes by the reaction $M + M^* \rightarrow (MM)^*$.

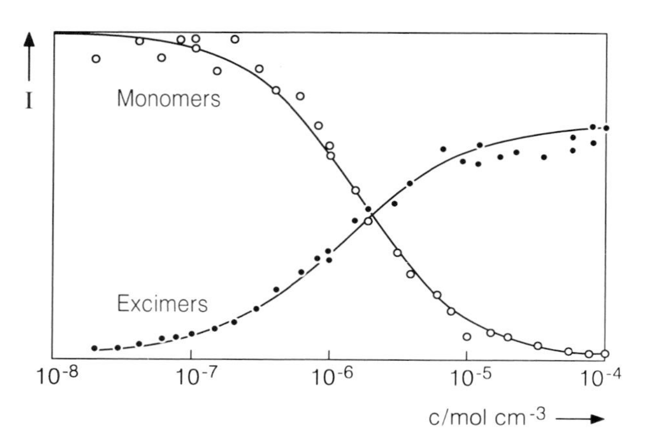

Fig. 14.17. The relative intensity of the fluorescence from pyrene in a benzene solution as a function of the concentration *c*; from Th. Förster and K. Kasper, Z. Elektrochem. **59**, 976 (1955). With increasing concentration, the probability for an excited and a ground-state molecule to collide and form an excited dimer (excimer) increases. The molecules are ordered as dimers in the pyrene crystal; its fluorescence is therefore pure excimer emission. This is already an example for spectroscopy of larger molecules

14.6 The Electronic Spectra of Larger Molecules

Our knowledge of the spectra of small molecules, particularly diatomics, and the information obtainable from them about molecular structure and bonding, can be extended in principle to larger molecules, which consist of more atoms. However, it then becomes more difficult to arrive at a reasonably complete description, because the number of possible excitations, the vibrations and rotations, the possibilities for rearrangements and dissociation within the molecule, and in general the number of allowed energetic and bonding states of the electrons increase strongly with increasing molecular size and complexity. For most polyatomic molecules, full knowledge of all the molecular data is therefore unobtainable. One has to be content with an understanding of the important aspects of the spectra and the bonding, and of the electrons in certain bonds. In the following, we therefore attempt to give only an overview of some particularly important and characteristic areas of information about larger molecules, especially in the almost infinite area of organic compounds.

The most important common characteristic of the electronic terms of polyatomic molecules is the fact that the excitation of a variety of different electrons within the molecule is possible. We can distinguish three typical limiting cases of possible excitations:

- Absorption by nonbonding electrons, which are not involved in the formation of chemical bonds in the molecule, but belong to a *localized* group within it, a so-called chromophore;
- light absorption by *bonding* electrons, which can lead to the dissociation of the molecule; and
- absorption by nonbonding but *delocalized* electrons which are spread over the entire molecule or a major part of it.

We consider first the absorption by groups within the molecule which are not primarily responsible for its chemical bonding. One refers to *chromophoric groups* (from the Greek for "carrier of colour"), with localized electronic orbitals, i.e. orbitals which extend over only a few nuclei; these groups exhibit a characteristic light absorption which causes little change in the rest of the molecule and which often produces a typical colour.

Examples from the area of inorganic molecules include the transition-metal complexes of elements such as Fe, Ti, or Co, where the excitation states of the atomic electrons remain intact in the molecule. These atomic-electronic spectra can, however, be modified in characteristic ways by the presence of the other electrons in the molecule with their typical symmetry properties. In the case of molecules containing atoms whose light absorption takes place through inner-shell electrons rather than valence electrons, such as the rare-earth salts, the observed spectra can be understood as atomic or ionic spectra which have been only slightly perturbed by chemical bond formation.

In the area of organic molecules, there are likewise many specific absorption spectra due to chromophores. The carbonyl group C=O, for example in the molecule

$$\begin{array}{c} CH_3-C-CH_3 \\ \| \\ O \end{array}$$

is thus characterised by its absorption around 290 nm, nearly independently of the position of the group within a larger molecular complex. This absorption is due to the excitation of an electron of oxygen into a previously empty π orbital of the carbonyl bond. An excitation of this kind is called a $(\pi^* \leftarrow n)$ transition, i.e. a transition from a nonbonding orbital into a π orbital. A C=C double bond can be excited by a $(\pi^* \leftarrow \pi)$ transition in the region of 180 nm. The symbol n stands for nonbonding, while π^* refers to an excited π orbital.

Furthermore, there can be absorption of radiation by *bonding electrons*, which play a decisive role in the chemical bonding of the molecule. Such excitations can lead to dissociation of the molecule. The absorptions of saturated hydrocarbons such as methane or ethane, which have σ bonds, lead to σ^* orbitals; they are thus $(\sigma^* \leftarrow \sigma)$ transitions. A relatively large amount of energy is needed for these transitions, and the absorption lies typically in the short-wavelength ultraviolet region around 120 nm.

Finally, absorption by electrons in *non-localized orbitals*, which are *not* primarily responsible for the bonding in the molecule, is especially interesting.

A well-known and important example of this type of absorption is benzene; cf. also Chap. 5. In the benzene molecule, not every electron pair can be associated with a particular bond between two atoms. Since each C atom in the molecule bonds to one H atom, three electrons per C atom are left over for the formation of additional bonds. Two of them form sp^2 hybrid orbitals and participate in localized σ bonds to the neighbouring C atoms. This hybridisation means that the formation of chemical bonds is accompanied by a reordering of the carbon electrons from $2s^2 2p^2$ to $2s2p^3$; the necessary energy of 4.2 eV is supplied by the binding energies as a result of the formation of the bonds. The remaining 6 valence electrons in the benzene molecule, one from each of the 6 C atoms, can form three localized double bonds. From the empirical fact that in benzene, all six bonds between the C atoms are equivalent, it was realised that these 6 electrons, as $2p_z$ electrons, form so-called π bonds, i.e. bonds in which the electrons are delocalized around the entire molecule, with a nodal plane in the plane of the benzene ring. These π bonds are weaker than the σ bonds and can be excited by quanta of lower energies than are necessary for the σ-bond electrons. Such excitations are denoted as $(\pi^* \leftarrow \pi)$ transitions, in which a π electron is lifted by absorption of radiation into a π^* orbital. They are the lowest-energy electronic excitations of aromatic molecules. These transitions exhibit a well-defined vibrational structure, since the excitation takes place between bound states of the molecule, in which the bonding remains intact, so that molecular vibrations are also preserved.

The absorption spectra of benzene (Fig. 14.18) thus exhibit the vibrational structure characteristic of the molecule in the various stages of electronic excitation. It remains intact when the molecules are observed in solution or in the solid state, as well as in the gas phase; however, the spectral lines are broadened in the condensed phases. In particular, no rotational structure is present there, since the whole rotational structure is smeared out or hidden within the linewidth or band width, insofar as the molecules are still able to rotate. In Fig. 14.18, the

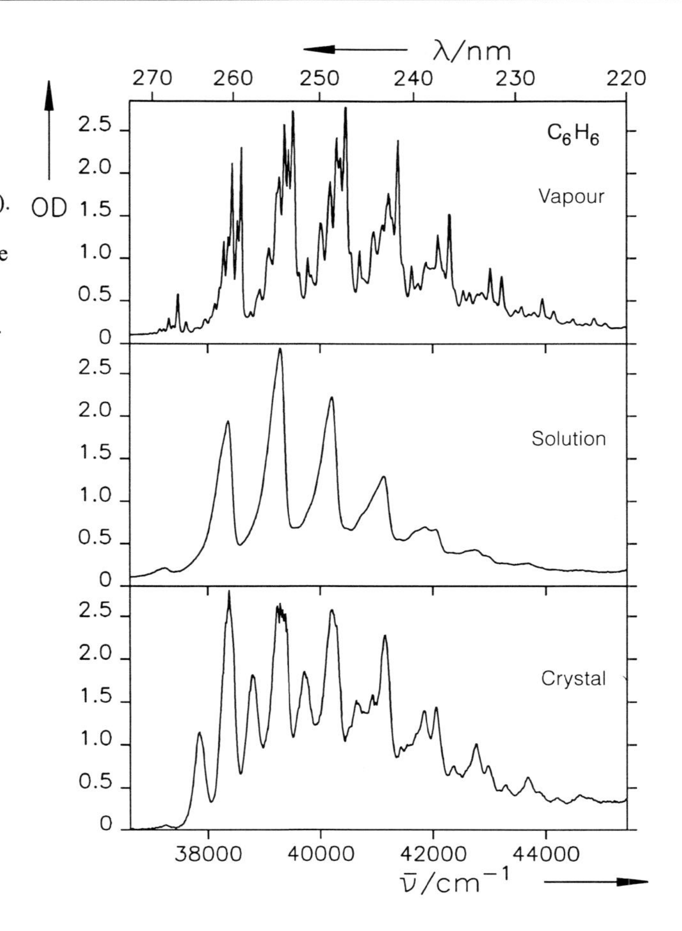

Fig. 14.18. The absorption spectrum of the long-wavelength transition ($S_2 \leftarrow S_0$) in benzene. The optical density OD (see p. 252) is plotted against the wavenumber (*lower scale*) and the wavelength (*upper scale*). (*Upper part*): Benzene in the gas phase at its room-temperature saturation vapour pressure, with an optical path length of 10 cm. The partially resolved rotational-vibrational structure can be seen. The strongest maximum in each progression corresponds to an excitation of the totally-symmetric vibration A_{1g} of benzene (breathing mode; cf. Fig. 10.14), $\bar{\nu} = 923$ cm^{-1}, by 1, 2, 3, 4, 5, 6, and 7 quanta. The 0,0 transition is forbidden, and the absorption between 37 000 and 37 500 cm^{-1} is therefore very weak. We cannot go into the details of the vibrational structure here; this is done in Steinfeld, p. 415. (*Centre*): Benzene dissolved in cyclohexane, at room temperature. The vibrational structure is similar, but considerably less detail can be resolved. (*Lower part*): The absorption spectrum of polycristalline benzene at 170 K. This spectrum is shifted somewhat relative to the spectrum in solution; in addition, a second progression of vibrational bands can be seen. The origin of this is the crystal symmetry, which changes the symmetry of the excited states of the molecules. At still lower temperatures (liquid helium temperature), the absorption lines in the crystal become considerably sharper

long-wavelength absorption of benzene in the solid state (crystal) is compared with the same spectrum in solution and in the gas phase; the latter two spectra were, however, recorded with relatively poor spectral resolution. The similarity of these spectra is obvious.

In Fig. 14.19, for comparison, we show the absorption spectra of the larger *aromatic molecules* illustrated in Fig. 14.20: napthalene, anthracene, tetracene, and pentacene. With increasing extension of the π electron system, i.e. with increasing size of the molecular rings, they are shifted to longer wavelengths. The same is true of the emission spectra (Fig. 14.21), which are mirror images of the absorption spectra around the longest-wavelength absorption transition. Referring to Fig. 14.19, we note here that the transition probabilities of the different electronic transitions are different; this means that the absorption strengths, measured by the extinction coefficient ε in different spectral ranges, have different magnitudes. We will not consider this fact in more detail at this point.

The benzene ring is a particularly well known and thoroughly-investigated example of conjugated double bonds, which are a common and important phenomenon in organic chemistry. In this case, successive C atoms in a ring or a chain are joined alternately by single and by double bonds; resonance occurs between these bonds, leading to an effective delocalisation of the electrons. An isolated, i.e. not conjugated C–C double bond absorbs light in the ultraviolet near 7 eV. Double bonds in conjugated chains absorb at longer wavelengths.

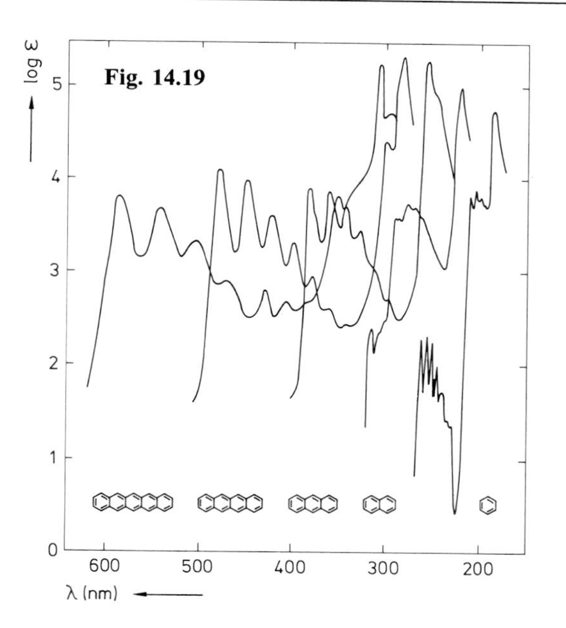

Molecule	Absorption
Benzene	2550 Å
Naphthaline	3150 Å
Anthracene	3800 Å
Tetracene	4800 Å
Pentacene	5800 Å

Fig. 14.20

Fig. 14.19. Absorption spectra of the aromatic ring molecules shown in Fig. 14.20 in the visible and the ultraviolet regions, in solution. The continuous shift of the spectra to smaller quantum energies, i.e. to longer wavelengths, with increasing size of the molecules can be readily seen

Fig. 14.20. The aromatic molecules, so-called polyacenes, whose absorption spectra are shown in Fig. 14.19, with the wavelength range of the longest-wavelength absorption indicated

Some important representatives of this group of molecules, which have more or less delocalized molecular orbitals, also referred to as "conjugated", are the *linear polyenes* known in organic chemistry. These are linear systems of n conjugated double bonds, –C=C–C= , with two end groups that close off the chain. Figure 14.22 shows the absorption spectra of the polyenes with $n = 1$ to 7 and two phenyl groups on the ends of the chain. The increasing shift of the absorption towards longer wavelengths with increasing conjugation number n can be readily seen. It can be explained in a model with a free electron which can move along the chain, as we shall show below.

A prominent member of this class of compounds is *retinal*, Fig. 14.23. This dye, bound to a certain protein, forms rhodopsin, which is responsible for the elementary visual process in the human eye. While retinal in solution absorbs at 380 nm, the spectrum of retinal bound to the protein is shifted to longer wavelengths and has its maximum near 500 nm. The absorption spectrum extends over the entire visible spectral region. Absorption of a photon in the eye by retinal [as a $(\pi^* \leftarrow \pi)$ or a $(\pi^* \leftarrow n)$ transition] leads to an isomerisation of the molecule from the cis- to the all-trans-conformation (see Fig. 14.23). This isomerisation produces the nerve impulse which is interpreted by the brain as "seeing". We shall have more to say about this topic in Sect. 20.4.

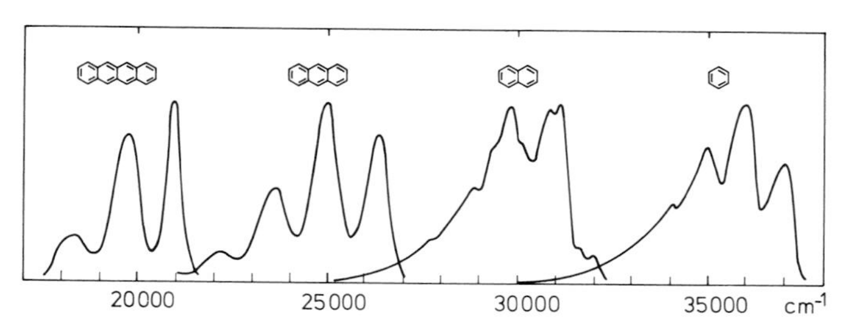

Fig. 14.21. Flourescence spectra of some polyacenes (in solution at room temperature). The spectra are, to a first approximation, mirror images of the longest-wavelength part of the absorption spectra shown in Fig. 14.19

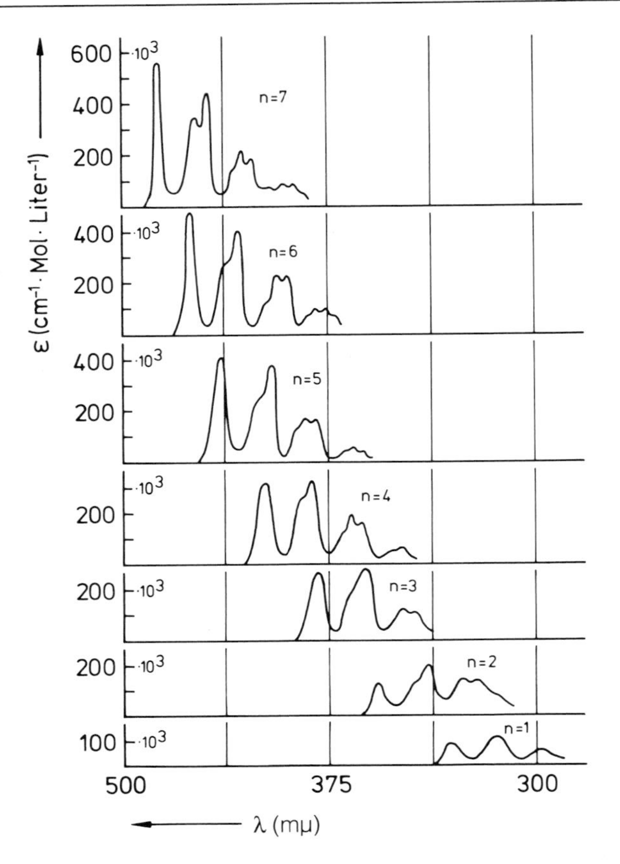

Fig. 14.22. Absorption spectra (long-wavelength part, lowest electronic transitions) of linear diphenyl polyenes, C_6H_5–$(CH{=}CH)_n$–C_6H_5 in solution at $-196\,^\circ$C, with $n = 1$ to 7. The absorption belongs to the polyene chain, –$(CH{=}CH)_n$, and can be understood to a good approximation on the basis of a model with an electron gas which is free to move along the chain. [After K.W. Hausser, R. Kuhn, and A. Smakula, Z. phys. Chem. B**29**, 371 (1935)]

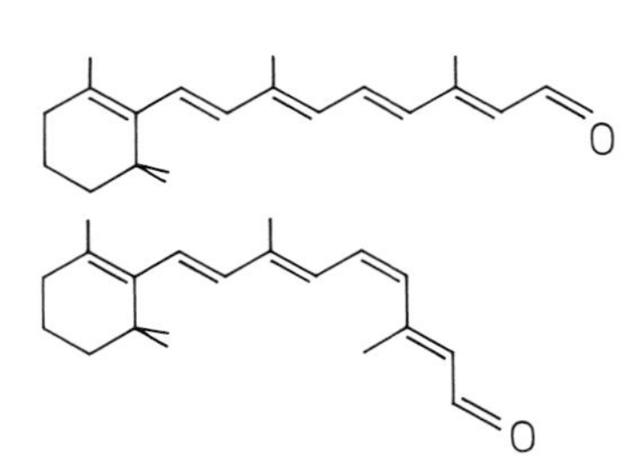

Fig. 14.23. The retinal molecule can be converted by light absorption from its cis-conformation (*above*) to trans-retinal (*below*). This is the first step in the visual process in the human eye. As a simplification, the CH_3-groups are indicated by lines and the H atoms along the conjugated chain system are not drawn in

Another molecule from the group of the polyenes which has considerable significance in biology is *β-carotene*, Fig. 14.24, with 11 conjugated double bonds; it gives carrots their yellow colour. This and related molecules, the carotenoids, play an important role in photosynthesis: the conversion of solar energy into biomass.

The electronic excitation states of the polyenes and molecules with similar structures can be understood in an intuitively clear first approximation by using the model of a particle in a box or in a one-dimensional potential well; see also Chap. 9 in I. For the energy eigenvalues of an electron of mass m_0 in a one-dimensional potential well of length a, the following relation holds, as we showed in I, Sect. 9.1, Eq. (9.14):

$$E_n = \frac{n^2h^2}{8m_0a^2}\,, \quad \text{with} \quad n = 1, 2, 3, \ldots\,. \tag{14.22}$$

If we populate this potential-energy scheme with the 22 electrons from the 11 double bonds of the conjugated chain of β-carotene and take into account the fact that each level n can hold only two electrons (with antiparallel spins), then the levels $n = 1$ to $n = 11$ are occupied in the ground state. The observed longest-wavelength absorption of the molecule at 450 nm would then correspond to a transition from $n = 11$ to $n = 12$, that is into the lowest unoccupied level. This implies an energy difference of

$$\Delta E = (E_{12} - E_{11}) = (12^2 - 11^2)\frac{h^2}{8m_0a^2}\,. \tag{14.23}$$

Inserting the observed absorption at $\Delta E = 450\,\mathrm{nm}$, we can calculate the length of the potential well, i.e. the length of the conjugated chain, $a = 17.7\,\text{Å}$. This is of the right order of magnitude for the length of the molecule, which can be determined more precisely by other methods. The agreement indicates that this model of an electron which is delocalized over the whole length of the molecule is a good approximation.

The delocalisation and free mobility of the π electrons along conjugated double bonds is related to the metallic conductivity of solids. It leads to a large and strongly anisotropic electrical polarisability of the molecules, and to a strongly anisotropic diamagnetism. The induced currents which are characteristic of diamagnetism on application of an external magnetic field are, for example, much stronger in the plane of an aromatic molecule than perpendicular to it. Correspondingly, the diamagnetic susceptibility of aromatic molecules for a magnetic field perpendicular to the plane of the rings is at least a factor of three larger than in the plane; cf. also Sect. 3.7.

We cannot describe here in detail the multiplicity of spectra and excited states which are encountered for large molecules. An important and, especially for physicists, interesting group of excitations are those referred to as charge-transfer, acceptor, or donor transitions. In these transitions, excitation by radiation causes an electron to be completely or partially (in the quantum-mechanical sense) transferred from one part of the molecule to another. This changes the whole charge distribution within the molecule, its dipole moment, and thus in general also its structure and its coupling to its surroundings. The spectra are often very broad, at least in the condensed phases, and unstructured as a result of this coupling. As an example from the area of inorganic chemistry, we mention the well-known deep violet MnO_4^- ion (the permanganate ion), with absorption between 430 and 700 nm. Here, the electronic transition occurs between the ligands, i.e. the O atoms surrounding the Mn ion, and the inner d orbitals of the central ion. Examples from the area of organic chemistry are the complexes between aromatic molecules such as anthracene and the tetracyano benzene molecule, which serves as electron acceptor. The latter is a benzene molecule in which four of the six H atoms are replaced by CN– groups.

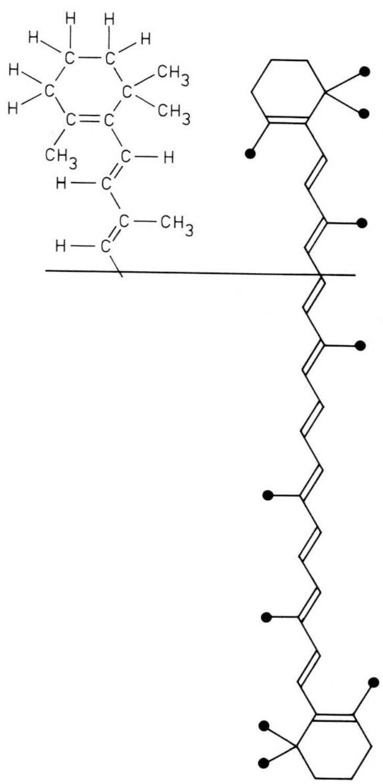

Fig. 14.24. The structure of the carotene molecule. At the left, we indicate how the complete molecular structure with all the C- and H-atoms as well as the CH_3-groups would look. In the structural sketch at the right, these symbols are left out for clarity

15. Further Remarks on the Techniques of Molecular Spectroscopy

In molecular spectroscopy, the energy levels of molecules which are connected by absorption, by radiationless processes, or by the emission of radiation are investigated (Sects. 15.1 through 15.3). The method of laser spectroscopy of cold molecules in a supersonic molecular beam has been the source of major advances in molecular spectroscopy (Sect. 15.4). With this technique, one can also study short-lived and weakly bound complexes and clusters; the new modification of carbon, the Fullerenes, were discovered in this way (Sect. 15.4). An important experimental tool is the tunable dye laser (Sect. 15.5). Employing a combination of these modern methods, it has proved possible to resolve the rotational structure of the vibronic bands of large molecules with two-photon absorption or excitation spectroscopy (Sect. 15.6). Using pulsed lasers, dynamic processes within and on molecules down to the femtosecond range can be investigated (Sect. 15.7).

Another experimental technique which is in a state of rapid development is photoelectron spectroscopy. Using this method, discussed in Sect. 15.8, the spectroscopy of inner-shell electrons as well as of valence electrons is possible. In the form of ZEKE spectroscopy, it can even resolve rotational lines in the spectra of molecular ions (Sect. 15.9).

15.1 The Absorption of Light

Aside from the energy difference between two molecular states which are connected by transitions corresponding to spectral lines, an important experimental quantity is the transition probability itself. In contrast to the case of atoms, where transitions can be classified as entirely "allowed" or "forbidden", in molecules all the intermediate stages between completely allowed and strongly forbidden transitions are found. The transition probabilities are determined by the electronic structures of the initial and final states. In Chap. 16, we shall concern ourselves with the quantum-mechanical derivation of these transition probabilities. Experimentally, they can be determined from the absorption strength or the lifetimes and quantum yields of the emission as fluorescence or as phsophorescence.

The *absorption of light* by molecules at a concentration C in a homogeneous sample of thickness x is governed by the Lambert-Beer law:

$$I = I_0 \mathrm{e}^{-\alpha C x} . \tag{15.1}$$

(We have used a capital C for the concentration in this expression in order to avoid confusion with the velocity of light, c.) In (15.1), I_0 is the incident light intensity and I the intensity transmitted by the sample, and α is the characteristic absorption coefficient of the molecules under consideration, which is defined by this equation. We have already met this law in a more general form as the definition of the interaction cross-section in I, Sect. 2.4.2. It gives the particular interaction cross-section for the absorption of light by molecules. The molecule-specific quantity ε, the molar absorption coefficient, is defined by (15.2):

$$\log\left(\frac{I_0}{I}\right) = \varepsilon C x \,. \tag{15.2}$$

The relation between ε and α is thus $\varepsilon = \alpha/\ln 10 = \alpha/2.303$. The dimension of ε is 1/(concentration·length) or, more commonly, $M^{-1}cm^{-1}$. M is the abbreviation for mol dm^{-3}; 1 M thus means 1 mole of a substance in one liter of solvent. The quantity ε is therefore also referred to as the molar decadic absorption coefficient or extinction coefficient. From the definition of M as moles per volume we can derive yet another unit for ε, namely cm^2 mol^{-1} or cm^2 $mmol^{-1}$, where mmol stands for 10^{-3} moles. We thus have 1 M cm^{-1} = 1 cm^2 $mmol^{-1}$; the latter unit expresses more clearly the fact that ε is a molar interaction cross-section for light absorption. The "optical density" OD is defined as the product $\varepsilon C x$, and the "transmissivity" as the ratio $T = I/I_0$.

The extinction coefficient ε is specific to the absorbing molecule and depends on the frequency of the light being absorbed. A transition between two electronic states of a molecule always includes a large number of vibrational and rotational levels, as we showed in Chap. 14; these levels belong to the potential curves of the initial and final states in the molecule, and give rise to a frequency range $\Delta\nu$ for the transition. The overall intensity of an electronic transition is therefore measured in terms of an integral:

$$A = \int \varepsilon(\nu)d\nu \,, \quad [\mathrm{M^{-1}\,cm^{-1}\,s^{-1}}] \quad \text{or} \quad [\mathrm{cm^2 mmol^{-1} s^{-1}}] \,, \tag{15.3}$$

which is called the integral absorption coefficient. It is related to the (dimensionless) oscillator strength f by the equation

$$f = \frac{4 m_0 c \varepsilon_0 \ln 10}{N_A e^2} A \tag{15.4}$$

(where m_0 is the mass and e the charge of the electron).

Inserting the numerical values, we find

$$f = 1.44 \cdot 10^{-19} A \tag{15.5}$$

with A measured in $cm^2\,mmol^{-1}s^{-1}$.

Very strong electronic transitions have $f = 1$, and ε lies between 10^4 and 10^5 [$M^{-1}cm^{-1}$]. The oscillator strength f depends directly on the transition matrix element Θ_{21} as defined in Sect. 16.3.6, i.e. on the wavefunctions of the initial and the final states. For an electric dipole transition between two states 1 and 2 with the dipole transition moment

$$\Theta_{21} = \int \psi_2^* e \boldsymbol{r} \psi_1 dV \tag{15.6}$$

(where ψ_1 and ψ_2 are the wavefunctions of the states, e the elementary charge, $\boldsymbol{r}$ is the distance between the centres of charge, and V is the volume), we find

$$f = \frac{8\pi^2}{3}\frac{m_0\nu}{he^2}|\Theta_{21}|^2 \,. \tag{15.7}$$

The oscillator strength can therefore be calculated if the wavefunctions are known. It is spread over a more or less large number of vibronic and rotational levels in the spectrum; cf. Chap. 16, especially Sect. 16.3.6.

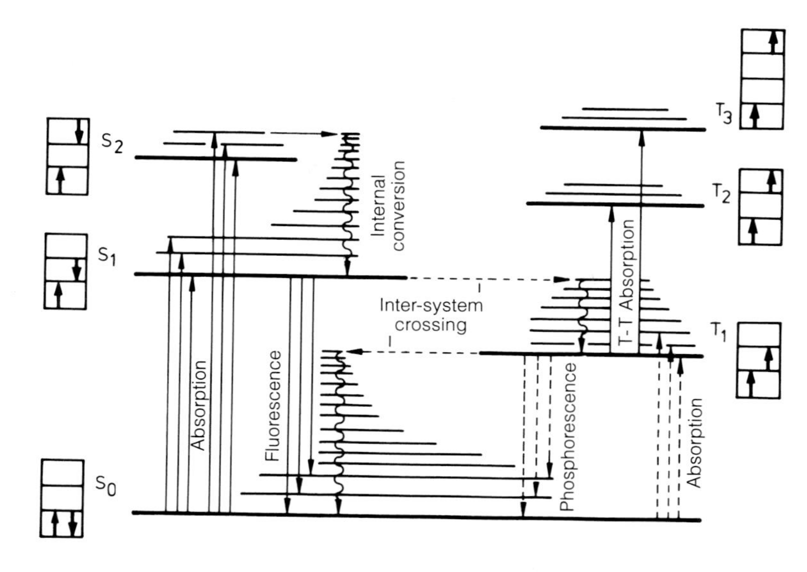

Fig. 15.1. A typical term diagram for a molecule with singlet and triplet systems, explaining the most important radiative and non-radiative processes. Intersystem crossing as well as a radiative transition between the two systems (phosphorescence, absorption) are more or less strongly forbidden. Internal conversion and intersystem crossing are radiationless processes ("Jablonski-Diagram")

The Lambert-Beer law is in general obeyed very well. Deviations are possible when the molecules interact with each other, but also when the light source is so intense that thermal equilibrium is disturbed by the radiation, or when a long-lived state is populated and thus the absorbing ground state has a reduced population during the irradiation.

The energy absorbed by a molecule can be released again in several quite different ways. If we first leave out photochemical processes, in which the molecules are modified or destroyed, then the energy can be released through radiationless processes or by fluorescence or phosphorescence. We will explain these latter two terms below. An overview of the most important radiative and radiationless or non-radiative transition processes in molecules is given in Fig. 15.1.

15.2 Radiationless Processes

Radiationless or non-radiative processes are in particular those processes in which electronic excitation energy is converted to vibrations, rotations, and translational motions of the molecules. Such processes are very important in the condensed phases, i.e. for molecules in liquids and in solids. They are for example the reason that molecules in general emit radiation only from their lowest excited electronic states, no matter what wavelength was used to excite them, and the emission gives a high yield only when the energy difference to the ground state is sufficiently large; it must be larger than several times the characteristic vibrational quantum energy in the electronic ground state. After excitation of the molecule, it can transfer the excitation energy via *internal conversion* processes (cf. Figs. 15.1 and 15.2) to vibrational and rotational quanta and thus release it to the molecular environment. Only when the possibility of energy release to the environment is eliminated, as is the case for example for isolated molecules in interstellar space, can these radiationless processes be avoided. Another important non-radiative process is the intersystem crossing between the singlet and triplet systems; it is enhanced by the spin-orbit coupling, and is also shown in Fig. 15.2.

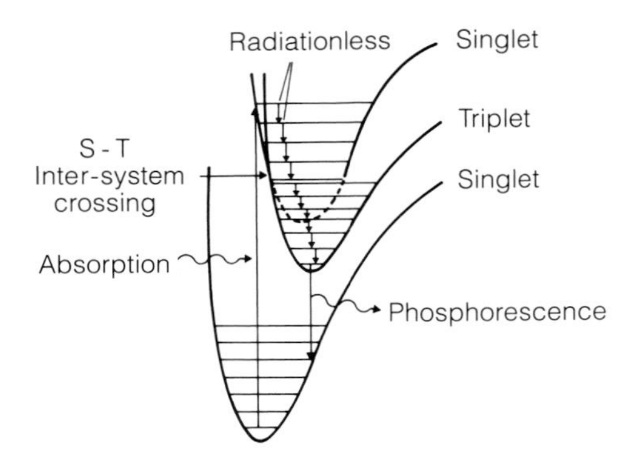

Fig. 15.2. Some non-radiative processes: an intersystem crossing between the singlet and the triplet systems is made possible by spin-orbit coupling and thus can occur more readily when several atoms with high atomic numbers are present in the molecule. Phosphorescence, which is a radiative intersystem crossing, produces emission with a long decay time

The fact that the spectral distribution in the fluorescence spectrum of molecules is usually independent of the excitation wavelength, but the fluorescence intensity is not, is employed in *excitation spectroscopy*. One measures the intensity of the fluorescence emission in a fixed wavelength range as a function of the wavelength or quantum energy of the light used for excitation. The fluorescence intensity then increases with the number of absorbed quanta. In this way, one can observe absorption spectra even in cases where transmission through the sample is not observable, i.e. when the absorption is very strong. To be sure, the quantitative determination of the absorption coefficients in this manner is usually not possible, because a quantitative relation between the number of quanta absorbed and the number emitted can be established only when perturbing effects such as saturation of the absorption and self-absorption of the fluorescence radiation can be avoided.

Excitation spectroscopy is particularly important as an extremely sensitive method for measuring weak absorptions. In such cases, in the usual absorption spectroscopy one is forced to determine the ratio of two quantities which are nearly the same, i.e. the incident and the transmitted intensities. However, in the excitation spectrum, only the emitted quantum flux is measured. In the case of weak absorption, this can be done with considerably improved accuracy. An example of an excitation spectrum is given later, in Fig. 21.8.

15.3 The Emission of Light

Radiative transitions from the vibronic levels (usually the lowest) of an excited electronic state (usually the first excited state) into the vibronic levels of the ground state are called *fluorescence* (see also Fig. 15.1 and Sect. 14.1). The Franck-Condon principle holds for these transitions just as for absorption of radiation; i.e. the transitions take place preferentially between levels which are directly above one another as in Fig. 15.2. Fluorescence occurs at quantum energies which are smaller than, or at most equal to, the energy of the absorbed radiation, as can be understood directly from Fig. 15.1. The only line common to both the absorption spectrum and the fluorescence spectrum, at least when the temperature is sufficiently low, is the 0,0 line, i.e. the transition between the states with $v'' = 0$ and $v' = 0$. Relative to this line, the absorption spectrum and the fluorescence spectrum extend in a mirror-image fashion towards higher and lower frequencies, respectively; cf. Figs. 14.8, 14.19, and 14.21. If higher vibronic levels of the electronic ground state, with $v'' > 0$, are populated thermally at higher temperatures, so that they can also provide initial states for absorption, then the region of overlap between the absorption and the fluorescence spectra becomes larger.

The *decay time* of the fluorescence following excitation by a short pulse of light is a measure of the radiative lifetime of the emitting state. In order to determine it, the fluorescence must be excited by a light pulse which is short compared to the decay time; the latter is then found from the time dependence of the emission intensity following the excitation. For allowed transitions, decay times in the nanosecond range are typical. The *quantum yield* η is defined by

$$\eta = \frac{\text{number of emitted quanta}}{\text{number of absorbed quanta}} . \tag{15.8}$$

If there are no competing radiationless processes, the quantum yield is 1. Values near 1 are in fact observed for e.g. the well-known and often-used fluorescence dyes such as rhodamine or anthracene. All competing non-radiative processes reduce the quantum yield and shorten the radiative lifetime.

Besides fluorescence, one often observes (especially in the case of organic molecules) an emission with a much longer decay time, which is called *phosphorescence*. In molecular physics, phosphorescence denotes emission from an excited triplet state, i.e. from a state with a total spin quantum number $S = 1$. Previously, and sometimes even today in solid-state physics, phosporescence refers in general to light emission with a long decay time. The longer decay time is a result of the forbidden intersystem crossing for a transition from an excited triplet state into the singlet ground state, i.e. the fact that spin-flip processes are forbidden in optical transitions. The triplet nature of the excited states can be verified by electron spin resonance; see Chap. 18.

We will now consider how radiative transitions between the terms T_1 and S_0 can take place in spite of the fact that they are forbidden. Most molecules are diamagnetic in their ground states, since all of the electron spins are paired to give an overall spin of zero. This is also true of many excited states, and the fluorescence which we have just described therefore corresponds to singlet-singlet transitions, that is transitions between states with $S = 0$. As we have already seen in atomic physics, however, spin-orbit coupling allows forbidden singlet-triplet transitions to occur; in these, an electronic spin flip accompanies a radiative optical transition. This is called *intercombination* in atomic physics, or, in molecular physics, where the process often occurs nonradiatively, *intersystem crossing*. There is thus a certain, usually very small probability for the excitation of molecules by light absorption from the singlet S_0 ground state into a triplet state T_1, or conversely for the emission from the triplet state into a singlet state. This probability is increased by the presence of atoms with high atomic numbers within the molecule or nearby, as we have learned in atomic physics (see I, Sect. 12.8). In molecular physics, this is termed the intramolecular or (if it is based on an interaction with the surroundings of the molecule) the intermolecular *heavy-atom effect*. It is, for example, responsible for the fact that the dibromonapthalene molecule, i.e. a naphthalene molecule in which two protons have been replaced by bromine atoms, has an $S_0 - T_1$ absorption which is several orders of magnitude stronger than that of unsubstituted naphthalene. Correspondingly, the lifetime of the metastable T_1 state, insofar as it is limited by radiative deactivation, is shorter in the Br-substituted naphthalene molecule.

A molecule which has been excited into a higher triplet state expends its excitation energy to its environment, if possible in a rapid series of non-radiative processes as rotational and vibronic quanta. When it then arrives at the lowest electronic triplet state T_1, it can release its remaining excitation energy, depending on the degree of spin-orbit coupling, via a forbidden, and therefore slow radiative transition as shown in Fig. 15.2. The triplet state is thus metastable; its lifetime and the radiative decay time for phosphorescence emission can be of the order of minutes.

A further spectroscopic quantity for the investigation of molecules is the determination of the *polarisation* of the absorption, fluorescence, and phosphorescence. The polarisation of any transition between two different electronic states can be computed from knowledge of the symmetries of the states. It can be measured by holding the molecules in a fixed spatial orientation or by carrying out measurements with polarised light relative to transitions of known polarisations. A loss of the degree of polarisation between absorption and emission, a so-called *depolarisation*, can for example be caused by motion of the molecule or of a part of the molecule during the time between the two processes. Measurements of the polarisation can thus be used to study molecular motions.

15.4 Cold Molecules

In condensed phases, molecular spectra consist of relatively broad inhomogeneous lines or bands without resolved rotational structure, and they therefore do not permit the determination of detailed spectral properties of the molecules. On the other hand, the analysis of the numerous lines of molecular spectra even in the gas phase, in particular for more complex molecules, with their enormous multiplicity of spectral lines and overlapping spectral regions from different series of rotational and vibrational levels, is nearly impossible. Simply cooling the molecules to low temperatures in order to reduce the number of occupied rotational and vibronic energy levels is possible only to a limited extent if the experiments are carried out in the gas phase. In the past decade, considerable progress has been made in this area by cooling molecules in a *supersonic molecular beam*.

If one allows a gas to expand into vacuum from a region of high pressure (several atmospheres) through a suitable nozzle at a high velocity (supersonic velocities), an adiabatic expansion accompanied by cooling occurs, whereby the thermal energy of the molecules is converted into directed kinetic energy in the direction of the expansion. This is explained in detail in Fig. 15.3.

In the region of the nozzle, energy is exchanged among the molecules by collisions, resulting in a translational cooling (i.e. a reduction of the width of the velocity distribution) and a cooling of the rotational and vibrational degrees of freedom accompanied by an increase in the velocity component in the direction of the beam. The width of the resulting velocity distribution is a measure of the temperature (Fig. 15.3).

In order to cool a molecular gas for spectroscopic measurements, one mixes a small amount of the desired gas with a large amount of a monatomic carrier gas, e.g. helium. On expansion of this mixture through the nozzle, the carrier gas is first translationally cooled. The molecules can then give up rotational and vibrational energy to the cold atoms through collisions, i.e. to the cold thermal reservoir of the monatomic carrier gas.

The degree of cooling obtainable depends on the interaction cross-sections for collisions with exchange of rotational and vibrational energy; it leads to a drastic reduction of the temperatures T_t, T_r, and T_v for the translational, rotational, and vibrational degrees of freedom, compared to the temperature of the environment. One can obtain $T_t = 0.5$ to 20 K, $T_r = 2$ to 50 K, and $T_v = 10$ to 100 K in this way; that is, those degrees of freedom of the molecules under investigation are "cooled" so strongly that they would correspond in thermal equilibrium to

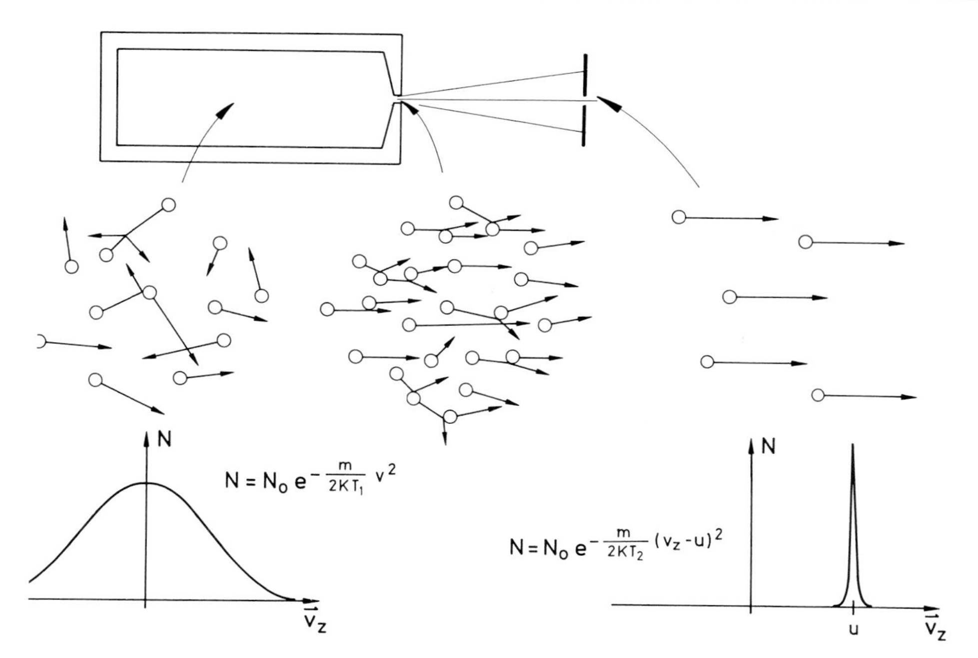

Fig. 15.3. Molecular spectroscopy in a supersonic molecular beam. The molecules are cooled by allowing them to stream out of a nozzle (*upper part*), whereby their disordered motion is converted into a directed motion with a selected translational velocity u (*centre*). The broad thermal velocity distribution at the initial temperature T_1 (*lower left*) is thus converted to a narrow distribution with a correspondingly reduced effective temperature T_2 for the molecular rotation and vibration. After Levy

the temperatures quoted. Excitation spectroscopy is then employed to measure the spectra, which still contain numerous lines. The absorption of the light from a tunable laser of narrow bandwidth is detected by means of the emission intensity from the molecular beam.

At such low rotational and vibrational temperatures, only the lowest rotational-vibrational levels in the electronic ground state are occupied. This reduces the number of possible initial states for absorption transitions and thus the number of lines in the spectrum by a considerable amount. The spectrum still contains a large number of lines, but it can be much more readily analyzed. Impressive examples for the increase in spectral resolution of molecular spectroscopy by means of this technique are shown in Figs. 15.4 and 15.5.

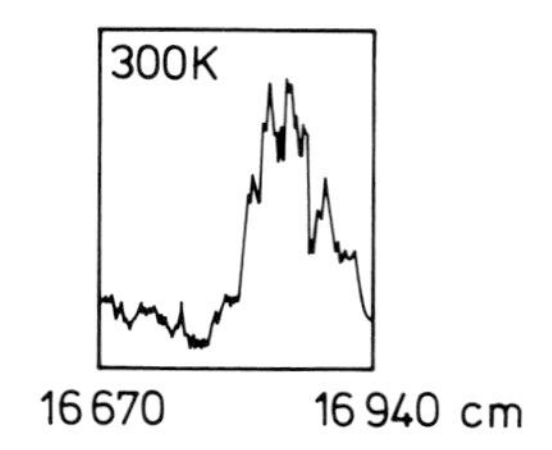

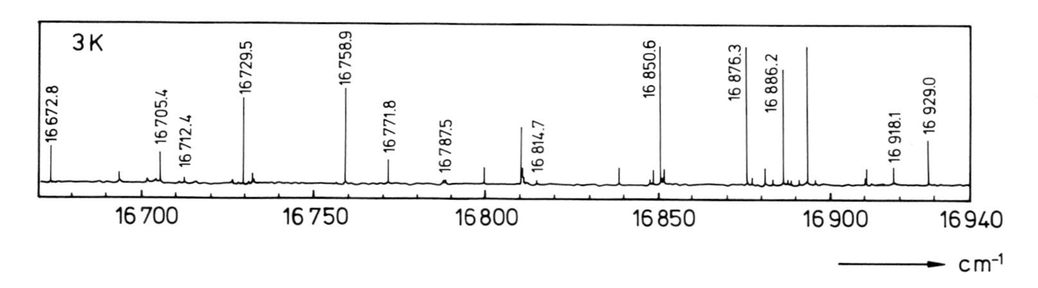

Fig. 15.4. The spectral resolution of the spectrum of NO_2 is drastically improved by adding the NO_2 gas to a helium supersonic beam. *Upper part of figure*: a section of the absorption spectrum of NO_2 in the gas phase, obtained by conventional absorption spectroscopy at 300 K. *Lower part of figure*: The excitation spectrum of NO_2 in a He supersonic beam; expanded wavenumber scale. After Levy

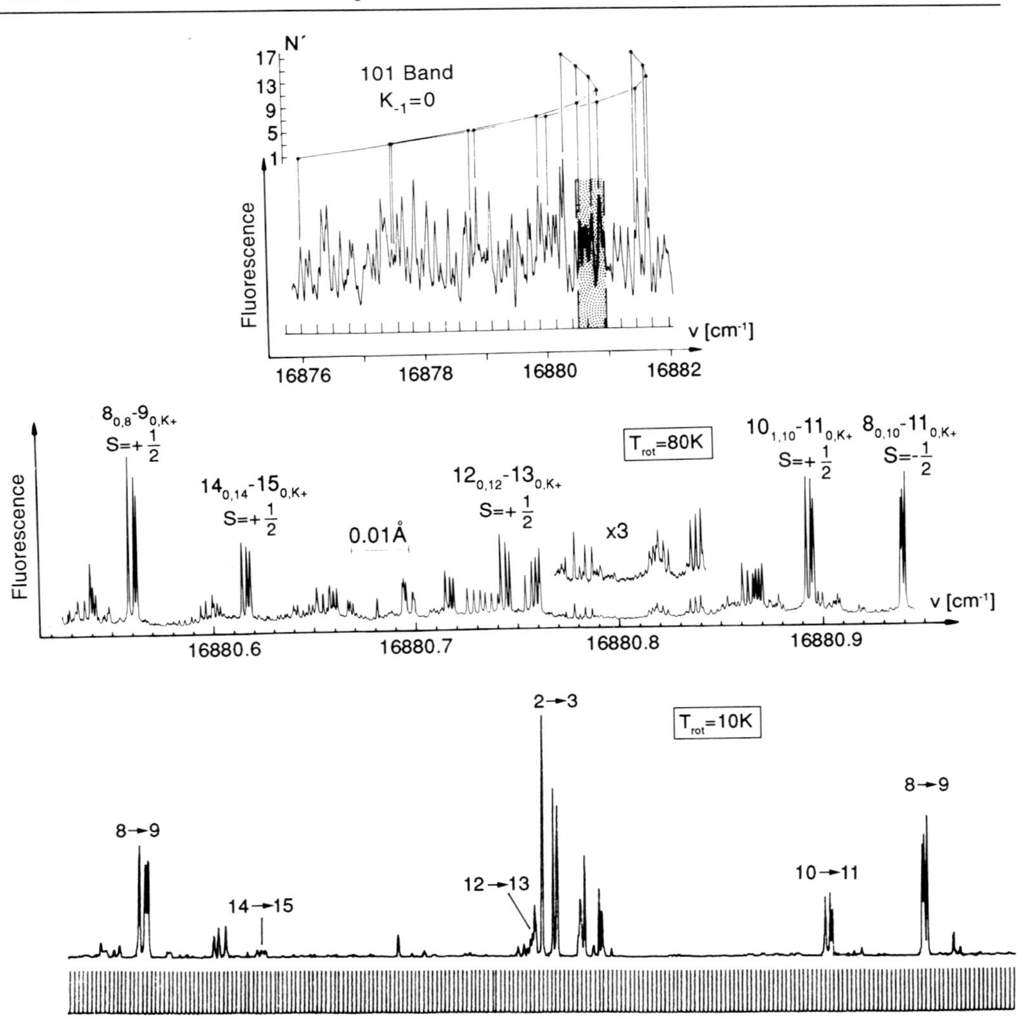

Fig. 15.5. A section of the NO_2 excitation spectrum, obtained using a narrow-band dye laser for excitation. *Above*: a Doppler-limited spectrum; NO_2 in an absorption cell at T = 300 K, p = 0.3 mbar. *Centre*: a Doppler-reduced spectrum showing the region which is shaded in the upper spectrum, obtained from a collimated molecular beam of pure NO_2. *Below*: as in the centre spectrum, but from a very cold molecular beam using 1 bar Ar carrier gas with 5% NO_2 added. It can be clearly seen that rotational lines corresponding to higher quantum numbers become less intense at the lower beam temperature. The $2 \to 3$ transition belongs to a different band system. The frequency markers are spaced at intervals of 63 MHz. Nomenclature: $8_{0,8} \to 9_{0,K_c'}$, $s = +1/2$ means a transition from the state with $N'' = 8$, $K_a = 0$, $K_c = 8$ to the state with $N' = 9$, $K_a' = 0$, K_c' variable, and $J = N+1/2$, where N is the quantum number of rotational angular momentum of the molecule and K_a and K_c are the projections of N on the molecular a or c axes, respectively. (Provided by W. Demtröder)

Figure 15.4 gives a section of the absorption spectrum of NO_2 measured using conventional absorption spectroscopy, and below it the highly resolved spectrum obtained with excitation spectroscopy from a supersonic molecular beam. Figure 15.5 explains a further refinement of the technique with a resulting large number of resolvable lines.

This supersonic-beam spectroscopy is very interesting for an additional reason: in the beam, one can also observe weakly bound molecular complexes, so-called Van der Waals molecules or *clusters*, which would dissociate immediately at higher temperatures (e.g. room temperature). These are associations of a few molecules to form weakly bound complexes. Clusters with up to 10^5 atoms or molecules as their components have also been found. This opens up a new possibility for investigating the interaction potentials between molecules and gives rise to a new transition region between molecular and solid-state physics.

The physics of clusters has a long history; in previous times, one referred to colloids. For example, it has long been known that finely dispersed colloidal gold can be used to colour glass ruby red. By changing the size distribution of finely divided metal clusters in glasses, one can colour them in various hues. This phenomenon was explained in 1908 by

Mie in terms of size-dependent plasma resonances of the metallic electrons. Since about 1980, interest in the study of such clusters has grown strongly. Important knowledge for various fields, such as catalysis, photography, the structure of amorphous substances, and the formation of large molecules in interstellar space can be obtained from cluster physics.

At present, supersonic molecular beams are an important preparation method, in particular for molecular clusters. Figure 15.6 shows an experimental set-up for the study of clusters.

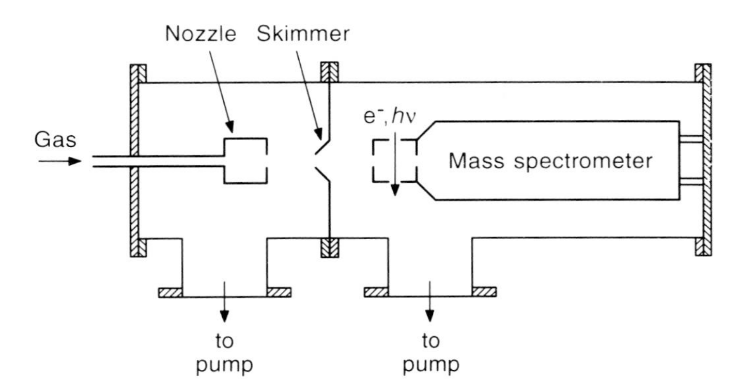

Fig. 15.6. An apparatus for the production and investigation of clusters. The clusters from the nozzle pass through an entrance diaphragm (skimmer) into the detection chamber, where they are ionised by an electron beam or by light. The analysis is performed in a mass spectrometer, which also detects the molecular fragments that are often formed on ionisation of the clusters

In the cold molecular beam, rather unusual molecules can be discovered. For example, at a beam temperature of 0.3 mK, it has proved possible to identify the helium dimer, He_2, cf. Fig. 13.3. Its binding energy corresponds to a temperature of 1 mK [J. Chem. Phys. **98**, 3564 (1993)]. This molecule has surprising properties: because of its small binding energy, the average internuclear distance is 55 Å, and it can exist only in its ground state, without rotational or vibrational excitations.

There are also stable clusters and large molecules which are produced in such cold molecular beams. A particularly notable molecule which was discovered in this way is Buckminster-Fullerene, C_{60}, consisting of 60 carbon atoms, which resembles a soccer ball and has icosahedral symmetry (see Fig. 4.18). This molecule can now be prepared in other ways, e.g. by vaporising graphite in a helium atmosphere. An especially sensational discovery was that C_{60} doped with alkali metal ions in the solid phase is superconducting, with an unexpectedly high transition temperature. Still larger molecules, such as C_{70}, have since also been prepared.

In the few years since the discovery of the Fullerenes, a number of new developments have been made; in addition to C_{60} and C_{70}, still other related but more complex forms have been found. The structures and the bonding of these molecules have been investigated with all the available spectroscopic methods. In particular, it could be shown that smaller atoms or molecules can be enclosed in the C_{60} or C_{70} balls; this has initiated a new chapter in the chemical physics of molecules.

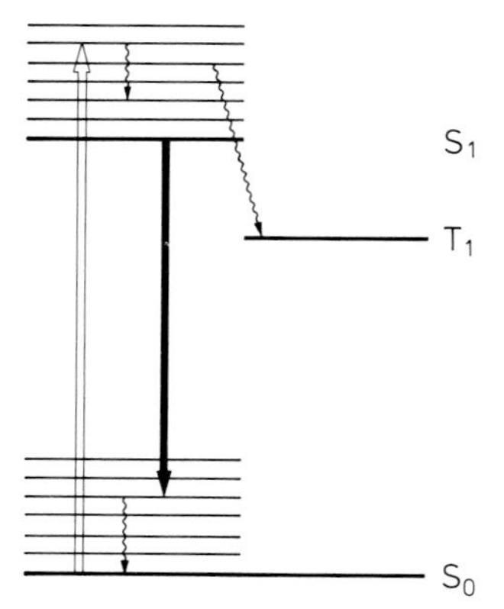

Fig. 15.7. The term diagram of an organic dye for a dye laser. The region of emission from the vibronic levels of the excited electronic state S_1 into the vibronic levels of the ground state S_0 extends over several 1000 cm^{-1}, depending on the dye. Intersystem crossing to the triplet state T_1 decreases the laser power. Further details are given in the text

15.5 Dye Lasers

For high-resolution molecular spectroscopy of cold molecules with very sharp energy terms, one requires spectrographs of high resolving power, or else tunable light sources with extremely small bandwidths. Using the latter, the molecular absorption can be measured with excitation spectroscopy by detecting the emission as a function of the excitation wavelength. If the spectral bandwidth of the light source is less than the energy width of the terms of the molecules, then they can be resolved in the spectrum.

The most important light source which can be tuned continuously over a large frequency range is the dye laser; see I, Sect. 21.1. In dye lasers, the laser medium consists of suitable organic dye molecules in solution. The corresponding term diagram is shown in Fig. 15.7.

The dye is optically pumped from its ground state S_0 into higher vibronic levels of the excited electronic state S_1. These levels, coupled to those of other molecules and to the solvent, are practically continuous in solution. Following a rapid radiationless relaxation to the lowest vibronic levels of S_1, emission from there into the likewise nearly continuous vibronic levels of the electronic ground state S_0 occurs. Since the latter are not thermally populated and their excess energy is quickly released to the environment by radiationless processes, the population inversion necessary for laser action is readily obtained. The desired laser frequency is selected from the broad spectral emission range of the dye, which can extend over several 1000 cm^{-1}, by tuning the laser resonator.

15.6 High-Resolution Two-Photon Spectroscopy

In larger molecules, the individual rotational lines in the electronic or vibronic-electronic transitions lie so close together that they can no longer be resolved by conventional spectroscopic methods. Larger molecules therefore do not exhibit line spectra, but instead broad bands; the Doppler broadening resulting from the thermal motions in the gas phase prevents the resolution of the lines and thus a precise measurement of the rotational and vibrational energies in the excited state. For the benzene molecule, C_6H_6, the Doppler width at room temperature is 1.7 GHz = 0.05 cm^{-1}, while the energy of the $S_1 \leftarrow S_0$ transition lies in the range of 40 000 cm^{-1}.

In such a case, the method of *Doppler-free two-photon spectroscopy* can be applied to advantage; it was first used in atomic spectroscopy and is described in I, Sect. 22.4. The light from an extremely narrow-band, tunable laser impinges as two oppositely-directed beams on the molecules to be studied. If a single molecule interacts with two photons which are travelling in opposite directions, the Doppler shifts are equal and opposite and compensate each other, assuming that both beams have exactly the same frequency. In that case, no Doppler broadening is observed and the determination of the transition frequency can be made much more exactly. Attainable precisions are of the order of 1 MHz.

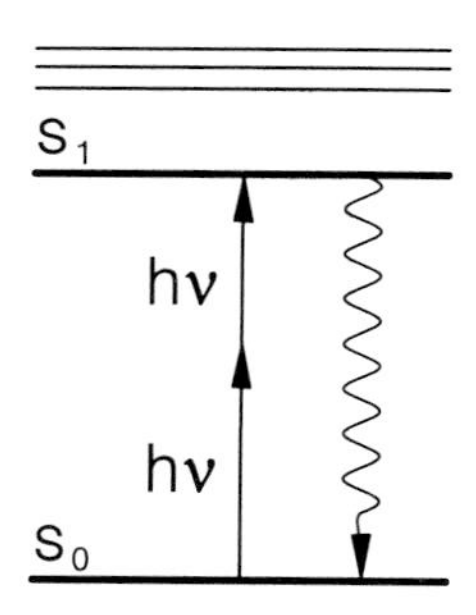

Fig. 15.8. The principle of two-photon absorption spectroscopy with fluorescence detection: the molecule is excited from its ground state S_0 by a two-photon transition, here into the state S_1. The absorption is detected by the resulting fluorescence

In this process, the molecule is raised to a state whose energy corresponds to the sum of the two photon energies. The absorption bands which lie in the ultraviolet region can thus be excited by blue light. To be sure, the selection rules for two-photon absorption are different from those for conventional single-photon processes (Sect. 17.3). Two-photon absorption is detected as an excitation spectrum using the UV fluorescence (see Fig. 15.8).

The power of this method is illustrated in Fig. 15.9 using the example of the $S_1 \leftarrow S_0$ absorption spectrum of the benzene molecule in the gas phase, as obtained in measurements

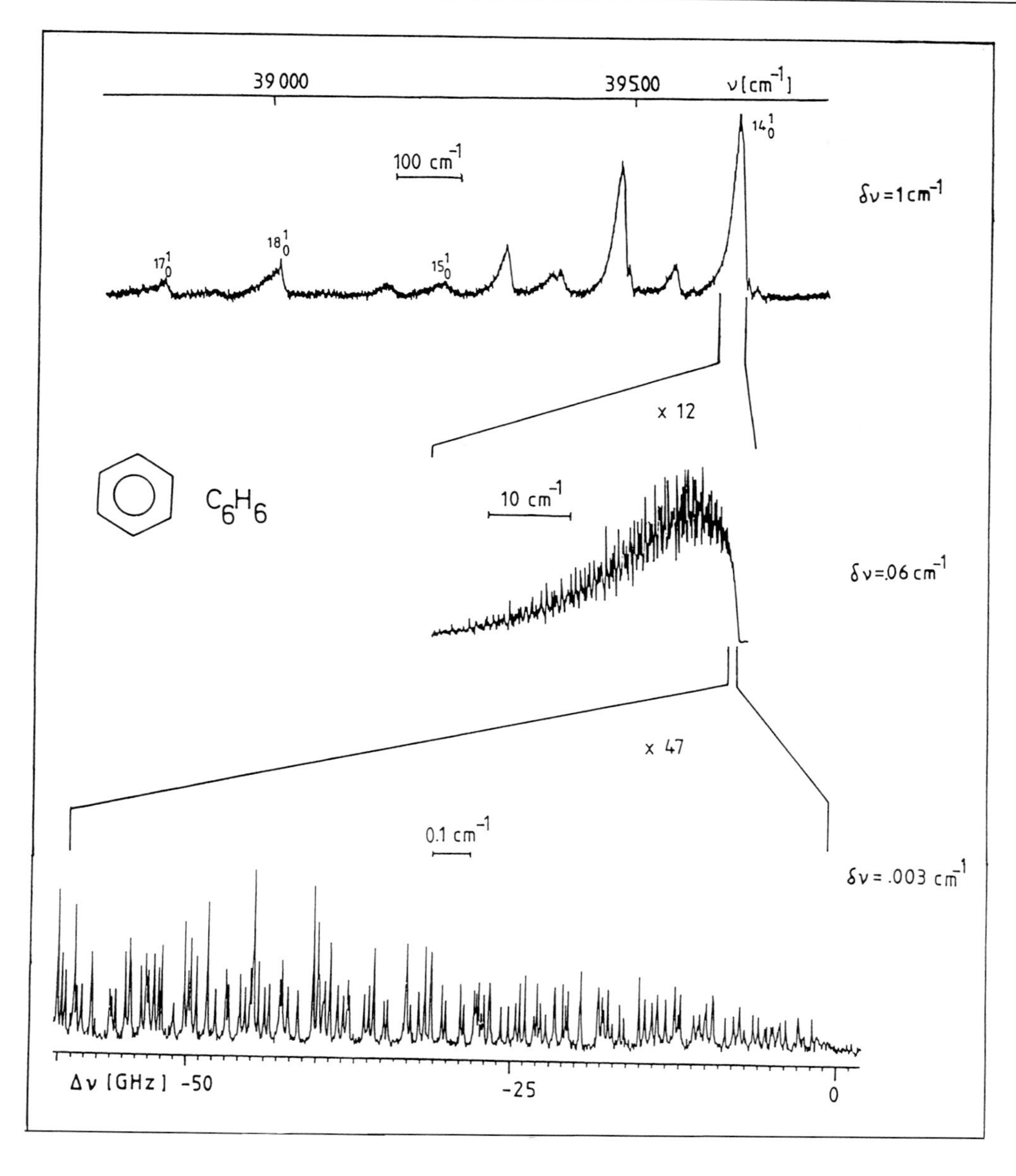

Fig. 15.9. The two-photon absorption spectrum of benzene from the electronic transition $S_1 \leftarrow S_0$, measured with varying spectral resolutions. In the upper part of the figure, taken at a resolution of 1 cm^{-1}, one can recognise the vibrational structure of the electronic transition. The middle part of the figure shows a spectrum with resolution limited by Doppler broadening to 0.06 cm^{-1}; it represents the Q branch of the strongest vibrational band. In the bottom spectrum, the resolution was improved to 0.003 cm^{-1} by using Doppler-free spectroscopy. The individual rotational lines can now be resolved. The following notation is used for the vibrational bands: 14_0^1 means the normal mode 14, which is excited in the final state (here S_1) by 1 quantum, but has 0 quanta of excitation in the initial state (ground state S_0). From H.J. Neusser and E.W. Schlag, Angew. Chem. **104**, 269 (1992)

with varying spectral resolutions. The resolution in the upper spectrum is 1 cm^{-1}, while it is Doppler limited to 0.06 cm^{-1} in the centre spectrum; only in the lower spectrum, where the Doppler broadening has been eliminated, giving a resolution of 0.003 cm^{-1}, can the individual rotational lines in the spectrum be distinguished.

There are thus numerous rotational lines hidden within the Doppler-broadened vibrational bands, which can be observed only when the Doppler broadening is eliminated. High-resolution spectroscopy makes it possible to determine the rotational constants and the centrifugal stretching constants for a molecule with a precision of 10^{-6} cm^{-1} in the ground state

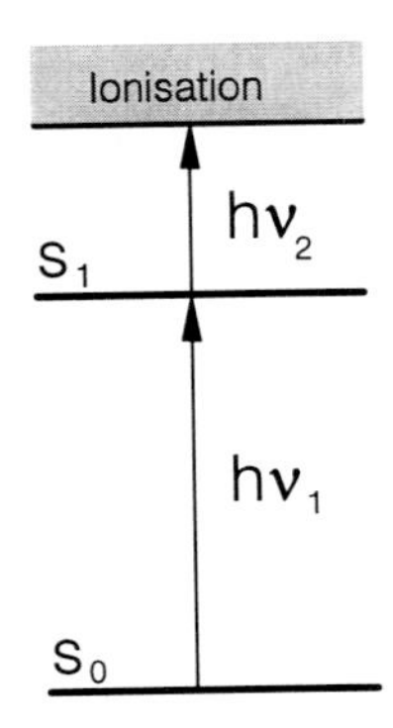

Fig. 15.10. An energy-level diagram to illustrate resonance-enhanced two-photon ionisation. A first light quantum $h\nu_1$ is tuned to a real rovibronic excitation level of the molecule being studied; a second light quantum $h\nu_2$ ionises the molecule. The ions, and thus the absorption process of the quantum $h\nu_1$ itself, can be detected by a mass spectrometer. The measurement can also be carried out in a Doppler-free manner. See also H.J. Neusser and E.W. Schlag, Angew. Chem. **104**, 269 (1992)

S_0 as well as in the excited state S_1. This in turn yields exact information on the molecular structure.

Another related technique for the optical spectroscopy of larger molecules at high resolution is *resonance-enhanced two-photon ionisation*. It is used in particular for the investigation of the weakly bound Van der Waals molecular complexes which can form in supersonic beams.

In this technique, the molecule or complex is first excited by a quantum $h\nu_1$ in the range of the $S_1 \leftarrow S_0$ transition, as shown in Fig. 15.10. This initial excitation quantum is produced by a tunable dye laser with a very narrow bandwidth. A second light quantum $h\nu_2$, which is emitted by a second dye laser with a relatively broad frequency distribution, ionises the excited molecule; thus the name resonance enhanced two-photon ionisation. The resulting ions are mass-selected and detected in a mass spectrometer. An example is shown in Fig. 15.11, which illustrates the structure of the Van der Waals complex formed by benzene and argon, C_6H_6–Ar_2, as found from an analysis of the vibrational and rotational structure of this complex; it can be studied only in a supersonic molecular beam.

15.7 Ultrashort Pulse Spectroscopy

The investigation of the velocities of molecular processes or reactions has progressed enormously in the past decades. The first studies of chemical rearrangements and reactions with optical spectroscopy were performed more than 40 years ago using flash lamps in the microsecond range; Nobel prizes were given to *Norrish* and *Porter* and to *Eigen* for this work.

In more recent times, it has become possible to produce light pulses of lengths in the range from pico- to femtoseconds using lasers. The current limit for short pulse times is about 5 ps, corresponding to a wave train only a few wavelengths long.

Pico- and femtosecond light pulses are necessary if one wishes to study ultrarapid dynamic processes of and on molecules, such as intra- and intermolecular energy transfer, relaxation between the rotational and vibrational levels of molecules following perturbation by light excitation or by disturbance of thermal equilibrium, or the rates of chemical reactions.

The most important experimental technique for these studies is the pump-probe method. A first light pulse raises the molecule into a short-lived excited state, and a second pulse probes the excitation, for example by measuring the short-lived absorption of the excited state. By varying the time interval between the first and the second light pulse, one can determine the lifetime of the excited state from the change in the probe signal as a function of this time interval. Available laser technology permits pulse lengths and pulse intervals of a few fs, so that dynamic processes can be investigated down to about 10 fs.

An impressive example is the photodissociation of the NaI molecule; cf. Fig. 15.12. Using an excitation pulse of 310 nm wavelength, the NaI molecule is excited into a state [NaI]*; in this state, it can dissociate into Na + I, but it can also remain in the [NaI]* state with a certain probability and dissociate only after one or more periods of oscillation. This temporal behaviour is observed with the second, or probe, pulse. If this pulse detects the absorption of the free Na atom at 589 nm – see the upper curve in Fig. 15.12 – then it can be used to follow the time development of the dissociation. The degree of dissociation increases with the period of oscillation of the [NaI]* complex, 1.25 ps $\equiv$ 27 cm^{-1}; the transition from the bound state to the dissociated state can take place with this period. The dissociation is complete after about 10 oscillation periods.

If one sets the probe pulse outside the Na resonance absorption wavelength, on the other hand, the absorption of Na which is still bound in the complex [NaI]* can be measured. The absorption frequency in this case depends on the momentary internuclear distance between Na and I, so that absorption is observed each time that the oscillation of the complex gives rise to the internuclear distance corresponding to the probe pulse light frequency. The result of such a measurement is shown as the lower curve in Fig. 15.12. The oscillatory decrease in this absorption with the vibrational period of the complex can be readily seen. The dissociation probability in each period is about 0.1.

Ultrashort pulse spectroscopy in the time range of pico- to femtoseconds thus allows the time development of chemical reactions to be observed. The term 'femtochemistry' has been suggested as a name for this field of research.

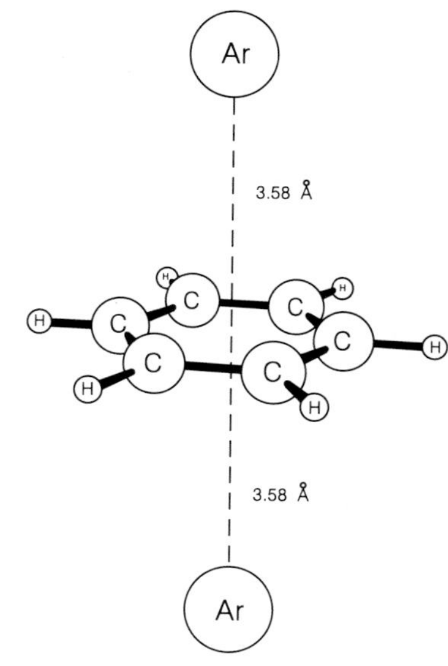

Fig. 15.11. The experimentally determined structure of the Van der Waals complex C_6H_6-Ar_2, as an example of the power of Doppler-free molecular spectroscopy. This (short-lived) complex is formed in a supersonic molecular beam. Its detection is performed with the aid of a mass spectrometer. See also H.J. Neusser and E.W. Schlag, Angew. Chem. **104**, 269 (1992)

15.8 Photoelectron Spectroscopy

With the spectroscopic methods in the infra-red, visible, and near ultraviolet spectral regions which we have thus far discussed, one can investigate more or less exclusively the outer, weakly bound electrons of a molecule. The methods of *photoelectron spectroscopy* have made great strides in recent years for the study of the inner electronic shells of molecules. In these methods, the photoelectric effect is employed, i.e. the fact that irradiation with light of a sufficiently high quantum energy can release electrons from their bound states; see I, Sect. 5.3. The sample is irradiated with monochromatic UV- or X-radiation, and the kinetic energy E_{kin} of the electrons released is measured in an electron velocity analyzer. In the simplest case, the energy is given by the basic equation of the photoeffect:

$$E_{\text{kin}} = h\nu - E_{\text{B}} , \tag{15.9}$$

where $h\nu$ is the quantum energy of the excitation light and E_{B} is the binding energy of the electrons; cf. Fig. 15.13.

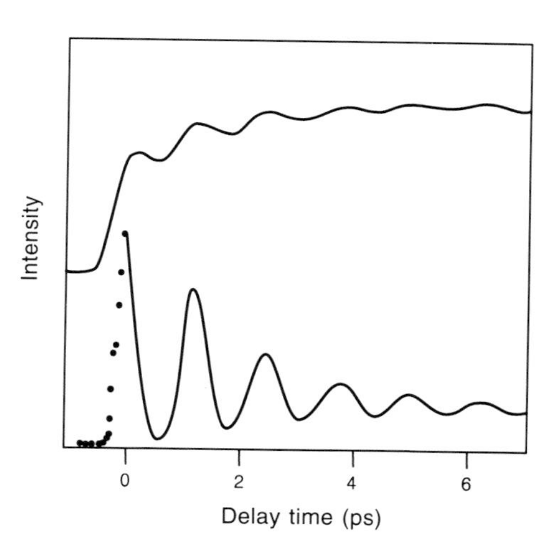

Fig. 15.12. A femtosecond-spectroscopic investigation of the dissociation reaction NaI → [NaI]* → Na+I. The absorption is plotted as a function of the delay time between the excitation light pulse and the probe pulse. The absorption is determined here as the intensity of the laser-induced fluorescence, i.e. as an excitation spectrum. *Upper curve*: the absorption of the free Na atom at its resonance frequency, 589 nm (Na *D* lines). *Lower curve*: the absorption of the complex, i.e. the Na absorption away from the resonance. Further details are given in the text. From A.H. Zewail, Science **242**, 1645 (1988)

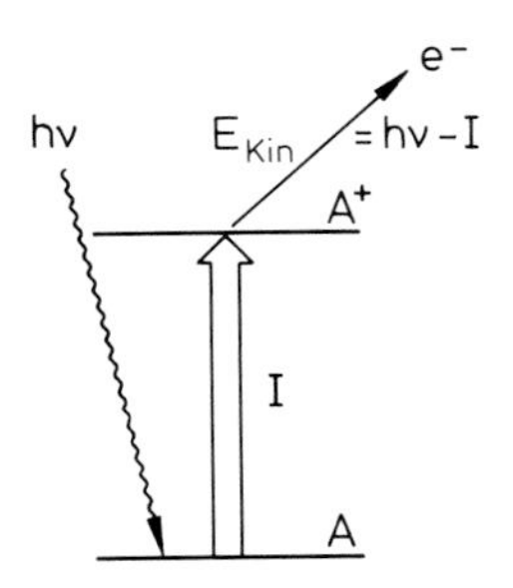

Fig. 15.13. The basis of photoelectron spectroscopy is the photoelectric effect. An incident photon with the energy $h\nu$ ionises an atom or molecule (ionisation energy I). The excess energy appears as the kinetic energy of the photoelectrons which are released in the process

For the excitation, one requires intense, monochromatic radiation. Since even for the outer valence electrons, the ionisation energies amount to several eV, the radiation must be in the ultraviolet or X-ray range. For this purpose, either the line spectrum of a gas discharge is used (e.g. the He(I) line from the $1s2p \rightarrow 1s^2$ transition at 58.43 nm or 21.22 eV), or, if more strongly bound electrons are to be studied, characteristic X-ray lines can be employed. Synchrotron radiation is especially suitable as a radiation source for this technique (see I, Sect. 5.1), since it is tunable (or selectable) over a wide frequency range.

The energy of the electrons is determined by deflecting them in an electric or magnetic analyzer, in a similar way as in an e/m_0 determination or in a mass spectrometer (I, Sects. 3.2 and 6.4). These energy analyzers can presently attain an energy resolution of better than 2 meV, i.e. 10 cm^{-1}. Figure 15.14 gives a schematic representation of an experimental set-up.

The term UPS (Ultraviolet Photoelectron Spectroscopy) is used when the excitation is produced by UV radiation, and XPS (X-ray Photoelectron Spectroscopy) when it is performed with X-radiation. In the process of releasing electrons with X-rays, one in the first approximation measures the atomic binding energies of the inner-shell electrons even in molecules, since they are influenced only weakly by chemical bonding and are determined for a given atom mainly by its ionisation state. The characteristic atomic electron binding energies can thus be used for the analysis of the atomic composition of a sample – hence the name "ESCA", which stands for 'Electron Spectroscopy for Chemical Analysis'. There are, however, some effects due to chemical bonding, especially on the outer electrons.

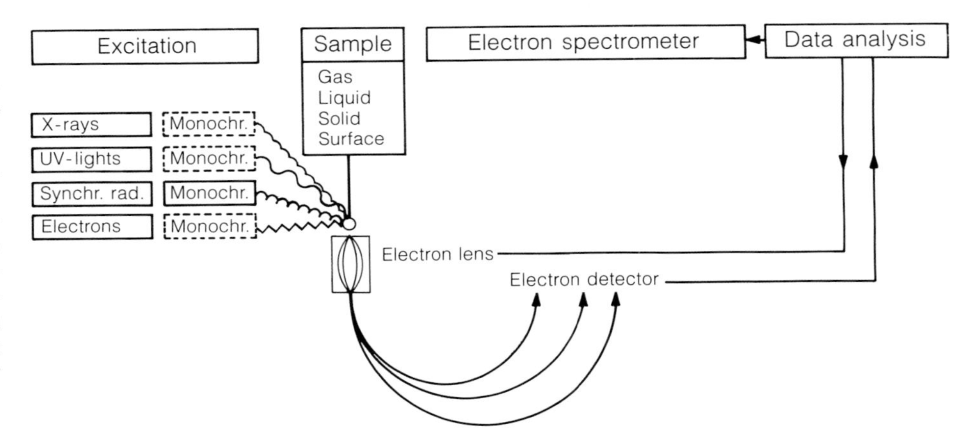

Fig. 15.14. Schematic drawing of an experimental set-up for the electron spectroscopy of atoms, molecules, or solids. The excitation is performed with radiation of various types, which must be as monochromatic as possible. The electrons released from the sample are focussed and analyzed with respect to their kinetic energies in an electron spectrometer, then amplified and detected. [After K. Siegbahn, Phys. Bl. **42**, 1 (1986)]

Figure 15.15 shows as an example the photoelectron spectrum of the N_2 molecule, and, for comparison, the molecular orbital diagram of the nitrogen molecule, which we have discussed earlier (Sect. 13.3). In the spectrum, photoelectrons from the reaction $N_2 + h\nu \rightarrow N_2^+ + e^- + E_{kin}$ from all the orbitals for the 14 electrons of the molecule can be analyzed, to be sure with different degrees of energy resolution. This makes the inner-shell orbitals, and not just the valence electrons, accessible to study. Figure 15.16 demonstrates with the example of the O_2 molecule that the ion resulting from photoionisation can also be excited into a vibrational level, following the Franck-Condon principle. Photoelectron spectroscopy thus also allows the measurement of the vibrational quanta of the ionisation products. Finally, Fig. 15.17 gives another example, this time of the photoelectron spectrum of a larger molecule, benzene, with photoelectrons from numerous orbitals.

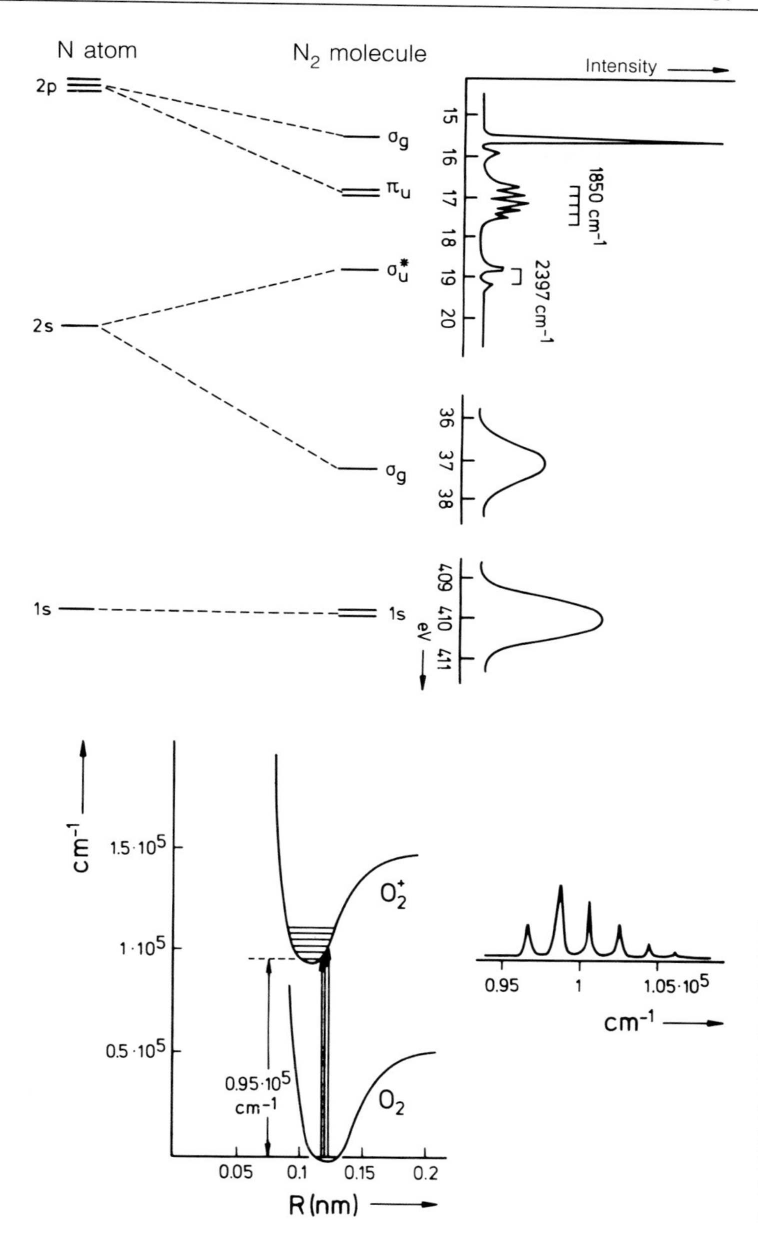

Fig. 15.15. The photoelectron spectrum of the N_2 molecule. The binding energies of the photoelectrons are computed from their measured kinetic energies and plotted in eV on the vertical scale. Comparison with the calculated molecular orbital scheme (cf. also Fig. 13.4) shows that the term energies of the inner-shell electrons of a molecule can be determined by photoelectron spectroscopy

Fig. 15.16. The photoelectron spectrum (*right*) and the term diagram for the O_2 molecule. The O_2^+ ion can be formed not only in its vibrational ground state, but also in excited vibrational states in the photoemission process. Then one can observe the vibrational structure of the O_2^+ molecule, here in cm^{-1}, relative to the vibrationless transition at $0.95 \cdot 10^5$ cm^{-1}

An important additional application of photoelectron spectroscopy is the study of the influence of chemical bond formation on the binding energies of inner-shell electrons in complex molecules. This can be explained using the example of ethyl propionate,

$$\mathrm{CH_2CH_3\underset{\underset{\displaystyle O}{\|}}{C}OCH_2CH_3}\ .$$

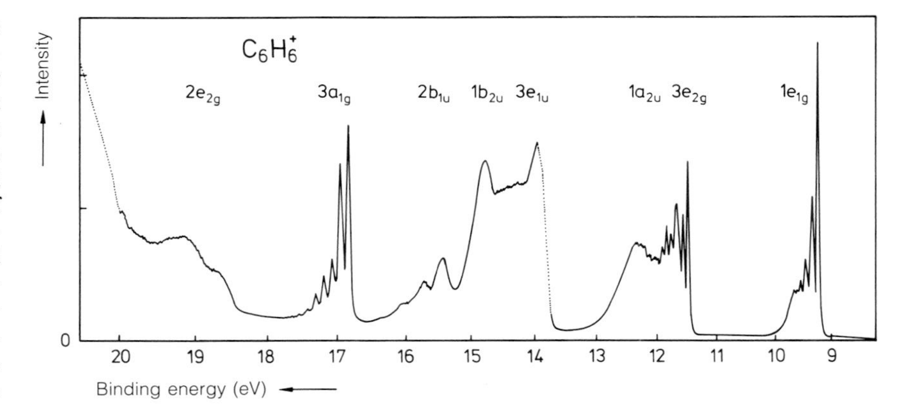

Fig. 15.17. The photoelectron spectrum of the valence electrons of benzene, C_6H_6, after excitation with the He I line at $h\nu = 21.22$ eV from a helium gas-discharge lamp. A unique identification of the measured binding energies in terms of the molecular orbitals is possible, but will not be explained in detail here. Vibronic structure can also be seen. [More information can be found in L. Karlsson, L. Mattson, R. Jadrny, T. Bergmark, and K. Siegbahn, Phys. Scr. **14**, 230 (1976)]

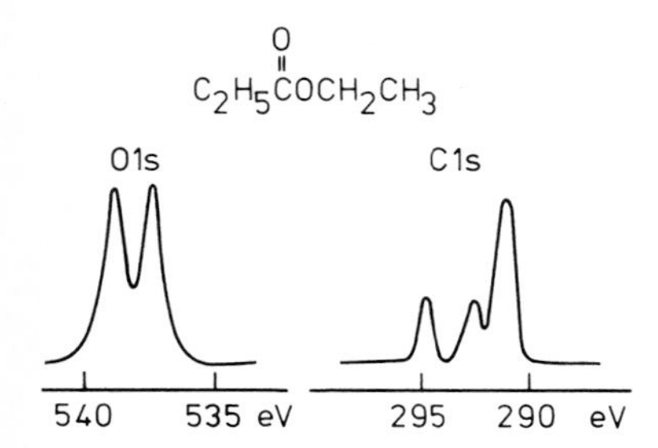

Fig. 15.18. The photoelectron spectrum of ethyl propionate in the gas phase. The spectrum shows the 1s core electrons of carbon (binding energy around 293 eV) and of oxygen (binding energy around 538 eV). The splitting of the two elemental lines corresponds to the different chemical shifts of the atoms, depending on their bonding in the molecule. Oxygen is present in the C=O group and in the chain; carbon as C=O, CH_2, and CH_3

In this molecule, O occurs in two different bonding states, and carbon in four, of which two are very similar. In the photoelectron spectrum of the 1s-electrons, one can indeed distinguish 2 groups of electrons from the O atoms and 3 groups from the C atoms in the spectrum, and these can be attributed to the various structural groups in the molecule; cf. Fig. 15.18. The concept of chemical shifts is used in photoelectron spectroscopy as well as in nuclear magnetic resonance spectroscopy (Chap. 18) and in Mössbauer spectroscopy, where it plays an important role. A further example is shown in Fig. 15.19.

Photoelectron spectroscopy thus provides an important complement to the optical spectroscopies for the investigation of the structure and the bonding of molecules.

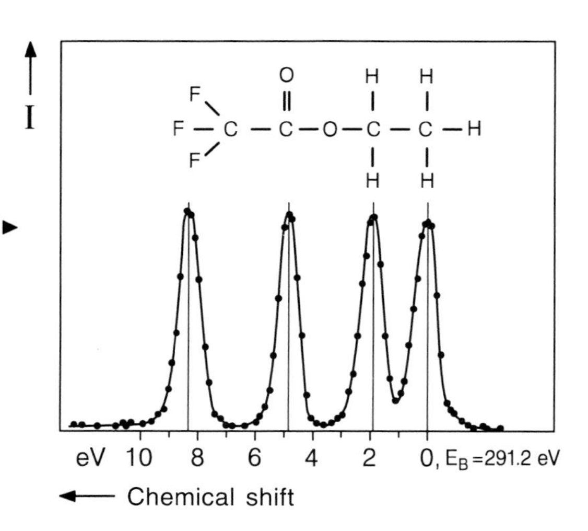

Fig. 15.19. The 1s electron lines of carbon in ethyl trifluoroacetate, as examples of the chemical shift in photoelectron spectroscopy. One can distinguish four lines of equal intensities, which originate from the four carbon atoms with different chemical environments in the molecule. After K. Siegbahn, J. Pure Appl. Chem. **48**, 77 (1976)

15.9 High-Resolution Photoelectron Spectroscopy

For the application of photoelectron spectroscopy to the spectroscopic analysis of molecules, its low spectral resolution proves to be a hindrance. In conventional photoelectron spectroscopy, the best attainable resolution is about 1 meV or 10 cm^{-1}. When the photoelectrons are produced by light from a narrow-band dye laser acting on molecules in a supersonic molecular beam, the energy resolution is limited by the precision with which the kinetic energy of the photoelectrons can be measured. Using conventional photoelectron spectroscopy, it is naturally not possible to resolve the rotational structure in the spectrum of larger molecules. How can this situation be improved?

A spectral resolution which is two orders of magnitude better (about 25 μeV, i.e. 0.2 cm^{-1}) can be attained with a new technique, that of Zero Kinetic Energy PhotoElectron Spectroscopy (ZEKE-PES). It thus permits the investigation of low-energy vibrations in cluster ions, such as that of phenol-water, or the rotationally resolved spectroscopy of large molecular ions, such as the benzene cation.

In the ZEKE method, light of a fixed wavelength is not used, as in conventional photoelectron spectroscopy, nor is the kinetic energy of the emitted photoelectrons analyzed to obtain information about the molecular states; instead, the light wavelength is varied and only those electrons are detected that are emitted with no kinetic energy (or a very small kinetic energy) from the sample. These extremely low-energy electrons (ZEKE or threshold electrons) are released when the energy of the photon just corresponds to the energy difference between an initial state of the neutral molecule and a final state of the ion, i.e. no excess kinetic energy is transferred to the emitted electron.

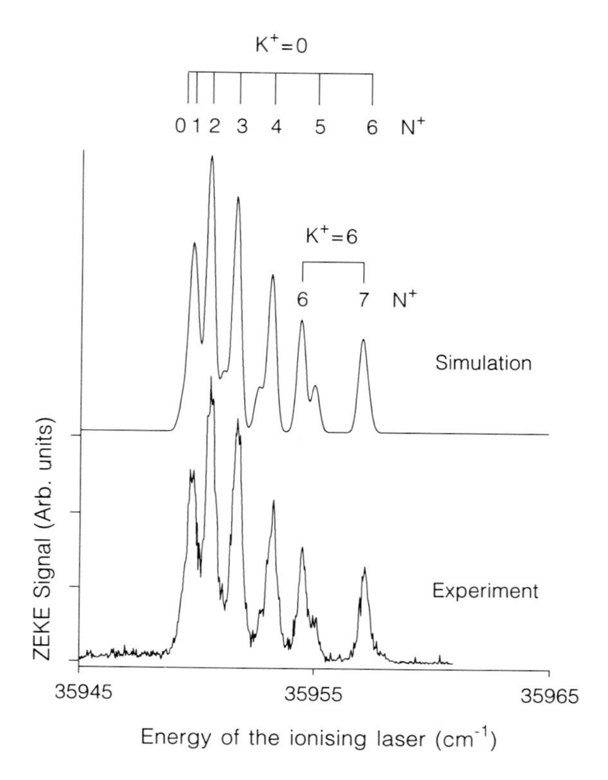

Fig. 15.20. A fully rotationally-resolved ZEKE spectrum of the benzene cation via the rovibronic state $S_1 6^1 (J' = 2, K' = 2, +1)$ as resonant intermediate state. For the case shown here, that of a symmetric top, the quantum numbers J and N denote the total angular momentum (N without electron spin) and K its projection on the symmetry axis. The spectrum shown, and similar spectra recorded via other intermediate states, clearly demonstrate that the benzene cation has D_{6h} symmetry. [For more details, see R. Lindner, H. Sekiya, B. Beyl, and K. Müller-Dethlefs, Angew. Chemie **105**, 631 (1993); also K. Müller-Dethlefs and E.W. Schlag, Ann. Rev. Phys. Chem. **42**, 109 (1992)]

The unwanted electrons with kinetic energy are distinguished from the "genuine" ZEKE electrons by means of an electric field pulse which extracts the emitted electrons after a time delay. During the field-free delay time, the kinetic electrons leave the region of the sample and can thus be discriminated from the ZEKE electrons, since the two groups arrive at the detector at different times. A second version, the field ionisation of highly excited Rydberg states, which converge to (ro)vibronic levels of the molecular ion (Pulsed Field Ionisation, PFI), is on the whole simpler to carry out and yields an even higher resolution. However, while the ZEKE method can be generally applied to neutral molecules and anions, the PFI technique cannot be used for the latter, since no Rydberg states exist for them. An example of the excellent energy resolution of ZEKE photoelectron spectroscopy is given in Fig. 15.20, which shows a fully resolved spectrum of the benzene cation, $C_6H_6^-$.

16. The Interaction of Molecules with Light: Quantum-Mechanical Treatment

Following an outline of time-dependent perturbation theory, we treat in detail the spontaneous and stimulated emission and the absorption of light by molecules. In particular, we derive the transition probabilities and the Einstein coefficients. The Franck-Condon principle and a discussion of the selection rules for transitions will again play an important role in our considerations.

16.1 An Overview

In the preceding chapters, we gained a general knowledge of a number of the spectroscopic properties of molecules. In the present one, we now want to lay the foundation for a strict quantum-mechanical treatment of those processes. As in atoms, electrons in molecules can undergo transitions when they absorb light, and can also emit light in spontaneous or stimulated (induced) emission processes. Molecules, like atoms, can furthermore scatter light. At high radiation intensities, such as those obtainable with lasers, nonlinear optical processes also occur, e.g. the absorption of two or more photons by a molecule. In the case of molecules, however, some additional important properties are present, due to the additional degrees of freedom of the rotations and vibrations; these molecular motions can also give rise to the absorption or the spontaneous or stimulated emission of radiation. Furthermore, transitions are possible in which a change in the electronic state is accompanied by a change in the molecular vibrations or rotations; in such electronic transitions, quanta of the molecular vibrations or rotations can thus be created or annihilated. A new effect, when compared to atoms, is Raman scattering by molecules, where the energy of the incident light quantum is split into a quantum of molecular vibration and a re-emitted light quantum. The same process can be observed with rotational quanta. Along with these processes, in which a light quantum is first absorbed and then another is emitted, i.e. where two light quanta are involved, there are other processes involving two or more light quanta. An example of these is two-photon emission or absorption.

The most important goal of the quantum-mechanical treatment which we will carry out in this chapter is the calculation of the transition probability per unit time, i.e. the mean number of transitions which a molecule undergoes in one second under the influence of light, or in emitting light quanta. As we shall see, this transition probability is determined in the main by the so-called optical transition matrix element. It follows from the quantum-mechanical treatment that all processes conserve energy, whereby the total energy of the molecule plus light before and after the corresponding absorption or emission is conserved. The selection rules can be derived from the optical matrix element, i.e. from the transition matrix element just mentioned. This matrix element contains the wavefunctions of the coupled motions of the electrons and the molecular vibrations and rotations. An important technique consists of evaluating the matrix elements by separating the electronic and nuclear coordinates in a suitable fashion. This involves the *Born-Oppenheimer approximation* and the *Franck-Condon principle* (cf. Sect. 14.1).

Our method of proceeding in this chapter will be the following: under the assumption that the light intensity is not too great, we develop time-dependent perturbation theory in first order to describe absorption as well as spontaneous and stimulated emission. We then write the Hamiltonian operator which describes the interaction between the radiation field on the one hand and the degrees of freedom of the molecule with its electronic and nuclear motions on the other. Thereafter, we will apply the initially quite general results of perturbation theory to this particular interaction of radiation with molecules; we discuss the various types of absorption and emission, derive the Einstein coefficients for these processes, and then turn to the Franck-Condon principle. Finally, we describe the methods which can be used to arrive at the selection rules.

16.2 Time-Dependent Perturbation Theory

In I, we treated *time-independent* perturbation theory in detail. For the treatment of the interaction of radiation with matter, however, we must make use of *time-dependent* perturbation theory, recalling in the process a number of ideas which we have already used for the time-independent theory.

The basic Schrödinger equation which underlies the problem can be written in the form

$$\mathrm{i}\hbar \dot{\Psi} = (H_0 + H^{\mathrm{S}})\Psi , \tag{16.1}$$

where H_0 is the unperturbed Hamiltonian, and H^{S} represents the perturbation operator. The exact meaning of H_0 and H^{S} will be specified later; here, it will suffice to mention some very general properties of H_0 and H^{S} in preparation for our further discussion. We first consider the Schrödinger equation containing the unperturbed operator H_0, i.e.

$$H_0 \psi_\nu^0 = \mathrm{i}\hbar \dot{\psi}_\nu^0 . \tag{16.2}$$

In (16.2), the Hamiltonian H_0 is supposed to be independent of time, so that we can write its solution in the form

$$\psi_\nu^0(t) = \exp\left(-\frac{\mathrm{i}}{\hbar}E_\nu t\right)\phi_\nu . \tag{16.3}$$

The energy eigenvalues E_ν and the unperturbed wavefunctions ϕ_ν, which are independent of time, are fixed by the time-independent Schrödinger equation

$$H_0\phi_\nu = E_\nu\phi_\nu , \quad \nu = 1, 2, \ldots . \tag{16.4}$$

The indices ν are, of course, the quantum numbers. In the following, we assume that the wavefunctions ϕ_ν and the corresponding eigenvalues E_ν are already known. We then represent the solution of (16.1) which we are seeking as a superposition of unperturbed wavefunctions in the form

$$\Psi(t) = \sum_{\nu=1}^{\infty} c_\nu(t)\psi_\nu^0 , \tag{16.5}$$

where the coefficients $c_\nu(t)$ depend explicitly on time and are still to be determined. Since the wavefunctions ϕ_ν form a complete basis set, the approach taken in (16.5) is mathematically exact. We must now compute the coefficients $c_\nu(t)$; to this end, we substitute (16.5) into (16.1) and, taking (16.3) into account, obtain the result

$$\mathrm{i}\hbar \sum_\nu \dot{c}_\nu \psi_\nu^0 + \sum_\nu c_\nu E_\nu \psi_\nu^0 = \sum_\nu c_\nu H_0 \psi_\nu^0 + \sum_\nu c_\nu H^{\mathrm{S}} \psi_\nu^0 \,. \tag{16.6}$$

Because of (16.4), the second term on the left-hand side and the first term on the right-hand side of (16.6) cancel each other. In order to arrive at equations for the coefficients c, we multiply Eq. (16.6) by ψ_ν^{0*} and integrate over the coordinates on which ψ depends, in complete analogy to the procedure used for time-independent perturbation theory. The coordinates are, in general, the electronic and the nuclear coordinates of the molecule. We thus find

$$\dot{c}_\mu(t) = \frac{1}{\mathrm{i}\hbar} \sum_\nu c_\nu(t) H_{\mu\nu}^{\mathrm{S}} \,, \tag{16.7}$$

where we have used the abbreviation

$$H_{\mu\nu}^{\mathrm{S}} = \int \psi_\mu^{0*} H^{\mathrm{S}} \psi_\nu^0 \, dV \,. \tag{16.8}$$

The integral $\int \ldots dV$ symbolises an integration over all the coordinates on which ϕ_ν depends. The solution of the system of Eqs. (16.7) is completely equivalent to the solution of the original Eq. (16.1) and therefore in general just as difficult to obtain.

We now consider a situation in which the system initially, at time $t = 0$, is in the unperturbed quantum state κ, i.e. $\psi(0) = \phi_\kappa$. We then have as initial condition for the coefficients

$$c_\mu(0) = \delta_{\mu\kappa} \,, \tag{16.9}$$

where

$$\delta_{\mu\kappa} = 1 \quad \text{for} \quad \mu = \kappa \,, \qquad \delta_{\mu\kappa} = 0 \quad \text{for} \quad \mu \neq \kappa \,.$$

We now assume that the perturbation is small. We can then expect that for times which are not too long, the coefficients $c_\mu(t)$ differ only by small amounts from their initial values (16.9). In this approximation, we may assume that we can replace the coefficients $c_\nu(t)$ on the right-hand side of (16.7) by those given in (16.9). We then immediately arrive at a new system of equations:

$$\dot{c}_\mu(t) = \frac{1}{\mathrm{i}\hbar} H_{\mu\kappa}^{\mathrm{S}} \,. \tag{16.10}$$

Since $H_{\mu\nu}^{\mathrm{S}}$, according to (16.8), contains the time-dependent wavefunctions ψ_μ^{0*} and ψ_ν^0, this matrix element is itself time dependent. We integrate both sides of (16.10) over time, employing the initial condition (16.9), and obtain the result:

$$c_\mu(t) = \frac{1}{\mathrm{i}\hbar} \int_0^t H_{\mu\kappa}^{\mathrm{S}}(\tau) \, dr + \delta_{\mu\kappa} \,. \tag{16.11}$$

In order to arrive at results which we can compare with experiment, it is expedient to specify the right-hand side of (16.11) more precisely. We insert the wavefunction (16.3) on the right-hand side of (16.8) and assume that H^S is time independent; then we can write (16.8) in the form

$$H^S_{\mu\kappa} = \exp(\mathrm{i}\,\omega_{\mu\kappa} t) H^S_{\mu\kappa}(0)\ , \tag{16.12}$$

where we have used the abbreviations

$$\omega_{\mu\kappa} = \frac{1}{\hbar}(E_\mu - E_\kappa) \tag{16.13}$$

and

$$H^S_{\mu\kappa}(0) = \int \phi^*_\mu H^S \phi_\kappa dV\ . \tag{16.14}$$

Since we are interested in finding out which states are repopulated as a result of the perturbation, we consider the case $\mu \neq \kappa$, i.e. a final state μ which is different from the initial state κ. With this assumption, we can immediately compute the integral over time in (16.11) by making use of (16.12–14), finding

$$c_\mu(t) = \frac{1}{\hbar\,\omega_{\mu\kappa}}[\exp(\mathrm{i}\,\omega_{\mu\kappa} t) - 1] H^S_{\mu\kappa}(0) \qquad \text{for} \qquad \mu \neq \kappa\ . \tag{16.15}$$

An exact measure of how strongly the state $\mu \neq \kappa$ is populated as a result of the perturbation is given by the square of the probability amplitude $c_\mu(t)$, i.e.

$$|c_\mu(t)|^2\ . \tag{16.16}$$

Inserting (16.15) into (16.16) and performing a minor rearrangement of the exponential function to give a sine function yields the result

$$|c_\mu(t)|^2 = \frac{4\sin^2(\omega_{\mu\kappa} t/2)}{\hbar^2(\omega_{\mu\kappa})^2} |H^S_{\mu\kappa}(0)|^2\ . \tag{16.17}$$

Instead of computing the occupation probability (16.16), it is often more expedient to determine the transition probability per unit time, i.e. the number of transitions per second. The transition probability per second is given by the time derivative of (16.16) or (16.17):

$$w_{\mu\kappa} = \frac{d\,|c_\mu(t)|^2}{dt}\ , \tag{16.18}$$

or, more explicitly,

$$w_{\mu\kappa} = \frac{2}{\hbar^2\,\omega_{\mu\kappa}} \sin(\omega_{\mu\kappa} t) |H^S_{\mu\kappa}(0)|^2\ . \tag{16.19}$$

If we apply this formula to the absorption of radiation by molecules, we find a peculiar difficulty: as we know, a certain number of molecules is transferred per unit time from the initial state κ to the final state μ, and the transition rate given by (16.18) is experimentally *independent of time*. However, if one plots the calculated transition probability (16.19) against time, it is seen to be a periodically oscillating function of t, which is therefore by no means independent of time. Here, it would seem that we have a contradiction between theory and experiment; it can be eliminated, however, if we think about the precise experimental conditions. One indeed finds such a periodic back-and-forth in the transition probability if the incident radiation is extremely monochromatic, i.e. coherent. This condition is often fulfilled in nuclear magnetic resonance, as well as in some experiments with laser radiation. In our present considerations, we have assumed that the radiation was emitted by conventional sources (thermal sources), which have a finite frequency bandwidth. The molecular states are also often broadened, e.g. due to the finite lifetime of excited energy levels. The assumption that the transition frequency $\omega_{\mu\kappa}$ is infinitely sharp is thus not justified; instead, we must assume that the initial state or the final state, or both, have a continuous spectrum of levels.

Let us first consider the case that the *final states* have a continuous spectrum. We shall proceed in two steps: in the first, we imagine the *continuous* energy spectrum to be approximated by a series of closely-spaced *discrete* energy levels. This is referred to as a *quasi-continuum*. In the second step, we make the transition to the limiting case of a true continuum. That is, we first take the sum over $w_{\mu\kappa}$ of the transition probabilities with final states in a quasi-continuum, i.e. we replace (16.19) by

$$W = \sum_{\mu\in\Omega} w_{\mu\kappa} \,. \tag{16.20}$$

The sum thus extends over all the quantum numbers μ of states which are located within the interval Ω of the quasi-continuum. It is often useful to replace the quantum number μ, which in general stands for a whole set of individual quantum numbers, by the energy E and a further set of quantum numbers q:

$$\mu \to E, q \,. \tag{16.21}$$

We now consider the number of states $\mu \in \Omega$. In particular, these states are supposed to belong to an energy continuum

$$E \ldots E + dE \,. \tag{16.22}$$

The number Z of states in this interval (16.22) can be expressed in the form

$$Z = \varrho(E)dE \,. \tag{16.23}$$

We take into account the fact that the density of states $\varrho(E)$ can also depend on the other quantum numbers q:

$$\varrho(E) = \varrho_q(E) \,. \tag{16.24}$$

With these considerations, we obtain instead of (16.20) the expression

$$W = \sum_q \int \varrho_q(E) \frac{2}{\hbar} \frac{\sin[(E - E_\kappa)t/\hbar]}{E - E_\kappa} |H^{S}_{q\kappa}(0)|^2 dE \,, \tag{16.25}$$

where we have used the abbreviation

$$E_\kappa = \hbar\,\omega_\kappa \,. \tag{16.26}$$

Let us look a bit more carefully at the integrand in (16.25), abbreviating $(E - E_\kappa)/\hbar$ as x. If we plot the function $\sin(xt)/x$ against x, then we obtain the curves shown in Fig. 16.1. They tell us that in the limit $t \to \infty$, the function $\sin(xt)/x$ becomes practically infinitely

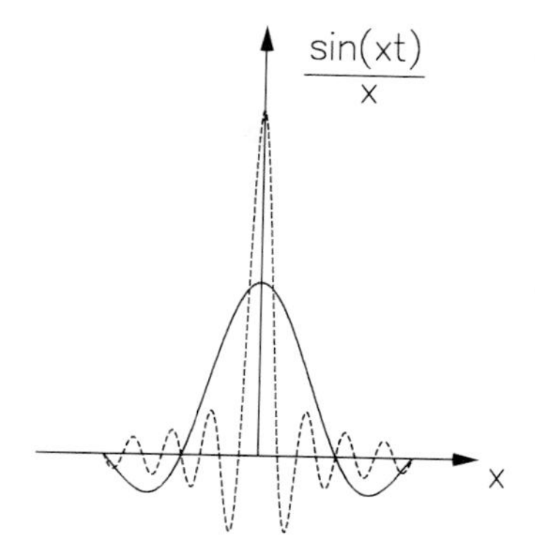

Fig. 16.1. The function $\sin(xt)/x$ is plotted against x for two values of the parameter t (*fully-drawn* and *dashed curves*). For increasing t-values, it becomes more and more concentrated around $x = 0$

sharply concentrated around the point $x = 0$; it thus acts like a δ-function, which can be proven mathematically to be in fact the case. Since, however,

$$\int_{-\infty}^{+\infty} \frac{\sin(xt)}{x} dx = \pi \tag{16.27}$$

and on the other hand

$$\int_{-\infty}^{+\infty} \delta(x) dx = 1 , \tag{16.27a}$$

we have to multiply the δ-function by a factor of π in order to make it quantitatively equivalent to the function $\lim_{t\to\infty} \sin(xt)/x$. Then, for sufficiently long times, we can replace the function $\sin(xt)/x$ by $\pi\,\delta(x)$. Using this replacement, we obtain as the final result, instead of (16.25), the following expression:

$$W = \sum_q \int \varrho_q(E) \frac{2\pi}{\hbar} \delta(E - E_\kappa) |H^{\mathrm{S}}_{q\kappa}(0)|^2 dE . \tag{16.28}$$

The δ-function in (16.28) ensures that the energy E of the final state will be equal to that of the initial state E_κ; it thus guarantees the conservation of energy. After carrying out the integration over E, we find

$$W = \sum_q \varrho_q(E_\kappa) \frac{2\pi}{\hbar} |H^{\mathrm{S}}_{q\kappa}(0)|^2 , \tag{16.29}$$

where the density of states $\varrho_q(E)$ is to be evaluated at the energy $E = E_\kappa$. In the literature, the result (16.29) is often called Fermi's Golden Rule. Taking the example of the emission of light by molecules, we shall see in Sect. 16.3 how the density of states $\varrho_q(E)$ can be calculated explicitly.

For many applications it is more practical not to replace the quantum numbers μ of the final states by E, q as in (16.21), but rather to let the sum over μ in (16.20) formally remain, remembering that the sum is finally to be expressed as an integral over continuously-variable quantum numbers. We then obtain the very simple formula

$$W = \frac{2\pi}{\hbar} \sum_{\mu\in\Omega} \delta(E_\mu - E_\kappa) |H^{\mathrm{S}}_{q\kappa}(0)|^2 . \tag{16.30}$$

Similar considerations may be persued in the case that the initial states are continuous. In this case, one does not sum over all the final states, but instead an average over the initial states is performed. We then find instead of (16.28) the following relation for the overall transition probability per second:

$$W = \frac{1}{Z} \sum_{\mu\in\Omega} w_{\mu\kappa} \tag{16.31}$$

with

$$Z = \sum_q \sum_{\Delta E} \varrho_q(E) dE , \tag{16.32}$$

where Z can be interpreted as the number of initial states.

Expression (16.30) for the transition probability per unit time will prove to be fundamental to the further developments in this chapter; we shall see explicitly how it is to be employed.

16.3 Spontaneous and Stimulated Emission and the Absorption of Light by Molecules

16.3.1 The Form of the Hamiltonian

In this section, we are interested in putting some physical meaning into the expressions (16.29) and (16.30) for the transition probability per unit time derived in the previous section. We first turn our attention to the evaluation of the matrix element $H^{\mathrm{S}}_{\mu\kappa}(0)$. In order to calculate it, we need the wavefunctions of the initial and final states and the explicit form of the perturbation operator. The starting point for our considerations is the Schrödinger equation (16.1), which we first write in the general form

$$\mathrm{i}\hbar\dot{\Psi} = H\Psi \, , \tag{16.33}$$

where we must still specify the Hamiltonian and the wavefunctions more precisely. We consider a molecule with N electrons having the coordinates $\boldsymbol{r}_1 \ldots \boldsymbol{r}_N$. If necessary, we could also include the spin variables explicitly; however, we will not do this in the following. Furthermore, the wavefunction depends upon the M nuclear coordinates, which we denote by $\boldsymbol{R}_1 \ldots \boldsymbol{R}_M$. The wavefunction must also contain the radiation field, which is described by the number of photons $n_{k.e}$ associated with light waves having a wavevector $\boldsymbol{k}$ and a polarisation direction $\boldsymbol{e}$. In the following, we combine $\boldsymbol{k}$ and $\boldsymbol{e}$ into a single index λ and write n_λ instead of $n_{k.e}$. Finally, Ψ also depends upon the time, so that in general, we must write

$$\Psi = \Psi(\boldsymbol{r}_1, \boldsymbol{r}_2, \ldots, \boldsymbol{r}_N; \boldsymbol{R}_1, \boldsymbol{R}_2, \ldots, \boldsymbol{R}_M; n_\lambda; t) \, . \tag{16.34}$$

The Hamiltonian H in (16.33) is given by

$$H = H_{\mathrm{el-n}} + H_{\mathrm{L}} + H^{\mathrm{S}} \, . \tag{16.35}$$

In this expression, $H_{\mathrm{el-n}}$ is the Hamiltonian which contains the kinetic energies of the electrons and the nuclei as well as all of the Coulomb interactions which occur between these particles. The spin-orbit coupling may also be included here, but we shall however not do this explicitly. H_{L} is the Hamiltonian of the radiation field, and H^{S} is the perturbation which results from the interaction of the radiation field with the molecule, i.e. its electrons and nuclei.

The unperturbed Hamiltonian H_0 includes those parts of (16.35) which refer to the electrons and the nuclei and their mutual interactions, added to the energy operator for the radiation field, i.e.

$$H_0 = H_{\mathrm{el-n}} + H_{\mathrm{L}} \, . \tag{16.36}$$

We have already described the Hamiltonian $H_{\mathrm{el-n}}$ in Chap. 11, and it should not be necessary to repeat it here. The Hamiltonian for the quantized radiation field was derived in I, Sect. 15.5; we remind the reader of the most important steps in that derivation. We begin with the electromagnetic field, representing the classical electric field vector $\boldsymbol{E}(\boldsymbol{r}, t)$ by an operator $\boldsymbol{E}(\boldsymbol{r})$ according to:

$$\boldsymbol{E}(\boldsymbol{r}, t) \Rightarrow \boldsymbol{E}(\boldsymbol{r}) \, . \tag{16.37}$$

In order to clarify its properties, we expand the position-dependent field strength as a series of partial waves with polarisation directions $\boldsymbol{e}_\lambda$ and wavevectors $\boldsymbol{k}_\lambda$:

$$\boldsymbol{E}(\boldsymbol{r}) = \sum_\lambda \boldsymbol{e}_\lambda N_\lambda [\mathrm{i} b_\lambda \exp(\mathrm{i}\boldsymbol{k}_\lambda \cdot \boldsymbol{r}) - \mathrm{i} b_\lambda^+ \exp(-\mathrm{i}\boldsymbol{k}_\lambda \cdot \boldsymbol{r})] \,. \tag{16.38}$$

The various symbols in (16.38) have the following meanings:

λ is an index which distinguishes the individual plane waves having wavevector $\boldsymbol{k}_\lambda$ and polarisation vector $\boldsymbol{e}_\lambda$;
ω_λ is the circular frequency of the light wave λ;
N_λ is a normalisation factor, with

$$N_\lambda = \sqrt{\frac{\hbar\,\omega_\lambda}{2\varepsilon_0 V}} \,; \tag{16.39}$$

ε_0 is the dielectric constant and V the normalisation volume in which the waves propagate, assuming periodic boundary conditions;
b_λ, b_λ^+ are creation and annihilation operators for light quanta with the index λ.

The vector potential enters in the form:

$$\boldsymbol{A}(\boldsymbol{r}) = \sum_\lambda \boldsymbol{e}_\lambda \frac{1}{\omega_\lambda} N_\lambda [b_\lambda \exp(\mathrm{i}\boldsymbol{k}_\lambda \cdot \boldsymbol{r}) + b_\lambda^+ \exp(-\mathrm{i}\boldsymbol{k}_\lambda \cdot \boldsymbol{r})] \,. \tag{16.40}$$

In a classical description, b_λ and b_λ^+ are time-dependent amplitudes, which, however, in the quantum-mechanical picture become operators, obeying the following commutation relations:

$$\begin{aligned} b_\lambda b_{\lambda'}^+ - b_{\lambda'}^+ b_\lambda &= \delta_{\lambda\lambda'} \,, \\ b_\lambda b_{\lambda'} - b_{\lambda'} b_\lambda &= 0 \,, \\ b_\lambda^+ b_{\lambda'}^+ - b_{\lambda'}^+ b_\lambda^+ &= 0 \,, \end{aligned} \tag{16.41}$$

which we have already seen in (7.47–49), but with different indices. They are defined in analogy to the commutation relations for the quantum-mechanical harmonic oscillator, which should be well known to us. In this formalism, the energy of the electromagnetic field can be represented as the sum of the energies of a series of uncoupled harmonic oscillators:

$$H_\mathrm{L} = \sum_\lambda \hbar\,\omega_\lambda b_\lambda^+ b_\lambda \,. \tag{16.42}$$

Equation (16.42) can be interpreted as a sum over the number operators $b_\lambda^+ b_\lambda$, each one multiplied by the energy of a quantum of radiation, $\hbar\,\omega_\lambda$.

We must now consider the form of the H^S in more detail. If we are dealing with an electron which is moving within an atom, then the dipole approximation is often sufficient:

$$H^\mathrm{S} = e\boldsymbol{r} \cdot \boldsymbol{E}(\boldsymbol{r}_0) \,, \tag{16.43}$$

where e is the elementary charge and $\boldsymbol{r}$ the coordinates of the electron, and $\boldsymbol{E}(\boldsymbol{r}_0)$ is the electric field vector of the radiation field at the position of the atom. In the case of larger molecules, however, this dipole approximation may no longer be valid; we therefore use the exact description here, based on the vector potential. As an example, we consider the

Hamiltonian of a single electron moving in the potential field of the atom and of the vector potential $\boldsymbol{A}$ of the radiation field. From I, Chap. 14, this Hamiltonian is given by:

$$H_{\mathrm{el}} = \frac{1}{2m_0}(\boldsymbol{p} - e\boldsymbol{A})^2 + V \,, \tag{16.44}$$

where V and $\boldsymbol{A}$ are functions of the electronic coordinates. In the following, we assume (as can always be done for the electromagnetic field) that the divergence of $\boldsymbol{A}$ vanishes:

$$\mathrm{div}\boldsymbol{A} = 0 \,. \tag{16.45}$$

Multiplying out the parentheses in (16.44), we obtain

$$H_{\mathrm{el}} = \frac{1}{2m_0}\boldsymbol{p}^2 + V - \frac{e}{m_0}\boldsymbol{A}\cdot\boldsymbol{p} + \frac{e^2}{2m_0}\boldsymbol{A}^2 \,, \tag{16.46}$$

where, owing to (16.45), we did not need to take account of the order of the operators $\boldsymbol{A}$ and $\boldsymbol{p}$. The first two terms in (16.46) are the operators of the kinetic and the potential energy, and the third and fourth terms give the interaction of the radiation field with the electron. If the intensity of the light field is not too great, we can neglect the term quadratic in $\boldsymbol{A}$, so that for the interaction operator between the radiation field and the electron, we obtain the expression:

$$H^{\mathrm{S}} = -\frac{e}{m_0}\boldsymbol{A}(\boldsymbol{r})\cdot\boldsymbol{p} \,. \tag{16.47}$$

In all these expressions, the momentum operator is naturally given by the usual prescription:

$$\boldsymbol{p} = \frac{\hbar}{\mathrm{i}}\nabla \,. \tag{16.48}$$

Up to now, we have considered the interaction of only a single electron with the radiation field. In a molecule, the electromagnetic field interacts with not only the whole set of electrons, but also with the nuclei. For this reason, we must replace the perturbation operator in (16.47) by a sum over the electronic indices and the nuclear indices. We then have for e.g. the electrons, the substitutions $\boldsymbol{r} \to \boldsymbol{r}_j, \boldsymbol{p} \to \boldsymbol{p}_j$. Our perturbation operator is thus expressed in terms of sums over the electronic indices j and the nuclear indices K:

$$H^{\mathrm{S}} = \sum_{j=1}^{N}\left[-\frac{e}{m_0}\boldsymbol{A}(\boldsymbol{r}_j)\cdot\boldsymbol{p}_j\right] + \sum_{K=1}^{M}\left[-\frac{eZ_K}{M_K}\boldsymbol{A}(\boldsymbol{R}_K)\cdot\boldsymbol{P}_K\right] \,. \tag{16.49}$$

In this equation, Z_K is the nuclear charge of the nucleus with index K. We have now determined all of the components of the Hamiltonians for the molecule (i.e. the electronic and nuclear motions), for the radiation field, and for their interactions.

16.3.2 Wavefunctions of the Initial and Final States

Before we apply the perturbation theory developed in Sect. 16.1, we must agree on what kinds of initial and final states we are going to deal with. We will need their wavefunctions for the evaluation of the matrix elements and for the determination of the corresponding energies. As the initial state, we take one in which the molecule and the radiation field have not yet begun to interact. Since the unperturbed Hamiltonian H_0 (16.36) is the sum of two terms, the overall initial-state wavefunction ϕ_κ (using the notation of Sect. 16.1) may be written as the product of the initial state of the molecule, $\phi_{Q_a}(\boldsymbol{r};\boldsymbol{R})$, with the initial state of the radiation field, Φ_a:

$$\phi_\kappa = \phi_{Q_a}(\boldsymbol{r};\boldsymbol{R})\Phi_a \,. \tag{16.50}$$

The index Q_a includes all the quantum numbers which determine the initial state of the molecule, both for the electrons and for the nuclei. The vector $\boldsymbol{r}$ summarizes all of the electronic coordinates, the vector $\boldsymbol{R}$ all of the nuclear coordinates in the molecule. The final state ϕ_μ can be written in a manner analogous to (16.50):

$$\phi_\mu = \phi_{Q_e}(\boldsymbol{r};\boldsymbol{R})\Phi_e \,. \tag{16.51}$$

16.3.3 The General Form of the Matrix Elements

Our task is now to compute the matrix elements $H^S_{\mu\kappa}(0)$ [cf. (16.14)]. We first take a closer look at the interaction operator H^S (16.49), by inserting (16.40) into (16.49) and employing the definition of N_λ, (16.39). Owing to the separation of (16.49) into two sums, we find two contributions to H^S, $H^S = H^S_{el} + H^S_n$, of which we give the first term explicitly, as an example:

$$H^S_{el} = \sum_\lambda b_\lambda \frac{1}{\sqrt{V}} O_\lambda(\boldsymbol{k}_\lambda) + \sum_\lambda b^+_\lambda \frac{1}{\sqrt{V}} O_\lambda(-\boldsymbol{k}_\lambda) \,. \tag{16.52}$$

In this expression, we have used the abbreviation

$$O_\lambda(\boldsymbol{k}_\lambda) = \sum_{j=1}^{N} \left[-\frac{e}{m_0}\sqrt{\frac{\hbar}{2\varepsilon_0\omega_\lambda}} \exp(\mathrm{i}\boldsymbol{k}_\lambda \cdot \boldsymbol{r}_j)\boldsymbol{e}_\lambda \cdot \boldsymbol{p}_j \right] \,. \tag{16.53}$$

Instead of $O_\lambda(-\boldsymbol{k}_\lambda)$, we can also write $O^+_\lambda(\boldsymbol{k}_\lambda)$.

The operator O_λ is a function of the coordinates $\boldsymbol{r}_j$ and the momentum operators $\boldsymbol{p}_j$ of all the electrons of the molecule, but not of the radiation field operators b^+_λ, b_λ. Equation (16.52) evidently consists of a sum over the following individual expressions:

$$b_\lambda \frac{1}{\sqrt{V}} O_\lambda(\boldsymbol{k}_\lambda) \tag{16.54}$$

and

$$b^+_\lambda \frac{1}{\sqrt{V}} O^+_\lambda(\boldsymbol{k}_\lambda) \,. \tag{16.55}$$

The perturbation operator H^S_n for the nuclei is constructed in a completely analogous manner.

As we shall see further below, the expression (16.54) describes the annihilation of a photon, while (16.55) describes its creation. In what follows, we consider those parts of the perturbation operator which act on the electronic coordinates, i.e. (16.54) and (16.55). As we know, (16.28) guarantees that energy is conserved. If we thus irradiate the molecules with light of a correspondingly high frequency, at which electronic transitions can take place, or if we observe light of such a frequency in emission, then these parts of the perturbation operator are acting. Let us insert the wavefunctions (16.50) and (16.51) and the parts (16.54) and (16.55) of the perturbation operator into the matrix element (16.14)! Since both the wavefunctions and the perturbation operator consist of products of functions or of operators which refer either to the radiation field or to the molecular coordinates (i.e. the electronic or nuclear coordinates), the matrix element (16.14) can also be split into factors containing the individual parts (16.54) or (16.55). For the factor with (16.54), we then find

$$\int \phi_\mu^* b_\lambda \frac{1}{\sqrt{V}} O_\lambda(\boldsymbol{k}_\lambda)\phi_\kappa \, dV = M_{\text{molecule}} M_{\text{radiation}} \, . \tag{16.56}$$

In this equation, M_{molecule} is defined by

$$M_{\text{molecule}} = \int \phi_{Q_e}^* \frac{1}{\sqrt{V}} O_\lambda(\boldsymbol{k}_\lambda)\phi_{Q_a} \, dV_{\text{el}} \, dV_{\text{n}} \, . \tag{16.57}$$

The integration which is indicated by $dV_{\text{el}}\, dV_{\text{n}}$ is to be performed over all the coordinates of the electrons and the nuclei of the molecule. We deal with the evaluation of this matrix element in Sect. 16.4.

We first discuss the second factor in (16.56), namely $M_{\text{radiation}}$. Here, we use the bra and ket notation, as we have done previously in the discussion of the harmonic oscillator (cf. I, Sect. 9.4 and Exercises). We then have:

$$M_{\text{radiation}} = \langle \Phi_{\text{e}} b_\lambda \, \Phi_{\text{a}} \rangle \, , \tag{16.58}$$

in the case that we take the perturbation operator (16.54), and

$$M_{\text{radiation}} = \langle \Phi_{\text{e}} b_\lambda^+ \, \Phi_{\text{a}} \rangle \, , \tag{16.59}$$

in the case of the perturbation operator (16.55).

In order to evaluate these matrix elements, we recall that the wavefunctions Φ_{e} and Φ_{a} are eigenfunctions of the unperturbed Hamiltonian, i.e. of (16.42). In the simplest case, the vacuum state is present, with no photon in the radiation field and $\Phi_{\text{a}} = \Phi_0$. In the matrix element (16.58), the annihilation operator b_λ then acts on the vacuum state, which as we know yields zero, so that (16.58) vanishes. In the matrix element (16.59), in contrast, a photon in the state λ (with the wavevector $\boldsymbol{k}_\lambda$ and the polarisation vector $\boldsymbol{e}_\lambda$) is created. As we learned in the quantum-mechanical treatment of the harmonic oscillator, the matrix element (16.59) is non-vanishing only when the final state Φ_{e} contains just this one photon:

$$\Phi_{\text{e}} = b_\lambda^+ \Phi_0 \, . \tag{16.60}$$

These results can be extended to generalized initial and final states; it will suffice for our purposes to consider a particular photon state λ, which is occupied by n_λ photons (where n_λ is an integer). The wavefunctions Φ_{a} or Φ_{e} then have the general form

$$\Phi = \frac{1}{\sqrt{n_\lambda!}} (b_\lambda^+)^{n_\lambda} \Phi_0 \, . \tag{16.61}$$

We can then make use of the following scheme:

Absorption: the number

$$n_\lambda \quad \text{in} \quad \Phi_{\text{a}} = \frac{1}{\sqrt{n_\lambda!}} (b_\lambda^+)^{n_\lambda} \Phi_0 \tag{16.62}$$

is decreased by one:

$$\Phi_e = \frac{1}{\sqrt{(n_\lambda - 1)!}} (b_\lambda^+)^{n_\lambda - 1} \Phi_0 . \tag{16.63}$$

As is known from the quantum theory of the harmonic oscillator, the matrix element responsible for absorption (16.58) is given by

$$M_{\text{radiation}} = \sqrt{n_\lambda} . \tag{16.64}$$

Emission: the number n_λ in (16.60) increases by one:

$$\Phi_e = \frac{1}{\sqrt{(n_\lambda + 1)!}} (b_\lambda^+)^{n_\lambda + 1} \Phi_0 . \tag{16.65}$$

The matrix element responsible for emission, (16.59), is then given by

$$M_{\text{radiation}} = \sqrt{n_\lambda + 1} . \tag{16.66}$$

Following these preliminary considerations, we can turn to the calculation of the matrix element (16.14), which contains the whole perturbation operator (16.49). In this process, we shall however continue to take only the electronic part of the matrix element, i.e. the first sum in (16.49), into account. The matrix element (16.14) is then to be computed with the whole perturbation operator for the electronic coordinates. We thus obtain a sum of the matrix elements having the index λ (16.56), which we have just finished calculating. This will permit us to determine the transition probabilities for the emission and the absorption of radiation directly.

16.3.4 Transition Probabilities and the Einstein Coefficients

As the first case, let us investigate *spontaneous emission*.

In this case, there are initially no radiation quanta present, i.e. $\Phi_a = \Phi_0$, and the molecule is in an excited electronic state. As we have already seen, the perturbation operator (16.55) creates a quantum of radiation of the type λ. We now sum over the index λ as in (16.52). Finally, we have to evaluate the matrix element, by inserting a particular final-state wavefunction $\Phi_e = b_{\lambda_0}^+ \Phi_0$ with a particular index λ_0. Since, however,

$$\langle (b_{\lambda_0}^+ \Phi_0) | b_\lambda^+ \Phi_0 \rangle = \delta_{\lambda\lambda_0} , \tag{16.67}$$

the only term in the sum over λ which remains is the term with $\lambda = \lambda_0$. If we also recall that $b_\lambda \Phi_0 = 0$, then that part of the matrix element which is derived from the first sum in (16.52) is seen to vanish completely.

Let us summarize all these results: the matrix element (16.14) of the perturbation operator (16.52) can now be written in the form

$$H_{\mu\kappa}^{\mathrm{S}}(0) = \frac{1}{\sqrt{V}} \int \phi_{Q_e}^*(\boldsymbol{r}, \boldsymbol{R}) O_{\lambda_0} \phi_{Q_a}(\boldsymbol{r}, \boldsymbol{R})\, dV_{\text{el}}\, dV_{\text{n}} . \tag{16.68}$$

On the left-hand side, the indices μ, κ are abbreviations for the quantum numbers of the initial and the final state of the molecule and the radiation field. We thus have the following correspondence:

	Molecule	Radiation	Energy	
Initial state	$\mu = Q_\mathrm{a}$,	Vacuum, 0	$E_\mu = E_{Q_\mathrm{a}.\mathrm{Mol}}$.	
Final state	$\kappa = Q_\mathrm{e}$,	Photon λ_0	$E_\kappa = E_{Q_\mathrm{e}.\mathrm{Mol}} + \hbar\,\omega_{\lambda_0}$.	(16.69)

$E_{Q_\mathrm{a}.\mathrm{Mol}}$ and $E_{Q_\mathrm{e}.\mathrm{Mol}}$ are the energies of the molecule in the corresponding quantum states. Because of (16.69), we denote the matrix element (16.68) more explicitly as

$$H^{\mathrm{S}}_{Q_\mathrm{e}.\lambda_0:Q_\mathrm{a}.0} = \frac{1}{\sqrt{V}} \int \phi^*_{Q_\mathrm{e}}(\boldsymbol{r},\boldsymbol{R})\,O_{\lambda_0}\phi_{Q_\mathrm{a}}(\boldsymbol{r},\boldsymbol{R})\,dV_\mathrm{el}\,dV_\mathrm{n}\ . \tag{16.70}$$

We now turn to the computation of the transition probability per unit time, W, as in (16.30). Since the quantum numbers Q_a, Q_e of the molecule are fixed, we need only to sum over the final states of the radiation field and to replace the index λ_0 in (16.70) by λ. If we now set $\Delta E = E_{Q_\mathrm{e}.\mathrm{Mol}} - E_{Q_\mathrm{a}.\mathrm{Mol}}$, then we obtain

$$W = \frac{2\pi}{\hbar}\sum_\lambda |H^{\mathrm{S}}_{Q_\mathrm{e}.\lambda_0:Q_\mathrm{a}.0}|^2\,\delta(\hbar\,\omega_\lambda - \Delta E)\ . \tag{16.71}$$

We must now evaluate the sum over λ by making use of the δ-function which it contains (while keeping the polarisation direction $\boldsymbol{e}$ of the emitted radiation quanta fixed). In fact, the wavevectors $\boldsymbol{k}_\lambda$ of the radiation quanta belong to a continuum: the vectors $\boldsymbol{k}_\lambda$ vary continuously in magnitude and direction. As we have seen in Sect. 16.1, energy is to be conserved. If the initial and final states of the molecule have discrete energy values, then the energy of the radiation quanta is likewise fixed at a particular discrete value $\hbar\,\omega$. Since, however, $\omega = ck$, the magnitude of $\boldsymbol{k}$ is also fixed. On the other hand, the *direction* of $\boldsymbol{k}$ remains continuously variable.

As we show explicitly in Appendix A.2, we can proceed with the calculation of the transition probability per second, W, as follows: we begin with light waves of the form $u_\lambda \propto \exp(\mathrm{i}\boldsymbol{k}\cdot\boldsymbol{r})$, which are normalized to 1 within a finite volume V through the prefactor $\frac{1}{\sqrt{V}}$, i.e.

$$\int_V |u_\lambda|^2 dV = 1\ . \tag{16.72}$$

The volume V may be supposed to have the shape of a cube of edge length L. The waves are subjected to periodic boundary conditions, i.e. for example

$$\mathrm{e}^{\mathrm{i}k_{\lambda.x}(x+L)} = \mathrm{e}^{\mathrm{i}k_{\lambda.x}x}\ . \tag{16.73}$$

This condition and the corresponding ones for the y- and z-directions require that

$$k_x = \frac{2\pi n_x}{L}\ ,\quad k_y = \frac{2\pi n_y}{L}\ ,\quad k_z = \frac{2\pi n_z}{L}\ , \tag{16.74}$$

where n_x, n_y and n_z are integers (and not to be confused with the radiation quantum numbers n_λ). We now consider the sum over the final states of the radiation quanta, which are denoted by the discrete numbers n_x, n_y, and n_z and their $\boldsymbol{k}$ vectors within an element of solid angle $d\Omega$ (cf. Fig. 16.2). We then take the limit $V \to \infty$ (i.e. $L \to \infty$), letting the $\boldsymbol{k}$ vectors become continuous and the sum $\sum_{n_x,n_y,n_z \in d\Omega}$ become an integral. From Appendix A.2, we then have the relation

$$\sum_\lambda \ldots \to \frac{V}{(2\pi)^3}\int \ldots k^2\,dk\,d\Omega\ . \tag{16.75}$$

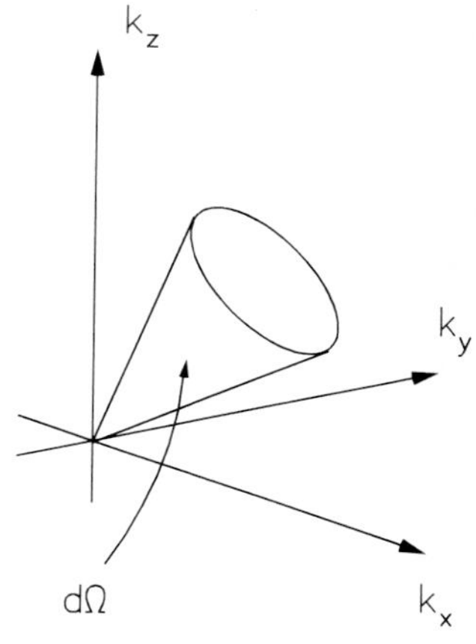

Fig. 16.2. The meaning of the solid angle element $d\Omega$

This limit is a special case of the limit represented by (16.25–30). In fact, for the case of radiation quanta, we can express the wavenumber k in terms of the energy E using the relation

$$E = \hbar\,\omega = \hbar\,ck\ ,$$

that is,

$$k = \frac{E}{\hbar c}\ ,$$

so that the right-hand side of (16.75) becomes

$$\frac{V}{(2\pi)^3}\frac{1}{(\hbar c)^3}\int \ldots E^2 dE d\Omega\ . \tag{16.76}$$

Within the integral in (16.30) or (16.76), however, there is a δ-function, which selects just that radiation quantum energy $\hbar\,\omega$ which is required by the law of conservation of energy:

$$E_{Q_\mathrm{a}} = E_{Q_\mathrm{e}} + \hbar\,\omega\ . \tag{16.77}$$

Using these results, we obtain

$$W = \frac{2\pi}{\hbar}\frac{V}{(2\pi)^3}\frac{(\hbar\omega)^2}{(\hbar c)^3}|H^{\mathrm{S}}_{Q_\mathrm{e}.\lambda_0:Q_\mathrm{a}.0}|^2\,d\Omega\ . \tag{16.78}$$

In order to arrive at the final result, we insert (16.68), (16.70) into (16.78) with (16.53). This yields

$$W = \frac{1}{\hbar}\frac{1}{8\pi^2}\frac{1}{c^3}\frac{e^2}{m_0^2}\frac{\omega}{\varepsilon_0}\left|\int \phi^*_{Q_\mathrm{e}}(\boldsymbol{r},\boldsymbol{R})\sum_j \exp(\mathrm{i}\boldsymbol{k}\cdot\boldsymbol{r}_j)\boldsymbol{e}\cdot\boldsymbol{p}_j\phi_{Q_\mathrm{a}}(\boldsymbol{r},\boldsymbol{R})\,dV_{\mathrm{el}}dV_{\mathrm{n}}\right|^2 d\Omega\ . \tag{16.79}$$

W gives the number of photons spontaneously emitted per unit time into the solid angle element $d\Omega$ and having the polarisation vector $\boldsymbol{e}$. This is, however, precisely the quantity which Einstein introduced phenomenologically for the spontaneous emission rate in his derivation of Planck's radiation formula.

The *Einstein coefficient* $a^1_{2.e}$ *for spontaneous emission* (with a polarisation vector $\boldsymbol{e}$) is found by comparison of

$$W = a^1_{2.e}d\Omega \tag{16.80}$$

with (16.78),(16.79).

In Sect. 16.3.6, we give the Einstein coefficients in the so-called dipole approximation, as well as carrying out the spatial average over all the polarisation directions.

Here, we consider the optical lifetime of an excited state having the quantum numbers Q_a. The transition probability W per unit time for the transition of a molecule with the quantum numbers Q_a to the final state Q_e with emission of a photon of polarisation vector $\boldsymbol{e}$ into the solid angle element $d\Omega$ is given in (16.78), (16.79). We denote it more precisely as $W(\mathrm{a} \to \mathrm{e}, \boldsymbol{e}, d\Omega)$. The optical lifetime τ of the (excited) initial state a can then be found immediately by taking the sum of the transition probabilities into all of the energetically lower-lying states of the molecule, as follows:

$$\frac{1}{\tau} = \sum W(\mathrm{a} \to \mathrm{e}, \boldsymbol{e}, d\Omega) .$$

The sum extends over all the final states of the molecule, as mentioned, over the total solid angle, and over the 2 possible polarisation states of the photons. The optical lifetime is related to the finite linewidth of the excited state (compare I, Sect. 16.2).

We now turn to *stimulated* or *induced emission.*

We assume initially that a particular number n of radiation quanta, which belong to a particular light wave having the index λ_0 (i.e. the wavevector $\boldsymbol{k}_{\lambda_0}$ and the polarisation vector $\boldsymbol{e}$), are present in the radiation field. The normalized initial state wavefunction of the radiation field, Φ_a, is then given by

$$\Phi_\mathrm{a} = \frac{1}{\sqrt{n!}} (b^+_{\lambda_0})^n \Phi_0 . \tag{16.81}$$

Let us allow the perturbation operator (16.52) to act on (16.81) in the matrix element $H^\mathrm{S}_{\mu\kappa}(0)$, (16.14), and take into account only those interactions which increase the number of photons; we then recognise that there are two types of final states: depending upon whether the index λ in the sum in (16.52) is equal to λ_0 or not, we obtain

$$(\alpha) \quad \lambda = \lambda_0 : \; \Phi_\mathrm{e} \propto (b^+_{\lambda_0})^{n+1} \, \Phi_0 \tag{16.82}$$

or

$$(\beta) \quad \lambda \neq \lambda_0 : \; \Phi_\mathrm{e} \propto b^+_\lambda \, (b^+_{\lambda_0})^n \, \Phi_0 . \tag{16.83}$$

In case (α), a light quantum of the same type as that in the initial state is added to the radiation field. In case (β), a light quantum of a different sort is emitted spontaneously. In order to calculate the overall transition probability into all the final states, we must set up the matrix element $H^\mathrm{S}_{\mu\kappa}$ for the cases (16.82) and (16.83) using the wavefunctions ϕ_{Q_a} and ϕ_{Q_e} of the molecule, take its absolute square, and sum over all the final states as in (16.30). In the process, we must keep track of the meaning of the energy difference $E_\mu - E_\kappa$ which occurs in the δ-function.

In case (α), E_κ, the energy of the initial state, is given by the sum of the energy of the molecule, $E_\mathrm{a.Mol}$, and the energy of all n light quanta in the initial state, i.e. $n\hbar\,\omega_{\lambda_0}$:

$$E_\kappa = E_\mathrm{a.Mol} + n\,\hbar\,\omega_{\lambda_0} . \tag{16.84}$$

The energy E_μ of the final state is given by the corresponding expression:

$$E_\mu = E_\mathrm{e.Mol} + (n+1)\,\hbar\,\omega_{\lambda_0} . \tag{16.85}$$

We thus obtain

$$E_\mu - E_\kappa = E_\mathrm{e.Mol} - E_\mathrm{a.Mol} + \hbar\,\omega_{\lambda_0} . \tag{16.86}$$

The δ-function ensures the conservation of energy:

$$\hbar\,\omega_{\lambda_0} = E_\mathrm{a.Mol} - E_\mathrm{e.Mol} \equiv \Delta E , \tag{16.87}$$

where the right-hand side is positive, since the molecule makes a transition from an excited state into an energetically lower-lying state. (Conversely, one could say that this is a precondition for the emission of a quantum of radiation.)

In case (β), we find

$$E_\kappa = E_{\mathrm{a.Mol}} + n\hbar\,\omega_{\lambda_0} \ . \tag{16.88}$$

$$E_\mu = E_{\mathrm{e.Mol}} + n\hbar\,\omega_{\lambda_0} + \hbar\,\omega_\lambda \tag{16.89}$$

and therefore

$$E_\mu - E_\kappa = \hbar\,\omega_\lambda - \Delta E \ . \tag{16.90}$$

Corresponding to the cases (α) and (β), the transition probability per unit time W, (16.30), consists of two parts:

$$W = \frac{2\pi}{\hbar}\delta(\hbar\,\omega_{\lambda_0} - \Delta E)(n+1)|H^{\mathrm{S}}_{Q_{\mathrm{e}}.\lambda_0:Q_{\mathrm{a}}.0}|^2 + \frac{2\pi}{\hbar}\sum_{\lambda\neq\lambda_0}|H^{\mathrm{S}}_{Q_{\mathrm{e}}.\lambda_0:Q_{\mathrm{a}}.0}|^2\delta(\hbar\,\omega_\lambda - \Delta E) \ . \tag{16.91}$$

The factor $(n+1)$ in the first sum is due to (16.66), since the matrix element occurs as its square. We can decompose this factor into n and 1 and add the expression which multiplies the 1 to the sum in the second line of (16.91); this, however, yields just the sum (16.71), which we have already found for the case of spontaneous emission. The remaining expression in (16.91),

$$W_{\mathrm{in}} = \frac{2\pi}{\hbar}|H^{\mathrm{S}}_{Q_{\mathrm{e}}.\lambda_0:Q_{\mathrm{a}}.0}|^2 n\delta(\hbar\,\omega_{\lambda_0} - \Delta E) \tag{16.92}$$

is the *induced* or *stimulated emission rate*.

In (16.92), no sum over the radiation quantum index λ occurs. On the other hand, it is necessary to integrate over a continuum in order to evaluate the function. The formalism thus forces us to start with a more realistic initial state, consisting of a wavepacket. We assume it to be made up of plane waves within a range Δk_x, Δk_y and Δk_z of wavenumbers and with a corresponding frequency width, $\Delta\omega = c\Delta k$. If M modes are present, the normalized wavefunction is given by

$$\Phi_{\mathrm{a}} = \frac{1}{\sqrt{M!}}\frac{1}{\sqrt{n!}}\sum_{\Delta k}(b_{\boldsymbol{k}}^+)^n\Phi_0 \ . \tag{16.93}$$

In order to be quite explicit, we have used the index $\boldsymbol{k}$ in (16.93) instead of λ. The sum over $\boldsymbol{k}$ in (16.93) becomes a sum over λ_0 in (16.92):

$$W_{\mathrm{in}} = \frac{2\pi}{\hbar}\frac{1}{M}\sum_{\lambda_0}|H^{\mathrm{S}}_{Q_{\mathrm{e}}.\lambda_0:Q_{\mathrm{a}}.0}|^2 n\delta(\hbar\,\omega_{\lambda_0} - \Delta E) \ . \tag{16.94}$$

Setting

$$M = m_x m_y m_z \ , \quad \Delta k_i = \frac{2\pi m_i}{L} \ , \quad i = x, y, z \ , \tag{16.95}$$

where L is the edge length of the normalisation cube, we find

$$M = \frac{L^3}{(2\pi)^3}\Delta k_x \Delta k_y \Delta k_z = \frac{V}{(2\pi)^3}k^2\Delta k d\Omega \ , \tag{16.96}$$

where, in a second step, we have introduced spherical polar coordinates with an element of solid angle $d\Omega$. Using (16.95), (16.96), and $\Delta E = \hbar\,\omega$, we obtain the following expression for the transition probability of stimulated emission of photons into a solid angle $d\Omega$:

$$W_{\text{in}} = n\frac{2\pi}{\hbar^2 \Delta\omega}|H^{\text{S}}_{Q_\text{e}.\lambda_0:Q_\text{a}.0}|^2 \,. \tag{16.97}$$

With (16.53) and (16.68), (16.97) can also be written in the form

$$W_{\text{in}} = \frac{\pi e^2}{m_0^2\varepsilon_0\Delta\omega\hbar\,\omega V}\left|\left(\sum_{j=1}^{N}[\exp(\mathrm{i}\boldsymbol{k}\cdot\boldsymbol{r}_j)\boldsymbol{e}\boldsymbol{p}_j]_{Q_\text{e}.Q_\text{a}}\right)\right|^2 \,, \tag{16.97a}$$

where the matrix element $(\ldots)_{Q_\text{e}.Q_\text{a}}$ is to be evaluated with the wavefunctions $\phi^*_{Q_\text{e}}$ and ϕ_{Q_a} (cf. 16.97).

We can write W_{in} in the form

$$W_{\text{in}}(d\Omega) = \varrho_e(\omega, d\Omega)b^1_{2.\text{e}}d\Omega \,, \tag{16.98}$$

in which

$$b^1_{2.e} = \frac{\pi e^2}{m_0^2\varepsilon_0(\hbar\,\omega)^2}\left|\left(\sum_{j=1}^{N}[\exp(\mathrm{i}\boldsymbol{k}\cdot\boldsymbol{r}_j)\boldsymbol{e}\boldsymbol{p}_j]_{Q_\text{e}.Q_\text{a}}\right)\right|^2 \tag{16.99}$$

is the *Einstein coefficient for the stimulated emission of photons with the polarisaton vector* ***e*** *into* $d\Omega$. The quantity

$$\varrho_\text{e}(\omega, d\Omega) = \frac{n\hbar\,\omega}{\Delta\omega V d\Omega} \tag{16.100}$$

is the total energy of the n photons divided by the frequency bandwidth, the solid angle, and the volume, or in other words, ϱ_e is the energy density per unit frequency interval, unit solid angle, and unit volume. A comparison of equations (16.78–80) with (16.99) yields the important *Einstein relation* for the ratio of spontaneous and stimulated emission probabilities as a function of the frequency [cf. also (5.22) in I; note that we used capital letters to denote the Einstein coefficients in I, but we use small letters in the present book]:

$$\frac{a^1_{2.e}}{b^1_{2.e}} = \frac{\hbar\,\omega^3}{(2\pi)^3c^3} \,. \tag{16.101}$$

An additional *comparison between spontaneous and stimulated emission* can be made as follows: the relation (16.92) can also be expressed in a different way. We determine the rate of spotaneous emission P per number of modes (not photons!) (cf. Appendix A.2) in the volume V, the solid angle $d\Omega$, and the frequency range $\Delta\omega(\omega\ldots\omega+\Delta\omega)$ which we have just been considering. Dividing (16.78) by this number,

$$N_m = \frac{k^2\Delta k V d\Omega}{8\pi^3} = \frac{\omega^2\Delta\omega V d\Omega}{8\pi^3c^3} \,, \tag{16.102}$$

we find

$$\tilde{W} = \frac{W}{N_m} = \frac{1}{n}W_{\text{in}} \,, \tag{16.103}$$

so that the ratio of the stimulated emission to spontaneous emission rates is equal to n, i.e. the total number of photons in this frequency range.

Absorption. The calculation of the transition probability W_{abs} for molecules which make transitions from their ground states to an excited state by absorption of a quantum of radiation from a wavepacket whose wavevectors lie in a solid angle element $d\Omega$ can be carried out analogously to that for stimulated emission. The essential difference lies in the fact that we are now dealing with a molecule initially in its ground state instead of in an excited state, and that a photon is removed from the wave with quantum index λ_0. The explicit form of E_μ and E_κ is thus changed:

$$E_\mu - E_\kappa = E_{\text{e.Mol}} - E_{\text{a.Mol}} - \hbar\,\omega_{\lambda_0}\ . \tag{16.104}$$

In addition, the factor $(n+1)$ in the first term of (16.91) is to be replaced by n, in accordance with (16.64) [instead of (16.66)], while the second term (sum) in (16.91) vanishes entirely. We again must average over the initial states of the radiation field. We then obtain for the transition probability of absorption per unit time

$$W_{\text{abs}} = \varrho_{\text{e}}(\overline{\omega}, d\Omega) b_{1.e}^2 d\Omega\ , \tag{16.105}$$

where $b_{1.e}^2 = b_{2.e}^1$, so that the Einstein coefficient for absorption is equal to the coefficient for stimulated emission (cf. p. 18 in I). The absorption rate is proportional to the energy density ϱ_{e} of the incident light as defined in (16.100).

16.3.5 The Calculation of the Absorption Coefficient

In order to compute the absorption coefficient α, we introduce the energy flux density $I(\omega) = \varrho_{\text{e}}(\omega, d\Omega)\, c d\Omega$ (energy flux per second per unit area) into (16.103), so that

$$W_{\text{abs}} = I(\omega)\frac{b_{1.e}^2}{c} \tag{16.106}$$

holds. The decrease of the number of photons n per second is equal to W_{abs} for a single molecule. If there are N molecules, then we find

$$\frac{dn}{dt} = -W_{\text{abs}} N\ . \tag{16.107}$$

By introducing the definition

$$I(\omega) = \frac{\overline{n}\hbar\,\omega c}{\Delta\omega} \tag{16.108}$$

(with $\overline{n} = n/V$: photon number density) and inserting it into (16.107), we obtain

$$\frac{dI(\omega)}{cdt} = -I(\omega) b_{1.e}^2 \frac{N\hbar\,\omega}{V\Delta\omega}\ . \tag{16.109}$$

Writing $dx = cdt$, we find the equation for the absorption as a function of sample thickness:

$$\frac{dI(\omega)}{dx} = -I(\omega)\alpha C\ , \tag{16.110}$$

with $C = (N/V)$ being the concentration of the molecules; then the absorption coefficient α is given by

$$\alpha = \frac{b_{1,e}^2 \hbar \omega}{c \Delta\omega} . \tag{16.111}$$

We have thus succeeded in expressing the absorption coefficient introduced in Sect. 15.1 in terms of the Einstein coefficient $b_{1,e}^2$. At the same time we can see that the Lambert-Beer law is simply the solution of the differential equation (16.110).

Since the transition probabilities W_{abs} and W_{in} play a completely symmetric role, we find quite generally for a system of noninteracting, partially inverted molecules

$$\frac{dI}{cdt} = I \frac{N_2 - N_1}{N} \alpha C , \tag{16.112}$$

where N_2 is the number of molecules in their excited states (inverted), and N_1 is the number in the ground state. Expression (16.112) is, incidentally, of fundamental importance for the laser action of molecules. When $N_2 - N_1 > 0$, then an *amplification* of the intensity I takes place and light is produced instead of being absorbed.

16.3.6 Transition Moments, Oscillator Strengths, and Spatial Averaging

a) Transition Moments and the Dipole Approximation

We will now explain the concept of the transition moment, which occurs in the expression for the optical transition probability per unit time. As we have seen, the transition probability per second is determined by the matrix element (16.70), wherein the operator O_λ is given by (16.53). Since all the essentials for the following discussion can be derived for a single electron, we consider only one term in the sum over j and thus leave off the index j. We assume that the light wavelength is much greater than the spatial extent of the molecule, so that the factor $\exp(\mathrm{i}\boldsymbol{k}_\lambda \cdot \boldsymbol{r})$ changes only slightly over this region. If we then choose the origin of our coordinate system to be at the centre of gravity of the molecule, we can replace the exponential function by 1. This is the so-called dipole approximation, which we now wish to examine in more detail. (If we instead expand the exponential function as a power series in $(\boldsymbol{k}_\lambda \cdot \boldsymbol{r})$, then we obtain the multipole series of matrix elements, in which both electric and magnetic multipoles can occur.)

We thus consider matrix elements having the form

$$\boldsymbol{p}_{\mu\kappa} = \int \psi_\mu^* \boldsymbol{p} \psi_\kappa dV , \tag{16.113}$$

where the wavefunctions and the integration over dV may refer to more than one particle. Equation (16.113) is thus an integral which is to be computed over the initial and the final states, in which the momentum $\boldsymbol{p}$ occurs as the operator. Matrix elements of the momentum operator may be reformulated in terms of matrix elements of the position operator $\boldsymbol{r}$ or of the dipole-moment operator

$$\boldsymbol{\Theta} = e\boldsymbol{r} , \tag{16.114}$$

as we shall now demonstrate. Here, e is the elementary charge and $\boldsymbol{r}$ is the coordinate vector of the particle. In order to make clear the relation between $\boldsymbol{r}$ and $\boldsymbol{p}$, we recall the classical relation between the momentum and the velocity of a particle, given by

$$\boldsymbol{p} = m_0 \dot{\boldsymbol{r}} \tag{16.115}$$

(m_0 is the mass of the particle). This well-known relation from classical mechanics can be interpreted quantum-mechanically as a relation between matrix elements in the following manner:

$$\boldsymbol{p}_{\mu\kappa} \equiv \int \psi_\mu^* \frac{\hbar}{\mathrm{i}} \nabla \psi_\kappa dV = m_0 \frac{d}{dt} \int \psi_\mu^* \boldsymbol{r} \psi_\kappa dV . \tag{16.116}$$

(As we have already mentioned, the integrals may extend over the coordinates of several particles.) We now assume that the time dependence of the wavefunctions enters into the integrals explicitly, i.e. that

$$\psi_\lambda(\boldsymbol{r}, t) = \mathrm{e}^{-\mathrm{i}E_\lambda t/\hbar} \phi_\lambda(\boldsymbol{r}) , \quad \lambda = \mu, \kappa . \tag{16.117}$$

We can then carry out the differentiation with respect to time directly and thereby obtain

$$\boldsymbol{p}_{\mu\kappa} = m_0 (\dot{\boldsymbol{r}})_{\mu\kappa} = m_0 \mathrm{i} \omega_{\mu\kappa} \boldsymbol{r}_{\mu\kappa} , \tag{16.118}$$

in which

$$\omega_{\mu\kappa} = \frac{1}{\hbar}(E_\mu - E_\kappa) .$$

The matrix element $\boldsymbol{r}_{\mu\kappa}$ is of course defined by

$$\boldsymbol{r}_{\mu\kappa} = \int \phi_\mu^* \boldsymbol{r} \phi_\kappa dV . \tag{16.119}$$

If we multiply $\boldsymbol{r}_{\mu\kappa}$ by the elementary charge e, we obtain the dipole moment $\boldsymbol{\Theta}$ (16.114). Because of (16.116), we can replace matrix elements containing the operator $\boldsymbol{p}$ everywhere by matrix elements containing $\boldsymbol{\Theta}$, using the relation

$$e\boldsymbol{p}_{\mu\kappa} = m_0 \mathrm{i} \omega_{\mu\kappa} \boldsymbol{\Theta}_{\mu\kappa} . \tag{16.120}$$

The matrix element $\boldsymbol{\Theta}_{\mu\kappa} = \int \phi_\mu^* e \boldsymbol{r} \phi_\kappa dV$ is referred to as the *transition moment* or as the *transition dipole moment.*

For computing the Einstein coefficients within the dipole approximation, we can therefore use the transition moment in place of the matrix elements of the momentum operator, remembering that in (16.120), $\omega_{\mu\kappa} = \omega$ as a result of energy conservation.

b) Oscillator Strengths

Along with the Einstein coefficients, the quantity "oscillator strength" is also used to describe optical transitions. This concept had already been defined before the introduction of quantum mechanics, for the description of the dispersion of light using oscillator models. The theory of dispersion can also be developed on a quantum-mechanical basis, but we shall not do that here; the result is that the quantum-mechanical expression for the atomic polarisability as a function of the radiation frequency has a form which is quite analogous to the classical formula, but the oscillator strengths $f_{Q_\mathrm{a} \to Q_\mathrm{e}}$ are now given as matrix elements for the transition $Q_\mathrm{a} \to Q_\mathrm{e}$, in the following manner:

$$f_{Q_\mathrm{a} \to Q_\mathrm{e}} = \frac{8\pi^2 m_0}{3he^2} \nu \, |\boldsymbol{\Theta}_{Q_\mathrm{e}, Q_\mathrm{a}}|^2 . \tag{16.121}$$

From (16.121), we can derive (15.4) as follows: we begin with the integral absorption coefficient A (15.3), which is obtained from $\varepsilon(\nu) = C\alpha(\omega)/(\ln 10)$ [cf. (15.3)] by integration over the entire absorption band:

$$A = \int \varepsilon(\omega) d\nu \, . \tag{16.121a}$$

If we now express α in (15.3) in terms of the Einstein coefficient $b^2_{1.e} = b^1_{2.e}$ as in (16.111) and carry out the integration over the interval $\Delta\nu \equiv \Delta\omega/2\pi$, then we find

$$A = C\hbar\, \omega b^2_{1.e}/(2\pi c \ln 10) \, . \tag{16.121b}$$

Using (16.99), we write the Einstein coefficient in terms of the absolute square of the dipole matrix element, i.e. $|\boldsymbol{\Theta}_{Q_e.Q_a}|^2$. Applying the dipole approximation, using (16.120), and carrying out the averaging over the spatial orientations of the molecules [see (16.126) below], which gives a factor of $\frac{1}{3}$ before $|\boldsymbol{\Theta}_{Q_e.Q_a}|^2$, we then obtain

$$A = C \frac{\omega}{6c\hbar\, \varepsilon_0 \ln 10} |\boldsymbol{\Theta}_{Q_e.Q_a}|^2 \, . \tag{16.121c}$$

If we solve this equation for $|\boldsymbol{\Theta}_{Q_e.Q_a}|^2$ and insert the result in (16.121), then on cancelling common factors we find

$$f = \frac{4m_0\, c\, \varepsilon_0 \ln 10}{Ce^2} A \, , \tag{16.121d}$$

i.e. (15.4), as we expected. With the same approximations, we can express a direct relation between the oscillator strength f and the Einstein coefficient b^2_1 (after spatial averaging) as follows:

$$f_{Q_a \to Q_e} = \frac{2\varepsilon_0 \omega m_0 \hbar}{\pi e^2} b^2_1 \, . \tag{16.122}$$

c) Spatial Averaging over the Polarisation Directions in Spontaneous Emission

The Einstein coefficient given in (16.79) and (16.80) refers to the emission of photons with a given polarisation vector $\boldsymbol{e}$. Frequently, one wishes to know the Einstein coefficient for the case of an average over all the possible polarisation directions of the emitted light. To obtain it, we apply the dipole approximation and consider the expression

$$\overline{|\boldsymbol{e} \cdot \boldsymbol{r}_{\mu\kappa}|^2} \, , \tag{16.123}$$

where the bar refers to a spatial average. To carry out this averaging, we imagine a coordinate system in which the z-axis is parallel to the direction of the vector $\boldsymbol{r}_{\mu\kappa}$. (The fact that the matrix element $\boldsymbol{r}_{\mu\kappa}$ represents a complex vector is of no consequence; its components are then simply complex numbers.) The initially randomly-oriented polarisation vector $\boldsymbol{e}$ thus makes an angle ϑ with $\boldsymbol{r}_{\mu\kappa}$, i.e. with the z-axis (Fig. 16.3). We can now carry out the spatial averaging using elementary geometrical considerations, by averaging over the solid-angle element $d\Omega$, which can be expressed as

$$d\Omega = \sin\vartheta\, d\vartheta\, d\phi \, . \tag{16.124}$$

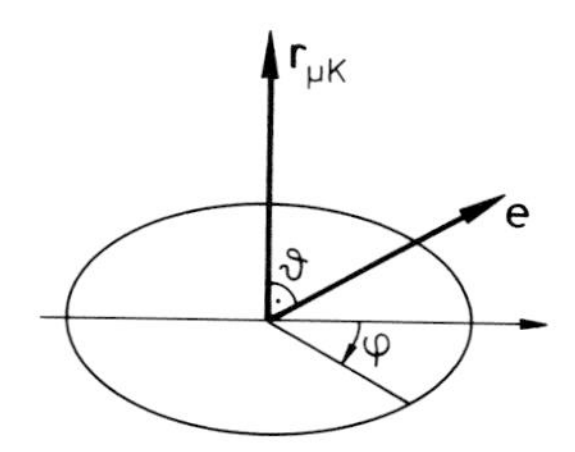

Fig. 16.3. The relative positions of $\boldsymbol{e}$ and $\boldsymbol{r}_{\mu\kappa}$

The average (16.123) is then given explicitly by

$$\overline{|\boldsymbol{e}\cdot\boldsymbol{r}_{\mu\kappa}|^2} = \frac{\int_0^{2\pi} d\phi \int_0^{\pi} |\boldsymbol{r}_{\mu\kappa}|^2 \cos^2\vartheta \sin\vartheta \, d\vartheta}{\int_0^{2\pi} d\phi \int_0^{\pi} \sin\vartheta \, d\vartheta} . \tag{16.125}$$

The absolute square of $\boldsymbol{r}_{\mu\kappa}$ does not depend on the angles and can be removed from the integrals. The integrals can then be computed in an elementary manner, and we obtain the final result

$$\overline{|\boldsymbol{e}\cdot\boldsymbol{r}_{\mu\kappa}|^2} = \tfrac{1}{3}|\boldsymbol{r}_{\mu\kappa}|^2 . \tag{16.126}$$

In order to find the Einstein coefficient averaged over all polarisation directions, we thus simply need to multiply (16.79) and (16.80) by the factor $\frac{1}{3}$.

16.4 The Franck-Condon Principle

Our treatment thus far applies to quite general motions of the electrons and the nuclei in a molecule. In order to proceed with the calculation of the corresponding matrix elements (16.68), we require some approximations, which however will make the physical content of the problem more clear. The goal of this section is to deepen our understanding within a quantum-mechanical framework of the Franck-Condon principle, which we introduced in Sect. 14.1.

Our first task is to separate the electronic motions from the nuclear motions. To this end, we apply the Born-Oppenheimer approximation, which was discussed in Sect. 11.1. We thus write the wavefunctions which refer to the electrons and the nuclei in the form of products:

$$\phi_{Q_\mathrm{a}}(\boldsymbol{r},\boldsymbol{R}) = \Psi_{q_\mathrm{a}}(\boldsymbol{r},\boldsymbol{R})\chi_{q_\mathrm{a}.K_\mathrm{a}}(\boldsymbol{R}) , \tag{16.127}$$

in which we make the fundamental assumption that the electrons can follow the nuclear motions immediately. The quantum numbers of the overall initial state, electrons and nuclei together, are denoted by Q_a, those of the electrons alone by q_a, and those of the nuclei by K_a. The electronic and nuclear coordinates are abbreviated as follows:

$$\begin{aligned} \boldsymbol{r} &= \boldsymbol{r}_1, \boldsymbol{r}_2, \ldots, \boldsymbol{r}_N , \\ \boldsymbol{R} &= \boldsymbol{R}_1, \boldsymbol{R}_2, \ldots, \boldsymbol{R}_M . \end{aligned} \tag{16.128}$$

In a similar manner, we denote the quantum numbers in the final state by Q_e, etc. Because of the separation (16.127), the matrix element (16.70) assumes the form (up to constant factors):

$$\propto \int \Psi_{q_\mathrm{e}}^*(\boldsymbol{r},\boldsymbol{R})\Big(\sum_j \mathrm{e}^{\mathrm{i}\boldsymbol{k}_\lambda\cdot\boldsymbol{r}_j}\boldsymbol{e}_\lambda\cdot\boldsymbol{p}_j\Big)\Psi_{q_\mathrm{a}}(\boldsymbol{r},\boldsymbol{R}) \int \chi_{q_\mathrm{e}.K_\mathrm{e}}^*(\boldsymbol{R})\chi_{q_\mathrm{a}.K_\mathrm{a}}(\boldsymbol{R})\, dV_\mathrm{nucl}\, dV_\mathrm{el} , \tag{16.129}$$

which we will use as a starting point. Within the integrals, there are two different types of wavefunctions, namely the electronic wavefunctions $\Psi(\boldsymbol{r},\boldsymbol{R})$, which contain the nuclear coordinates as parameters, and the wavefunctions $\chi(\boldsymbol{R})$ of the nuclear motions. We now want to make it clear that the spatial variation of $\Psi(\boldsymbol{r},\boldsymbol{R})$ as a function of $\boldsymbol{R}$ is much more gradual than that of $\chi(\boldsymbol{R})$. We recall the specific form of Ψ, which we saw in Chap. 4 in the treatment of H_2^+ and of H_2: there, the electronic wavefunctions were found to depend upon the difference vectors $\boldsymbol{r} - \boldsymbol{R}_j$, where $\boldsymbol{R}_j$ is the position coordinate of nucleus j. The atomic orbitals which occur in the LCAO method (cf. Chap. 4) vary over the spatial extent r_0 of the corresponding wavefunctions. This extent is, however, much greater than the vibrational amplitude of the nuclear wavefunctions $\chi(\boldsymbol{R})$.

The integral over dV_{nucl} which occurs in (16.129) thus in fact contains the product of a function which varies gradually with position, i.e. the expression under the first integral in (16.129), with a strongly varying function, i.e. that within the second integral. This allows us to take the first integral at a fixed position $\boldsymbol{R} = \boldsymbol{R}_0$, at which the second integrand $\chi^*\chi$ has its maximum value. Then we can rewrite (16.129) as

$$\int \Psi^*_{q_e}(\boldsymbol{r}, \boldsymbol{R}_0)\Big(\sum_j \mathrm{e}^{\mathrm{i}\boldsymbol{k}_\lambda \cdot \boldsymbol{r}_j} \boldsymbol{e}_\lambda \cdot \boldsymbol{p}_j\Big)\Psi_{q_a}(\boldsymbol{r}, \boldsymbol{R}_0)\, dV_{\text{el}}$$

$$\cdot \int \chi^*_{q_e.K_a}(\boldsymbol{R})\chi_{q_a.K_a}(\boldsymbol{R})\, dV_{\text{nucl}} \,. \tag{16.129a}$$

We can therefore replace the integration over electronic and nuclear coordinates by a product of integrals over the electronic coordinates alone and over the nuclear coordinates alone. The transition from (16.129) to (16.129a) may be justified in a strict mathematical sense by applying the mean value theorem of integral calculus.

We now begin by investigating the integral which describes the nuclear motions:

$$\int \chi^*_{q_e.K_e}(\boldsymbol{R})\chi_{q_a.K_a}(\boldsymbol{R})\, dV_{\text{nucl}} \,. \tag{16.130}$$

Here, we must recall a result of the Born-Oppenheimer approximation concerning the nuclei. We saw that the force between the nuclei consists not only of their mutual Coulomb repulsion, but also has a contribution due to the electrons. The electronic energy W which determines this force (11.9) depends on which of the electronic wavefunctions are occupied. This implies in particular that the nuclear coordinates before and after an optical transition must not necessarily have the same equilibrium positions, but instead may be shifted, depending on which electronic states are present. This shift in the equilibrium positions of the nuclei can be represented in an intuitively clear manner by using a so-called configuration coordinate [cf. Fig. 16.4, as well as (14.3) and (14.4)]. Even if the oscillator wavefunctions of the nuclei before and after the transition are, for example, both in their ground states, the integral in (16.130) is still in general smaller than 1, since the equilibrium positions are shifted and the overlap of the wavefunctions is thus incomplete. Owing to this shift of the nuclear equilibrium positions, the wavefunctions in (16.130) are no longer mutually orthogonal, even when they have different quantum numbers K_e and K_a. Different quantum numbers, however, mean that the number of vibrational quanta in the initial and final states are not the same; this means that in the course of an optical transition, vibrational quanta can be created or annihilated. There is thus no strict selection rule for the change in the vibrational quantum number during an electronic transition, as we noted already in Chap. 14.

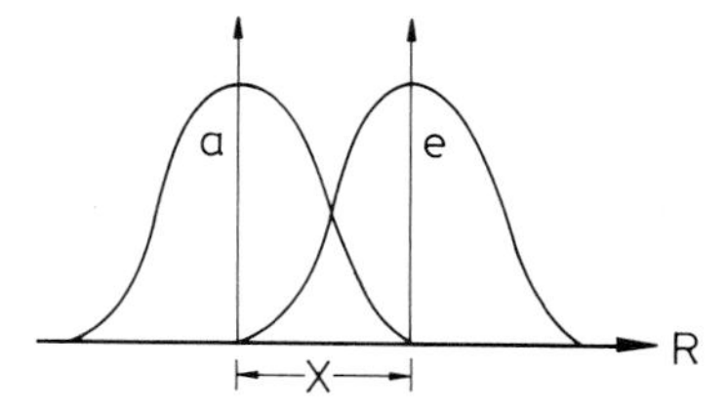

Fig. 16.4. The displacement of the equilibrium positions of the nuclei (and of their wavefunctions) before and after an optical transition of the electrons, illustrated by means of the configuration coordinate X

Let us now have a look at the part of the matrix element (16.129a) which refers to the electronic coordinates, i.e.

$$\int \Psi^*_{q_e}(\boldsymbol{r}, \boldsymbol{R}_0)\Big(\sum_j \mathrm{e}^{\mathrm{i}\boldsymbol{k}_\lambda \cdot \boldsymbol{r}_j} \boldsymbol{e}_\lambda \cdot \boldsymbol{p}_j\Big)\Psi_{q_a}(\boldsymbol{r}, \boldsymbol{R}_0)\, dV_{\text{el}} \,. \tag{16.131}$$

We assume that the wavelength of the light is much greater than the extent of the molecular orbitals, so that we can ignore the spatial variation of the exponential function $\exp(\mathrm{i}\boldsymbol{k} \cdot \boldsymbol{r})$

and replace it by $\exp(i\boldsymbol{k}\cdot\boldsymbol{r}_0)$, with $\boldsymbol{r}_0$ referring to an average coordinate within the molecule. (This, as we have already seen, is the dipole approximation.) We can then extract this constant factor from the integral and finally leave it off entirely, since it cancels on taking the absolute square of the matrix element. The remaining part of the matrix element (16.131) is then given by

$$\int \Psi_{q_e}^*(\boldsymbol{r},\boldsymbol{R}_0)\Big(\sum_j \boldsymbol{e}_\lambda\cdot\boldsymbol{p}_j\Big)\Psi_{q_a}(\boldsymbol{r},\boldsymbol{R}_0)\,dV_{\mathrm{el}}\ . \tag{16.132}$$

This matrix element clearly does not depend on the vibrational quantum numbers, but only on the average nuclear coordinate $\boldsymbol{R}_0$ (compare Sect. 14.1).

The matrix element (16.132) is still difficult to evaluate, because the wavefunctions Ψ^* and Ψ refer to all of the electrons in the molecule. In order to proceed, we make use of the Hartree-Fock approximation, according to which the initial-state wavefunction (as well as that of the final state) can be written in the form

$$\Psi_{q_a}^* = \frac{1}{\sqrt{N!}}\begin{vmatrix} \psi_{q_1}(\boldsymbol{r}_1) & \psi_{q_2}(\boldsymbol{r}_1)\ldots & \psi_{q_N}(\boldsymbol{r}_1) \\ \psi_{q_1}(\boldsymbol{r}_2) & \psi_{q_2}(\boldsymbol{r}_2)\ldots & \\ \vdots & & \\ \psi_{q_1}(\boldsymbol{r}_N) & \psi_{q_2}(\boldsymbol{r}_N)\ldots & \psi_{q_N}(\boldsymbol{r}_N) \end{vmatrix}\ . \tag{16.133}$$

For simplicity, we have written this Slater determinant in terms of the spatial wavefunctions, without the spin parts. As in Sect. 7.1.2, we could however readily take the spin dependence into account. In (16.133), we have split up the overall quantum numbers of the electrons, q_a, into the individual quantum numbers $q_1, q_2, \ldots$ of the states occupied by *single* electrons. The quantum numbers of the final state are distinguished from those of the initial state by a prime:

$$\begin{aligned} q_a &= (q_1, q_2, \ldots q_N)\ , \\ q_e &= (q_1', q_2', \ldots q_N')\ . \end{aligned} \tag{16.134}$$

As we know (cf. also the Appendix), a determinant can be represented as the sum of the products of all the possible combinations of the indices, with a + or − sign depending on whether the permutation is even or odd. According to (16.133) and the corresponding expression for Ψ_{q_e}, two determinants occur in (16.132) and must be multiplied, and then the integration over the electronic coordinates is to be carried out. This evaluation of the matrix element gives a very limited physical insight, and so we shall simply present its result and refer the reader who is interested in the details to the Appendix. We find the following: the set of quantum numbers q_a (16.134) must agree with the set q_e except for one pair, where $q_l \neq q_l'$, with l an integer from the series $1\ldots N$ (N is the total number of electrons in the molecule). In other words, in an optical transition only one electron changes its state and all the others remain in their respective initial states. Due to the indistinguishability of electrons, the identity of this one electron is arbitrary. The matrix element (16.132) is then reduced to the expression

$$(16.132) = \boldsymbol{e}_\lambda\cdot\boldsymbol{p}_{q_l',q_l} = \boldsymbol{e}\cdot\int \psi_{q_l'}^*(\boldsymbol{r})\boldsymbol{p}\psi_{q_l}(\boldsymbol{r})dV\ . \tag{16.135}$$

We can now apply group-theoretical considerations to (16.135), which will show whether the matrix element is in principle nonvanishing or if it is zero due to symmetry properties.

If we drop the Hartree-Fock approximation, then it is possible that several electrons will be excited during an optical transition, and the excitation energy can then be spread in a complicated way over the excitation states of the various electrons.

16.5 Selection Rules

Whether or not an optical transition can occur depends on the optical matrix elements. Applying group theory, it is possible to determine which matrix elements vanish, i.e. which transitions are forbidden, or, conversely, which matrix elements can be non-zero (but do not have to be!). Group theory can, however, make no predictions about the magnitudes of the non-zero matrix elements.

In order to become familiar with this method, we first consider the so-called *direct product*. In Chap. 6, we introduced the wavefunctions Ψ_j as a basis for the representation of a transformation group. We now consider, along with one such basis, which we denote as $\Psi_j^{(1)}$, a second basis denoted as $\Psi_j^{(2)}$. We can also permit the set of the $\Psi_j^{(2)}$ to be identical with the $\Psi_j^{(1)}$. We then form a new set of basis functions from $\Psi_j^{(1)}$, $\Psi_k^{(2)}$ by taking the direct product, $\Psi_j^{(1)}\Psi_k^{(2)}$. If we then apply a symmetry operation to such a product, each of the factors is transformed into a linear combination of the basis functions belonging to its set, and the product itself becomes a linear combination of the products $\Psi_j^{(1)}\Psi_k^{(2)}$. It may be shown in detail that the product $\Psi_j^{(1)}\Psi_k^{(2)}$ also provides a set of basis functions for the representation of the group. In particular, applying the matrix rules, we can show that *the characters of the representation of a direct product are equal to the products of the characters of the representations of the original functions*. The direct product of two irreducible representations is, as stated, a new representation, which is itself either irreducible or reducible to irreducible representations. Tables 16.1 and 16.2 show some examples.

Table 16.1. Examples of irreducible representations of direct products of the group C_{2v}

C_{2v}	E	C_2	σ_v	σ_v'	
A_1	1	1	1	1	
A_2	1	1	−1	−1	
B_1	1	−1	1	−1	
B_2	1	−1	−1	1	
A_1A_1	1	1	1	1	$= A_1$
A_1A_2	1	1	−1	−1	$= A_2$
A_2B_1	1	−1	−1	1	$= B_2$
A_2B_2	1	−1	1	−1	$= B_1$
B_1B_2	1	1	−1	−1	$= A_2$

Table 16.2. Examples of irreducible representations of direct products of the group C_{3v}

C_{3v}	E	$2C_3$	$3\sigma_v$	
A_1	1	1	1	
A_2	1	1	−1	
E	2	−1	0	
A_2A_2	1	1	1	$= A_1$
A_2E	2	−1	0	$= E$
EE	4	1	0	$= A_1 + A_2 + E$

In the same manner, products of three or more functions, i.e.

$$\Psi_i^{(1)}\Psi_j^{(2)}\Psi_k^{(3)} \tag{16.136}$$

may be used as basis functions. These functions need not be just wavefunctions, but could also be the variables x, y, z which occur, for example, in dipole matrix elements (16.119). Finally, operators are also transformed linearly under symmetry operations and thus can form the basis for a representation; an example is the momentum operator, $p_x = (\hbar/\mathrm{i})\partial/\partial x$, $p_y = (\hbar/\mathrm{i})\partial/\partial y$, $p_z = (\hbar/\mathrm{i})\partial/\partial z$. We could, for example, replace $\Psi_j^{(2)}$ in (16.136) by the components of the momentum operator.

The *optical matrix elements* can be written as integrals having the form

$$\int \Psi_i^{(1)} \Psi_j^{(2)} \Psi_k^{(3)} dV , \tag{16.137}$$

where the integration extends over all the variables which occur in the Ψ's, e.g. over the coordinates of one or more electrons. We can now decompose the basis (16.136) with respect to the symmetry operations of the given molecule into its irreducible representations. It can be shown mathematically that integrals of the form (16.137) vanish with respect to such symmetry operations unless the integrand is invariant under all the operations of the point group. We demonstrate this with two simple examples:

1) In the one-dimensional integral

$$\int_{-a}^{+a} f(x)dx \tag{16.138}$$

we take $f(x)$ to be noninvariant with respect to the reflection $x \to -x$; let us assume that $f(-x) = -f(x)$. We replace x by $-x$ in (16.138), which leaves the value of the integral unchanged, and obtain

$$\int_{-a}^{+a} f(x)dx = \int_{-a}^{+a} f(-x)dx = -\int_{-a}^{+a} f(x)dx ,$$

from which it follows that

$$2\int_{-a}^{+a} f(x)dx = 0 ,$$

i.e. the integral vanishes.

2) In the two two-dimensional integrals (with limits $\pm\infty$)

$$I_j = \int\int f_j(x, y)dxdy , \quad j = 1, 2 \tag{16.139}$$

let the functions transform as follows under a rotation operation:

$$\begin{pmatrix} f_1 \\ f_2 \end{pmatrix} \quad \text{as} \quad \begin{pmatrix} \cos\alpha & \sin\alpha \\ -\sin\alpha & \cos\alpha \end{pmatrix} \begin{pmatrix} f_1 \\ f_2 \end{pmatrix} .$$

Then, on the one hand, the values of the integrals (16.139) remain unchanged under the rotation, but on the other hand, the following equation holds:

$$\begin{pmatrix} I_1 \\ I_2 \end{pmatrix} = \begin{pmatrix} \cos\alpha & \sin\alpha \\ -\sin\alpha & \cos\alpha \end{pmatrix} \begin{pmatrix} I_1 \\ I_2 \end{pmatrix} .$$

Since the determinant

$$\begin{vmatrix} \cos\alpha - 1 & \sin\alpha \\ -\sin\alpha & \cos\alpha - 1 \end{vmatrix}$$

is non-vanishing (except for $\cos\alpha = 1$, i.e. $d = 2\pi n$, $n = 0, \pm 1, \ldots$), it follows that $I_1 = I_2 = 0$.

In order to determine whether integrals of the form (16.137) can differ from zero for given basis functions (and operators), we can proceed as follows: we decompose the product (16.136) into its irreducible representations. Then the integrals can be non-vanishing only if the identical representation is among the irreducible representations. The following mathematical theorems are useful:

Theorem: The representation of a direct product (of 2 basis sets) contains the complete symmetrical representation only if the original basis functions belong to the same irreducible representation of the point group. In the case of a triple product, the integral may be nonzero only if the representation of the product of two functions is the same as the representation of the third function or contains the representation of the third function. In this case, the functions or operators can be grouped together in expedient ways. For example, the representation of $\Psi_j^{(1)} \cdot \text{operator} \cdot \Psi_k(2)$ can be first examined with respect to $\Psi_j^{(1)}\Psi_k^{(2)}$ to find out if it contains a representation of the operator (e.g. p_x, p_y, p_z).

We show the application of these theorems using the example of the point group C_{2v}, which we have met in the treatment of the H_2O molecule in Sect. 6.7, and consider the dipole matrix element

$$\int \Psi_\mu^*(\boldsymbol{r})\boldsymbol{r}\Psi_\kappa(\boldsymbol{r})dV \,. \tag{16.140}$$

We thus identify Ψ_μ^* with $\Psi_i^{(1)}$ in (16.137), $\boldsymbol{r}$ with $\Psi_j^{(2)}$, and Ψ_κ with $\Psi_k^{(3)}$. We assume that Ψ_μ^* and Ψ_κ already belong to an irreducible representation whose symmetry properties are listed in Table 16.1. Now we still need the irreducible representations of $\boldsymbol{r} = (x, y, z)$, which we determine here as a little exercise (but can also read off directly from Table 16.3). As one can readily see from Fig. 6.5, the following transformation rules hold for the symmetry operations (see Table 16.3):

Table 16.3

E	$\begin{pmatrix} x \\ y \\ z \end{pmatrix} = \begin{pmatrix} x \\ y \\ z \end{pmatrix} = \begin{pmatrix} 1 & 0 & 0 \\ 0 & 1 & 0 \\ 0 & 0 & 1 \end{pmatrix} \begin{pmatrix} x \\ y \\ z \end{pmatrix}$,	$\chi = 3$
C_2	$\begin{pmatrix} x \\ y \\ z \end{pmatrix} = \begin{pmatrix} -x \\ -y \\ z \end{pmatrix} = \begin{pmatrix} -1 & 0 & 0 \\ 0 & -1 & 0 \\ 0 & 0 & 1 \end{pmatrix} \begin{pmatrix} x \\ y \\ z \end{pmatrix}$,	$\chi = -1$
σ_v	$\begin{pmatrix} x \\ y \\ z \end{pmatrix} = \begin{pmatrix} -x \\ y \\ z \end{pmatrix} = \begin{pmatrix} -1 & 0 & 0 \\ 0 & 1 & 0 \\ 0 & 0 & 1 \end{pmatrix} \begin{pmatrix} x \\ y \\ z \end{pmatrix}$,	$\chi = 1$
σ_v'	$\begin{pmatrix} x \\ y \\ z \end{pmatrix} = \begin{pmatrix} x \\ -y \\ z \end{pmatrix} = \begin{pmatrix} 1 & 0 & 0 \\ 0 & -1 & 0 \\ 0 & 0 & 1 \end{pmatrix} \begin{pmatrix} x \\ y \\ z \end{pmatrix}$,	$\chi = 1$

In the last column, the characters are given, i.e. the traces of the representation matrices which can be found directly from the matrices themselves. How these representations Γ decompose into the irreducible representations can be easily seen by using these characters

Table 16.4. *Upper row*: characters for the representation of $\boldsymbol{r}$; *Lower rows*: characters of the irreducible representations which contribute to Γ

	E	C_2	σ_v	σ_v'
Γ	3	-1	1	1
A_1	1	1	1	1
B_1	1	-1	1	-1
B_2	1	-1	-1	1

Table 16.5. *Left column*: $\boldsymbol{r}$ and its irreducible representations; *Right column*: $\Psi_\mu^* \Psi_\kappa$ and its representation products, which yield A_1, B_1, and B_2

$\boldsymbol{r}$	$\Psi_\mu^* \Psi_\kappa$
A_1	$A_1 A_1$
B_1	$A_2 B_2$
B_2	$A_2 B_1$ (or $B_1 A_2$)

χ and formula (6.47), or by trial and error. The latter method can be easily applied using Table 16.3 and the characters it contains, as one can see from Table 16.4. The result is

$$\Gamma = A_1 + B_1 + B_2 \ . \tag{16.141}$$

Let us now return to our original goal and apply the theorem quoted above to (16.140). We first combine Ψ_μ^*, Ψ_κ; then we can use Table 16.1 and investigate which products of the lower left columns yield one of the irreducible representations in (16.141), i.e. A_1, B_1, or B_2. We thus obtain Table 16.5.

The second column of the table shows the wavefunctions between which an optical dipole transition is allowed. All other transitions are forbidden. In Sect. 6.7, we can find examples of LCAO wavefunctions having the corresponding symmetry properties.

Using Table 6.14, we can read off the facts that z belongs to the representation A_1, x to the representation B_1, and y to the representation B_2. With these correspondences, we can determine which polarisation directions are allowed (or forbidden) in optical dipole transitions. Thus, for example, only the polarisation x is allowed in the transition $A_2 \rightarrow B_2$. We leave it to the reader as an exercise to show that x, y, z each forms a basis by itself for the irreducible representations A_1, B_1, and B_2, respectively.

16.6 Summary

In this relatively long chapter, we have treated the spontaneous and stimulated emission as well as the absorption of light using first-order perturbation theory, with the creation or the annihilation of one photon in each case. From the transition probabilities (per unit time), we obtained the Einstein coefficients for emission and absorption; they are determined by the optical matrix elements. The latter may be simplified to give the dipole matrix elements or transition moments and related to the oscillator strengths. The Born-Oppenheimer approximation and the Franck-Condon principle allow us to separate the optical matrix elements into the product of a matrix element for the electronic transition with the nuclear coordinates held constant, and a matrix element for the vibrational transition of the nuclei. Finally, we have shown how it is possible to derive selection rules for the optical transitions by making use of group theory.

17. Theoretical Treatment of the Raman Effect and the Elements of Nonlinear Optics

We first introduce time-dependent perturbation theory in higher orders, and then apply it to a quantum-mechanical treatment of the Raman effect and of two-photon absorption.

17.1 Time-Dependent Perturbation Theory in Higher Orders

The methods developed in the preceding Chap. 16 will now permit us to deal with the Raman effect on a quantum-theoretical basis, having discussed it from the experimental standpoint in Chap. 12. Furthermore, we are now in a position to treat, for example, the two-photon processes of nonlinear optics. For the mathematical description, it will prove expedient in the following sections that we at first do not specify the interaction, but instead develop the formalism on a general basis. However, to gain an intuitive picture, we can consider the following problem: we imagine an electron which is moving in a fixed potential field, e.g. that of an atomic nucleus or of a whole molecule. This electron is then exposed to a radiation field. If the radiation field is weak, then it can be considered as a small perturbation; we can then describe the emission and absorption of single photons by the electron in a molecule, as we showed in Sect. 16.3.

In nuclear magnetic resonance, it has long been possible to detect multiple-quantum transitions, as well; more recently, lasers have provided a light source which is so intense that the observation of multiple-quantum transitions in the visible or the UV spectral regions is also feasible. For multiple-quantum processes, which also occur in light scattering, it no longer suffices to employ only the first-order approximation of time-dependent perturbation theory (cf. Sect. 16.2). Instead, one must systematically take the higher orders of the perturbation into account. We shall now demonstrate just how this can be done in general, keeping Sect. 16.2 in mind and using the same notation that we introduced there. Our starting point is again the Schrödinger equation (16.1). We expand the desired solution in terms of the solutions of the unperturbed Schrödinger equation (16.2) in the form given by (16.5), i.e.

$$\Psi(t) = \sum_{\nu=1}^{\infty} c_\nu(t)\, \Psi_\nu^0 \,. \qquad (17.1)$$

The coefficients are found to obey (16.7), which we repeat here:

$$\dot{c}_\mu(t) = \frac{1}{\mathrm{i}\hbar} \sum_\nu c_\nu(t) H^{\mathrm{S}}_{\mu\nu} \qquad (17.2)$$

with the matrix elements $H^{\mathrm{S}}_{\mu\nu}$ as defined in (16.8). At the initial time $t = t_0$, only the state having the quantum number κ is occupied, i.e.

$$c_\nu(t_0) = \begin{Bmatrix} 1 & \text{for} & \nu = \kappa \\ 0 & \text{for} & \nu \neq \kappa \end{Bmatrix} = \delta_{\nu\kappa} \; . \tag{17.3}$$

Equation (17.3) at the same time defines the zeroth order of our perturbation method. The coefficients in this approximation are denoted by a zero as superscript, that is we define

$$c_\nu^{(0)} \equiv c_\nu(t_0) \; . \tag{17.4}$$

The basic idea of perturbation theory is the following: since the perturbation H^{S} is assumed to be relatively small, the coefficients c_ν will not differ too much from those in (17.4) at times which are not too long. It is then apparent that we can use (17.4) as trial values for c_ν on the right-hand side of (17.2), in order to obtain as the result an improved value, $c_\nu^{(1)}$, on the left-hand side of (17.2). Using this approach and (17.3), (17.4) as starting values, we arrive at the following relation for the c_ν's on the right-hand side:

$$\dot{c}_\mu^{(1)}(t) = (-\mathrm{i}/\hbar) H_{\mu\kappa}^{\mathrm{S}}(t) \; . \tag{17.5}$$

By integrating over time, we obtain immediately an explicit expression for the c_μ's in the first-order approximation:

$$c_\mu^{(1)}(t) = (-\mathrm{i}/\hbar) \int_{t_0}^{t} H_{\mu\kappa}^{\mathrm{S}}(\tau)\, d\tau + \delta_{\mu\kappa} \; . \tag{17.6}$$

Here, we have still written the initial time in the general form t_0. The Kronecker delta $\delta_{\mu\kappa}$ guarantees that the initial condition (17.3) will be obeyed. In evaluating the integral (17.6), we must keep in mind that $H_{\mu\kappa}^{\mathrm{S}}$ is time dependent [compare (16.8) with (16.3)]. The coefficients (17.6) yield an improvement as compared to those of (17.4).

It is apparent that this process can be continued to higher orders, by inserting the now improved values of $c_\nu^{(1)}$ on the right-hand side of (17.2) and thus obtaining still further improved coefficients $c_\nu^{(2)}$ on the left-hand side. By imagining this process to proceed still further, we obtain after the $(l+1)$–th step the relation:

$$c_\mu^{(l+1)}(t) = c_\mu^{(0)} + (-\mathrm{i}/\hbar) \int_{t_0}^{t} H_{\mu\nu}^{\mathrm{S}}(\tau)\, c_\nu^{(l)}(\tau)\, d\tau \; . \tag{17.7}$$

(We could also obtain this expression in a mathematically well-defined manner by equipping the perturbation H^{S} with an "infinitesimal parameter" ε and expanding c_μ as a series in ε, i.e.

$$c_\mu = \sum_{l=0}^{\infty} \varepsilon^l c_\mu^{(l)} \; ,$$

then inserting this series into (17.2), carrying out a comparison of the coefficients, and integrating the result over time from t_0 to t.) Relation (17.7) is a recursion formula; applying it, we can compute $c_\mu^{(l+1)}$ whenever we have determined $c_\mu^{(l)}$ in a previous step.

We can see most of the important features by examining the case $l = 1$; then, of course, we have instead of (17.7) the expression

$$c_\mu^{(2)}(t) = (-\mathrm{i}/\hbar) \sum_{\mu_1} \int_{t_0}^{t} H_{\mu\mu_1}^{\mathrm{S}}(\tau)\, c_{\mu_1}^{(1)}(\tau)\, d\tau + c_\mu^{(0)} \; . \tag{17.8}$$

For $l = 0$, we arrive back at the case (17.6) which we have already treated, given here in a slightly more generalized notation:

$$c_\mu^{(1)}(t) = (-\mathrm{i}/\hbar) \sum_{\mu_2} \int_{t_0}^{t} H^{\mathrm{S}}_{\mu_1\mu_2}(\tau)\, d\tau\, c_{\mu_2}^{(0)} + c_{\mu_1}^{(0)} \,. \tag{17.9}$$

As we have already seen, we have to take $c_\mu^{(0)}$ as given at the time $t = t_0$, so that these quantities can be assumed to be known. We then have the task of expressing the coefficients in second order, $c_\mu^{(2)}$, in terms of the $c_\mu^{(0)}$. This is possible if we first express the coefficients $c_\mu^{(1)}$ which occur in (17.8) in terms of the $c_\mu^{(0)}$ by using (17.9); accordingly, we insert (17.9) into (17.8) and find

$$\begin{aligned} c_\mu^{(2)}(t) = {} & c_\mu^{(0)} + (-\mathrm{i}/\hbar) \sum_{\mu_1} \int_{t_0}^{t} H^{\mathrm{S}}_{\mu\mu_1}(\tau)\, d\tau\, c_{\mu_1}^{(0)} \\ & + (-\mathrm{i}/\hbar)^2 \sum_{\mu_1\mu_2} \int_{t_0}^{t} d\tau_1\, H^{\mathrm{S}}_{\mu\mu_1}(\tau_1) \int_{t_0}^{\tau_1} d\tau_2\, H^{\mathrm{S}}_{\mu_1\mu_2}(\tau_2)\, c_{\mu_2}^{(0)} \,. \end{aligned} \tag{17.10}$$

In a corresponding fashion, we can employ the general expression (17.7), successively eliminating all the intermediate results $c_\mu^{(l)}$.

In the $(l+1)$–th approximation, $c_\mu^{(l+1)}$ is given by

$$\begin{aligned} c_\mu^{(l+1)}(t) = {} & c_\mu^{(0)} + (-\mathrm{i}/\hbar) \sum_{\mu_1} \int_{t_0}^{t} H^{\mathrm{S}}_{\mu\mu_1}(\tau)\, d\tau\, c_{\mu_1}^{(0)} \\ & + (-\mathrm{i}/\hbar)^2 \sum_{\mu_1\mu_2} \int_{t_0}^{t} \int_{t_0}^{\tau_2} d\tau_1 d\tau_2 H^{\mathrm{S}}_{\mu\mu_2}(\tau_2) H^{\mathrm{S}}_{\mu_2\mu_1}(\tau_1)\, c_{\mu_1}^{(0)} \\ & + \ldots + \ldots + \ldots \\ & + (-\mathrm{i}/\hbar)^{(l+1)} \sum_{\mu_1\mu_2\ldots\mu_{l+1}} \int_{t_0}^{t} \int_{t_0}^{\tau_{l+1}} \int_{t_0}^{\tau_l} \cdots \int_{t_0}^{\tau_2} d\tau_1 \\ & \ldots d\tau_{l+1} H^{\mathrm{S}}_{\mu\mu_{l+1}} H^{\mathrm{S}}_{\mu_{l+1}\mu_l} \cdots H^{\mathrm{S}}_{\mu_2\mu_1} c_{\mu_1}^{(0)} \,. \end{aligned} \tag{17.11}$$

The coefficients we were originally seeking, $c_\mu(t)$, are obtained by letting the perturbation series go to infinite order:

$$c_\mu(t) = \lim_{l\to\infty} c_\mu^{(l)}(t) \,. \tag{17.12}$$

17.2 Theoretical Description of the Raman Effect

We have described the Raman effect in Chap. 12, in that chapter primarily from the experimental standpoint. As we saw there, incident light with a frequency ν_0 is scattered inelastically by molecules, leading to a new frequency ν_1. In Chap. 12, we also denoted ν_0 by ν_p (p for "primary") and ν_1 by $\nu_\mathrm{p} \pm \nu_\mathrm{vib}$ or $\nu_\mathrm{p} \pm \nu_\mathrm{rot}$. In this section, we treat the vibronic Raman effect; the description of the rotational Raman effect is analogous.

From the quantum-mechanical point of view, the effect can be explained as follows: a photon of frequency ν_0 produces a transition in a molecule from the electronic ground state with quantum numbers q_a into an excited intermediate state with quantum numbers q_i; during the transition, the vibronic state, which has, e.g. v quanta initially, changes also to a state with v'' vibrational quanta. This transition is a so-called virtual transition, in which energy conservation is not necessarily fulfilled. The molecule then makes a second transition from the virtual intermediate state to the final state, which has the same electronic quantum numbers as the initial state, but a different number of vibronic quanta, v' (see Fig. 17.1). If the vibronic energy is now greater than in the initial state, then the increase in energy must be obtained from the difference of the absorbed and emitted photons. The emitted photon thus has a smaller quantum energy $\hbar\,\omega$, i.e. a lower frequency than the incident photon, and $\nu_1 < \nu_0$; this is termed the Stokes shift. However, especially through the availability of lasers it has now become possible also to produce anti-Stokes Raman lines, which have $\nu_1 > \nu_0$.

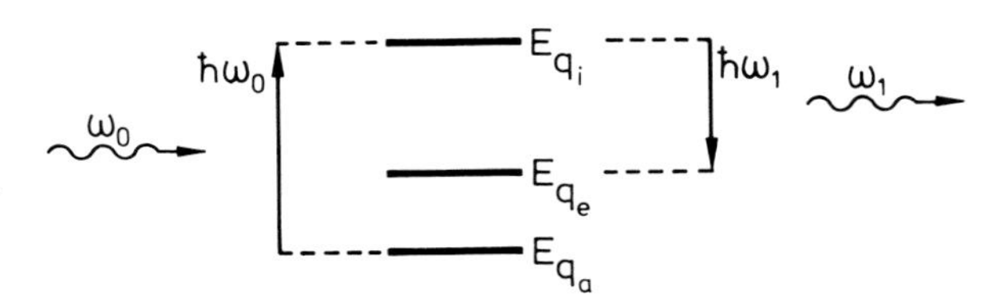

Fig. 17.1. The Raman effect: absorption of a light quantum with the energy $\hbar\,\omega_0$ and emission of another light quantum of energy $\hbar\,\omega_1$

In order to describe the Raman effect theoretically, we make use of time-dependent perturbation theory of higher order, which we have developed in the previous section. We consider a second-order process and, as a first step, we calculate the time integrals which occur in (17.10). Using the same notation, we write

$$\omega_{\mu\kappa} = \frac{1}{\hbar}(E_\mu - E_\kappa)\,. \tag{17.13}$$

As a result of the form of the unperturbed wavefunctions according to (16.3), we can express the matrix element (16.8) as

$$H^{\mathrm{S}}_{\mu\nu}(t) = H^{\mathrm{S}}_{\mu\nu}(0)\exp(\mathrm{i}\,\omega_{\mu\nu}t)\,. \tag{17.14}$$

For simplicity, we take the initial time $t_0 = 0$. We are now interested in the coefficient which describes a second-order transition, i.e. for example the absorption of a photon and the emission of a second photon. In the notation of (17.13) and (17.14), this coefficient is given by the third term on the right-hand side of (17.10), which we denote by $c_\mu^{(2)'}$:

$$c_\mu^{(2)'} = -\frac{1}{\hbar^2}\sum_{\mu_1}\int_0^t \exp(\mathrm{i}\,\omega_{\mu\mu_1}\tau_1)\,d\tau_1 \int_0^{\tau_1} \exp(\mathrm{i}\,\omega_{\mu_1\kappa}\tau_2)\,d\tau_2 H^{\mathrm{S}}_{\mu\mu_1}(0)H^{\mathrm{S}}_{\mu_1\kappa}(0)\,. \tag{17.15}$$

The integrals over time can be readily evaluated, yielding

$$\int_0^t\int_0^{\tau_1}\ldots d\tau_1 d\tau_2 = -\frac{\exp(\mathrm{i}\,\omega_{\mu\kappa}t)-1}{\omega_{\mu\kappa}\omega_{\mu_1\kappa}} + \frac{\exp(\mathrm{i}\,\omega_{\mu\mu_1}t)-1}{\omega_{\mu\mu_1}\omega_{\mu_1\kappa}}\,. \tag{17.16}$$

The two terms which occur in (17.16) have quite different meanings and magnitudes. We have previously seen the factor $(\mathrm{e}^{\mathrm{i}\omega_{\mu\kappa}t} - 1)/\omega_{\mu\kappa}$ within the first term of (16.15); as we saw there, this expression in the end leads to a δ-function, which guarantees that energy is conserved. In the neighbourhood of $\omega_{\mu\kappa} = 0$, a singularity occurs, and, if we do not integrate over a frequency continuum, it leads to an increase of the coefficient $c_\mu^{(2)'}$, proportional to the time. In the factor $1/\omega_{\mu_1\kappa}$, the denominator remains nonzero for the processes we are considering. The second term in (17.16) contains transition frequencies between the virtual states μ_1 and the final state μ, in which energy is with certainty not conserved. This term in (17.16) thus exhibits oscillatory behaviour but does not increase with time, and can therefore be neglected. Using these results from (17.16), we obtain the following expression for (17.15):

$$c_\mu^{(2)'} = \frac{1}{\hbar} \frac{\exp(\mathrm{i}\,\omega_{\mu\kappa} t) - 1}{\omega_{\mu\kappa}} H_{\mu\kappa}^{\mathrm{S.eff}} , \tag{17.17}$$

where we have introduced the abbreviation

$$H_{\mu\kappa}^{\mathrm{S.eff}} = \sum_{\mu_1} \frac{H_{\mu\mu_1}^{\mathrm{S}}(0) H_{\mu_1\kappa}^{\mathrm{S}}(0)}{(E_{\mu_1} - E_\kappa)} . \tag{17.18}$$

This abbreviation permits us to make a direct comparison between the coefficients (17.17) and the coefficients which we have seen earlier in (16.15). Clearly, the matrix element we calculated there is to be substituted by (17.18), but all other considerations remain the same. We can thus immediately write down the transition probability per unit time in the form

$$W = \frac{2\pi}{\hbar} \sum_{\mu \in \Omega} \delta(E_\mu - E_\kappa) \left| H_{\mu\kappa}^{\mathrm{S.eff}} \right|^2 , \tag{17.19}$$

in which the quantum numbers μ of the final state (or the quantum numbers κ of the initial state) are spread over a continuum Ω.

In order to apply this general formula to the Raman effect, we must consider in more detail the individual states and the energies which occur here, first directing our attention to the states of the radiation field. The wavefunction of the initial state with the index κ can be written (as we have already shown in Chap. 16) as a product of the wavefunction ϕ_{Q_a} of the molecule (i.e. electrons and nuclei) with that of the radiation field in the initial state, Φ_a:

$$\kappa : \phi_{Q_\mathrm{a}} \Phi_\mathrm{a} . \tag{17.20}$$

The index Q_a in this expression summarizes all of the quantum numbers of the electrons and the nuclei of the molecule in the initial state. The initial state of the radiation field is in general occupied by n photons of the type λ_0:

$$\Phi_\mathrm{a} = \frac{1}{\sqrt{n!}} (b_{\lambda_0}^+)^n \Phi_0 . \tag{17.21}$$

The energy belonging to (17.20) can, in a readily understandable way, be written in the form:

$$E_\kappa = E_{Q_\mathrm{a}.\mathrm{Mol}} + n\hbar\,\omega_{\lambda_0} , \tag{17.22}$$

where $\omega_{\lambda_0} \equiv 2\pi\,\nu_0$.

In order to evaluate the matrix elements, we require the perturbation operator which we have already given in (16.52) but will repeat briefly here:

$$H^\mathrm{S} = \sum_\lambda b_\lambda \frac{1}{\sqrt{V}} O_\lambda + \sum_\lambda b_\lambda^+ \frac{1}{\sqrt{V}} O_\lambda^+ . \tag{17.23}$$

In this equation, V is the normalisation volume for the waves in the radiation field, while O_λ was defined in (16.53) and is an operator which refers to the electronic motions. The final state, with index μ, is written as

$$\mu : \phi_{Q_e} \Phi_e \, , \tag{17.24}$$

in which the final state of the radiation field differs from that of (17.21) in that now one photon fewer of type λ_0 is present, while one photon of type λ_1 has been created. This is precisely the effect of Raman scattering. The perturbation operator H^S includes both the creation and the annihilation of photons. Since the pertrubation operator however enters (17.18) two times, it can cause two different processes, which differ in the order of photon absorption and emission.

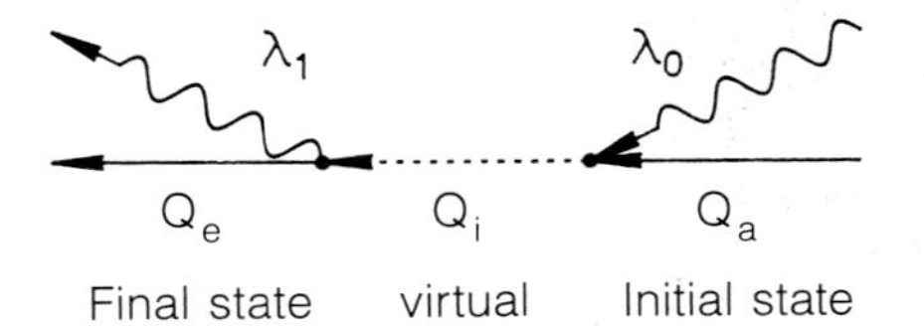

Fig. 17.2. Feynman diagrams (to be read from *right* to *left*!) for the absorption and emission of a quantum of radiation. The molecule is initially in the state associated with the quantum numbers Q_a, in which it absorbs a photon denoted by the quantum numbers λ_0. It thereby makes a transition into the intermediate state Q_i. Finally, it emits the photon λ_1 and in the process makes a transition into the final state Q_e

These two paths are best made clear in so-called Feynman diagrams (Figs. 17.2 and 17.3), and we recommend that the reader compare the following text with the corresponding figures. The first path (Fig. 17.2) takes the following course: in the initial step, a photon of type λ_0 is absorbed, leading to the intermediate virtual state with index μ_1, which takes the form

$$\mu_1 : \phi_{Q_i} \Phi_i \, , \tag{17.25}$$

where Q_i summarizes the quantum numbers of the molecule in the intermediate state, while the wavefunction of the radiation field is distinguished from that in the initial state by having one photon fewer of type λ_0:

$$\Phi_i = \frac{1}{\sqrt{(n-1)!}} (b^+_{\lambda_0})^{n-1} \Phi_0 \, . \tag{17.26}$$

(If only one photon was initially present in the radiation field, then of course we now have the vacuum state.)

For this process, the associated matrix element is given by

$$H^S_{\mu_1 \kappa}(0) = \sqrt{n} \int \phi^*_{Q_i} \frac{1}{\sqrt{V}} O_{\lambda_0} \phi_{Q_a} dV_{el} dV_n = \sqrt{n} \frac{1}{\sqrt{V}} H^S_{Q_i, \lambda_0 : Q_a, 0} \, . \tag{17.27}$$

In this equation, we have used the abbreviations of (16.68) and (16.70). If we read the indices which occur in (17.15, 18) from right to left, then the matrix element which refers to this process occurs first. The second matrix element is the one which refers to the second step in the process, namely the emission of a photon of type λ_1, where λ_1 is supposed to be different from λ_0. This matrix element is then given by:

$$H^S_{\mu \mu_1}(0) \equiv \frac{1}{\sqrt{V}} \int \phi^*_{Q_e} O_{\lambda_1} \phi_{Q_i} dV_{el} dV_n = \frac{1}{\sqrt{V}} H^S_{Q_e, \lambda_1 : Q_i, 0} \, . \tag{17.28}$$

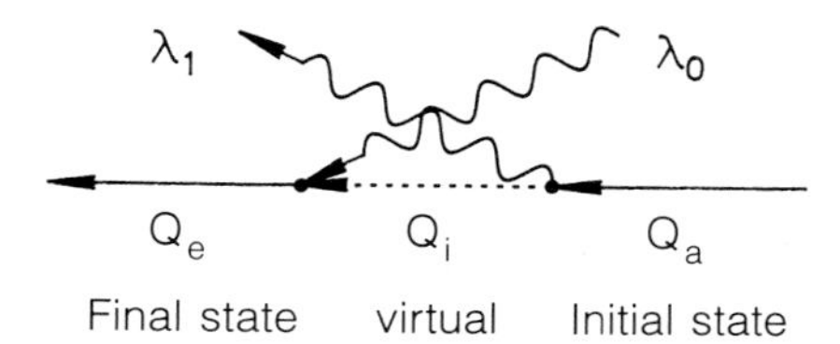

Fig. 17.3. As in Fig. 17.2, but now the emission of one quantum occurs before the absorption of the other

Combining these results, we find the contribution to the effective matrix element (17.18) for the first path to be as follows:

$$H^{\mathrm{S.eff}}_{Q_e.\lambda_1:Q_a.\lambda_0} = \sum_{Q_i} \frac{H^{\mathrm{S}}_{Q_e.\lambda_1:Q_i.0} H^{\mathrm{S}}_{Q_i.\lambda_0:Q_a.0}}{E_{Q_i.\mathrm{Mol}} - E_{Q_a.\mathrm{Mol}} - \hbar\,\omega_{\lambda_0}} \sqrt{n}\,. \tag{17.29}$$

The structure of (17.18), however, contains a second path (cf. Fig. 17.3), since the operator which enters into the matrix element $H^{\mathrm{S}}_{\mu\kappa}$ can both create and annihilate photons. In this second path, the first step is the creation of a photon of type λ_1; it is then present in the final state, so that the intermediate state with index μ_1 is given as

$$\mu_1 : \phi_{Q_i}\Phi_i\,, \tag{17.30}$$

with the radiation field in the intermediate state represented by

$$\Phi_i = b^+_{\lambda_1} \frac{1}{\sqrt{n!}} (b^+_{\lambda_0})^n \Phi_0\,. \tag{17.31}$$

The associated matrix element is then

$$H^{\mathrm{S}}_{\mu_1\kappa}(0) = H^{\mathrm{S}}_{Q_i.\lambda_1:Q_a.0}\,. \tag{17.32}$$

In the second step, one of the photons of type λ_0 which was originally present is annihilated. This is described by the matrix element

$$H^{\mathrm{S}}_{\mu\mu_1}(0) = H^{\mathrm{S}}_{Q_e.\lambda_0:Q_i.0}\sqrt{n}\,. \tag{17.33}$$

The energy denominator is found to be

$$\begin{aligned} E_{\mu_1} - E_\kappa &= E_{Q_i.\mathrm{Mol}} + n\hbar\,\omega_{\lambda_0} + \hbar\,\omega_{\lambda_1} - E_{Q_a.\mathrm{Mol}} - n\hbar\,\omega_{\lambda_0} \\ &= E_{Q_i.\mathrm{Mol}} - E_{Q_a.\mathrm{Mol}} + \hbar\,\omega_{\lambda_1}\,. \end{aligned} \tag{17.34}$$

Both paths must be taken into account in the overall expression (17.18), so that we obtain as the final result for the effective matrix element:

$$\begin{aligned} |H^{\mathrm{S.eff}}_{\mu\kappa}|^2 = n \Bigg| &\sum_i \frac{H^{\mathrm{S}}_{Q_e.\lambda_1:Q_i.0} H^{\mathrm{S}}_{Q_i.\lambda_0:Q_a.0}}{E_{Q_i.\mathrm{Mol}} - E_{Q_a.\mathrm{Mol}} - \hbar\,\omega_{\lambda_0}} \\ &+ \sum_i \frac{H^{\mathrm{S}}_{Q_e.\lambda_0:Q_i.0} H^{\mathrm{S}}_{Q_i.\lambda_1:Q_a.0}}{E_{Q_i.\mathrm{Mol}} - E_{Q_a.\mathrm{Mol}} + \hbar\,\omega_{\lambda_1}} \Bigg|^2\,. \end{aligned} \tag{17.35}$$

The sums in (17.35) run over all the intermediate levels in the molecule. This result can now be inserted into (17.19), with summation of the final states and averaging over the initial states. The summation of the final states may be carried out exactly as in the case of spontaneous emission, which we treated in Sect. 16.3.4, summing the wavevectors within a solid-angle element $d\Omega$. The average (or summation) over the initial states just corresponds to the computation of the average which we carried out for the case of the absorption of radiation, also in Sect. 16.3.4.

Up to this point, our treatment has been kept quite general, without any specific assumptions about the internal structure of the electronic and nuclear states of the molecule. However, in order to obtain results which may be compared with experiment, we must now specify the exact meaning of these states and of the quantum numbers Q_a, Q_e, and Q_i which characterize them. To help us in this examination of the molecular states, we employ the Born-Oppenheimer approximation, which we have already used in Sect. 16.4. It allows us to separate the molecular wavefunctions, which depend on all of the electronic coordinates and on all of the nuclear coordinates, into a particular product of individual wavefunctions:

$$\phi_{Q_\mathrm{a}}(\boldsymbol{r},\boldsymbol{R}) = \psi_{q_\mathrm{a}}(\boldsymbol{r},\boldsymbol{R})\chi_{q_\mathrm{a}.v}(\boldsymbol{R}) \,, \tag{17.36}$$

$$\phi_{Q_\mathrm{i}}(\boldsymbol{r},\boldsymbol{R}) = \psi_{q_\mathrm{i}}(\boldsymbol{r},\boldsymbol{R})\chi_{q_\mathrm{i}.v''}(\boldsymbol{R}) \,, \tag{17.37}$$

$$\phi_{Q_\mathrm{e}}(\boldsymbol{r},\boldsymbol{R}) = \psi_{q_\mathrm{e}}(\boldsymbol{r},\boldsymbol{R})\chi_{q_\mathrm{e}.v'}(\boldsymbol{R}) \,. \tag{17.38}$$

The first factor ψ in each case refers to the electronic motions with the nuclear coordinates $\boldsymbol{R}$ held fixed, while the functions χ refer to the nuclear motions with respect to the coordinates $\boldsymbol{R}$. As we recall, the wavefunctions χ and their associated energies are functions of both the electronic state (characterized by its quantum numbers q_a, etc.) and of the vibrational quantum number v which determines the vibronic excitation level. The energies associated with (17.36–38) can be written in the form

$$E_{Q_\mathrm{a}.\mathrm{Mol}} = E_{q_\mathrm{a}.\mathrm{el}} + E_{q_\mathrm{a}.v} \tag{17.39}$$

$$E_{Q_\mathrm{i}.\mathrm{Mol}} = E_{q_\mathrm{i}.\mathrm{el}} + E_{q_\mathrm{i}.v''} \tag{17.40}$$

$$E_{Q_\mathrm{e}.\mathrm{Mol}} = E_{q_\mathrm{e}.\mathrm{el}} + E_{q_\mathrm{e}.v'} \,. \tag{17.41}$$

The quantum numbers Q have been separated here, as in (17.36–38), into those of the electronic states, q, and those of the vibronic states, v, v'', and v'; the latter are the vibrational quantum numbers of the initial, the intermediate, and the final states, respectively, as defined at the beginning of this section. The energy E_Mol is the total energy of the molecule, E_el that of its electrons, and E_v the molecular vibrational energy. With the specifications (17.36–41), we can rewrite the effective matrix element as follows:

$$\begin{aligned} H^{\mathrm{S.eff}}_{Q_\mathrm{e}.Q_\mathrm{a}} &= \sqrt{n}\sum_{q_\mathrm{i}.v''}\frac{(\boldsymbol{e}_1\cdot\int\chi^*_{q_\mathrm{e}.v'}\boldsymbol{\Theta}_{q_\mathrm{e}q_\mathrm{i}}\chi_{q_\mathrm{i}.v''}dV_\mathrm{n})(\int\chi^*_{q_\mathrm{i}.v''}\boldsymbol{\Theta}_{q_\mathrm{i}q_\mathrm{a}}\chi_{q_\mathrm{a}.v}dV_\mathrm{n}\boldsymbol{e}_0)}{E_{q_\mathrm{i}.\mathrm{el}}-E_{q_\mathrm{a}.\mathrm{el}}+E_{q_\mathrm{i}.v''}-E_{q_\mathrm{a}.v}-\hbar\,\omega_{\lambda_0}} \\ &+ \sqrt{n}\sum_{q_\mathrm{i}.v''}\frac{(\boldsymbol{e}_0\cdot\int\chi^*_{q_\mathrm{e}.v'}\boldsymbol{\Theta}_{q_\mathrm{a}q_\mathrm{i}}\chi_{q_\mathrm{i}.v''}dV_\mathrm{n})(\int\chi^*_{q_\mathrm{i}.v''}\boldsymbol{\Theta}_{q_\mathrm{i}q_\mathrm{a}}\chi_{q_\mathrm{a}.v}dV_\mathrm{n}\boldsymbol{e}_1)}{E_{q_\mathrm{i}.\mathrm{el}}-E_{q_\mathrm{a}.\mathrm{el}}+E_{q_\mathrm{i}.v''}-E_{q_\mathrm{a}.v}+\hbar\,\omega_{\lambda_1}} \,. \end{aligned} \tag{17.42}$$

In this equation, dV_n is the volume element in nuclear coordinate space, i.e. dV_nuclear, and $\boldsymbol{e}_0$, $\boldsymbol{e}_1$ are the polarisation vectors of the incident and the scattered light waves, respectively. The abbreviations $\boldsymbol{\Theta}_{q_\mathrm{i}q_\mathrm{a}} = \boldsymbol{\Theta}^*_{q_\mathrm{a}q_\mathrm{i}}$ stand for the electronic matrix element which we have already discussed in Sect. 16.3.4,

$$\boldsymbol{\Theta}_{q_\mathrm{i}q_\mathrm{a}} = -\frac{e}{m_0}\sqrt{\frac{\hbar}{2\varepsilon_0\omega}}\int\phi^*_{q_\mathrm{i}}(\boldsymbol{r},\boldsymbol{R})\sum_j\boldsymbol{p}_j\phi_{q_\mathrm{a}}(\boldsymbol{r},\boldsymbol{R})dV_\mathrm{el} \,, \tag{17.43}$$

where we have set

$$2\pi \nu_0 \approx 2\pi \nu_1 = \omega \,, \tag{17.44}$$

and have replaced the exponential function in the operator O_λ by 1; this, as we have seen earlier, is a good approximation when the size of the molecule is small compared to the wavelength of the radiation. In this approximation, (17.43) can also be expressed in terms of the transition moment (16.118) (cf. Sect. 16.3.6).

To arrive at useful *selection rules*, we can often apply another approximation. The change of the energy between the vibronic levels is in general much smaller than the energy difference between the electronic states. For this reason, we can set

$$E_{q_\mathrm{i}.v''} - E_{q_\mathrm{a}.v} \approx 0 \tag{17.45}$$

and, in addition, employ the approximation of (17.44). Because of (17.45), the quantum number v'', which is a summation index for the sums in (17.42), drops out of the expressions.

This permits us to carry out a very pleasing simplification: as can be proven mathematically, a complete set of wavefunctions implies a so-called completeness relation, which in our case takes the form

$$\sum_{v''} \chi_{q_\mathrm{i}.v''}(\boldsymbol{R})\chi_{q_\mathrm{i}.v''}(\boldsymbol{R}') = \delta(\boldsymbol{R} - \boldsymbol{R}') \,. \tag{17.46}$$

The δ-function on the right-hand side can be more precisely expressed as

$$\delta(\boldsymbol{R} - \boldsymbol{R}') = \delta(\boldsymbol{R}_1 - \boldsymbol{R}_1')\,\delta(\boldsymbol{R}_2 - \boldsymbol{R}_2')\ldots\delta(\boldsymbol{R}_M - \boldsymbol{R}_M') \,, \tag{17.47}$$

where $\boldsymbol{R}_j, \boldsymbol{R}_j'$ are the nuclear coordinates. We can, however, choose these coordinates from the beginning to be normal coordinates, and we will make use of this fact in the following.

With the aid of (17.46), we can reduce the double integrals which occur in the sums in (17.42) to single integrals, so that this equation can be put into the very simple form

$$H^{\mathrm{S.eff}}_{Q_\mathrm{e}\cdot Q_\mathrm{a}} = \boldsymbol{e}_1 \cdot \int \chi^*_{q_\mathrm{a}.v'}\alpha\chi_{q_\mathrm{a}.v} dV_\mathrm{n}\,\boldsymbol{e}_0 \,. \tag{17.48}$$

In (17.48), we have introduced the polarisability tensor α, which, as can be seen from a comparison with (17.42), has the following components:

$$\alpha_{jk} = \sum_{q_\mathrm{i}} \left(\frac{(\boldsymbol{\Theta}_{q_\mathrm{a}q_\mathrm{i}})_j(\boldsymbol{\Theta}_{q_\mathrm{i}q_\mathrm{a}})_k}{E_{q_\mathrm{i}.\mathrm{el}} - E_{q_\mathrm{a}.\mathrm{el}} - \hbar\,\omega} + \frac{(\boldsymbol{\Theta}_{q_\mathrm{a}q_\mathrm{i}})_k(\boldsymbol{\Theta}_{q_\mathrm{i}q_\mathrm{a}})_j}{E_{q_\mathrm{i}.\mathrm{el}} - E_{q_\mathrm{a}.\mathrm{el}} + \hbar\,\omega} \right) , \tag{17.49}$$

where j, k refer to the spatial coordinates x, y, and z.

This polarisability tensor was derived in the theory of dispersion, which we review briefly here. If a static or oscillating electric field $\boldsymbol{E}$ is allowed to act on an atom or a molecule, the electronic charge clouds (and possibly also the nuclei) are displaced so that an electric dipole moment $\boldsymbol{p}$ is induced as a result. If the field strength is not too great, there is a linear relation between $\boldsymbol{p}$ and $\boldsymbol{E}$, i.e.

$$\boldsymbol{p} = \alpha\boldsymbol{E} \,.$$

Since the directions of $\boldsymbol{p}$ and $\boldsymbol{E}$ are not necessarily collinear, α is in general a tensor quantity. As a detailed quantum-mechanical calculation within the theory of dispersion shows, the expression (17.49) for α can be obtained by applying perturbation theory. Since the electronic wavefunctions, which enter into the computation of α, are functions of the nuclear coordinates $\boldsymbol{R}$, i.e. $\phi_Q = \phi_Q(\boldsymbol{r}, \boldsymbol{R})$, α of course also depends on these coordinates.

To obtain selection rules for vibrational Raman transitions, we expand the polarisability tensor in the normal coordinates $\boldsymbol{R}_j$ of the nuclei about their equilibrium positions in the molecule. This yields

$$\alpha = \alpha_0 + \sum_j \left(\frac{\partial \alpha}{\partial \boldsymbol{R}_j}\right)_0 \boldsymbol{R}_j + \frac{1}{2}\sum_{jk}\left(\frac{\partial^2 \alpha}{\partial \boldsymbol{R}_j \partial \boldsymbol{R}_k}\right)_0 \boldsymbol{R}_j \boldsymbol{R}_k + \dots \; . \tag{17.50}$$

If we insert this relation into (17.48), then the selection rules are determined by whether or not the expression

$$\int \chi^*_{q_a.v'} \alpha \chi_{q_a.v} dV_n = \alpha_0 \delta_{vv'} + \sum_j \int \chi^*_{q_a.v'} \left(\frac{\partial \alpha}{\partial \boldsymbol{R}_j}\right)_0 \boldsymbol{R}_j \chi_{q_a.v} dV_n + \dots \tag{17.51}$$

goes to zero or remains finite. The Kronecker delta on the right-hand side of this equation, $\delta_{vv'}$, is due to the orthogonality of the wavefunctions $\chi_{q_a.v}$, $\chi_{q_a.v'}$.

Let us examine the terms on the right-hand side of (17.51). The first term, $\alpha_0\delta_{vv'}$, requires that the vibrational quantum number remain unchanged, i.e. it does not describe the Raman effect, but instead the elastic scattering of photons. We therefore look at the second term and consider a typical summand, which we can rewrite in the form

$$\left(\frac{\partial \alpha}{\partial \boldsymbol{R}_j}\right)_0 \int \chi^*_{q_a.v'} \boldsymbol{R}_j \chi_{q_a.v} dV_n \; , \tag{17.52}$$

since the first factor does not depend on the nuclear coordinates and can therefore be extracted from the integral. The functions $\chi_{q_a.v'}$, $\chi_{q_a.v}$ depend on all the normal coordinates of the nuclei. As we have seen in Sect. 11.3, the normal coordinates allow us to represent the Hamiltonian for the molecular vibrations as the sum of Hamiltonians for individual harmonic oscillations (of the various normal modes). As a consequence, the wavefunctions, e.g. $\chi_{q_a.v}(\boldsymbol{R})$, can be expressed as products (we leave off the index q_a in the following expressions):

$$\chi_v = \chi_{v_1}(\boldsymbol{R}_1)\, \chi_{v_2}(\boldsymbol{R}_2) \dots \chi_{v_M}(\boldsymbol{R}_M) \; , \tag{17.53}$$

where M is the number of vibrational degrees of freedom of the molecule. Correspondingly, we have

$$\chi'_v = \chi_{v'_1}(\boldsymbol{R}_1)\, \chi_{v'_2}(\boldsymbol{R}_2) \dots \chi_{v'_M}(\boldsymbol{R}_M) \; . \tag{17.54}$$

Substituting (17.53) and (17.54) into (17.52) and separating the integral into integrals over the individual nuclear coordinates, we find

$$\begin{aligned} &\int \chi^*_{v'_1}(\boldsymbol{R}_1)\, \chi_{v_1}(\boldsymbol{R}_1) dV_1 \dots \int \chi^*_{v'_j}(\boldsymbol{R}_j) \boldsymbol{R}_j \chi_{v_j}(\boldsymbol{R}_j) dV_j \\ &\qquad \dots \int \chi^*_{v'_M}(\boldsymbol{R}_M)\, \chi_{v_M}(\boldsymbol{R}_M) dV_M \; . \end{aligned} \tag{17.55}$$

Apart from the integral with the index j, these are all orthogonality integrals, so that

$$v'_k = v_k \; , \quad k \neq j$$

must hold, i.e. these vibrational quantum numbers do not change. Inserting the usual oscillator functions for χ in the remaining integral

$$\int \chi^*_{v'_j}(\boldsymbol{R}_j)\boldsymbol{R}_j \chi_{v_j}(\boldsymbol{R}_j) dV_j \ , \tag{17.56}$$

we obtain on applying the theory of the quantum-mechanical harmonic oscillator the following selection rule:

$$v'_j = v_j \pm 1 \ .$$

The corresponding transitions are called the *fundamental transitions*.

The next term in (17.51), which is due to the third term on the right-hand side of (17.50), can be discussed in a similar manner. In the double sum, we have either $j \neq k$ or $j = k$. In the case that $j \neq k$, two integrals similar to (17.56) occur, leading to the selection rules

$$v'_l = v_l \ , \quad l \neq j, k$$

and

$$v'_j = v_j \pm 1 \ ,$$

$$v'_k = v_k \pm 1 \ .$$

In the case that $j = k$, the integral

$$\int \chi^*_{v'_j}(\boldsymbol{R}_j)\boldsymbol{R}^2_j \chi_{v_j}(\boldsymbol{R}_j) dV_j$$

is found from a generalisation of (17.56). For the oscillator wavefunctions, we then find the selection rules

$$v'_j = v_j \pm 2 \ ,$$

i.e. the creation or annihilation of two vibrational quanta, or else

$$v'_j = 0 \ ,$$

i.e. no Raman effect at all.

In this manner, one can include higher and higher terms in the expansion (17.50), (17.51), of course keeping in mind the relative size of the various terms. As is shown by a detailed treatment, the size of the terms goes as $(\xi/r_0)^n$, where ξ is the average vibrational amplitude of the nuclei, and r_0 is a measure of the spatial extent of the electronic wavefunctions. Due to $\xi \ll r_0$, the intensity of higher vibrational quantum transitions decreases rapidly.

Along with the selection rules which we have discussed here, some others can be derived on the basis of symmetry considerations. Just as there are irreducible representations for the *electronic* wavefunctions, one can also find them for the molecular vibrations and apply the results of Chap. 6 to these. This is true in particular of the symmetry selection rules (cf. Sect. 16.5). We can take into account the fact that quantities such as $(\partial\alpha/\partial R_j)R_j$ and $(\partial^2\alpha/\partial R_j \partial R_k)R_j R_k$ transform like α itself under the symmetry operations of the point groups. Every component α_{jk} of α transforms like $x_j x_k$; as a result, the fundamental transitions in the Raman effect are due in general only to those normal modes R_i which transform under the operations of the point groups as linear combinations of $x_j x_k$, e.g. $x^2 - y^2, z^2, xy$. Such vibrational Raman transitions are based only on the vibrational modes which change the molecular polarisability; otherwise, the right side of (17.51) would be reduced to a quantity proportional to the Kronecker delta $\delta_{v'v}$, and no Raman effect would occur. There are also transitions, with a small probability, in which the vibrational quantum number changes by ± 2, ± 3. Therefore, the vibrational modes which exhibit infra-red and Raman fundamentals belong to mutually-exclusive sets when the molecule has a centre of symmetry. This fact is used to identify particular molecular geometries.

A special situation, termed *resonant Raman scattering*, occurs when the frequency of the incident laser light, ν_0, is near to the resonance frequency of one of the molecular eigenstates (i.e. electronic and vibronic). The energy denominator in (17.42) then becomes very small compared to its value for normal Raman scattering, and the transition probability becomes extremely large. In this limit, the vibrational energy differences (17.45) can no longer be neglected relative to the other quantities in the energy denominator, so that the transition amplitude (17.42) can no longer be reduced to the symmetrical form (17.48), (17.49). As a result, the usual selection rules for the normal Raman effect do not apply to resonant Raman transitions. One finds that certain transitions which are forbidden in the normal Raman effect are allowed in the resonant Raman effect.

We mention also that time-resolved resonant Raman scattering has been developed to the point of being a useful technique for the observation of the populations of large molecules in electronically excited states.

17.3 Two-Photon Absorption

If we consider the second-order perturbations which occur in the interactions between radiation and molecules, then we find not only the Raman effect discussed in the previous section, but also the processes of two-photon absorption and emission (compare the Feynman diagrams of Fig. 17.4a,b). As an example, we consider here two-photon absorption, limiting our treatment to the case of the absorption of two photons from the same light wave. In this case, the perturbation operator (17.23) first annihilates one photon, and then, in a second step, a second one.

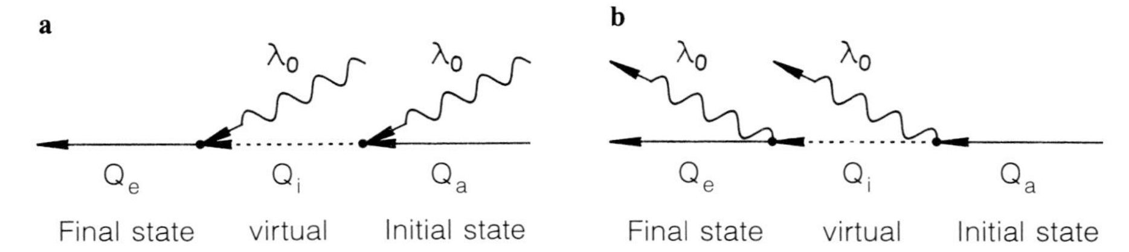

Fig. 17.4a,b. Feynman diagrams for **(a)**: Two-photon absorption; and **(b)** two-photon emission

We can carry out the calculations in exact analogy to those of the previous section, so that we need only point out the decisive differences here. The initial state κ again consists of a product of the molecular state wavefunction ϕ_{Q_a} with that of the radiation field, Φ_a:

$$\kappa : \phi_{Q_a} \Phi_a \, . \tag{17.57}$$

In the radiation field, there are initially n photons present:

$$\Phi_a = \frac{1}{\sqrt{n}}(b_{\lambda_0}^+)^n \Phi_0 \,. \tag{17.58}$$

The energy of the initial state is thus given by

$$E_\kappa = E_{Q_a.\mathrm{Mol}} + n\hbar\,\omega_0 \,. \tag{17.59}$$

In the first step, the perturbation operator (17.23) gives rise to the annihilation of a photon, so that we go from the initial state (17.57) to an intermediate state of the form

$$\phi_{Q_i}\Phi_i \,, \tag{17.60}$$

where the intermediate state of the radiation field is represented by

$$\Phi_i = \frac{1}{\sqrt{(n-1)!}}(b_{\lambda_0}^+)^{n-1}\Phi_0 \,. \tag{17.61}$$

The energy of the intermediate state is then

$$E_{\mu_1} = E_{Q_i.\mathrm{Mol}} + \hbar\,\omega_0(n-1) \,. \tag{17.62}$$

The evaluation of the matrix element is identical to that in the previous section, so that we need only repeat the result here:

$$H^{\mathrm{S}}_{\mu_1\kappa}(0) = \sqrt{n}\frac{1}{\sqrt{V}}\,H^{\mathrm{S}}_{Q_i.\lambda_0:Q_a.0} \,. \tag{17.63}$$

In the second step, a final state is produced in which the annihilation operator b_{λ_0} in the perturbation operator has acted on the intermediate state, giving

$$\phi_{Q_e}\Phi_e \,, \tag{17.64}$$

a state which contains only $n-2$ photons:

$$\Phi_e = \frac{1}{\sqrt{(n-2)!}}(b_{\lambda_0}^+)^{n-2}\Phi_0 \,. \tag{17.65}$$

The matrix element can be evaluated in a manner corresponding to (17.63); it is found to be

$$H^{\mathrm{S}}_{\mu\mu_1}(0) = \frac{1}{\sqrt{V}}\sqrt{n-1}\,H^{\mathrm{S}}_{Q_e.\lambda_0:Q_i.0} \,. \tag{17.66}$$

If we insert all these results into (17.18), we obtain

$$H^{\mathrm{S.eff}}_{Q_e.\lambda_0:Q_a.\lambda_0} = \sqrt{n}\sqrt{n-1}\sum_{Q_i}\frac{H^{\mathrm{S}}_{Q_e.\lambda_0:Q_i.0}H^{\mathrm{S}}_{Q_i.\lambda_0:Q_a.0}}{E_{Q_i.\mathrm{Mol}} - E_{Q_a.\mathrm{Mol}} + \hbar\,\omega_0} \,. \tag{17.67}$$

Since the transition probability per unit time, W, is proportional to the *absolute square* of (17.67), we obtain the especially important result:

$$W \propto n(n-1) \propto I^2 \,. \tag{17.68}$$

The last proportionality is a result of the fact that the intensity I of the incident light wave is proportional to the number of photons the wave contains, and that for n sufficiently large, we can neglect the 1 relative to n. Relation (17.68) states that the number of transitions per second which a molecule undergoes in two-photon absorption is proportional to the *square* of the incident intensity. This fact was employed by *Kaiser* and *Garrett* to measure the length of ultra-short laser pulses. Furthermore, it is important to note that different selection rules hold for two-photon absorption from those for normal single-photon absorption. In the latter case, in the dipole matrix elements ("transition moments"), the parity of the initial-state and final-state wavefunctions must be different (even → odd or odd → even), but in two-photon transitions, the parity of the wavefunctions remains unchanged. This is immediately clear from (17.67), where the parity changes in going from the initial state to the intermediate state, and then changes back again in the transition from the intermediate state to the final state; overall, it thus remains unchanged.

18. Nuclear Magnetic Resonance

In this and the next chapter, we illustrate some of the contributions which can be made to molecular physics by the methods of *magnetic resonance spectroscopy*. These methods occupy a place at the low end of the energy scale (see Fig. 8.1) of the spectroscopic techniques. In the magnetic resonance methods, one makes use of the spins and magnetic moments of nuclei and electrons as probes to study the electronic structure, dynamics and reactivity of molecules. The investigations are usually carried out in the condensed phases, i.e. in solutions or on solid samples.

In nuclear magnetic resonance, the nuclear spin is the probe which samples the structure and dynamics of the electron clouds around it, as well as the coupling to other nuclei in its neighbourhood. Since its first demonstration in 1946, this spectroscopic technique has been developed into what is now perhaps the most important and powerful method of all the various molecular spectroscopies. Here, we can describe only the fundamentals of the method (Sect. 18.1) and the most important basic aspects of nuclear resonance on hydrogen nuclei (proton resonance) in molecules (Sect. 18.2). The study of dynamic processes (Sect. 18.3), resonance using other nuclei besides H (Sect. 18.4), and applications of two-dimensional and spatially-resolved spectroscopies (Sects. 18.5 and 18.6) illustrate the far-reaching capabilities of this *spectroscopic method*.

18.1 Fundamentals of Nuclear Resonance

We first treat nuclear magnetic resonance, NMR, in the present chapter. The fundamentals of this spectroscopy are described in detail in I, Chap. 20. In order to explain the significance of the method for obtaining structural information on molecules, we remind the reader again of those basic aspects of the technique.

18.1.1 Nuclear Spins in a Magnetic Field

Atomic nuclei can have a magnetic moment

$$\boldsymbol{\mu}_I = \gamma \boldsymbol{I} , \tag{18.1}$$

where $\boldsymbol{I}$ is the angular momentum (spin) of the nucleus and γ is the magnetogyric ratio. If we introduce the nuclear magneton μ_N as the unit of the nuclear magnetic moment (it is smaller than the Bohr magneton μ_B by the ratio of the electron mass to the proton mass), then we obtain

$$\boldsymbol{\mu}_I = \frac{g_I \mu_\mathrm{N}}{\hbar} \boldsymbol{I} , \quad \text{with} \quad \mu_\mathrm{N} = 0.505 \cdot 10^{-26}\,\mathrm{Am}^2 . \tag{18.2}$$

This equation defines the nuclear g-factor, $g_I = (\gamma\hbar/\mu_N)$; it is a dimensionless number which one can also use to describe the magnetic moments of nuclei. The factor g_I, in contrast to the Landé g_J-factor of the electronic shells, cannot be computed from a combination of quantum numbers, but instead must be measured experimentally for each nucleus with non-vanishing spin $\boldsymbol{I}$.

The nuclear spin vector obeys the relation

$$|\boldsymbol{I}| = \sqrt{I(I+1)}\,\hbar\ . \tag{18.3}$$

The nuclear spin quantum number I (also called "nuclear spin") can be an integer or a half-integer. It is a characteristic property of the nuclide and takes on values between 0 and 8 for stable nuclei (the value 8 is for the metastable nuclide $^{178}_{72}$Hf).

Only the component of the spin and the magnetic moment in the direction of a quantisation axis, the so-called z-component, can be experimentally observed. This quantisation axis can be defined for example by the direction of an external magnetic field $\boldsymbol{B}_0$. The z-components obey

$$(\boldsymbol{I})_z = m_I\hbar \quad \text{with} \quad m_I = I, I-1, \ldots -I \tag{18.4}$$

and

$$(\boldsymbol{\mu})_z = \gamma\hbar m_I = g_I\mu_N m_I\ . \tag{18.5}$$

There are thus $2I+1$ possible quantized orientations for the nuclear angular momentum vector and the nuclear magnetic moment relative to the quantisation axis. For simplicity, one frequently refers to the largest possible value of $(\boldsymbol{\mu})_z$ as the nuclear moment, i.e. $\mu_I = g_I I \mu_N$.

The magnetic moment $\boldsymbol{\mu}_I$ and the g_I-factor can have a positive or a negative sign. A positive sign means that $\boldsymbol{\mu}_I$ and $\boldsymbol{I}$ have the same direction, as would be expected from classical electrodynamics for a rotating positive charge. A negative sign means that the two vectors are antiparallel. Table 18.1 gives the values of I, g_I, and μ_I for some nuclides which are important to nuclear resonance spectroscopy.

Table 18.1. Properties of some atomic nuclei that are important in the nuclear spin resonance spectroscopy of molecules

Nucleus	I	g_I	μ_I (in μ_N)
^{1}H	1/2	5.5856912	2.7928456
^{2}H	1	0.8574376	0.8574376
^{13}C	1/2	1.40482	0.70241
^{14}C	0	0	0
^{14}N	1	0.4037607	0.4037607
^{17}O	5/2	−0.757516	−1.89379
^{19}F	1/2	5.257732	2.628866
^{31}P	1/2	2.26320	1.1316

Owing to its magnetic moment, a nucleus in an external magnetic field $\boldsymbol{B}_0$ has the magnetic interaction energy

$$V = -\boldsymbol{\mu}_I \cdot \boldsymbol{B}_0 = -g_I \mu_\mathrm{N} B_0 m_I \; , \tag{18.6}$$

with the magnetic quantum number $m_I = I, I-1, \ldots - I$.

The energy difference between two adjacent orientations of the magnetic moment in a field $\boldsymbol{B}_0$, i.e. for transitions with $\Delta m_I = \pm 1$, is

$$\Delta E = g_I \mu_\mathrm{N} B_0 \; . \tag{18.7}$$

If electromagnetic radiation of frequency

$$\nu = \frac{\Delta E}{h} = \frac{g_I \mu_\mathrm{N}}{h} B_0 \tag{18.8}$$

is applied to a sample containing the corresponding nuclei in a direction perpendicular to the direction of $\boldsymbol{B}_0$, this radiation can be absorbed by the sample and can give rise to transitions between the possible orientations of the nuclear spin as found from (18.6). This process is called nuclear spin resonance. As shown in detail in I, Chap. 20, resonance refers to an equality of the frequency of the applied radiation to the Larmor precession frequency of the nuclear spins in the field B_0.

Equation (18.8) is fulfilled for protons in a field of 1 T (= 10 kG) at the frequency $\nu = 42.578$ MHz. This corresponds to radiation with a wavelength λ of about 7 m, and the energy quanta ΔE are equivalent to about $1.8 \cdot 10^{-7}$ eV.

In general, one can also write the resonance condition in the convenient form

$$\nu = 762.3 \frac{\mu_I}{I} B_0 \; [\mathrm{s}^{-1}] \; ,$$

when ν is quoted in s^{-1} and B_0 in Gauss.

Nuclear spin resonance is often abbreviated as NMR (for Nuclear Magnetic Resonance).

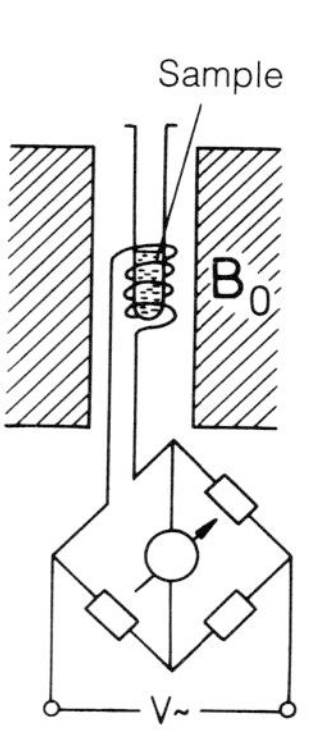

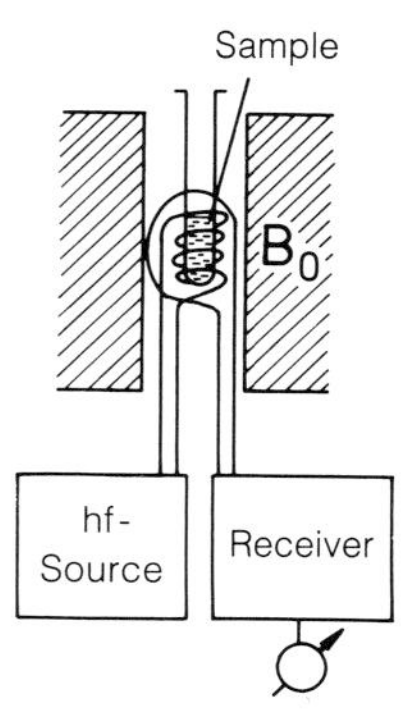

Fig. 18.1. Diagrams illustrating the principles of a nuclear magnetic resonance apparatus which measures the absorbed high-frequency power. The resonance can be detected at a fixed magnetic field and variable hf-frequency, or with a fixed frequency and variable magnetic field. *Upper part*: The impedance of the hf-coil changes at resonance, because energy is transferred from the oscillating hf field to the nuclear spin system. This can be detected by a bridge circuit. *Lower part*: This drawing shows the so-called induction method: at resonance, the precessing nuclear magnetisation produces an alternating magnetic field perpendicular to the direction of the static applied field B_0 and to the applied hf field. Detection is accomplished by means of induction in a receiver coil. Still other methods of NMR detection are described in I, Sect. 20.6

18.1.2 Detection of Nuclear Resonance

The principle of a nuclear resonance experiment in its simplest form is shown in Fig. 18.1. More details are given in I, sect. 20.6. At present, superconducting magnets with $B = 10$ T and higher are available for NMR measurements; still higher fields with correspondingly higher resonance frequencies are extremely interesting for improving the sensitivity of the measurements. The NMR signal is, indeed, due to only the small fraction of the transitions by which the population of the energetically higher state, N_2, differs from that of the lower state, N_1 in thermal equilibrium. At resonance, the electromagnetic radiation (hf field) stimulates transitions in both directions; the nett absorption is due to the excess of absorption transitions over stimulated emission transitions. The difference in the populations, relative to the total number of nuclei present, is given by:

$$\frac{N_1 - N_2}{N_1 + N_2} = \frac{1 - \mathrm{e}^{-g_I \mu_\mathrm{N} B_0 / kT}}{1 + \mathrm{e}^{-g_I \mu_\mathrm{N} B_0 / kT}} \approx g_I \mu_\mathrm{N} B_0 / 2kT \; . \tag{18.9}$$

For protons at room temperature and $B_0 = 1.4$ T, that is [using (18.8)] at $\nu \approx 60$ MHz, this relative population difference is $2.6 \cdot 10^{-6}$. There is thus only a small fraction of spins in excess in the lower state over those in the upper state, and only these provide a detectable absorption signal. The remaining fraction of the nuclei undergoes the same number of transitions in absorption as in stimulated emission under the influence of the radiation field, and thus yields no nett effect for the measurement. An increase in the strength of the applied static magnetic field and thus of the resonance frequency by a factor of 10 therefore increases the detection sensitivity by the same factor.

The actual construction of a nuclear resonance spectrometer is described in I, Chap. 20, in more detail. There, we explain also why most nuclear resonance experiments today are carried out using pulsed-hf methods, in which the signal is then Fourier transformed from the time domain to the frequency domain; this is called Fourier-Transformed NMR (FT-NMR). We shall not repeat this discussion here.

Nuclear magnetic resonance spectroscopy differs from optical spectroscopy in several characteristic ways:

- The energy quanta are very small ($10^{-4} - 10^{-8}$ eV);
- Magnetic dipole transitions and not electric dipole transitions are observed;
- The wavelength of the radiation used is large compared to the dimensions of the sample; this means that all the nuclei in the sample can be excited coherently.

Nuclear magnetic resonance of protons, i.e. employing the nuclei of hydrogen atoms in molecules, is particularly important, especially in view of its application to the investigation of the enormous variety of organic molecules. It can be applied to permit the use of nuclear spins as probes of the structure and bonding in molecules. Two measurable quantities are especially important in this process: the chemical shift and the coupling constants. These quantities will be discussed in the following section, taking proton resonance as an example.

18.2 Proton Resonance in Molecules

18.2.1 The Chemical Shift

The reason for the great importance of nuclear magnetic resonance in molecular physics lies in the fact that one can obtain extremely detailed structural data on molecules with its aid. We illustrate this here using the example of proton resonance in ethyl alcohol, CH_3CH_2OH. In what follows, we shall see that many of the most important properties of nuclear resonance on molecules and the value of the technique in molecular physics can be understood by referring to the ethanol spectrum. Figure 18.2 shows the proton resonance spectrum of this compound. The 3 groups of protons in the molecular functional groups CH_3, CH_2, and OH give rise to three groups of resonance signals with relative intensities (i.e. the areas under the absorption curves) of 3:2:1 at somewhat different resonance frequencies; or, if the measuring frequency is held fixed, at somewhat different magnetic fields. This is due to the *chemical shift*, which refers to the following phenomenon:

The resonance frequency in (18.8) is not, in fact, determined by the applied static magnetic field B_0 alone; instead, the field entering this equation is the local field at the position of the nucleus being studied. It is not equal to B_0, because the application of an external field B_0 to an atom or a molecule induces a current, and thereby a magnetic moment in the electronic shells, which is directed antiparallel to the applied field, according to Lenz's rule. One therefore refers to *diamagnetic shielding*; cf. Fig. 18.3. For the local or effective field which acts on the nucleus, we then have

$$B_{\text{local}} = B_0 - B_{\text{induced}} ,$$

and, since the strength of the induced field is proportional to the applied field,

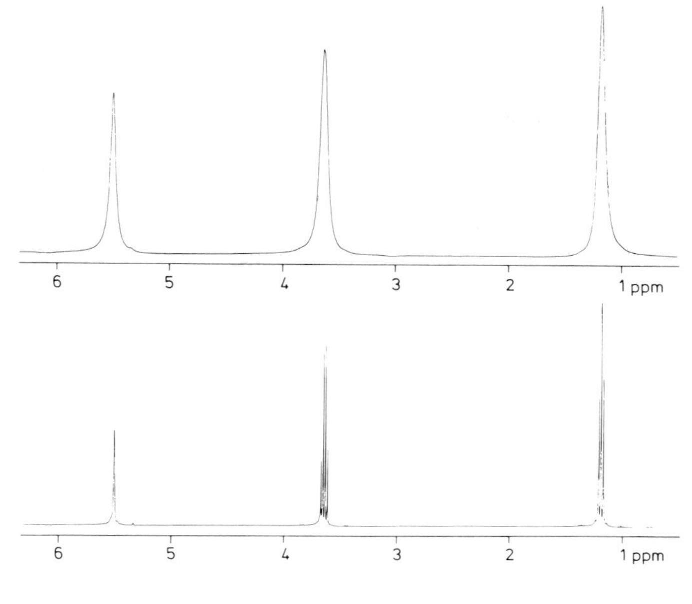

Fig. 18.2. The nuclear magnetic resonance spectrum of ethanol, CH_3CH_2OH. The signal intensity in the receiver is plotted as a function of the transition frequency or of the magnetic field strength at resonance. This is measured in ppm (parts per million) relative to a suitable standard compound. In the *upper spectrum*, one can see three signals having areas with the ratios 1:2:3. They are due to the proton spins in the OH, the CH_2, and the CH_3 groups, with 1, 2, and 3 protons, respectively. Owing to the differences in chemical bonding, the resonance frequencies or the resonance field strengths of the protons in the different groups differ by a few ppm. The lower spectrum is similar, but was measured at a higher resolution. The CH_2 signal is now seen to be split into a quartet of lines due to the indirect spin-spin interaction with the protons of the CH_3 group, while the CH_3 signal is split into a triplet by the indirect interaction with the CH_2 protons. The protons of the OH group are rapidly exchanged between different molecules; the indirect interactions are thereby averaged out and the line remains unsplit

$$B_{\text{local}} = B_0 - \sigma B_0 \ . \tag{18.10}$$

The magnitude of the diamagnetic shielding depends on the density and the bonding state of the electrons in the neighbourhood of the probe nucleus. Protons in different molecular groups have different shielding constants σ as a result of their different electronic environments. The nuclear resonance frequency is then given by

$$\nu = g_I \mu_N B_{\text{local}} = g_I \mu_N B_0 (1 - \sigma) \ ; \tag{18.11}$$

that is, nuclei in differing chemical environments and thus with differering shielding constants σ have different resonance frequencies ν in a given field B_0. Figure 18.4 shows the different resonance fields for the OH protons and the CH_3 protons in methanol, CH_3OH, at a given resonance frequency, as a result of their differing shielding factors. The product σB_0 measures the *chemical shift* and is often denoted by δ. This explains the three groups of lines seen in Fig. 18.2 for the alcohol CH_2CH_3OH, with the intensity ratios 1:2:3 corresponding to the numbers of protons in the OH, CH_2, and CH_3 groups: these groups of lines represent the resonance signals of the protons in the molecular groups, and the areas of the absorption lines are proportional to the numbers of nuclei taking part in each resonance.

The significance of the chemical shift for understanding chemical bonding can be made clear by considering the difference between the C–H and the O–H bonds. The O atom is a better electron acceptor than the C atom; therefore, the electron density on the H atom in an O–H bond is smaller than that in a C–H bond. This leads to a larger diamagnetic shielding for the CH protons, and thus to a larger chemical shift. For the local fields, we find the relation:

$$B_{\text{CH}} = B_0(1 - \sigma_{\text{CH}}) < B_{\text{OH}} = B_0(1 - \sigma_{\text{OH}}) \ .$$

At a fixed frequency of the hf field B_1, the resonance condition for the CH protons therefore occurs at a somewhat larger applied static field B_0 than that of the OH protons; cf. Fig. 18.4.

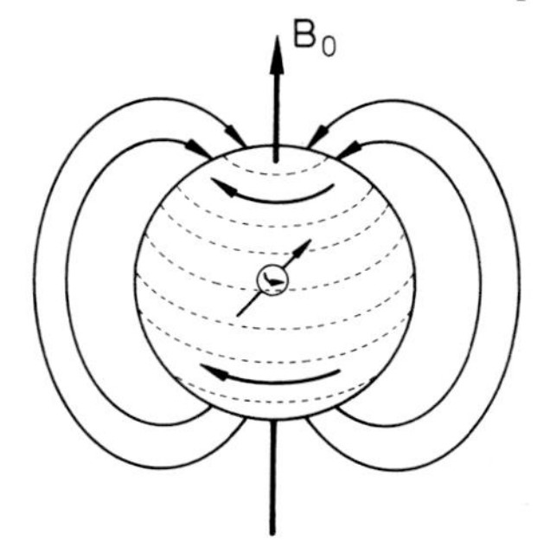

Fig. 18.3. Diamagnetic shielding as the source of the chemical shift: the diamagnetic electron clouds in the neighbourhood of the nucleus produce a magnetic field at the nucleus which is oppositely directed to the applied field B_0

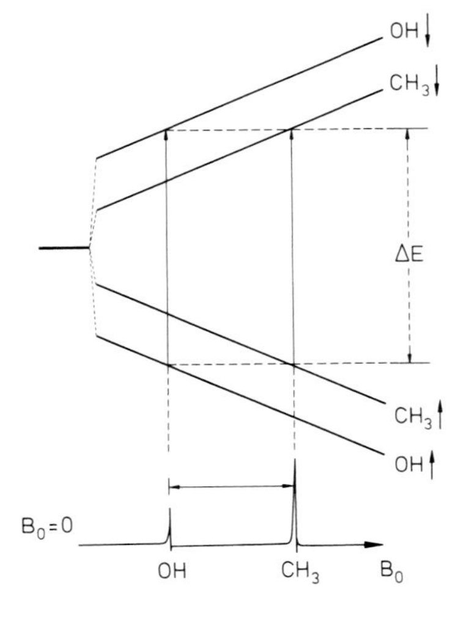

Fig. 18.4. The energy levels of the protons of the methyl group and of the OH group of methyl alcohol, CH_3OH, in an applied field B_0. The different chemical shifts of these groups have the result that the NMR signals from the protons in the two groups appear at differing values of B_0. At a fixed frequency of 100 MHz, the resonant field strength is about 2.35 T, and the interval between the signals from OH and CH_3 protons is about 3.2 μT; after Banwell. The shielding constant for protons in the OH group is smaller than for protons in the CH_3 group, so that the resonance signal from the OH protons appears at a lower applied field strength than that of the CH_3 protons

A diamagnetic shielding and resulting chemical shift can be produced not only by the binding electrons in the immediate neighbourhood of the probe nucleus, as in our example, but can be caused also by other electrons near the nucleus. The induced magnetic field at the position of the nucleus can be either antiparallel or parallel to the applied field B_0 and can thus amplify the latter. Some examples are shown in Fig. 18.5. In the acetylene molecule, the protons are shielded by the electrons of the triple C≡C bonds; in benzene, the delocalized π-electrons in the plane of the ring induce a field at the positions of the protons which is parallel to the applied field B_0, and thus effectively increases it.

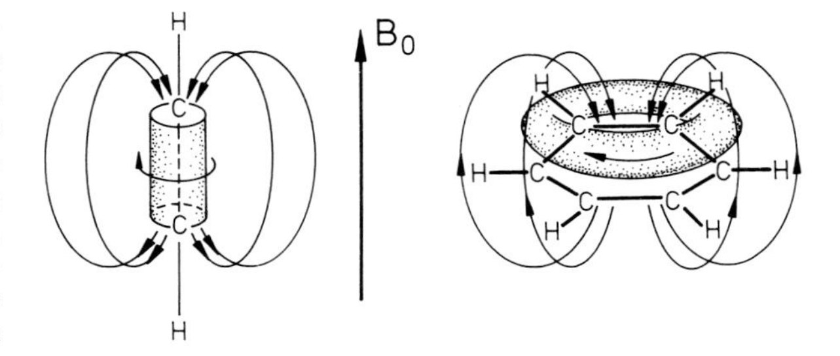

Fig. 18.5. The principle of diamagnetic shielding. The protons in the acetylene molecule (*left*) experience a reduction of the applied field B_0 due to the field-induced electronic currents. In the benzene molecule, C_6H_6 (*right*), these can lead in contrast to an additional magnetic field at the proton sites which *adds* to the applied field B_0. In this case, B_0 is not shielded but is instead amplified. For clarity only the upper ring of π-electrons is shown; below the plane of the molecule, a second indentical ring is present, with its currents in the same sense as those of the upper ring

Although the shielding constant σ is a quantity which is a property of the molecule and does not depend on the applied field, the shift itself is proportional to the field. This is an additional reason for choosing high magnetic fields in NMR: the spectral resolution is increased, by increasing the spacing between resonance lines from different molecular subgroups with differing chemical shifts, when a greater magnetic field strength is used.

According to its definition, the chemical shift is referred to the chemical shift of a completely free, nonshielded proton without an electronic environment as standard; in this case, σ would be zero. Since, however, it would be experimentally difficult to carry out measurements with such a standard, one uses instead a compound with narrow, reproducible NMR lines as standard for the chemical shift. The shift is defined as the ratio of the frequency displacement $\Delta\nu$ of a proton group, relative to the resonance frequency ν_0 of the standard

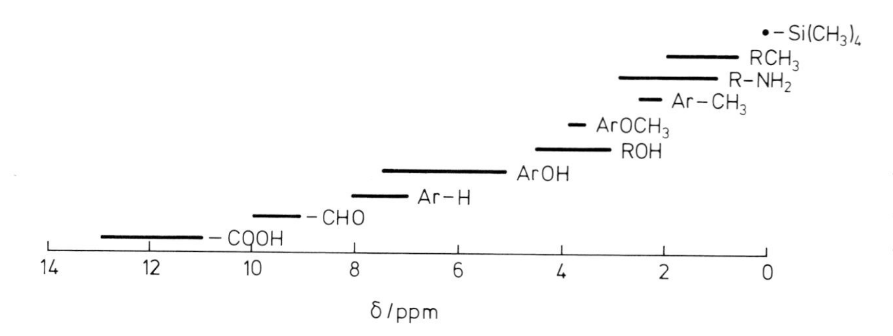

Fig. 18.6. Ranges of the values of the chemical shift for the proton resonance in the case of different bond types. In this figure, *Ar* stands for aromatic and *R* for non-aromatic subgroups. The shifts δ are quoted relative to $Si(CH_3)_4$, TMS, for which $\delta = 0$ is assumed

compound; this very small quantity is measured in ppm (i.e. units of 10^{-6}). Usually, a fixed hf frequency is employed; then the resonance field strength and its shift are measured. Denoting the resonance field of the probe nuclei in the sample by B_{probe} and that of the same type of nuclei in the standard compound as B_{standard}, we define δ by:

$$\delta = \frac{B_{\text{standard}} - B_{\text{probe}}}{B_{\text{standard}}} \cdot 10^6 \text{ [ppm]} . \tag{18.12}$$

According to (18.12), δ is a dimensionless number. The usual standard is tetramethyl silane, $Si(CH_3)_4$ (TMS), a compound which is readily soluble in many solvents (not water, however) and has 12 protons in 4 equivalent groups. The absolute magnitude of σ is of lesser interest; it can, to be sure, be calculated if one has sufficient knowledge of the electron density distribution around the probe nucleus. There are, however, hardly any exact calculations to be found in the literature.

More important for purposes of analysis and for the determination of molecular structures are relative measurements which can be compared to a standard compound, since the values of the shifts are specific for various groups and bond types and are known, at least empirically. When protons having a particular chemical shift are observed in a sample of unknown composition, then some conclusions can be drawn from the measured shifts about the bonds or groups which are present in the sample. Figure 18.6 shows the values of the chemical shift for some groups which are of importance in organic chemistry. Other compounds as well, such as NH_3, SiH_4, or H_2, also have their characteristic chemical shifts δ.

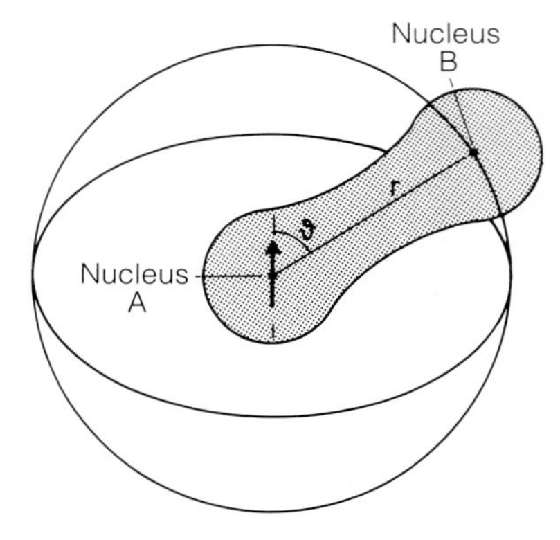

Fig. 18.7. Two nuclei A and B in a spherical polar coordinate system, defining the angles for the magnetic dipole-dipole interaction. The magnetic field which is produced at the position of nucleus B by the magnetic moment of nucleus A depends on the angle ϑ between the line joining the two nuclei and the applied magnetic field B_0 [see (18.13)]. If the molecule is rotating, ϑ passes through all possible values and the time-averaged orientation factor $(1 - 3\cos^2\vartheta)$ is equal to 0. This interaction is then averaged away

18.2.2 Fine Structure and the Direct Nuclear Spin-Spin Coupling

The additional splittings which can be seen in the chemically-shifted lines in the spectrum of ethanol (lower spectrum, Fig. 18.2) are called fine structure –not to be confused with the "fine structure" in the optical spectra of atoms. In the case of atomic spectra, the fine structure is caused by the magnetic interactions of the electronic spins with the orbital motions; here, in NMR spectra, the cause is the magnetic interaction between the nuclear magnetic moments. This can be either a direct dipole-dipole interaction between the moments of two nuclei, or an indirect interaction which arises through the polarisation of the electronic shells around the nuclei. The interaction is in both cases referred to simply as the spin-spin coupling.

The *direct magnetic dipole-dipole interaction* between the moments of two nuclei A and B can be readily calculated. The field B_A which nucleus A with an orientation corresponding to a quantum number m_I produces at the position of nucleus B at a distance r depends on the direction ϑ of the line joining the two nuclei relative to the applied field $\boldsymbol{B}_0$ (see Fig. 18.7). It has the magnitude

$$B_A = -\frac{\mu_0}{4\pi} g_I \mu_N m_I \left(\frac{1}{r^3}\right)(1 - 3\cos^2\vartheta) . \tag{18.13}$$

If we assume that nucleus A is a proton with spin 1/2, then there are two possible orientations, $m_I = \pm 1/2$, i.e. α (in the direction of the field) and β (opposing the field). Nucleus B therefore experiences an effective field of $B_0 \pm B_A$, and it thus has two resonance fields with a spacing of $2B_A$; the resonance signal becomes a doublet. The same is true in reverse for the resonance signal of nucleus A. The energy spacing or the frequency spacing of the two components of the doublet is denoted by J, and this symbol also defines the spin-spin coupling constant, which is normally quoted in frequency units, i.e. Hz.

Inserting the numerical values for typical nuclear moments, we obtain values for B_A of order 10^{-4} T at an internuclear distance of 0.2 nm. Such large spin-spin interactions J are indeed observed in the solid phase. In solid samples, one can even use the direct spin-spin interaction to determine the internuclear distances according to (18.13).

In the liquid phase, however, the angle ϑ between the interacting nuclei and $\boldsymbol{B}_0$ usually varies rapidly relative to the nuclear resonance frequency or Larmor frequency. This causes an averaging of the function $(1 - 3\cos^2\vartheta)$ to zero, and as a result the field B_{nucl} at the position of another nucleus vanishes when averaged over the precession period of the nuclei; the direct dipole-dipole interaction is thus averaged out. The spatial average over $\cos^2\vartheta$, as we saw in Chap. 16, gives just the value 1/3. This may be different in the case of molecules, however, which may move sufficiently slowly in solution that the interaction is not averaged away; this can be true for example of large, biologically active molecules.

18.2.3 Fine Structure and the Indirect Nuclear Spin-Spin Coupling Between Two Nuclei

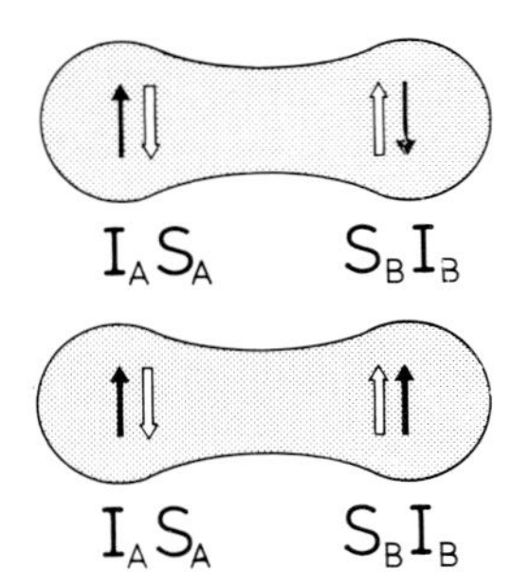

Fig. 18.8. The indirect magnetic spin-spin coupling, for example in the H_2 molecule or the ^{13}C–^{1}H bond. The proton with spin I_A polarizes the spin S_A of a binding electron in its neighbourhood. According to the Pauli principle, the spin orientation of the second binding electron S_B is then antiparallel to S_A, and thus the orientation of the second nuclear spin, I_B, is energetically favoured when it is antiparallel to I_A (*Upper sketch*) relative to the parallel orientation (*Lower sketch*). The energy spacing between the two orientations is measured as the coupling constant J, which is positive in the present case

On the other hand, the *indirect magnetic spin-spin coupling* of the nuclei, which is mediated by the binding electrons between them, is not averaged away by motions. The reason for this is the *Fermi contact interaction* between electrons and nuclei, which we treated in I, Sect. 20.3. It produces an indirect spin-spin coupling of the nuclei which is 10^2 to 10^4 times smaller than the direct interaction, and is isotropic.

This coupling can be most simply understood for a system of two nuclei with spins of 1/2 and two binding electrons with paired spins. Figure 18.8 shows the spin arrangement. The two nuclei A and B with spins I_A and I_B may be different nuclides, for example ^{13}C and ^{1}H in a CH- group, or they may be identical nuclei with different chemical shifts, so that they would give rise to two resonance lines in the NMR spectrum without coupling; see Fig. 18.9. We can take as an example the ^{13}C–^{1}H bond. Without spin-spin coupling, one resonance line is observed for the proton and another for the ^{13}C nucleus, at a quite different resonance field strength. The binding electron S_A which is nearest to the ^{13}C nucleus will have its spin antiparallel to that of the nucleus in the most favourable energy state. Then, following the Pauli exclusion principle, the spin of the other binding electron, S_B, is fixed in the direction antiparallel to S_A, and one of the two possible spin orientations of the proton spin I_B is energetically more favourable, namely the orientation antiparallel to S_B and thus also antiparallel to the spin I_A of the ^{13}C nucleus. The energy spacing between the two orientations of the nuclear spins is quoted in terms of the spin-spin coupling constant J in Hz. Without the coupling, each nucleus, ^{13}C and ^{1}H, exhibits a single NMR line; these single lines split in the presence of the spin-spin coupling mediated by the binding electrons into a doublet with a spacing of J between its two components, according to the scheme explained in Fig. 18.8. In contrast to the chemical shift, this splitting, which arises from an intramolecular magnetic field, is independent of the applied static field strength B_0. This indirect spin-spin coupling is also found in the H_2 molecule, where the spins of the two protons are coupled via the two binding electrons.

However, one should note here that the relative orientation of the spins I_A and I_B is not fixed, e.g. in the form ↑↓, by the indirect spin-spin coupling. The coupling merely has the effect that the two orientations of the nuclear spins which are sketched in Fig. 18.8, antiparallel, ↑↓, and parallel, ↑↑, differ in energy by a very small amount. Both configurations are thus found with equal probabilities in the molecule, but there is a small energy splitting between them and thus the possibility of observing the indirect spin-spin coupling.

If, as is the case in the CH_2-group in ethanol, two protons are bound to the same C atom, then there is likewise a coupling of the two proton spins via the C atom; however, it plays only an indirect role through the electrons. This is explained in Fig. 18.10: one proton tends to orient the spin of the binding electron which is closest to it in the C–H bond in such a way that their spins are antiparallel. The second electron in the C–H bond, which is closer to the C atom, has its spin antiparallel to the first. This has the effect that the electron near the C atom in the second C–H bond tends to align its spin parallel to that of the first electron near the C atom, as is required by Hund's rule: it states that the parallel spin orientation for two otherwise equivalent electrons on the same atom is energetically more favourable. This then gives a favoured orientation for the spin of the second proton, as can be seen in Fig. 18.10. Overall, this indirect spin-spin coupling leads to an energetically favoured parallel orientation of the two proton spins relative to the antiparallel orientation. The coupling constant J is therefore negative for this spin-spin interaction. The remarks about the probability of different spin orientations in the previous paragraph apply here as well.

The interaction between two protons that is mediated by binding electrons can also be measurable over several intervening bonds. Thus, for the proton-proton coupling in the configuration H–C–H, a coupling constant of $J = -10$ to -15 Hz is found, but for H–C–C–H, it is $J = +5$ to $+8$ Hz. In the configuration H–C–C–C–H, i.e. over three intervening C atoms, the proton spin-spin coupling becomes unmeasurably small.

Some numerical values for the coupling constants J are given in Table 18.2.

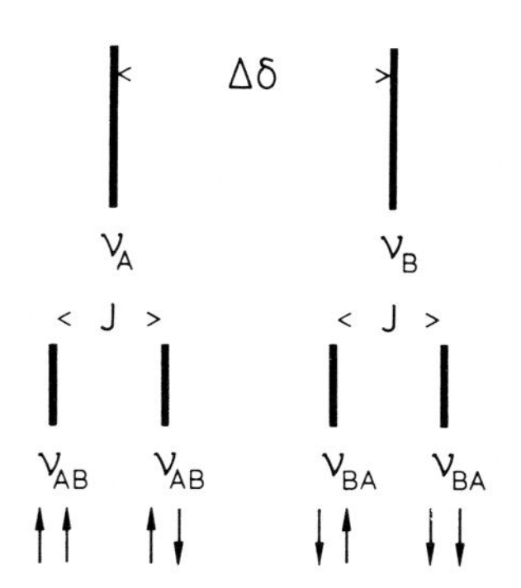

Fig. 18.9. The magnetic coupling between two nuclei A and B: their resonance lines are shifted relative to one another by the chemical shift difference $\Delta\delta$ (*Upper drawing*). The coupling then splits each resonance line into two components. From the magnitude of the splitting, one can obtain the coupling constant J. Here, the case of $J < \delta$ is shown

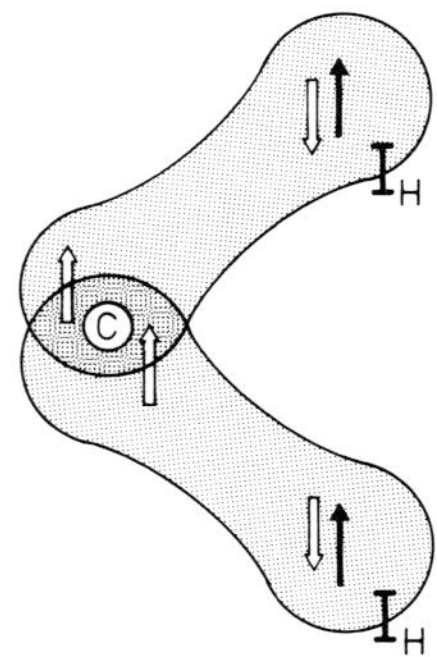

Fig. 18.10. The spin-spin coupling between two protons bound to the same C atom, which has a negative coupling constant; for example, in a CH_2 group. The parallel orientation of the two proton spins I_H is energetically more favoured than the antiparallel orientation. This coupling is mediated by the binding electrons; the nucleus of the C atom is not directly involved. According to Hund's rule, a parallel spin orientation for the two electrons as indicated on the C atom is more favourable than the antiparallel orientation. This explains the reversal of the nuclear spin coupling in comparison with Fig. 18.8

Table 18.2. Coupling constants J in Hz for some nuclei in various types of bonds or different molecules

Nuclei	Molecule	$J\,[s^{-1}]$
H–H	H_2	276
H–C	CH_4	125
H–O	H_2O	73
H–Si	SiH_4	−202
C–C	CH_3CH_3	35
C–F	CF_4	−259
H...H	H–C–H	−(10 − 15)
	H–C–C–H	5 − 8
	H–C–C–C–H	≈ 0

18.2.4 The Indirect Spin-Spin Interaction Among Several Nuclei

We have thus far assumed that the two coupled nuclei A and B have different chemical shifts whose difference is large compared to the coupling constant J. The spectrum of two coupled nuclei A and B with spins 1/2, which we have so far assumed for simplicity both to be protons, thus consists of two doublets with the same intensities, as indicated in Fig. 18.9; the spacing of the two component lines in each doublet corresponds to the coupling constant J, and the centres of the doublets are separated by the difference of the chemical shifts δ. If the difference in the chemical shifts (typically a few hundred Hz for H in different types of bonds) is no longer large compared to the coupling constant (typically a few Hz), then a superposition of the perturbing fields can occur, resulting in (especially) differing intensities, but also in differing splittings of the component lines relative to the unperturbed lines. Finally, if δ approaches 0, the two innermost lines of the doublets meet and form a single line, while the outermost lines vanish. In spite of the coupling, the spectrum exhibits only single lines. Such nuclei are termed *equivalent*.

The behaviour of equivalent nuclei in NMR is especially simple. One refers to chemical equivalence when the same type of nuclei have the same chemical shifts, while magnetic equivalence means that the angular dependence of the energy terms in an applied field due to the direct spin-spin interaction is the same for both nuclei. The two protons in the CH_2 group, or the three protons in the CH_3 group of CH_3CH_2OH (Fig. 18.2) are each chemically equivalent. Magnetic equivalence is found when the groups are sufficiently free to rotate, e.g. in solution, so that the anisotropy can be averaged out.

Here, at least for only moderate spectral resolution, one observes only one line for each of the two proton groups. An isolated CH_3 group, or an isolated CH_2 group, thus each has only one proton resonance line. The coupling within a group of magnetic and chemically equivalent nuclei has no influence on the spectrum and therefore does not lead to a splitting as described above (Sect. 18.2.3). In this case, there is a rapid exchange between the equivalent nuclei by means of so-called spin diffusion or flip-flop processes, in which only the spins are exchanged, not the particles themselves. This causes the splitting, which is in principle present, to be averaged out and to become unobservable. An isolated CH_3 or CH_2 group thus gives rise to only one resonance line each.

The situation is different when there are several mutually inequivalent groups within the molecule, e.g. in ethanol, the mutually equivalent CH_2 protons on the one hand and the mutually equivalent CH_3 protons on the other. These groups are then no longer isolated, but rather can interact with each other. The protons in the CH_2 groups differ, as shown above, from those in the CH_3 groups by their differing chemical shifts; the two groups of protons are thus not equivalent to one another. The resonance of the CH_3 protons in Fig. 18.2 is split into three lines with intensity ratios 1:2:1 by the spin-spin coupling with the CH_2 protons. The coupling with *one* CH_2 proton namely splits the line from the CH_3 protons into two component lines, and the coupling with the *second* CH_2 proton causes another splitting with the same coupling constant, so that the centre lines fall together. This produces the splitting pattern of the CH_3 protons; cf. Fig. 18.11. One can summarize this as follows: the spin configurations of the two CH_2 protons in their interaction with the CH_3 protons can be $\uparrow\uparrow$, $\uparrow\downarrow$, $\downarrow\uparrow$, $\downarrow\downarrow$. The two middle configurations are energetically equivalent.

Now, in order to understand the splitting of the line from the CH_2 protons, we must, in turn, investigate the possible spin configurations of the 3 protons in the CH_3 group. Here, there are 8 possibilities, of which only 4 are energetically distinct: $\uparrow\uparrow\uparrow$; $\uparrow\uparrow\downarrow$; $\uparrow\downarrow\uparrow$; $\downarrow\uparrow\uparrow$; $\uparrow\downarrow\downarrow$; $\downarrow\uparrow\downarrow$; $\downarrow\downarrow\uparrow$; $\downarrow\downarrow\downarrow$ This explains the intensity ratio of 1:3:3:1 for the four component

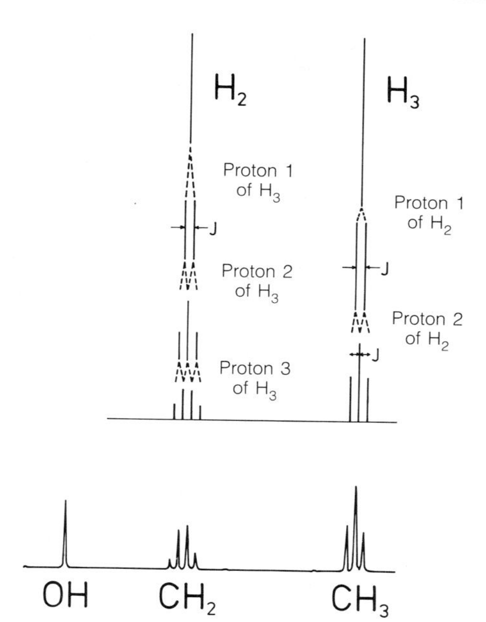

Fig. 18.11. The splitting pattern for the interactions with several equivalent nuclei. The nuclear resonance signal of the two equivalent protons in the CH_2 group is split into four component lines by the three CH_3 protons; conversely, the CH_3 signal is split into three lines by the two CH_2 protons. The reason why the OH signal is not split and also does not contribute to the other splittings is explained in the text

lines in the CH_2 proton resonance in Fig. 18.2, as illustrated in the diagram of Fig. 18.11. In general, it is found that N equivalent protons split the resonance line of a neighbouring group into $N + 1$ component lines. The intensity ratios are given by the number of ways of producing each of the energetically different configurations, i.e. the binomial coefficients, which can be represented by using Pascal's triangle, Fig. 18.12. This is an arrangement of numbers in which the entries in a particular row are obtained by adding their two nearest neighbour numbers in the preceding row.

N	Pascal's triangle
0	1
1	1 1
2	1 2 1
3	1 3 3 1
4	1 4 6 4 1
5	1 5 10 10 5 1
6	1 6 15 20 15 6 1

Fig. 18.12. Pascal's triangle: N equivalent protons split the resonance line of a neighbouring group into $N + 1$ lines; their intensity ratios can be read off the triangle. The numbers in a particular row are found by adding the two neighbouring entries in the row above

Finally, the resonance of the OH-group protons in Figs. 18.2 and 18.11 is still lacking. We should expect from the above discussion that this proton would give rise to a doublet splitting of the lines of the CH_2 and CH_3 protons, due to its two possible spin orientations, and that its own resonance line would be split into a triplet by the interaction with the CH_2 protons, while the three components would again be split into quartets by the CH_3 protons. These splittings are in general not observable, as can be seen in Fig. 18.2, because ethanol usually contains a small amount of water admixture.

The vanishing of the expected splittings is due to a chemical *exchange* of the hydroxyl protons with the protons of water. When the protons are exchanged between different molecules, the lifetime of a given spin configuration is limited by the exchange rate. The protons donated by water molecules, which undergo a rapid chemical exchange with the OH protons of the ethanol, have statistically distributed spin orientations and therefore shorten the lifetime of a particular OH–proton spin orientation. Since the fine structure splittings discussed above have magnitudes of the order of $J = 1$ Hz, an exchange frequency of $J/2\pi \approx 0.1\ \mathrm{s}^{-1}$ is already sufficient to make the splitting $\delta\nu$ unobservable by lifetime broadening, according to the relation $I_{\mathrm{exch}} < 1/2\pi\delta\nu$. Here, by exchange we are referring to chemical exchange. Only when a proton remains on the molecule without exchange for a time longer than $\tau = 1/2\pi\delta\nu$ can it produce a splitting of the lines of other proton groups in the molecule. If water is used as the solvent for ethanol, the exchange occurs much more rapidly; only in extremely anhydrous ethanol does the exchange become sufficiently slow that the splitting of the proton resonance lines from the OH group and the splitting of the lines of the other groups in the spectrum of Fig. 18.2 by the OH protons become observable. The occurrence or nonoccurrence of line splittings, or, more generally, of line broadening, can thus contribute to knowledge of a proton exchange rate. This is at the same time an example of the fact that NMR is also useful for the observation of dynamic processes and the determination of their rates.

While the chemical shift is proportional to the applied magnetic field strength B_0, as shown in Sect. 18.2.1, the fine structure which results from the direct or indirect dipole-dipole interactions of the nuclear moments is independent of the applied field. This simplifies the analysis of NMR spectra, by allowing a comparison of spectra measured at different field strengths to distinguish the two effects. It is still another reason for the desirability of high magnetic field strengths in NMR spectroscopy. The various mechanisms which are responsible for the line frequencies in a high-resolution spectrum can thus be more easily separated and analysed by carrying out the measurements at several fields B_0 and thus in several different frequency ranges. However, the chemical shifts and the spin-spin couplings are often of the same order of magnitude; then the analysis becomes more difficult, as indicated above.

18.3 Dynamic Processes and Relaxation Times

In the preceding sections, we have seen that the linewidths in nuclear resonance spectra can yield information about the rates of processes which take place in the molecules under investigation or in the sample. As we have already mentioned, the resolvable structure of an NMR spectrum depends on whether and how rapidly the molecules are moving. If, for example, we place a system of similar molecules in a disordered fashion into a matrix, then the anisotropic direct dipole-dipole interaction will depend on the orientation and distance of each molecule relative to its neighbours, as seen from (18.12), and the resonance lines will have a corresponding inhomogeneous linewidth, which can be as much as 1 mT for small intermolecular distances. The smaller, isotropic indirect spin-spin interaction is in general hidden within this linewidth and therefore not observable. If we now transform the rigid matrix into a liquid and thereby give the molecules the freedom to move and change their orientations rapidly, then the resonance frequency of each molecule will become time-dependent through the anisotropic spin-spin interaction. This can lead to an averaging out of the interaction and thus to a narrowing of the spectral lines.

This effect can be understood as follows: when the exchange frequency ν is small compared to the splitting $\delta\nu$, expressed as a frequency, then the splitting corresponds to two separated states, A and B. If ν is of the same order of magnitude as $\delta\nu$, the lifetimes of the separated states become shorter and the linewidth correspondingly greater (lifetime broadening of A and B). If, however, ν becomes large compared to $\delta\nu$, the two previously separate states are now no longer distinguishable; instead, one obtains a new state AB, in which the different fields or frequencies are averaged out due to the rapidity of the exchange. The resonance line appears narrow and unsplit, as in a homogeneous field. Depending on the origin of the rapid exchange, this phenomenon is called *exchange narrowing* or *motional narrowing*. This is explained more fully in Fig. 18.13 with the aid of a simulated example.

The time-energy uncertainty holds in classical physics, also; in order to distinguish a resonance frequency $\nu + \Delta\nu$ from a frequency ν, the number n of oscillations measured in the time τ must differ by at least 1 for the two frequencies, i.e. τ must be at least long enough that

$$n = \tau\nu \quad \text{and} \quad n + 1 = \tau(\nu + \Delta\nu), \tag{18.14}$$

that is,

$$\tau = \frac{1}{\Delta\nu} \,. \tag{18.15}$$

If now the resonance frequency changes in times shorter than $1/\Delta\nu$, that is if the molecule moves more rapidly, then the variation of the resonance frequency can no longer be detected; instead, one measures a time-averaged intermediate frequency. The broad resonance line becomes narrow. This is referred to as *motional narrowing*.

For protons, at $\Delta B = 1$ mT, we find

$$\Delta\nu = \frac{\gamma}{2\pi}\Delta B \approx 4.3 \cdot 10^4 \ \mathrm{s}^{-1} \tag{18.16a}$$

and

$$(\Delta\nu)^{-1} \approx 2.5 \cdot 10^{-5}\mathrm{s} \,. \tag{18.16b}$$

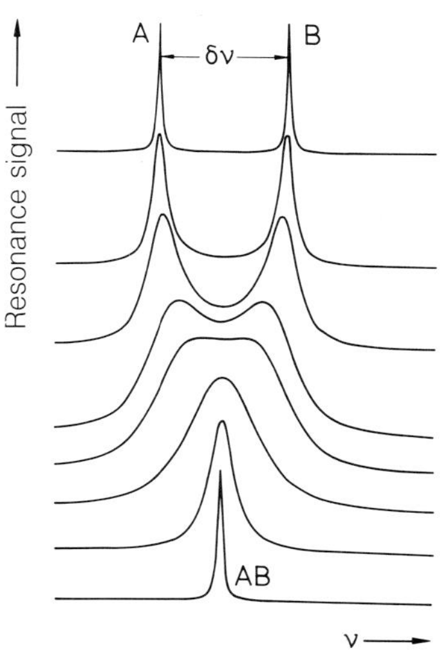

Fig. 18.13. The exchange narrowing of magnetic resonance lines: two nuclei, A and B, differ in their resonance frequencies by $\delta\nu$. The *uppermost curve* shows the resonance spectrum without exchange (schematically). The exchange frequency increases in going from the *upper* to the *lower curves*. The lines are at first broadened due to the shortening of the lifetimes of the states, then they merge together and finally, for very rapid exchange, there is only a single sharp line in the centre between the two original lines (from *top* to *bottom*). We quote the numerical values as an example: $\delta\nu = 30$ Hz, the exchange rate $A \rightleftharpoons B = 1\mathrm{s}^{-1}$ for the *uppermost curve*, 10^2 s^{-1} for the *middle curve*, and 10^4 s^{-1} for the *bottom curve*

The reorientation times for molecules in low-viscosity liquids are much shorter, typically 10^{-10} s. The anisotropic spin-spin coupling is thus averaged out, and the NMR spectrum consists of sharp lines as shown in the lower spectrum in Fig. 18.2.

In a similar manner, an anisotropic interaction can be averaged away by a rapid exchange of nuclei between different molecules, e.g. of protons between water and the OH-groups of alcohols as in Sect. 18.2.3. If this exchange takes place quickly enough, i.e. more rapidly than would correspond to the frequency range which it is averaged over, then with increasing exchange rate it at first leads to lifetime broadening, and then, at a sufficiently rapid rate, to an *exchange narrowing* of the signals. The velocities of configurational changes in molecules, that is of the rearrangement of groups within the molecule, can be determined in an analogous way, when they lead to an averaging out of the different chemical shifts of the differing configurations. Figure 18.13 illustrates exchange broadening and narrowing in an NMR spectrum as a function of the exchange rate.

In general, the linewidths of NMR signals open up the possibility of studying motional or exchange processes in molecules. Such processes are also the origin of many temperature dependencies in the measured resonance spectra. In many cases, for example, the averaging out of anisotropies through motions is not possible at low temperatures, but becomes so with increasing temperature and thus leads to a narrowing of the resonance lines in the spectrum.

In solids, where motional narrowing is not possible at all due to the rigid lattice structure, a dynamic averaging-out of the anisotropic interactions can be produced by *multiple resonance* and *pulse methods*. For example, using multiple resonance techniques with two or more differing frequencies or resonance fields, the spins of one type of nuclei can be selectively resonated and caused to flip rapidly between their allowed orientations. This allows them to be decoupled from a second type of nuclei. With special pulse sequences, one can therefore supress the anisotropic interactions with other nuclei and the line broadening which they cause.

This is the basis of the high-resolution NMR technique known as *spin decoupling*. With its use, for example, the coupling of the CH_3 protons to the CH_2 protons in the spectrum of ethanol (cf. Figs. 18.2 and 18.11) can be removed by applying an intense second radiofrequency signal which causes one of the groups of protons to flip its spins so rapidly that the other group experiences only its average orientation. The quartet or the triplet of lines in Fig. 18.11 then collapses to a single line. Such spin-decoupling methods in many variations find application in high-resolution NMR spectroscopy.

Linewidths in resonance spectra can frequently be described in terms of the two *relaxation times* T_1 and T_2. These times were introduced and defined in I, Sect. 14.5. The time T_1 measures the length of time in which an excited spin state returns to thermal equilibrium through interactions with the environment (production of heat) or with other spins. The time T_2 is a measure of the rate of loss of phase coherence among the spins within a particular spin system. Relaxation times can be determined quite directly using pulse methods, such as the spin-echo method described in I, by following the time development of the nuclear magnetisation after a brief disturbance of the thermal equilibrium magnetisation.

Longitudinal or spin-lattice relaxation, with the time constant T_1, is produced in the main by fluctuations of neighbouring molecules or of paramagnetic impurities which generate fluctuating magnetic fields at the positions of the relaxing nuclei. It measures the spin-flip rate of the nuclei relative to the quantisation axis z determined by B_0, and with it, the transfer of energy to the nuclear environment. Typical T_1 times in NMR are in the range between 10^{-4} and 10 s in liquids and between 10^{-2} and 10^3 s in solids.

Transverse or spin-spin relaxation, with time constant T_2, is a measure of the change in the phase relation between the spins of otherwise equivalent nuclei in the xy-plane perpendicular to the direction of the static field, z. This can be caused by somewhat different local fields acting at different nuclear sites, or else a phase exchange can be produced by the spin-spin interaction of equivalent nuclei. Both processes contribute to the effective relaxation time T_2'. Since the longitudinal relaxation naturally also destroys the phase relation, the overall transverse relaxation rate of a single type of nuclei with spin 1/2 is given by:

$$\frac{1}{T_2} = \frac{1}{T_2'} + \frac{1}{2T_1} \,. \tag{18.17}$$

Typical values of T_2 in solids lie in the range of 10^{-4} s. In liquids, T_2 has similar values to T_1; see above.

If the sample is irradiated with such a high radiofrequency power level that the population difference between the resonant levels given by the thermal energy kT can no longer be maintained by T_1-relaxation processes, the signal intensity no longer increases proportionally to the rf power, and the resonance lines are broadened. Except in the case of this saturation broadening, the linewidth is determined by T_2, as long as no additional mechanisms contribute to an inhomogeneous line broadening. The uncertainty relation for the energy width of a resonance line limited by T_2 gives $\Delta E = h/T_2$. From this it follows that the frequency width of a homogeneous line is given by

$$\delta\nu = \frac{1}{2\pi T_2} \,. \tag{18.18}$$

When T_2 has a value of 1 s, the linewidth is thus of the order of 0.1 s^{-1}.

Measurement of relaxation times opens up a broad field for the investigation of dynamic processes of spins and thus of the molecules to which they belong.

18.4 Nuclear Resonance with Other Nuclei

Of course, nuclear magnetic resonance in molecules is not limited to protons, about which we have mainly spoken thus far. All nuclei with non-vanishing nuclear spins can in principle be used for nuclear magnetic resonance spectroscopy. The same interaction mechanisms apply here as to protons, i.e. the chemical shifts and spin-spin coupling.

Both quantities are, to be sure, usually considerably larger for other nuclei and thus more readily measurable than for protons. This is due to the fact that in larger atoms, there are more electrons and they are spread over a larger volume than in hydrogen. The possibilities for diamagnetic shielding and spin polarisation are thus increased.

For the investigation of organic molecules, the nuclide ^{13}C (natural abundance 1 %) is of importance; for biologically relevant molecules, ^{31}P (abundance 100 %) is significant.

Other nuclei with $I > 1/2$, such as ^{14}N ($I = 1$), or Cl and Br ($I = 3/2$) split the resonance lines of neighbouring nuclei with which they have spin-spin interactions into $(2I+1)$ components, since these nuclei have $(2I+1)$ possible spin orientations in an applied field.

In addition, nuclei with $I > 1$ have an *electric quadrupole moment*. This moment is subjected to an orienting interaction in an electric field gradient; such field gradients occur in molecules, owing to the directed chemical bonds. The orienting tendency of the field gradient acts in competition with the magnetic orientation of the nuclear spins in the applied field and leads to an additional relaxation mechanism, shorter relaxation times, and thus to line broadening in the spectrum.

If the nuclear quadrupole moment is known, the electric field gradients in molecules and solids can be determined by means of *Nuclear Quadrupole Resonance* (NQR); see also I, Sect. 20.8.

18.5 Two-Dimensional Nuclear Resonance

Modern nuclear resonance spectroscopy consists almost exclusively of Fourier transformed nuclear resonance with computer-aided data collection and data treatment, as described in I, Sect. 20.6. This is carried out as follows: a series of resonant hf pulses of a certain length, a certain delay time and a certain intensity is applied to the sample and produces a well-defined initial state, with the nuclear spin system having a particular value of its magnetisation (in z-direction) and its motion (in the xy-plane); thereafter, the time evolution of the state is observed as an induction signal in the receiver. This signal as a function of time is Fourier transformed to give the corresponding frequency distribution; the time dependence contains in fact all of the spectral information. One thus obtains data about the energy levels of the sample nuclei and their dynamic behaviour.

In the case of one-dimensional nuclear resonance, as described in the previous sections, the resonance signal is observed as a function of *one* hf frequency which is applied to the sample. Information such as the chemical shifts and the spin-spin coupling is then contained in the spectrum. In recent years, in particular through the work of R.R. *Ernst*, an enormous methodological development in the direction of two- or multidimensional nuclear resonance has taken place. This branch of NMR spectroscopy, which is in a period of very rapid development and has become possible only through the use of modern, powerful electronic computers, deserves at least a mention here.

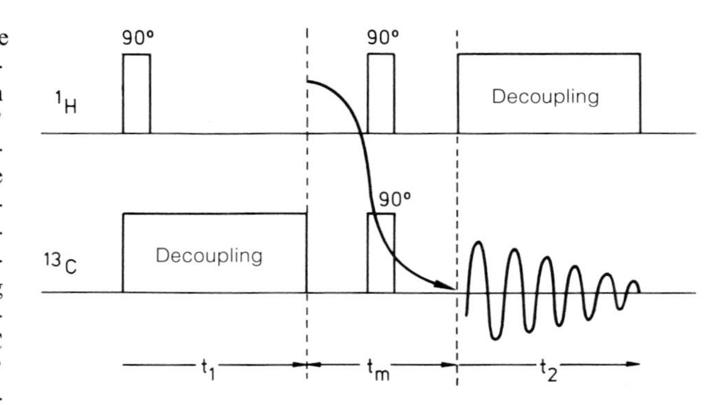

Fig. 18.14. A sequencing scheme for heteronuclear two-dimensional NMR spectroscopy with spin decoupling. The first 90° pulse flips the proton magnetisation into the xy-plane, where it precesses about the z-axis during the time t_1 with the frequencies characteristic of the various groups of protons. During the mixing time (t_m), the magnetisation of the ^1H and the ^{13}C nuclei is transferred by two 90° pulses to the system of ^{13}C nuclei which are directly bound to the protons affected by the first pulse. During the detection period t_2, the free induction decay is recorded. When the protons are decoupled, the magnetisation precesses at the frequencies characteristic of the ^{13}C nuclei

In two-dimensional NMR, the time evolution of the nuclear magnetisation is observed as a function of *two* hf frequencies which are both applied to the sample, for example the resonant frequencies for protons and for the ^{13}C nuclei of molecules in which the interactions of these two types of nuclei are of interest. The resonance signal is observed e.g. as a function of the time, the so-called evolution time, during which these two nuclear species are interacting. They can be decoupled by the application of suitable pulse sequences, as mentioned above. The sequencing of such a measurement is illustrated with an example in Fig. 18.14. Fourier transformation with respect to two frequencies then yields *two-dimensional* resonance diagrams in which for example the chemical shift is plotted along the x axis and the spin-spin coupling on the y axis, or the chemical shifts of two different nuclear species are plotted on the two axes. There are many variations of this technique, which is highly developed both experimentally and theoretically, that we cannot treat in detail in this introduction. An example of the result of such a measurement is shown by Fig. 18.15.

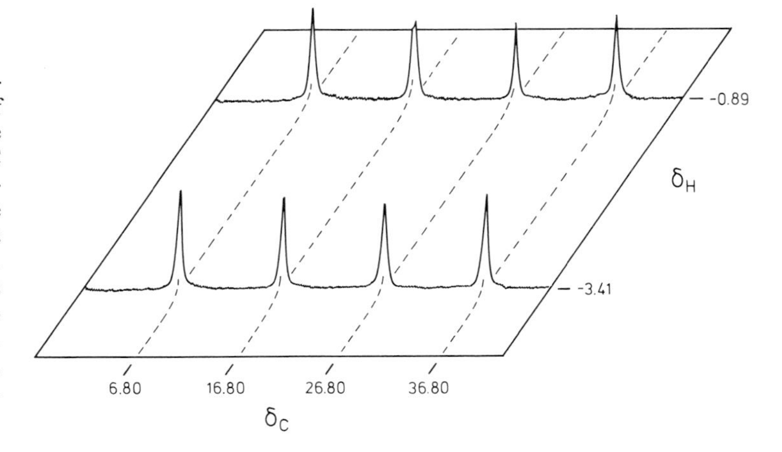

Fig. 18.15. A heteronuclear two-dimensional NMR spectrum of ^{13}C-methyl iodide showing the correlation of the ^1H and ^{13}C nuclei. The axis labelled δ_H indicates the chemical shift of the ^1H nuclei, while the other axis gives the shift of the ^{13}C nuclei. Note that here, the intensities cannot be predicted by the coupling model described in Sect. 18.2.4. [After A.A. Maudsley, L. Müller, and R.R. Ernst, J. Magn. Res. **28**, 463 (1977)]

18.6 Applications of Nuclear Magnetic Resonance

Nuclear spin resonance spectroscopy has been developed in the past decades with the help of very sophisticated experimental techniques to a method with which one can analyse the *structures* and the *bonding* in molecules in more detail than with all of the other methods available to physicists and chemists. This is true not only of small molecules, but also of polymers and large, biophysically and biochemically relevant functional units. This method is indispensable both for analysis and for structural investigations, and it belongs among the most important tools of the chemist.

It can furthermore be applied to the study of dynamic processes, of the *motions* of molecules or parts of molecules, and of molecular reactions. In recent years, spatially resolved nuclear spin resonance, called *spin tomography*, and in general the *in vivo* resonance methods, have gained considerable importance in biology and medicine. As explained in more detail in I, Sect. 20.7, in tomography the spatial positions of the nuclear spins in an inhomogeneous magnetic field are marked by their splittings and thus their resonance frequencies. It thus becomes possible to investigate molecules and their reactions in the interior of living organisms without damaging them. Important applications are the study of metabolic processes and of the mechanisms of action of pharmaceutical compounds *in vivo*.

In addition to protons, as we have already mentioned in Sect. 18.4, also the nuclides ^{13}C, ^{19}F, and ^{31}P are often used as probes for NMR studies. However, the method allows the investigation of any nucleus which has a non-vanishing nuclear spin. Additional information can be obtained from nuclear quadrupole resonance using nuclei with $I > 1$ in the electric field gradients associated with chemical bonds.

19. Electron Spin Resonance

The use of the electron spin in resonance spectroscopy allows important insights to be gained into the structure and dynamics of paramagnetic molecules (Sects. 19.1 through 19.4). This is particularly true of molecules in triplet states, with a total spin quantum number of $S = 1$ (Sect. 19.5). Especially useful, in part due to their great detection sensitivities and excellent spectral resolutions, are the various multiple resonance techniques (Sects. 19.6 through 19.8); they were to some extent developed for problems in molecular physics and have since become important in other areas, for example in solid-state physics.

19.1 Fundamentals

Electron Spin Resonance spectroscopy (ESR) is less important to molecular physics than is NMR spectroscopy, because molecules are in general diamagnetic and thus do not give rise to an ESR signal; there are, in contrast, only a very few molecules which do not contain at least one nucleus with a non-zero nuclear spin and magnetic moment, and therefore most molecules are accessible to NMR methods. ESR spectroscopy is limited to those molecules which contain an unpaired electron and are paramagnetic; in these cases, ESR is a very important experimental method, from which one can learn a great deal about the structure, bonding, and dynamics of the molecules.

Which molecules are paramagnetic? The most important groups are

- Molecules containing paramagnetic atoms as integral parts of their structures, in particular the case when, as for the rare earth atoms or transition-element atoms, the paramagnetism arises through inner electrons. Examples are the ions Fe^{3+} or $[Fe(CN_6)]^{3-}$. The outer or valence electrons may in these cases have their spins paired, and thus be diamagnetic.
- Molecules with an unpaired outer electron, called radicals. There are stable radicals, for example DPPH (diphenyl-picryl hydrazyl), which is often used as a standard for the calibration of resonance fields due to its well-known and precisely measurable g-factor (see below). There are also radicals which are formed from diamagnetic molecules under the influence of a solvent, through chemical reactions, or upon irradiation, which vanish after a short time by recombining with the split-off molecular fragments.
- Molecules in the triplet state, whether it be their ground state, as for O_2, NO, or NO_2, or a metastable excited triplet state, as in napthalene (cf. also Sect. 13.4). When the lifetimes are 10^{-6} s or longer, these excited states may also be studied using stationary-state ESR.

The fundamentals of ESR spectroscopy and the experimental methods used to carry it out were already described in I, Chap. 13. An electron with its magnetic moment $\mu_s = \sqrt{s(s+1)}\, g_s \mu_B$ (s = spin quantum number, i.e. 1/2, μ_B = Bohr magneton, g_s = electronic

g-factor = 2.0023 for the free electron) has two possible orientations in a magnetic field $\boldsymbol{B}_0$, given by $m_s = \pm\frac{1}{2}$, with the energy splitting $\Delta E = g_s \mu_B B_0$. Here, m_s is the magnetic spin quantum number. The relation

$$h\nu = \Delta E = g_s \mu_B B_0 \tag{19.1}$$

is the fundamental equation of ESR. If electromagnetic radiation having this frequency ν is applied perpendicular to the direction of $\boldsymbol{B}_0$, then the resonance condition is fulfilled for a free electron and transitions between the two allowed orientations of the electronic spin, i.e. ESR, can be observed. The numerical value of ν is

$$\nu = 2.8026 \cdot 10^{10} \boldsymbol{B}_0 \frac{[\mathrm{Hz}]}{[\mathrm{T}]} \tag{19.2}$$

for the frequency of the allowed magnetic dipole transitions with $\Delta m_s = \pm 1$. The field B_0 is usually chosen to have a strength of the order of 0.1 to 1 T, so that the resonance frequency lies in the microwave range, i.e. is of the order of GHz.

The signal intensities in ESR spectra are proportional to the number of unpaired spins in the sample, as long as saturation effects are avoided (i.e. if the relaxation times T_1 are not too long and the microwave power is not too great). As in NMR, only the temperature-dependent difference of populations between the spin orientations parallel and antiparallel to the applied field B_0 contribute to the signal. The detection limit is inversely proportional to the linewidth and is of the order of 10^{10} spins for a linewidth of $1\ \mathrm{G} = 10^{-4}$ T when using conventional ESR spectrometers.

Spin-lattice relaxation, similarly to the case of NMR, is produced by time-varying magnetic perturbing fields with a correlation time of the order of the Larmor frequency. Such perturbing fields can be due to the motions of neighbouring magnetic moments in solution or in a solid. A typical value of T_1 at room temperature for molecules in solution is $T_1 \approx 10^{-7}$ s; this relaxation time becomes longer at lower temperatures. The transverse relaxation time T_2 takes on similar or shorter values.

19.2 The g-Factor

The simplest quantity which can be determined in an ESR measurement is the g-factor of the paramagnetic electron, as in (19.1). In molecules, the g-factor is usually anisotropic; however, its anisotropy is often hidden due to motions of the molecules in solution or to a disordered molecular arrangement in a solid host. Surprisingly, nearly all radicals, as well as the triplet states of organic molecules and even the paramagnetic electronic states in many ionic crystals have g-factors which differ from that of the free electron by a very small amount, i.e. by a few tenths of a percent. This demonstrates that one is dealing for the most part with electrons which appear to have no orbital angular momentum, i.e. with $l = 0$. This can result from their being in fact s electrons, or else they can be electrons in orbitals which belong to the entire molecule and can be classified as nonlocalized orbitals at the location of an atom in the molecule, which likewise have no orbital angular momentum.

Molecules which contain atoms with paramagnetic electrons in their inner shells, and also some ionic crystals, can have g-factors which result from a combination of spin and orbital quantum numbers and which may be much greater than 2. In this case, the paramagnetic electrons are localized on one atom, and the coupling between their spins and orbital moments must be taken into account. Just how the spin and orbital moments couple to give a total magnetic moment, and how the corresponding g_J-factor is defined, has already been discussed in I, Sects. 12.7 and 12.8. However, even such molecules can have electronic states with $g = 2$, that is with practically pure spin magnetism. This can be due to the fact that the atom or ion has an electronic configuration with $L = 0$, as is the case of Fe^{3+}, with 5 unpaired d electrons and $S = 5/2$, $L = 0$, and the configuration $^6S_{5/2}$. Another possibility is that Russell-Saunders coupling has been broken by the strong internal electric field of the chemical bonds in the molecule or the crystal, so that the quantum number L is no longer valid.

The g-factor as defined here in (19.1), by the way, includes the "chemical shift" caused by local currents induced by the applied field at the location of the magnetic moment.

19.3 Hyperfine Structure

In addition to the electronic g-factor, in ESR one can measure the interaction between the magnetic moment of the paramagnetic electron and the nuclear moments of nuclei with non-vanishing spins $\boldsymbol{I}$. The energy terms and the resonance line of the electrons are split by this *hyperfine interaction* with the nuclear spin I into a multiplet of $2I + 1$ terms or a corresponding number of component lines. This can be understood as follows: the nuclear spin produces a magnetic field which adds to or subtracts from the applied field B_0, depending on its orientation. A dipolar contribution to the interaction will be averaged out in solution by the molecular motions, and only the scalar or contact interaction remains. The field at the position of the electron is then given by

$$B_{\text{loc}} = B_0 + a\, m_I \,, \tag{19.3}$$

where m_I is the magnetic spin quantum number of the nucleus, i.e. $m_I = I, I - 1, \ldots, -I$, and a is the hyperfine coupling constant for the particular electron/nucleus configuration, which here is measured in units of magnetic field.

This is illustrated in Fig. 19.1 for a proton, with $I = \frac{1}{2}$ and $m_I = \frac{1}{2}$. The resonance condition is now fulfilled for two values of the field, and it is given by

$$h\nu = g\mu_{\text{B}} \left(B \pm \frac{a}{2}\right) . \tag{19.4}$$

Due to the selection rules $\Delta m_s = \pm 1$, $\Delta m_I = 0$, a proton splits the ESR line of an electron into two lines, as we have already seen for the case of a hydrogen atom in I, Chap. 20. The energy splitting is equal to

$$\Delta E_{\text{hfs}} = a \,, \tag{19.5}$$

if we denote the hyperfine coupling as an energy by the same symbol, a. The conversion can be carried out with the aid of (19.1) (1 Gauss corresponds to 2.8 MHz). Each electronic term is displaced by an amount

$$\Delta E_{\text{hfs}} = a\, m_I \,. \tag{19.6}$$

For the interaction of a paramagnetic molecular electron with a nucleus of spin I, the number of hyperfine levels with equal statistical weights is $2I + 1$, since such a nucleus has $2I + 1$

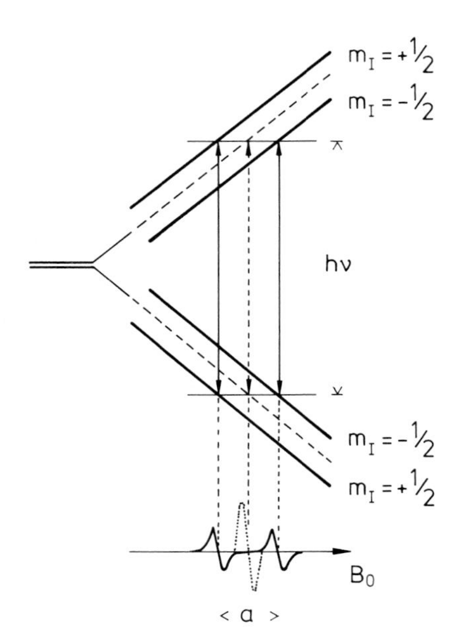

Fig. 19.1. The hyperfine interaction between an electron and a nucleus with spin $I = 1/2$ gives four levels with two ESR lines of equal intensities. The hyperfine splitting is denoted by a. The two-fold degeneracy of the magnetic states of an electron in the absence of a magnetic field is lifted by the field B_0 to give two levels with $m_s = \pm 1/2$; this gives one ESR line (*dashed* transition). A proton has in addition two allowed orientations corresponding to $m_I = \pm 1/2$, leading to a displacement of the terms as in (19.3) and thus to two ESR lines, each with half the original intensity, split by a

allowed orientations relative to the magnetic moment of the electron or to the applied field. A typical order of magnitude for this hyperfine interaction is $10^{-3} - 10^{-4}$T, if we quote it in magnetic-field units. Figure 19.2 shows the case of a nucleus with $I = 3/2$.

When the paramagnetic electronic state interacts with several equivalent nuclei, then the number and relative probability of the spin orientations of these nuclei must be taken into account in order to understand the observable hyperfine structure. For N equivalent protons, one finds $N + 1$ hyperfine component lines. Their intenstiy distribution is given by Pascal's triangle, just as for the component lines in the nuclear spin-spin coupling; see Fig. 18.12.

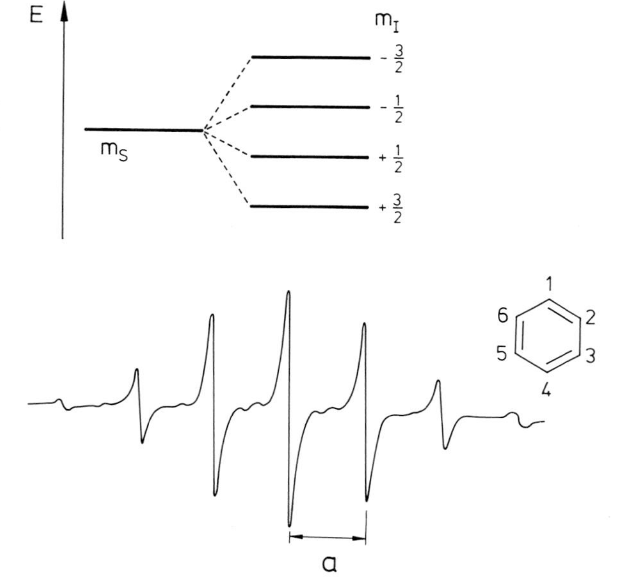

Fig. 19.2. The hyperfine splitting of the term m_s of an electron by a nucleus with $I = 3/2$; it is split into 4 components, corresponding to $m_I = 3/2, 1/2, -1/2, -3/2$

Fig. 19.3. The ESR spectrum of the benzene radical anion, $C_6H_6^-$, in solution. The ESR absorption (for experimental reasons shown as the derivative signal) is plotted against the magnetic field B_0. One observes 7 hyperfine component lines with a splitting $a = 0.375$ mT

Figure 19.3 shows as an example the ESR spectrum of the benzene radical anion, $(C_6H_6)^-$, which can be readily produced by electron transfer from alkali metal atoms to neutral benzene molecules in solution. The unpaired electron is uniformly distributed around the benzene ring, as can be seen from the hyperfine splitting pattern of its ESR signal, which we discuss here. Its hyperfine interactions with the 6 protons lead to 7 lines having the intensity ratios 1:6:15:20:15:6:1 (Fig. 19.4). Benzene radical cations, $(C_6H_6)^+$, can also be produced electrolytically or by electron abstraction using sulphuric acid; the ESR spectrum of the cation is very similar to that of the anion. An additional electron is thus distributed around a benzene ring in a similar way as a missing electron.

In the napthalene radical anion (Fig. 19.5), the spin distribution of the additional electron is no longer uniformly spread over all the C atoms, and the hyperfine interaction is thus no longer the same with all 8 protons. The C atoms at the α positions, i.e. at the positions

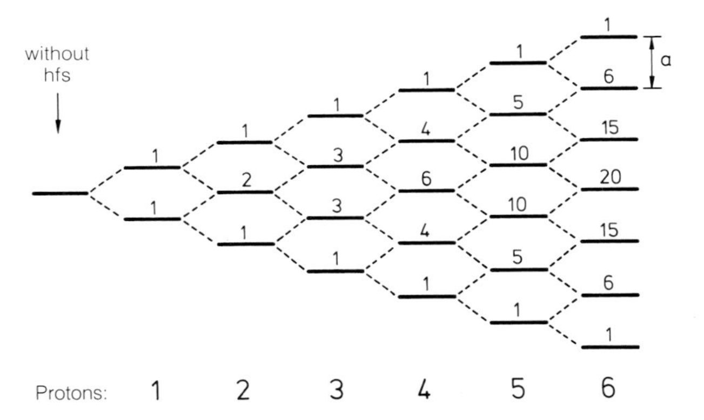

Fig. 19.4. Pascal's triangle, for determining the number and relative weight of the hyperfine components from the interaction of an electron with 6 equivalent protons, i.e. in the benzene radical. The interaction with N equivalent protons leads to $N+1$ equivalent lines having the intensity ratios given in the diagram. See also Fig. 18.12

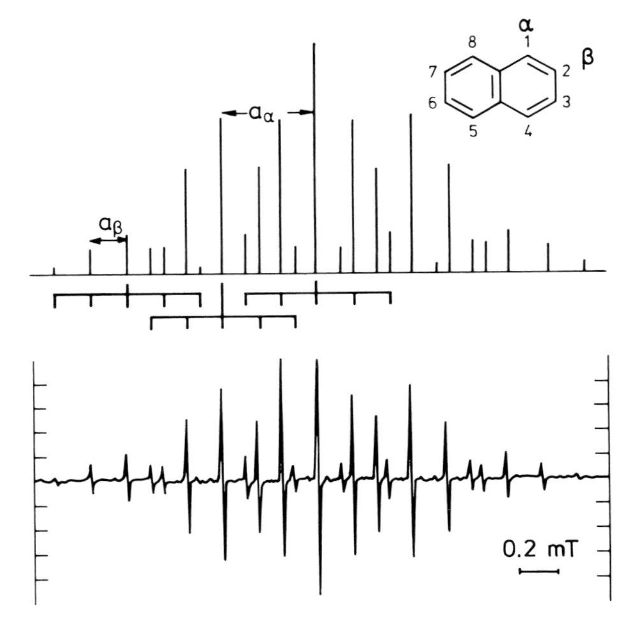

Fig. 19.5. The ESR spectrum of the napthalene radical anion in solution, similar to Fig. 19.3. Here, there are two groups of protons with differing hyperfine constants, since the probability density of the unpaired electron is larger at the α positions (1,4,5,8) than at the β positions (2,3,6,7). The measured values of the hyperfine splitting constants are $a_\alpha = 0.495$ mT and $a_\beta = 0.186$ mT

1,4,5, and 8, are equivalent to one another, as are the β atoms at positions 2,3,6, and 7, but the probability density of the electron near the α positions is larger than that near the β positions. The protons on the 4 equivalent α C atoms thus lead to a splitting of the ESR line into 5 lines with the intensity ratios 1:4:6:4:1. Each of these component lines is then split again by the (smaller) hyperfine interaction with the mutually equivalent β protons into 5 sublines. The resulting spectrum is shown in Fig. 19.5. The coupling constants a_α and a_β are a measure of the probability density of the unpaired electron at the corresponding positions.

These examples show how ESR hyperfine structure can be used to determine the electron density distribution, or, more precisely, the spin density distribution in a molecule, and thus to characterize the molecular orbitals in more detail.

Still more information can be obtained from a quantitative analysis of the hyperfine splitting. The hyperfine interaction between an electron and a nucleus has, in general, an anisotropic part and an isotropic part. The *anisotropic* part can be understood as a magnetic *dipole-dipole interaction* between the magnetic moment of the nucleus and that of the electron, and it has the well-known angular dependence of the dipole-dipole interaction [cf. (18.13)]. It is found for example for electrons in p orbitals. Its magnitude and sign depend on the orientation of the molecule in the applied field. When molecules move rapidly in solution, this interaction is averaged over time to zero and cannot be observed. For this reason, the ESR lines of paramagnetic molecules in solution are usually much sharper than those of the same molecules in the solid phase. If only this anisotropic hyperfine interaction existed, no hyperfine structure could be measured in liquid solutions.

The other, isotropic part of the hyperfine interaction is more important in molecular physics. The *isotropic* or *Fermi contact interaction*, which was introduced in I, Sect. 20.3, is the magnetic interaction between the magnetic moments of electrons at the location of the nucleus, and the nuclear moment. It is independent of the orientation of the radical or molecule and is therefore observed even for molecules which are in rapid and disordered motion, as is the case in solution. It is non-vanishing only for those electronic orbitals which do not have a node at the position of the nucleus, i.e. mainly for s electrons, which have a spherically symmetric distribution around the nucleus. This interaction energy between a proton and an electron has the magnitude [cf. (20.11) in I]

$$E = \frac{8\pi}{3} g_I \mu_N g_e \mu_B |\psi(0)|^2 \frac{1}{4} = \frac{a}{4} , \tag{19.7}$$

where $|\psi(0)|^2$ is the electron density at the nucleus and a the observable hyperfine splitting energy.

For the ground state of the hydrogen atom, the measured value of the hyperfine splitting is 50 mT, if we again quote a as a magnetic field. That is, the s electron of the H atom is acted on by a magnetic field of 50 mT due to the proton. This is, by the way, the largest known hyperfine splitting for protons. We take the density of the hydrogen $1s$ electron at the nucleus to be 1 for purposes of normalisation, and can then use the general relation

$$a = \varrho R , \qquad R = 50\ \text{mT} \tag{19.8}$$

to determine the electron densities ϱ at the nuclei of other atoms from the measured values of the hyperfine splitting a for protons. Strictly speaking, ϱ is not an electron or spin density, but rather the probability density of the electron at the nucleus, and is thus dimensionless.

We will discuss this further using the example of the methyl radical anion, CH_3^-. The ESR spectrum of this radical consists of 4 lines with the intensity ratios 1:3:3:1 and a separation of 2.3 mT. We conclude from this that the unpaired electron has hyperfine interactions with 3 equivalent protons. The spin density at each of the protons is, from (19.8), equal to

$$\varrho = \frac{2.3}{50} = 0.046\,, \qquad \text{that is, about } 5\%\,.$$

The electron thus spends about 5% of its time at each of the protons as a $1s$ electron, and the remaining 85% in the neighbourhood of the C atom.

The measured coupling constant a is, according to (19.8), proportional to the spin density and thus to the electron density at the position of a nucleus; i.e. the electron density is greater, the larger the measured splitting a. From a measurement on the benzene radical, for example, with the assumption of a uniform distribution of the unpaired electron around the ring with its six C atoms, one finds that 1/6 electron leads to a coupling constant of -0.375 mT. A whole electron as a π electron in the neighbourhood of a C atom then gives for the proton bound to the atom $a = 6(-0.375)$ mT $= -2.25$ mT. For C–H bonds, the *McConnell relation* can be applied: it states that the isotropic hyperfine structure coupling constant a_i of the proton in the C–H bond and the associated spin density ϱ_i of the unpaired π electron on the neighbouring C atom are proportional to each other, with

$$\begin{aligned} a_i &= Q\varrho_i\,, \qquad Q = -2.25 \text{ mT} \\ \varrho_i &= \text{spin density, normalized to 1 for a whole electron, i.e.} \\ \Sigma\,\varrho_i &= 1, \text{ where the sum runs over all } i \text{ nuclei involved.} \end{aligned} \tag{19.9}$$

Using this important relation, one can compute the electron densities at the positions of the various C atoms in a molecule from the measured hyperfine splitting constants of the protons, taking electron density and spin density to be equal. The C atoms themselves do not appear in the resonance spectrum, if they are present as the most abundant form of carbon, ^{12}C. This isotope has a nuclear spin of $I = 0$. The stable isotope ^{13}C, with a natural abundance of 1% and $I = 1/2$, gives rise to a hyperfine splitting, however. Its hyperfine interaction with the paramagnetic electron of the radical cannot be seen in Figs. 19.3 and 19.5, due to the small natural abundance of ^{13}C.

At this point, we should mention an additional important interaction mechanism, which is responsible for the fact that the hyperfine splitting in C–H groups treated in the previous paragraphs can occur at all. The unpaired electrons which we considered in the methyl radical or the benzene radical are electrons from the p orbitals of the C atom; cf. Fig. 19.6. Electrons in the p orbitals should, however, have no hyperfine interactions with the C atoms or with the H atoms which lie in the plane perpendicular to the p orbitals. The hyperfine structure shown in Figs. 19.3 and 19.5 nevertheless *is* observed; to be sure indirectly, through *spin polarisation*.

We shall explain this phenomenon using the methyl radical as an example, as in Fig. 19.6. The unpaired electron is in a p_z orbital of the C atom and therefore has a spin density

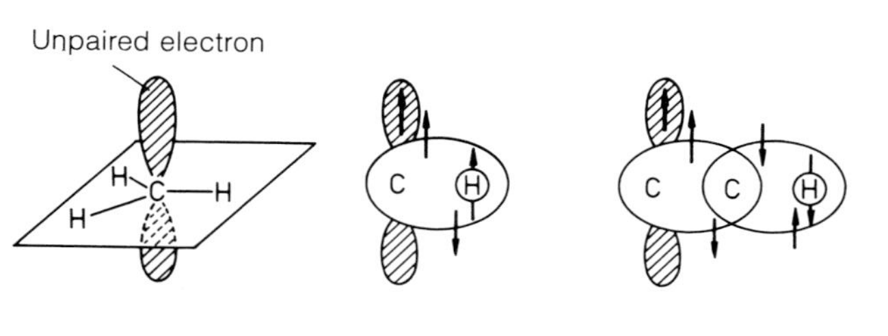

Fig. 19.6. The indirect coupling between the spin of an unpaired electron and a nuclear spin in its neighbourhood, using the methyl radical as an example. The unpaired electron in the p_z orbital polarizes the spin of one of the two σ electrons of the C–H bond; therefore, both σ electrons are polarized, since their spins must remain antiparallel. This leads to a coupling with the proton spin. In the centre, the case of "positive" coupling is indicated; at the right, the "negative" coupling via two C atoms is shown. In the latter case, one also refers to a negative spin density

of zero at the sites of the H nuclei; in this case, one should expect no hyperfine structure. There is, however, an indirect coupling, similar to that which we discussed in Sects. 18.2.3 and 18.2.4. The unpaired electron in the p_z orbital has a tendency to favour the parallel orientation of the spin of one of the two σ electrons of the C atom in the C–H bond; this follows from Hund's rule. It also means that the second σ electron in each of the three C–H bonds has a preferred orientation in the opposite sense, which follows from the Pauli exclusion principle. The nuclear spins of the protons orient preferentially antiparallel to the neighbouring electronic spins, and therefore, finally, parallel to the unpaired spin in the p_z orbital. This indirect interaction can thus be described by assuming that an electron experiences a contact interaction having a certain probability with a given H atom. The probability can be expressed in terms of a spin density of the unpaired electron at the location of the carbon atom or the protons, as in (19.9). The corresponding hyperfine interaction is then observable.

Here, again, it is true that the interaction energy which leads to a mutual spin orientation is small compared to the quantum energy of the electron spin resonance, similarly to the case of the indirect nuclear spin-spin interaction discussed in Chap. 18. Thus, when we say that "the spins have a tendency to orient parallel or antiparallel to each other"(which is also referred to as spin polarisation), then we mean that both spin orientations occur, but that they have a (small) difference in energy.

This example is intended to show how electron spin resonance can contribute to the elucidation of the electron distribution in a molecule and thus to a better understanding of molecular structure and chemical bonding. The quantity ϱ in (19.8) and (19.9) is, as we mentioned, strictly speaking a spin density and not an electron density. The indirect mechanism of the interaction also makes it clear that the spin density can take on negative values: if the coupling extends over an additional C atom, as is shown on the right in Fig. 19.6, this leads to a further application of Hund's rule and thus to an antiparallel orientation between the proton spin and the spin of the unpaired electron. Clearly, this spin polarisation mechanism is closely related to that of the indirect nuclear spin-spin coupling which we treated in Sect. 18.2.3.

In any case, the observation of hyperfine structure in the ESR spectra of molecules opens a way to the experimental determination of the electron density distributions in molecules, and thus of the spatial distribution of the molecular orbitals.

19.4 Fine Structure

The paramagnetic state of a molecule can, as we have thus far assumed, be based on the existence of an electron with an unpaired spin $s = 1/2$ within the molecule. This is a doublet state, since the electron has two possible orientations in the field $\boldsymbol{B}_0$. However, as we have already pointed out, there are also molecular states in which *two* electrons have their spins parallel; such *triplet states* with a spin quantum number $S = 1$ occur for example as metastable excited states in organic molecules, cf. Fig. 15.1.

For the purposes of ESR, we can treat this triplet state as one particle with spin $S = 1$, i.e. $|\boldsymbol{S}| = \sqrt{S(S+1)}\hbar$ with $S = 1$. In the applied field $\boldsymbol{B}_0$, such a state will split into three substates with $m_S = 0$ and ± 1. The substates are equidistant, and for transitions with $\Delta m = \pm 1$, we would expect only a single resonance line. However, in addition to the splitting by the field $\boldsymbol{B}_0$, the substates are shifted by the dipole-dipole interaction of the magnetic moments of the two electrons which form the $S = 1$ state. In a somewhat simplified description, this leads to an additional magnetic field $\boldsymbol{D}$ which each electron experiences as a result of the interaction with the other electron, as illustrated in Fig. 19.7. A more precise treatment is given in Sect. 19.5.

In the state $S_z = m_S = +1$, i.e. both electrons with their spins parallel to the applied field, the additional dipole field adds to $\boldsymbol{B}_0$; in the state with $S_z = m_S = -1$, the applied field is reduced by the same amount; cf. Fig. 19.7. The middle sublevel remains unshifted, because the dipolar field is oriented perpendicular to $\boldsymbol{B}_0$ in this case.

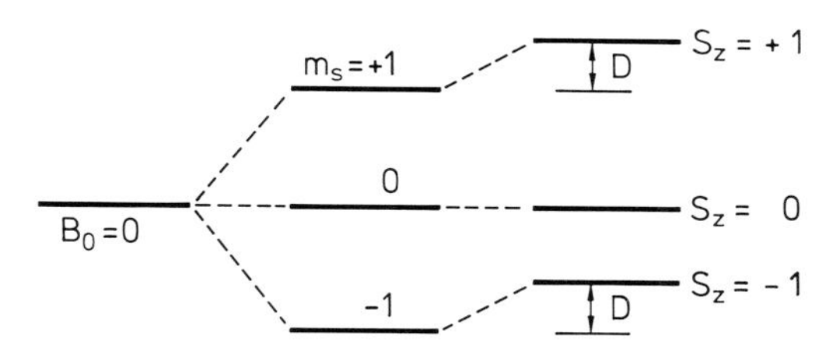

Fig. 19.7. For a state with the overall spin quantum number $S = 1$, i.e. with two parallel electron spins, there are three possible spin orientations in an applied magnetic field (*centre* and *right parts of figure*), $m_S = +1, 0, -1$. This gives three energy terms in a field. The terms $S_z = m_S = \pm 1$ are shifted by the dipole-dipole interaction D, while the term $S_z = 0$ remains unshifted

If the applied field $\boldsymbol{B}_0$ is smaller than the dipolar field $\boldsymbol{D}$, and in the limit as it is reduced to zero, the energy difference between the substates with $S_z = \pm 1$ and $S_z = 0$ persists, and the term diagram behaves as shown in Fig. 19.8. The degeneracy for the case $\boldsymbol{B}_0 = 0$ is lifted by the dipole-dipole interaction. If $\boldsymbol{B}_0 \neq 0$, one expects two resonance lines, and they are in fact observed in the spectrum. Their splitting allows the determination of the dipolar field $\boldsymbol{D}$ and thus of the interaction energy. When the electrons are localized, the interaction energy of the two electronic moments at a distance r can be calculated in terms of the dipole-dipole interaction. The magnetic field which the two electrons produce at each other's locations is given by:

$$D = \frac{3}{2} g_e \mu_B \frac{1}{r^3} \qquad [\mathrm{Vsm^{-2}}] \, . \tag{19.10}$$

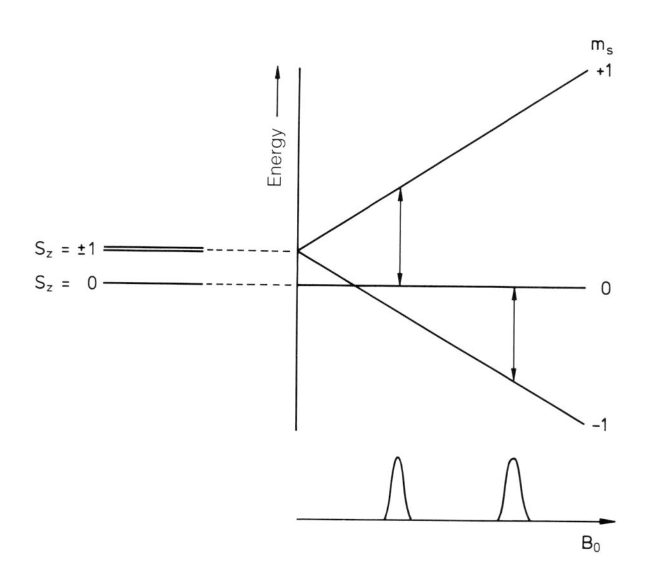

Fig. 19.8. The zero-field splitting of a triplet state (*left*) between $S_z (= m_S) = \pm 1$ and 0, and the splitting and allowed transitions in an applied field $\boldsymbol{B}_0$. In a field $\boldsymbol{B}_0 \neq 0$, one observes two ESR transitions. Without the applied field, no quantisation axis z is defined; the symbols X, Y and Z are then used to characterize the states, instead of S_z and m_S, as in Fig. 19.10

A measurement of D thus gives the distance r between the two electrons which form the triplet state in a molecule with localized electrons, corresponding to the so-called point dipole model.

The fine structure thus leads to a splitting of the triplet electronic state even without an external field, the so-called *zero-field splitting*. The model described in this section and the previous one, with two localized electrons at a fixed distance r, is oversimplified to the degree that the electrons are delocalized. A calculation of the fine structure interaction is therefore carried out in detail in Sect. 19.5.

19.5 Calculation of the Fine Structure Tensor and the Spin Wavefunctions of Triplet States

In this section, we wish to treat the energies of the triplet state levels in detail. The basis for this treatment is naturally the Hamiltonian operator, which contains the orbital motions of the electrons with coordinates $\boldsymbol{r}_1$ and $\boldsymbol{r}_2$, the energy of the spins in the applied magnetic field, and the dipole interaction between the spins. This Hamiltonian has the form

$$H(\boldsymbol{r}_1, \boldsymbol{r}_2, 1, 2) = H_{0.\mathrm{space}} + H_{0.\mathrm{spins}} + H^{\mathrm{S}} , \tag{19.11}$$

where the first term on the right-hand side refers to the orbital motions and $H_{0.\mathrm{spins}}$ is given explicitly by

$$H_{0.\mathrm{spins}} = g_{\mathrm{e}}\mu_{\mathrm{B}}\boldsymbol{B}_0 \cdot \hat{\boldsymbol{S}} . \tag{19.12}$$

In this expression, $\boldsymbol{B}_0$ is the applied magnetic field, while the spin operator $\hat{\boldsymbol{S}}$ contains the vector sum of the spin operators of the two electrons, $\hat{\boldsymbol{S}} = \hat{\boldsymbol{S}}_1 + \hat{\boldsymbol{S}}_2$, and has the usual three components

$$\hat{\boldsymbol{S}} = (\hat{S}_x, \hat{S}_y, \hat{S}_z) . \tag{19.13}$$

The hat over $\boldsymbol{S}$ denotes an operator. The dipole interaction operator in (19.11) is given explicitly by

$$H^{\mathrm{S}} = g_{\mathrm{e}}^2\mu_{\mathrm{B}}^2 \left(\frac{\hat{\boldsymbol{S}}_1 \cdot \hat{\boldsymbol{S}}_2}{r_{12}^3} - \frac{3(\hat{\boldsymbol{S}}_1 \cdot \boldsymbol{r}_{12})(\hat{\boldsymbol{S}}_2 \cdot \boldsymbol{r}_{12})}{r_{12}^5} \right) . \tag{19.14}$$

It is obtained in a straightforward manner from the classical interaction energy of two magnetic dipoles by replacing the magnetic moments by the products of the spin operators with $g_{\mathrm{e}}\mu_{\mathrm{B}}$. Without the dipole interaction operator (19.14) in (19.11), the eigenfunctions of the Hamiltonian H can be readily found. They take the form of products:

$$\Psi(\boldsymbol{r}_1, \boldsymbol{r}_2, 1, 2) = {}^3\Psi(\boldsymbol{r}_1, \boldsymbol{r}_2)\,\sigma(1, 2) , \tag{19.15}$$

where the first factor refers only to the orbital motions of the triplet-state electrons and the second factor σ to their spins 1 and 2.

It is our goal to derive a Hamiltonian which refers only to the spin wavefunctions, and not to the orbital wavefunctions of the electrons. As can be shown in detail, the orbital motions are influenced only weakly by the spin interactions, so that the hypothesis (19.15) remains a good approximation and it is sufficient to replace the interaction energy H^S (19.14) by its quantum-mechanical average:

$$\overline{H^S} = \int {}^3\Psi^*(\boldsymbol{r}_1, \boldsymbol{r}_2) H^{S3}\Psi(\boldsymbol{r}_1, \boldsymbol{r}_2)\, dV_1 dV_2 \,. \tag{19.16}$$

Equation (19.16) thus means that the distance $\boldsymbol{r}_{12}$ between the two electrons can be averaged over by using the probability distribution ${}^3\Psi^{*3}\Psi$. Considering the expressions which result from inserting (19.14) into (19.16) in more detail, we initially look at the expression obtained from the first term in parentheses in (19.14). The integral is a scalar quantity, so that we can just as well write it within the scalar product of $\hat{\boldsymbol{S}}_1$ and $\hat{\boldsymbol{S}}_2$, obtaining

$$g_e^2 \mu_B^2 \hat{\boldsymbol{S}}_1 \hat{\boldsymbol{S}}_2 \int {}^3\Psi^* \frac{1}{r_{12}^3} {}^3\Psi\, dV_1 dV_2 = \hat{\boldsymbol{S}}_1 \cdot \int \ldots dV_1 dV_2 \hat{\boldsymbol{S}}_2 \,. \tag{19.17}$$

In order to be able to extract the spin operators from the second term in parentheses in (19.14), we write the corresponding part of (19.16) in the form

$$\underbrace{-3\, g_e^2 \mu_B^2 \hat{\boldsymbol{S}}_1 \cdot \int}_{(\hat{\boldsymbol{S}}_1 \cdot \hat{\boldsymbol{r}}_{12})} {}^3\Psi^* \frac{\boldsymbol{r}_{12} \cdot \boldsymbol{r}_{12}}{r_{12}^5} \underbrace{{}^3\Psi\, dV_1 dV_2 \hat{\boldsymbol{S}}_2}_{(\hat{\boldsymbol{r}}_{12} \cdot \hat{\boldsymbol{S}}_2)} \,. \tag{19.18}$$

In this expression, we have indicated by horizontal brackets how the terms are to be interpreted. First, the scalar product of $\hat{\boldsymbol{S}}_1$ and $\boldsymbol{r}_{12}$ is taken, and likewise that of $\boldsymbol{r}_{12}$ and $\hat{\boldsymbol{S}}_2$. The product $\boldsymbol{r}_{12} \cdot \boldsymbol{r}_{12}$ in the centre has the properties of a tensor, as can be shown mathematically, and it can be written explicitly in the form

$$\boldsymbol{r}_{12} \cdot \boldsymbol{r}_{12} = \begin{pmatrix} x_{12} \cdot x_{12} & x_{12} \cdot y_{12} & x_{12} \cdot z_{12} \\ y_{12} \cdot x_{12} & y_{12} \cdot y_{12} & y_{12} \cdot z_{12} \\ z_{12} \cdot x_{12} & z_{12} \cdot y_{12} & z_{12} \cdot z_{12} \end{pmatrix} . \tag{19.19}$$

Using this formalism, we can rewrite (19.15) in the simple way

$$\overline{H^S} = \hat{\boldsymbol{S}}_1 \underbrace{g_e^2 \mu_B^2 \int {}^3\Psi^* \left(\frac{1}{r_{12}^3} - \frac{3\boldsymbol{r}_{12} \cdot \boldsymbol{r}_{12}}{r_{12}^5} \right) {}^3\Psi\, dV_1 dV_2}_{2F} \hat{\boldsymbol{S}}_2 \,, \tag{19.20}$$

where the spin operators referring to electrons 1 and 2 are extracted to the left or to the right, respectively. The integral remaining in the centre has the properties of a tensor, which we have abbreviated as $2F$. For clarity, we write this tensor with its components once more:

$$F_{ij} = \frac{1}{2} g_e^2 \mu_B^2 \int {}^3\Psi^* \left(\frac{\delta_{ij}}{r_{12}^3} - \frac{3\boldsymbol{r}_{12.i} \cdot \boldsymbol{r}_{12.j}}{r_{12}^5} \right) {}^3\Psi\, dV_1 dV_2 \,, \tag{19.21}$$

where we define

$$i, j = x, y, z \tag{19.22}$$

and

$$r_{12.x} = x_{12} \,, \quad r_{12.y} = y_{12}, \; \ldots \,. \tag{19.23}$$

As one can readily see from this explicit representation, F_{ij} is symmetric. A tensor of this type can be brought into diagonal form by a suitable choice of the coordinate system, so that the non-diagonal elements vanish:

$$F_{ij} = 0 \quad \text{for} \quad i \neq j \,. \tag{19.24}$$

In what follows, we will always assume that F_{ij} has already been diagonalized. In this coordinate system, the diagonal elements of F take on the following form:

$$F_{xx} = \frac{1}{2} g_e^2 \mu_B^2 \int {}^3\Psi^* \left(\frac{r_{12}^2 - 3x_{12}^2}{r_{12}^5} \right) {}^3\Psi \, dV_1 dV_2 \,, \tag{19.25}$$

$$F_{yy} = \frac{1}{2} g_e^2 \mu_B^2 \int {}^3\Psi^* \left(\frac{r_{12}^2 - 3y_{12}^2}{r_{12}^5} \right) {}^3\Psi \, dV_1 dV_2 \,, \tag{19.26}$$

$$F_{zz} = \frac{1}{2} g_e^2 \mu_B^2 \int {}^3\Psi^* \left(\frac{r_{12}^2 - 3z_{12}^2}{r_{12}^5} \right) {}^3\Psi \, dV_1 dV_2 \,. \tag{19.27}$$

As can be readily verified by direct calculation, the trace, i.e. the sum of the diagonal elements, is zero:

$$F_{xx} + F_{yy} + F_{zz} = 0 \,. \tag{19.28}$$

For the spin-Hamiltonian (19.16), we obtain finally the form

$$\overline{H^S} = 2\hat{\boldsymbol{S}}_1 F \hat{\boldsymbol{S}}_2 = 2F_{xx}\hat{S}_{1x}\hat{S}_{2x} + 2F_{yy}\hat{S}_{1y}\hat{S}_{2y} + 2F_{zz}\hat{S}_{1z}\hat{S}_{2z} \,. \tag{19.29}$$

We now want to show that (19.29) can also be expressed directly in terms of the overall spin

$$\hat{\boldsymbol{S}} = \hat{\boldsymbol{S}}_1 + \hat{\boldsymbol{S}}_2 \,, \tag{19.30}$$

that is, that we can write

$$\overline{H^S} = \hat{\boldsymbol{S}} F \hat{\boldsymbol{S}} \,. \tag{19.31}$$

To show this, we insert (19.30) into (19.31) and multiply out the terms of the sum in (19.30), leading immediately to

$$\overline{H^S} = \hat{S}_1 F \hat{S}_1 + \hat{S}_2 F \hat{S}_2 + \hat{S}_1 F \hat{S}_2 + \hat{S}_2 F \hat{S}_1 \,. \tag{19.32}$$

Considering the first term on the right-hand side, we see that it can be written in the form:

$$\hat{S}_1 F \hat{S}_1 = F_{xx}\hat{S}_{1x}^2 + F_{yy}\hat{S}_{1y}^2 + F_{zz}\hat{S}_{1z}^2 \,. \tag{19.33}$$

However, as we know from the spin matrices, the following relation holds:

$$\hat{S}_{1x}^2 = \hat{S}_{1y}^2 = \hat{S}_{1z}^2 = \tfrac{1}{4} \,. \tag{19.34}$$

Using these values, (19.33) becomes

$$(F_{xx} + F_{yy} + F_{zz})\tfrac{1}{4} = 0 \,. \tag{19.35}$$

This expression is equal to zero, as indicated, since the trace of F is zero. Similarly, one can show that the second term on the right side of (19.32) is zero. Due to the symmetry of the tensor F, the last two terms are equal and yield just the expression (19.29); this ends our little ancillary computation, showing that we can replace (19.29) by (19.31).

We now come to the important concept of the *fine structure constants*. Since the tensor F is traceless, we can characterize it by two constants. We choose these constants to have the following values, for reasons which will become clear later:

$$D = F_{zz} - \tfrac{1}{2}(F_{xx} + F_{yy}) \tag{19.36}$$

and

$$E = \tfrac{1}{2}(F_{xx} - F_{yy}) \; . \tag{19.37}$$

Using the expressions (19.25–27) for the elements on the main diagonal of F, we can write D and E explicitly in the forms

$$D = \tfrac{3}{4} g_e^2 \mu_B^2 \int {}^3\Psi^* \left(\frac{r_{12}^2 - 3z_{12}^2}{r_{12}^5} \right) {}^3\Psi \, dV_1 dV_2 \tag{19.38}$$

and

$$E = \tfrac{3}{4} g_e^2 \mu_B^2 \int {}^3\Psi^* \left(\frac{y_{12}^2 - x_{12}^2}{r_{12}^5} \right) {}^3\Psi \, dV_1 dV_2 \; . \tag{19.39}$$

As can be derived from the definitions of D and E, the order of magnitude of the fine structure interaction corresponds to the dipole-dipole interaction energy of two electrons, i.e. two Bohr magnetons, at a distance which is of the order of the dimensions of the molecule. For the triplet state in napthalene, measurements yield $D = 0.1012\ \mathrm{cm}^{-1}$ and $E = 0.0141\ \mathrm{cm}^{-1}$. As one can also see from the quantity in parentheses in (19.38) for D, it is a *measure of the deviation from spherical symmetry*, while in E, the quantity in parentheses is a measure of *the difference in spatial extent of the wavefunctions in the y- and x-directions*. Clearly, equations (19.38) and (19.39) are related to the spatial extent of the wavefunctions and thus to the shape of the molecule.

The fine structure constants D and E, which can be obtained from ESR spectra, thus give information about the mean values of the squared distances of the electrons, which are described by the spatial wavefunctions; see also (19.10). The principal axes x, y, and z of the fine structure tensor are, in the case of symmetrical molecules, identical to the molecular symmetry axes x, y, and z. These axes are determined by the wavefunctions, which are in turn oriented by the molecular skeleton. We abbreviate the integral $\int {}^3\Psi^* \Omega\, {}^3\Psi \, dV_1 dV_2$ as $\langle \Omega \rangle$, where the operator Ω corresponds to x^2, y^2, z^2; this makes Table 19.1 readily understandable.

Following these preparations, we can turn to our further task, namely the computation of the spin wavefunctions and the corresponding energies when both the dipole interaction and an applied magnetic field are present. The basic Hamiltonian for the spins in that case takes on the form

$$H_{\mathrm{spin}} = g_e \mu_B \boldsymbol{B}_0 \cdot \hat{\boldsymbol{S}} + D\hat{S}_z^2 + E(\hat{S}_x^2 - \hat{S}_y^2) \; , \tag{19.40}$$

where $\hat{\boldsymbol{S}}$ is the spin operator for the overall spin of the two electrons. We have assumed in writing (19.40) that the coordinates refer to the principal axes of the fine structure tensor. We have neglected the term

$$-\tfrac{1}{3}\hat{\boldsymbol{S}}^2 \tag{19.41}$$

Table 19.1. The relation between the fine structure constants D and E and the symmetry and spatial extension of the wavefunctions

Fine structure constants	Wavefunctions	Examples
$D = 0$, $E = 0$	spherically symmetric $\langle x^2 \rangle = \langle y^2 \rangle = \langle z^2 \rangle$	all atoms
$D \neq 0$, $E = 0$	three- or more-fold axis of symmetry $\langle x^2 \rangle = \langle y^2 \rangle$	triphenyl, coronene
$E > 0$	stretched in the y-direction $\langle y^2 \rangle > \langle x^2 \rangle$	carbenes (BPG)
$E < 0$	stretched in the x direction $\langle y^2 \rangle < \langle x^2 \rangle$	napthalene, anthracene
$D > 0$	plate or disk-shaped	carbenes –Ċ̣
$D < 0$	dumbbell or club-shaped	biradicals, e.g. –Ċ=C=Ċ–

in the Hamiltonian (19.40), since it is simply a constant for the triplet state. We now have the task of solving the Schrödinger equation for the spin wavefunctions σ. It is given by

$$H_{\text{spin}}\sigma = \varepsilon\sigma \ . \tag{19.42}$$

In order to solve this problem, we use the so-called zero-field functions, which correspond to the principal axes of the fine structure tensor. We first give these wavefunctions and then examine their properties using some examples, which will permit us to recognize their suitability. For single electrons, we use the notation α, β to denote the spin wavefunctions, as already defined in Chap. 4. We introduce the following functions:

$$\tau_x = \frac{1}{\sqrt{2}}[\beta(1)\beta(2) - \alpha(1)\alpha(2)] \ , \tag{19.43}$$

$$\tau_y = \frac{\mathrm{i}}{\sqrt{2}}[\beta(1)\beta(2) + \alpha(1)\alpha(2)] \ , \tag{19.44}$$

$$\tau_z = \frac{1}{\sqrt{2}}[\alpha(1)\beta(2) + \beta(1)\alpha(2)] \ . \tag{19.45}$$

As can be readily verfied with the help of relations between the spin functions as defined in I, the following equations hold for the τ functions:

$$\begin{aligned} \hat{S}_x\tau_y &= \mathrm{i}\tau_z \ , \\ \hat{S}_y\tau_z &= \mathrm{i}\tau_x \ , \\ \hat{S}_z\tau_x &= \mathrm{i}\tau_y \ ; \end{aligned} \tag{19.46}$$

$$\begin{aligned} \hat{S}_y\tau_x &= -\mathrm{i}\tau_z \ , \\ \hat{S}_z\tau_y &= -\mathrm{i}\tau_x \ , \\ \hat{S}_x\tau_z &= -\mathrm{i}\tau_y \ , \end{aligned} \tag{19.47}$$

as well as

$$\hat{S}_x \tau_x = 0, \quad \hat{S}_y \tau_y = 0 . \tag{19.48}$$

Clearly, the application of a component of the spin operator to one of the spin wavefunctions causes it to be transformed into another of the functions.

Let us first consider the case that the applied magnetic field $\boldsymbol{B}_0$ is zero. We claim that the wavefunctions introduced in (19.43–45) are then already eigenfunctions of the Hamiltonian (19.40), which enters the Schrödinger equation (19.42). To verify this claim, we ask how the operators $\hat{S}_x^2, \ldots$ act upon the individual spin wavefunctions. As can be easily calculated using the relations (19.46) and (19.47), one finds that

$$\hat{S}_z^2 \tau_x = \hat{S}_z(\hat{S}_z \tau_x) = \hat{S}_z \mathrm{i} \tau_y = \tau_x , \tag{19.49}$$

$$\hat{S}_y^2 \tau_x = \hat{S}_y(\hat{S}_y \tau_x) = \hat{S}_y(-\mathrm{i} \tau_z) = \tau_x , \tag{19.50}$$

or, using also (19.50), we find the result

$$H_{\text{spin}} \tau_x = (D - E) \tau_x . \tag{19.51}$$

This shows immediately that the spin operator applied to τ_x reproduces the wavefunction, i.e. that this wavefunction is an eigenfunction of the spin-Hamiltonian operator with the energy eigenvalue

$$\varepsilon = (D - E) . \tag{19.52}$$

In a similar manner, one finds for the wavefunctions τ_y and τ_z the corresponding eigenvalues:

$$\tau_y : \varepsilon = D + E , \tag{19.53}$$

and

$$\tau_z : \varepsilon = 0 . \tag{19.54}$$

The resulting term diagram for the zero-field states is given in Fig. 19.9.

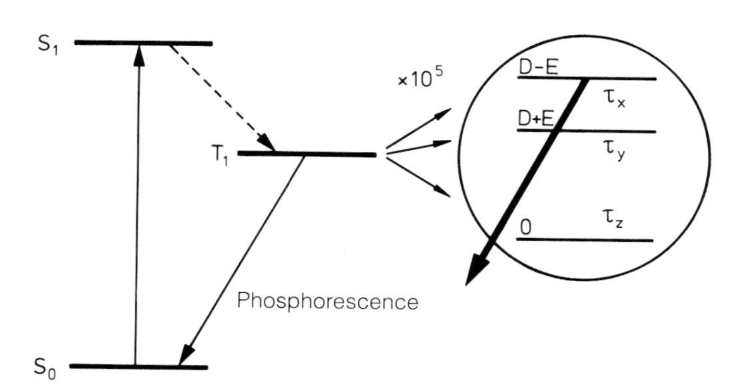

Fig. 19.9. An explanation of the zero-field spin functions and of optically-detected magnetic resonance, ODMR. The three triplet substates τ_x, τ_y and τ_z of a molecule have the energy spacings $D+E$, $D-E$, and $2E$. They differ, in general, in terms of their probabilities of being populated or depopulated by transitions, as a result of symmetry properties. The spin-orbit interaction, which weakens the selection rule forbidding intercombination transitions between the singlet and the triplet systems, has differing strengths for the different substates. In this example, we have assumed that the state τ_x has the largest radiative transition probability. By inducing transitions in the microwave spectral range, one can pump population from the τ_y or τ_z substates into τ_x, thereby increasing the phosphorescence intensity. In this way, in particular the zero-field resonance can be detected. The energy scale within the circle is magnified ca. 10^5 times relative to that of the left-hand part of the figure

Let us now turn to the general case of a non-zero applied magnetic field having the components

$$\boldsymbol{B}_0 \neq 0 \qquad B_0 = (B_{0x}, B_{0y}, B_{0z}) . \tag{19.55}$$

As can be seen by inserting the wavefunctions τ_x, τ_y and τ_z, they are no longer eigenfunctions of the spin operator. We therefore need to take linear combinations of these wavefunctions, i.e.

$$\sigma = c_x \tau_x + c_y \tau_y + c_z \tau_z \ . \tag{19.56}$$

In this equation, c_x, c_y and c_z are constants which are still to be determined. If we now substitute (19.56) into the Schrödinger equation (19.42) with the Hamiltonian (19.40) and multiply the resulting equation by τ_x or τ_y or τ_z, thus taking the spin matrix elements, we obtain a secular equation system, as we have often seen in applying perturbation theory; it can be written in the form:

$$\begin{bmatrix} D-E-\varepsilon & -\mathrm{i}g_e\mu_B B_{0z} & \mathrm{i}g_e\mu_B B_{0y} \\ \mathrm{i}g_e\mu_B B_{0z} & D+E-\varepsilon & -\mathrm{i}g_e\mu_B B_{0x} \\ -\mathrm{i}g_e\mu_B B_{0y} & \mathrm{i}g_e\mu_B B_{0x} & -\varepsilon \end{bmatrix} \begin{bmatrix} c_x \\ c_y \\ c_z \end{bmatrix} = 0 \ . \tag{19.57}$$

Setting the determinant which is defined by the matrix on the left of (19.57) equal to zero, we can determine the eigenvalues ε. In the special case that the applied magnetic field $\boldsymbol{B}_0$ is parallel to one of the principal axes, (19.57) is easy to solve, since then the secular determinant leads to a quadratic equation in ε. The results are given in Table 19.2 and in Fig. 19.10 for the cases:

$$\boldsymbol{B}_0 || x \ , \quad \boldsymbol{B}_0 || y \ , \quad \text{and} \quad \boldsymbol{B}_0 || z \ . \tag{19.58}$$

A term diagram such as the one shown in Fig. 19.10 for the example of the triplet state of napthalene is obtained. It should be clear that due to the zero-field splitting of a triplet state, two ESR lines at different resonance field strengths occur. A measurement of the ESR spectra as a function of the angle between the applied magnetic field and the principal axes of the molecule permits the determination of the fine structure tensor and thus an investigation of the distribution of the two triplet-state electrons over the molecule.

Finally, we should also mention that a complete determination of the fine structure tensor is possible only when the molecules are oriented – that is, in a solid matrix.

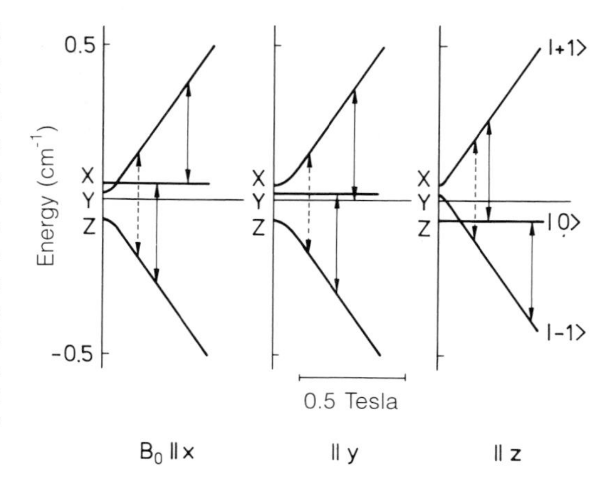

Fig. 19.10. The term diagram for the triplet terms of napthalene as a function of the applied field strength B_0. The direction of $\boldsymbol{B}_0$ is – from left to right – parallel to the x-, y-, and z-axes of the molecule, respectively; x is the long and y the short axis in the molecular plane, while z is the axis perpendicular to the plane. One observes two anisotropic ESR transitions with $\Delta m_S = \pm 1$. The (forbidden) transition with $\Delta m_S = 2$ is also indicated by *dashed arrows*

Table 19.2. The energy eigenvalues ε of the Schrödinger equation (19.42) with the Hamiltonian (19.40) for $\boldsymbol{B}_0||x$, $\boldsymbol{B}_0||y$, $\boldsymbol{B}_0||z$

	$\boldsymbol{B}_0\|\|x$	$\boldsymbol{B}_0\|\|y$	$\boldsymbol{B}_0\|\|z$
ε_{+1}	$(D+E)/2+\sqrt{(D+E)^2/4+(g_e\mu_B B_0)^2}$	$(D-E)/2+\sqrt{(D-E)^2/4+(g_e\mu_B B_0)^2}$	$D+\sqrt{E^2+(g_e\mu_B B_0)^2}$
ε_0	$(D-E)$	$(D+E)$	0
ε_{-1}	$(D+E)/2-\sqrt{(D+E)^2/4+(g_e\mu_B B_0)^2}$	$(D-E)/2-\sqrt{(D-E)^2/4+(g_e\mu_B B_0)^2}$	$D-\sqrt{E^2+(g_e\mu_B B_0)^2}$

19.6 Double Resonance Methods: ENDOR

In experimental molecular physics, various double and multiple resonance methods have considerable importance, in particular for increasing sensitivities and spectral resolutions. Here, we will briefly describe two of these, ENDOR and ODMR.

The technique of *Electron Nuclear Double Resonance* (ENDOR for short) can be explained using the example of a hydrogen atom, with one electron and one proton; cf. Fig. 19.11. In molecular physics, we could just as well be considering an unpaired electron in a radical, which is coupled to a proton. This system has four states in an applied magnetic field, which we denote by the symbols ↓↑ ↓↓ ↑↑ ↑↓ (in order of increasing energy). The first arrow refers in each case to the direction of the electronic spin and the second to the direction of the proton spin. The energy diagram in Fig. 19.11 shows, going from left to right, the energy of the electronic moment in the field $\boldsymbol{B}_0$, that of the nuclear moment, and the energy of the hyperfine interaction. In an ESR spectrum, transitions with $\Delta m_S = \pm 1$ are allowed, i.e. the two transitions 1–3 and 2–4 which are indicated in the figure. Their observation, and the determination from their difference of the hyperfine interaction with the proton, is not possible with sufficient precision in every case, for example when the ESR lines are inhomogeneously broadened by interactions with additional nuclei and thus consist of a superposition of many lines with differing hyperfine interactions.

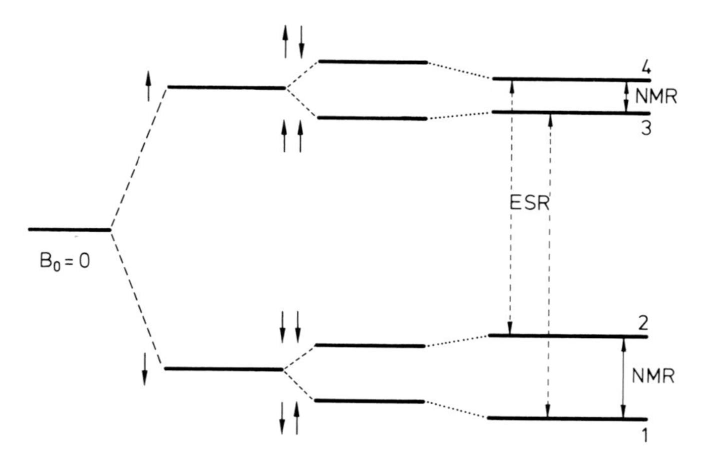

Fig. 19.11. An energy level diagram illustrating the coupling between electronic and nuclear spins (spin $I = 1/2$) with ESR and NMR transitions. The electron spin can be parallel or antiparallel to the direction of the applied field (*left*). For the nuclear spin (*right-hand arrows*), the same is true (*centre*). The hyperfine interaction is superposed on this energy splitting (*right*). In considering the energetic ordering of the terms, one must keep in mind that the magnetogyric ratio of electrons has the opposite sign from that of protons. For more details of the ENDOR technique, see the text; cf. also Sect. 20.5 and Fig. 20.14 in I

A direct observation of the nuclear resonance transitions with $\Delta m_I = \pm 1$, that is the transitions from 1 to 2 or from 3 to 4 in Fig. 19.11, which would give the same information, is often difficult or impossible. Owing to the much smaller detection sensitivity for NMR, one would require many more spins than are needed for the ESR measurement, and the resonance lines can also be strongly broadened due to nuclear spin relaxation under the influence of the electronic moments.

ENDOR consists in using the intensity of the ESR signal to detect the nuclear resonance. To achieve this, resonant microwaves are applied to the sample at such a high power that an ESR line, for example the 1–3 transition, is partially saturated; that is, it is reduced in intensity, because the population difference between the terms 1 and 3 has been decreased. If now resonant radiation corresponding to one of the two proton resonance transitions, e.g. 3–4, is applied, it will produce a change in the populations of the two participating nuclear substates and a reduction of the population of state 3. This, in turn, leads to a de-saturation of the ESR signal, since now the population difference between the states 1 and 3 is increased. This causes the intensity of the ESR absorption signal for the transition 1–3 to increase correspondingly.

In ENDOR spectroscopy, one thus observes nuclear resonance – here the resonance transition 3–4 – by using the intensity of the ESR signal – here 1–3 – as detector. This allows an enhancement of the detection sensitivity for NMR by many orders of magnitude; also, the hyperfine interactions of the electron with different nuclei can by individually observed by irradiating the sample with the corresponding resonance frequencies, which in the ESR spectrum are hidden within the inhomogeneously broadened resonance line.

19.7 Optically Detected Magnetic Resonance (ODMR)

Another important double resonance method, *Optically Detected Magnetic Resonance* (abbreviated ODMR), is based on the use of the intensity of an electronic transition in the optical spectral range as detector for electron spin (or also nuclear spin) resonances, which are induced by simultaneous irradiation using microwaves or high-frequency radiation of the appropriate frequency. This method was described in I, Sect. 13.5, with an example from atomic physics.

Here, we take an example from molecular physics, namely the investigation of the metastable triplet state T_1 in organic molecules (Sect. 15.3), to illustrate the method. The energy splitting diagram of an electronic molecular term, as in Fig. 19.9, applies to a metastable excited state T_1 that lies roughly 20 000 cm^{-1} above the ground state S_0, and from which an emission transition of long lifetime, i.e. a phosphorescence transition, can lead to S_0. Now, for reasons of symmetry which we shall not discuss here, the transition probability for emissions from T_1 to S_0 differs for the three magnetic substates, which are denoted by τ_x, τ_y, and τ_z and are indicated in Fig. 19.9. If one observes the emission and at the same time pumps population from a substate with a smaller transition probability into one with a larger probability by inducing transitions between two of the three spin levels τ_x, τ_y and τ_z, this increases the intensity of the phosphorescence observed. Then electron spin resonance can be optically detected through the change in intensity of an emission in the visible or the UV range. An example of an experimental set-up for zero-field measurements is shown in Fig. 19.12. If the observed emission spectrum belongs as a superposition to two different types of molecules, then one observes a change in the phosphorescence intensity at a particular wavelength only when the frequency of the hf-radiation corresponds to a zero-field transition of the triplet state T_1 in the particular molecule to which that emission line belongs. If the emission intensity of a line belonging to the other molecule is observed, it will change at the frequencies characteristic of the zero-field resonance transitions of that molecule.

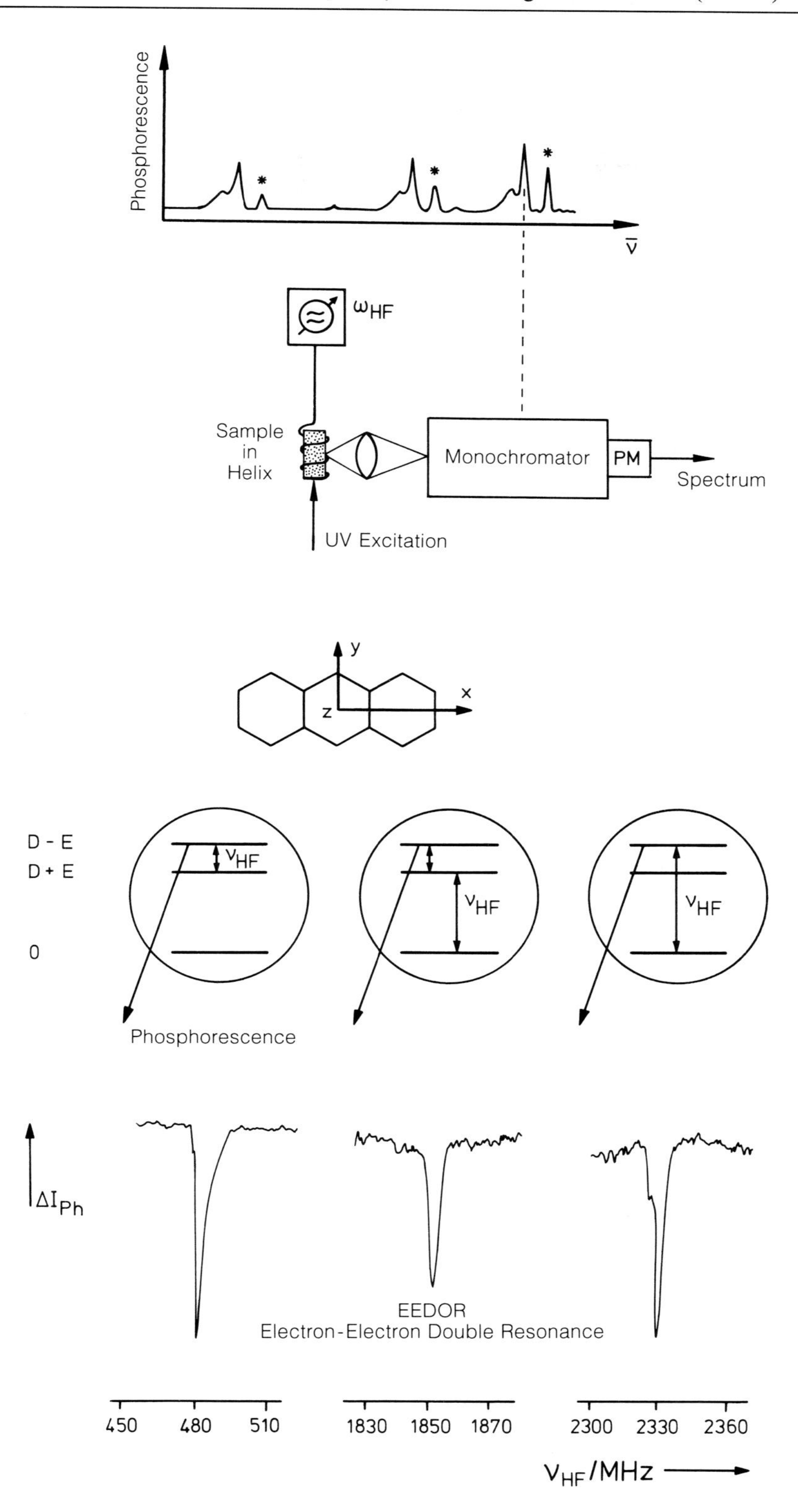

Fig. 19.12. A schematic representation of an experimental set-up for ODMR. The phosphorescence spectrum here is the superposition of the spectra from two different molecules, denoted by the presence or absence of an asterisk in the figure. The ODMR apparatus allows a change in the phosphorescence intensity to be produced selectively for both components of the spectrum by irradiation with hf power, and thus yields a correlation between the optical spectrum and the hf resonance frequencies. It is obtained by simultaneously monitoring the intensity change of individual phosphorescence lines when irradiating with microwave power of frequencies corresponding to the resonant transitions in the particular molecular triplet state

Fig. 19.13. ODMR in zero applied field, from the anthracene molecule. *Upper part:* the molecular principal axes x, y, and z. *Centre:* the zero-field splitting of the triplet state T_1 into states with relative energies 0, $D + E$, and $D - E$, with the three allowed resonance transitions. Optical emission takes place with a higher intensity from the uppermost substate in each case.; thus, high-frequency transitions between this substate and the other two states increase the emission probability and thereby the observed phosphorescence intensity. This change in intensity is used to detect the resonance due to the corresponding hf transitions. *Lower part:* the ODMR lines observed on irradiation with hf power at quantum energy $2E$, $D + E$, and $D - E$. The transition $D + E$ is observed as electron-electron double resonance by simultaneous irradiation with $2E$ quanta. [After J.-U. von Schütz, F. Gückel, W. Steudle, and H.C. Wolf, Chem. Phys. **53**, 365 (1980)]

Figure 19.13 shows an example of such a measurement, the three zero-field resonance transitions of the anthracene molecule with optical detection. From the observed frequencies (ODMR frequencies in zero field), one obtains directly the fine structure constants D and E of the molecule. The three zero-field resonance transitions shown are allowed magnetic dipole transitions.

There are numerous variations on this method. They all have as a common characteristic an enhancement of the detection sensitivity for small spin concentrations. Furthermore, when the optical spectrum consists of a superposition of spectra from several different molecules, one can attribute the different spectral lines to the different molecules; cf. Fig. 19.12. The microwave frequencies at which one observes ODMR signals as a change in the optical emission intensity are namely molecule-specific properties and therefore represent a sort of fingerprint of the participating molecule. The resolution of such overlapping spectra is thus improved, and a clearcut correlation is obtained between the optical spectrum, the hf-resonance spectrum, and the particular molecule. Figure 19.14 shows an especially instructive example of how ODMR allows the individual analysis of the spectra of a number of molecules which overlap to give a fluorescence spectrum with very little structure. This figure represents the fluorescence spectrum of photosynthesizing bacteria; it consists of relatively broad, overlapping bands. When the ODMR spectra are observed at discrete wavelengths, it can be shown that the total optical spectrum is a superposition of the emission from numerous different molecules. Eleven of these are detected individually in the ODMR spectra shown in Fig. 19.14. Employing the measured fine structure constants which are characteristic of the different molecules, one can analyse them. They are in the main chlorophylls and their precursors, which are produced in intermediate stages of biosynthesis and are all present simultaneously in the bacteria.

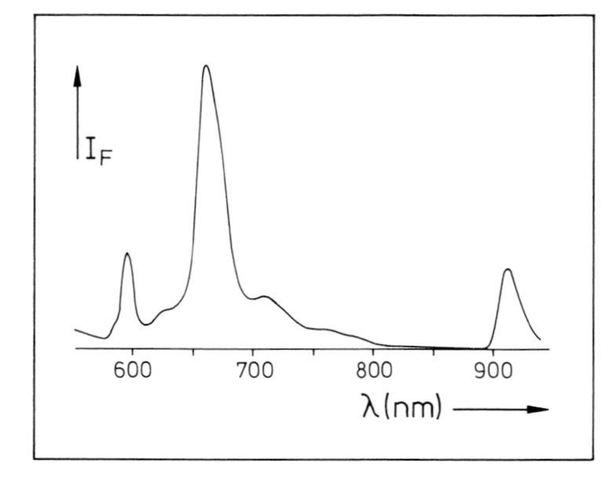

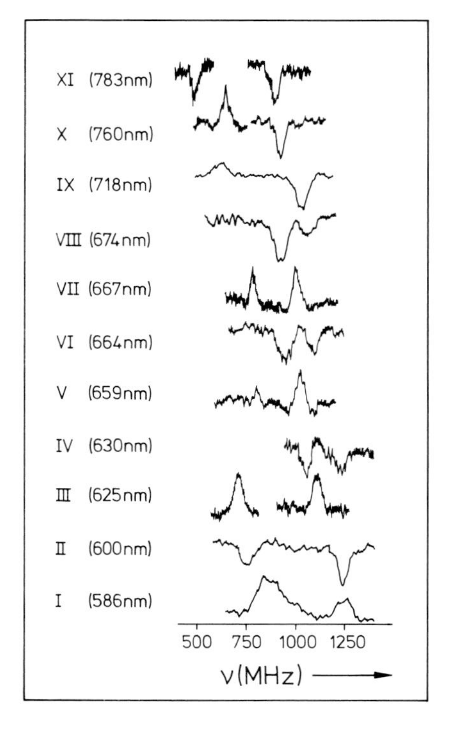

Fig. 19.14. An example of the power of the ODMR method for the analysis of complex spectra. In the fluorescence spectrum of photosynthesizing bacteria (*left-hand part of figure*), between the wavelengths 586 and 783 nm alone, 11 different fluorescing molecules can be detected by ODMR. In the right-hand part of the figure, the corresponding ODMR spectra are plotted, i.e. the intensity of the optical emission at the given wavelength is drawn as a function of the frequency of the hf-power applied to the sample. Using the fine structure constants D and E determined from these spectra, the individual molecules can be identified. They consist of various protoporphyrines, phaeophorbides, and chlorophyllides. [After J. Beck, J.-U. von Schütz, and H.C. Wolf, in *Photochemistry and Photobiology*, ed. by A.H. Zewail, Harwood Acad. Publ. (1983)]

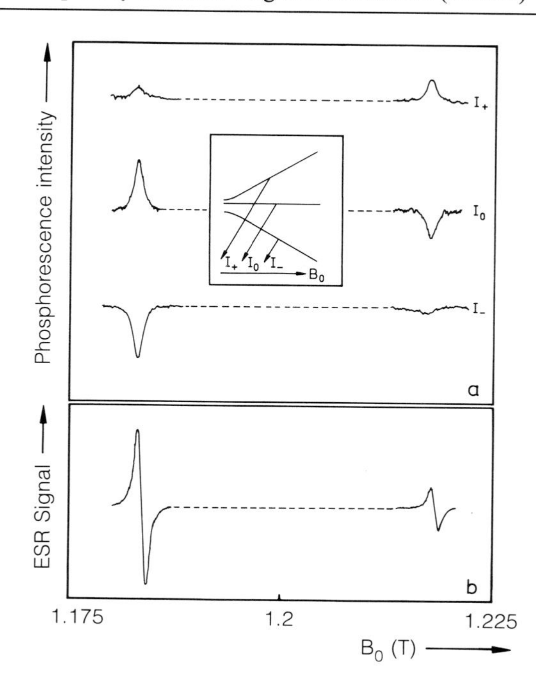

Fig. 19.15. An ODMR spectrum taken in a high magnetic field and measured at 35 GHz, of triplet excitons in a 1,4-dibromonapthalene crystal. In the *upper part of the figure*, the ODMR spectrum of the triplet terms $m_S = +1, 0, -1$ is plotted; the *lower part of the figure* shows, for comparison, the ESR spectrum with conventional detection. In the ODMR spectrum, the spectrally resolved phosphorescence emssion from the three triplet sublevels +, 0, and − is observed as a function of the applied magnetic field, with simultaneous irradiation at the microwave resonance frequency of 35 GHz. The resulting ESR transitions change the populations of the three sublevels and thus the intensities of the three phosphorescence components. In this way, one can determine the fine structure constants as well as the relative population and depopulation rates of the levels. [After R. Schmidberger and H.C. Wolf, Chem. Phys. Lett. **16**, 402 (1972)]

In the examples of optically detected resonance which we have given so far, the resonance detection was accomplished by observation of the triplet–singlet emission, i.e. phosphorescence. One can readily understand by referring to Fig. 19.9 that a change of the level populations and thus of the effective lifetime of the triplet state T_1 can also lead to a change in the steady-state population of the ground state S_0 and the excited state S_1, since these three states are strongly coupled together in the steady state, so that any change in the population of one of them has an influence on the other two. This is the basis for the *detection of magnetic resonance using* $S_1 \rightarrow S_0$ *fluorescence* (FDMR, Fluorescence Detected Magnetic Resonance) or using the *absorption* $S_1 \leftarrow S_0$ (ADMR, Absorption Detected Magnetic Resonance). These methods become especially important when the intensity of the phosphorescence $T_1 \rightarrow S_0$ is too weak to permit its use for detection.

Finally, this double resonance technique can also be inverted, by measuring the optical absorption spectrum $S_1 \leftarrow S_0$ or the phosphorescence spectrum $T_1 \rightarrow S_0$ with simultaneous irradiation of one of the microwave transitions between τ_x, τ_y, and τ_z in Fig. 19.9, or between the states $|+1\rangle$, $|0\rangle$, and $|-1\rangle$ in Fig. 19.10. If the detection apparatus is modulated at this microwave frequency, one observes only the absorption or phosphorescence spectrum of that particular molecule which has its microwave resonance at the frequency being applied. A superposition of the absorption or phosphorescence spectra of a number of different molecules can then be resolved into individual spectra; one thus obtains the correspondence between the microwave frequency and each particular molecule. These methods are termed *Microwave Induced Absorption* (MIA) or *Phosphorescence Microwave Double Resonance* (PMDR).

The ODMR method can, by the way, be applied not only in zero field, but also with success in an external field B_0. An example of such high-field ODMR measurements is shown in Fig. 19.15, which represents spectra obtained from the triplet state T_1 of 1,4-dibromonapthalene, taken at a microwave frequency of about 35 GHz using an ESR spectrometer and a monochromator for the optical detection.

19.8 Applications of ESR

In summary, we can say that electron spin resonance is an important method for the determination of the electronic and the geometric structures of paramagnetic molecules, and thus is a significant technique of molecular physics. These structural determinations are accomplished especially through the observation of the hyperfine structure and fine structure. From linewidths and relaxation times, one obtains information on the motions of spins, molecular groups, and whole molecules, as well as on molecular reactions. Finally, the method is particularly suitable for the simple and exact determination of the concentrations of unpaired spins, and thus also for the measurement of spin susceptibilities, as we have already pointed out in Sect. 3.6.

20. Macromolecules, Biomolecules, and Supermolecules

This chapter is intended to widen the horizons of physicists with regard to the enormous variety of large molecules, and to their significance for physics, chemistry, and biology (Sect. 20.1), as well as to the question of what molecular physicists have already contributed, and can in the future contribute, to this field. Our overview here must necessarily be limited and thus rather superficial. It is intended in the main for students of physics, who might otherwise be left with the impression after studying a text on molecular physics that diatomic molecules are the most important. The present chapter is intended to make it clear that this is not so; and that, on the contrary, the investigation of complex molecular functional units (Sects. 20.5 to 20.7) offers many fascinating and unanswered questions to challenge the molecular physicist.

20.1 Their Significance for Physics, Chemistry, and Biology

In molecular physics, small molecules have initially occupied the most prominent place, for good reasons. The physicist wishes to determine, understand and calculate the physical properties of the objects investigated in the most precise and complete manner possible. This is incomparably more straightforward in the case of a small diatomic molecule such as HCl than for a large molecule like chlorophyll, not to mention a macromolecule from among the proteins, for example.

The variety of molecules is however enormous, and it becomes even more so when one considers larger molecules. An essential component of this variety is the formation of hybrid bonds by the carbon atom and the resulting ability to enter into multiple covalent bonding. Macromolecules are everywhere; they form the material basis of all biological structures and processes. Life and all biological processes, the occurrences within living cells, within organisms, and their interactions with the environment, are all due essentially to the chemistry of macromolecules. Along with natural macromolecules such as proteins, DNA, or cellulose, we find in our modern environment a great number of synthetic macromolecules like polyethylene, polystyrene, or Teflon. One can hardly imagine our present-day life without these materials.

When macromolecules are formed out of many, usually identical small molecular units, called monomers, then we term them *polymers*. Macromolecules can also be formed when identical or different molecules group together to give new units under the influence of intermolecular forces, especially Van der Waals forces; one then obtains *supermolecules, molecular clusters*, and *inclusion compounds*. Molecules can also form larger units through covalent binding, yielding *supramolecules*. When their functions, e.g. in biological systems, can be carried out only through the form of these specifically organized macromolecules, they are also called *molecular functional units*.

The investigation and understanding of such molecular functional units is still in its early stages. The contributions of physicists have initially been in the area of structure determinations using the physical methods of X-ray and neutron diffraction and magnetic resonance techniques. Physicists can develop appropriate spectroscopic methods and apply them in order to understand the statics, the dynamics, and the reactive behaviour of these functional units. Physics has an important role to play in understanding the functions of these molecules and how they are determined by molecular organisation, and in the investigation of their conformation and configuration, that is the construction of the molecules from atomic units and their spatial structures, as well as the possible reactions of the molecules with each other and with other molecules. It can be confidently predicted that in these fields, the interdisciplinary, collaborative work of physicists, chemists, and biologists will have a great and increasing significance in the future. In these sections, we attempt to show which contributions can be made specifically by physicists, using as examples a small and subjective selection of supramolecules, macromolecules, and functional units.

20.2 Polymers

Among the largest artificially produced "molecules" are the so-called polymers. A typical polymer, which is familiar from daily life, is *polyethylene*, a saturated hydrocarbon with the formula $(CH_2)_n$ (Fig. 20.1, upper part), where n lies between 5 000 and 50 000. The polymer molecules consist of long saturated hydrocarbon chains with differing chain lengths. In the polymeric material, there is no strict mutual ordering of these chains, aside from molecules which are oriented parallel to one another over limited regions of space.

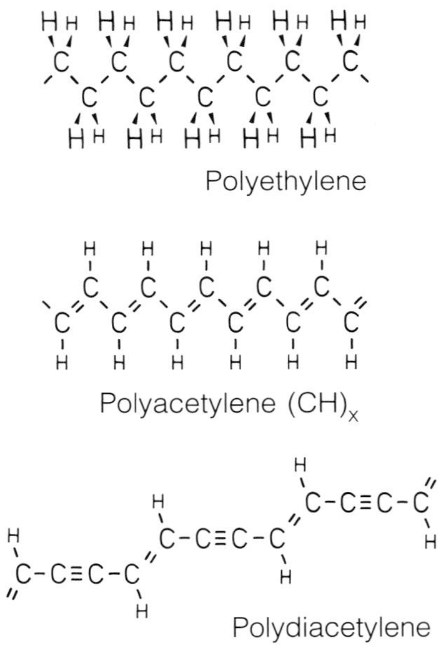

Polyethylene

Polyacetylene $(CH)_x$

Polydiacetylene

Fig. 20.1. The molecular structures of three important classes of polymers

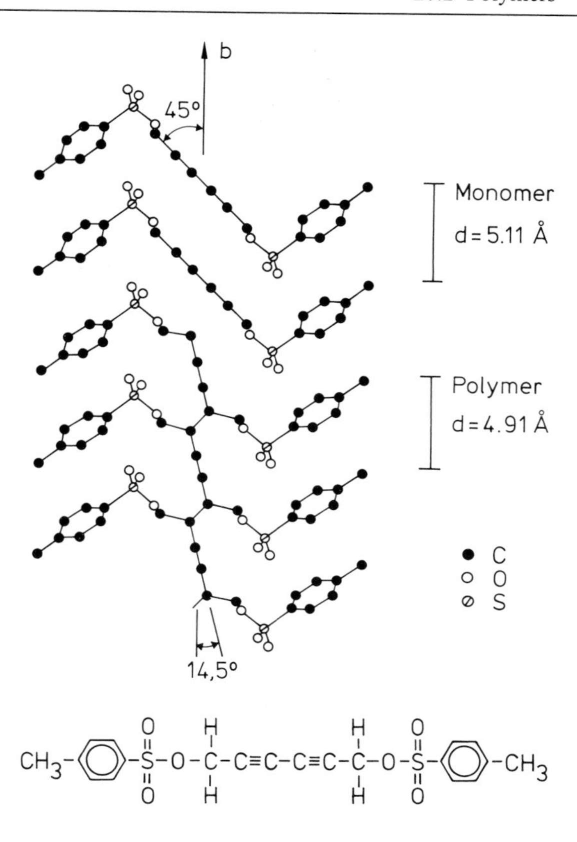

Fig. 20.2. The polymerisation of polydiacetylene. In the upper part of the figure, the X-ray structure of the monomeric diacetylene TS6 is shown (the notation refers to the particular side group, in general denoted by *R*); in the lower part of the figure, the structure of the polymerized crystal is indicated. The protons in the molecule are not shown. The complete formula of the TS6 monomer is given at the bottom of the figure

For physicists, the polymer *polydiacetylene* is of particular interest (Fig. 20.1, lower part), since the polymerisation reaction can take place within a single crystal of the subunits. It is formed from the monomer R–C≡C–C≡C–R′ following the mechanism illustrated in Fig. 20.2. Here, R and R′ stand for two of many possible substituent groups, of which Fig. 20.2 shows one example. The groups R and R′ can also be the same.

On polymerisation, the neighbouring monomers, which are stacked one above the other in the crystal, bind together to form chains by means of a so-called 1,4 addition, forming various reaction intermediates in the process (1 and 4 refer to the numbering of the 4 central C atoms in the monomers). To induce the polymerisation and to maintain its progress, radiation and/or thermal energy are necessary. Accompanied by reorientation of the molecules and the formation of new bonds, first dimers, then trimers, tetramers (Fig. 20.3), and finally polymers with a large value of n are formed. The special feature of this reaction is that it takes place in the solid state in a single crystal, retaining the crystalline order; a single crystal of the monomer molecules is therefore converted into a single polymer crystal. The uniform orientation of the polymer molecules in the ordered crystal allows the investigation of the initial and intermediate stages of the polymerisation using spectroscopic methods, which is not possible for other polymerisation reactions. The intermediate products of the polymerisation are diradicals, that is molecules with an unpaired electron at each end, as well as carbenes, i.e. molecules with reactive end groups containing a C atom, and two non-saturated electrons ("dangling bonds"). Excitation by optical radiation is necessary to initiate

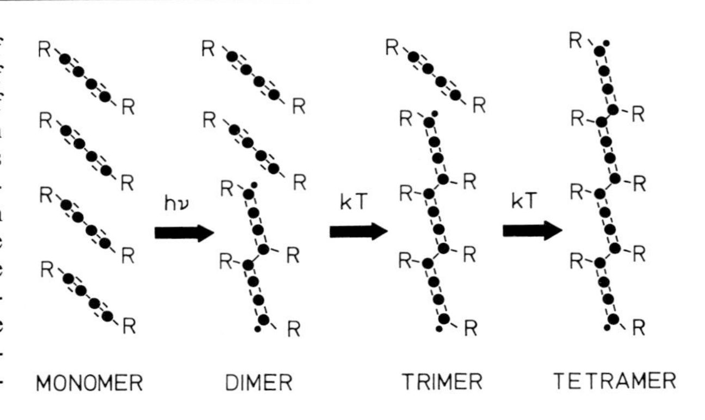

Fig. 20.3. The mechanism of the solid-state polymerisation of diacetylene. The formation of dimers is initiated by irradiation with light; the further progress of the polymerisation is thermally activated as an addition reaction. At the ends of the oligomer chains, the reactive radical electrons of the diradicals which are formed in the course of the polymerisation reaction are indicated as dots. After Sixl

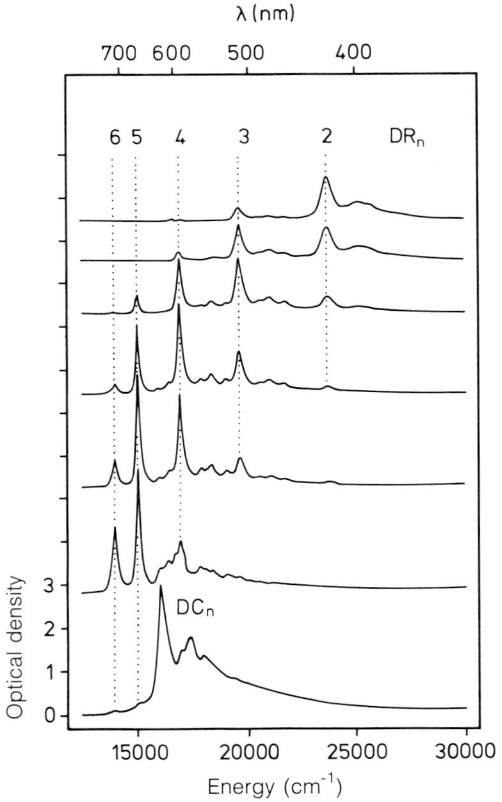

Fig. 20.4. The absorption spectrum of a diacetylene crystal after the initiation of the polymerisation reaction (*upper curve*) and at various times during the following thermally activated addition reaction. The tempering times at 100 K were (from *top* to *bottom*) 0, 15, 35, 65, 80, 150, and 240 min. The numbers 2, 3, 4, 5, and 6 refer to the intermediate diradical stages DR_n going from the dimer to the hexamer; for $n \geq 6$, one sees essentially the dicarbenes DC_n. These are molecules with carbene-like (–C–) reactive end groups at each end. After Sixl

the transformation steps, and thermal energy is required to maintain the addition reaction, i.e. to continue the polymerisation. These processes can be followed step by step using the methods of optical and ESR spectroscopy and can thus be investigated in detail. Figure 20.4 shows as an example the absorption spectrum of the oligomers during the course of the polymerisation reaction. Oligomeric molecules, i.e. molecules which are made up of only a few monomeric units, thus become accessible to investigation and can at the same time be used as probes for the elucidation of the polymerisation process.

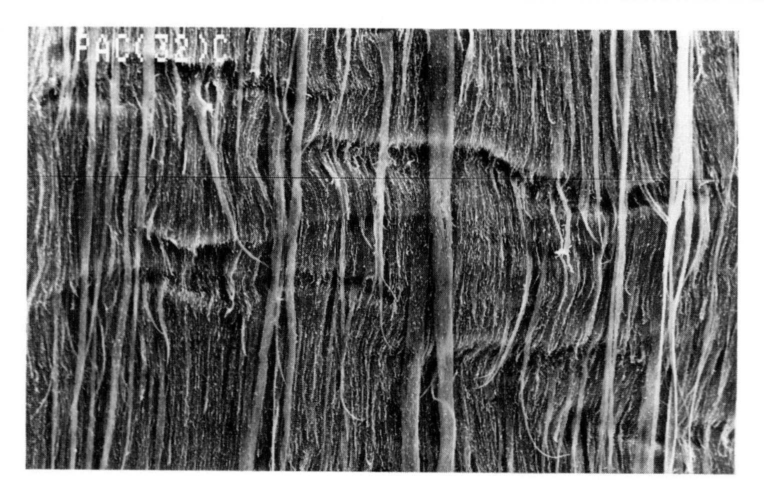

Fig. 20.5. A scanning electron microscope image of a polyacetylene foil. The foil contains more or less disordered strands of polymer molecules. A certain degree of preferred orientation was obtained by stretching the foil. The length of the picture corresponds to 125 μm. (Kindly placed at our disposal by M. Schwoerer)

The solid-state polymerisation of diacetylene is furthermore a good example of the "topochemical principle". This states that solid-state reactions of molecules proceed with a minimum of molecular rearrangements. The monomer molecules must be ordered within the crystal in such a way that they can form chemical bonds with their neighbours with only a small deviation from their equilibrium shapes and positions. This is made possible in diacetylene by attaching suitable substituent groups. During the reaction, the lattice parameters of the crystal may change only slightly; only then is it possible that the single crystal will remain intact throughout the solid-state polymerisation.

Another molecule which has been of particular interest to molecular physics is polyacetylene, $(CH)_n$ (Fig. 20.1, centre). This long-chain polyene has alternating double and single bonds and can, like most polymers, be prepared only in disordered or partially ordered form. This is illustrated by Fig. 20.5. Polyacetylene is a polymer which can be made electrically conducting by "doping": the addition of oxidants such as AsF_5 or $FeCl_3$, or of reducing agents such as alkali metals, i.e. the production of an excess or a deficiency of electrons, increases the electrical conductivity by many orders of magnitude to values of order 10^5 $S\,cm^{-1}$. The "record" at present is 800 000 $S\,cm^{-1}$; in this way, one can obtain conducting polymers.

To understand the electrical conduction in polyacetylene, the concept of conductivity through "defects" has been developed; they are referred to as "solitons" by theoretical physicists. In this concept, so-called alternating-bond defects are important, such as those found in molecules of the polyacetylene structure, or as formed by breaking a double bond and thus creating an electron-hole pair. Figure 20.6 shows as an example the removal of an unpaired electron by a dopant, giving a missing electron or hole, at such a defect. It is distributed over several carbon atoms in the molecule and can propagate as a wave along the molecular chain owing to the quantum-mechanical energy equivalence; exchange with neighbouring chains is also possible. This mechanism makes the conductivity of such molecular systems understandable. The search for the optimum conditions under which one can obtain conducting polymers or, in general, molecular systems with a high electrical conductivity is one of the currently important research areas of molecular physics.

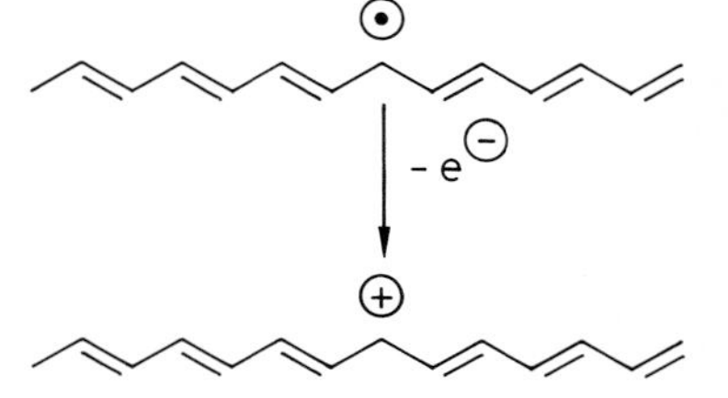

Fig. 20.6. A neutral and a charged "soliton" defect in the conjugated π-electron chain of polyacetylene. The upper part of the picture shows a neutral soliton, consisting of an alternating-bond defect and an unpaired electron (i.e. a free radical), which can move along the conjugated chain. In the lower part of the picture, an electron has been removed from the neutral soliton by oxidation. The remaining positively-charged soliton is likewise delocalized along the chain and is mobile

20.3 Molecular Recognition, Molecular Inclusion

Can one molecule "recognize" another? Can it choose a specific molecule from within a group of others, identify it and bind to it? These questions occupy the attention of a modern branch of chemistry, which is referred to as supermolecular chemistry. Molecular interactions are indeed the basis of many highly selective recognition, transport, and regulatory processes in biological systems. As examples we mention the binding of a biologically important molecule to a protein complex, enzymatically regulated reactions, antigen-antibody association, or, as the most spectacular example, the replication and transcription of the genetic code in the reproduction and genetics of living organisms.

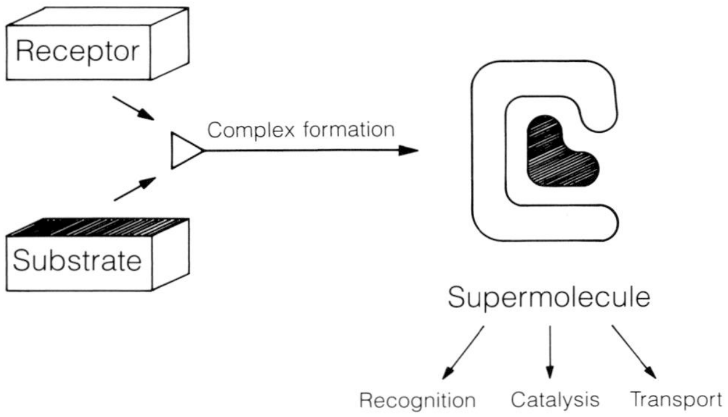

Fig. 20.7. The "receptor" and the "substrate" bind together to form a supermolecule or supermolecular complex. This forms the basis for the concept of of "molecular recognition", important in the catalysis and transport processes of molecular units. After J.M. Lehn

Molecular recognition can be carried out by *supermolecules*. According to K.L. *Wolf*, these are molecules whose covalent bonds are saturated and which are held together by intermolecular forces. The smaller component is termed the "substrate", the larger the "receptor"; "supermolecules" are thus formed, as illustrated schematically in Fig. 20.7. The binding of the substrate to the receptor must be selective and specific. Molecular recognition is then based upon the storage and readout of information with the aid of such supermolecules, formed by the selective interactions of suitable molecules. The fitting together of the substrate and the receptor must/can be determined either electronically or geometrically, corresponding to the famous postulate of E. *Fischer* (1894), according to which the molecules fit together like a key into a lock.

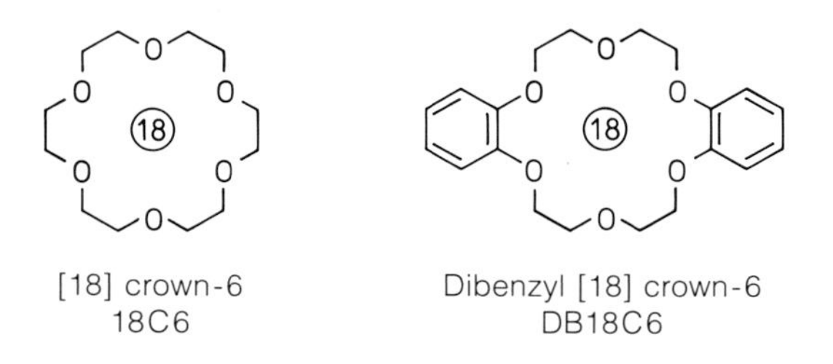

Fig. 20.8. Two crown ether molecules; the C atoms are not drawn in. *Notation*: the ring size, that is the number of bonds in the ring, is first given in square brackets, followed by "crown" or "C" for the class of molecules, and finally the number of donor positions is noted. In the ethers shown, these are the 6 oxygen atoms

As an example, in Fig. 20.8 we show a molecule from the series of the crown ethers, which play an important role in this field of molecular physics. In this molecule, the ring of six O atoms and twelve C atoms forms a cavity which is just the right size to permit a

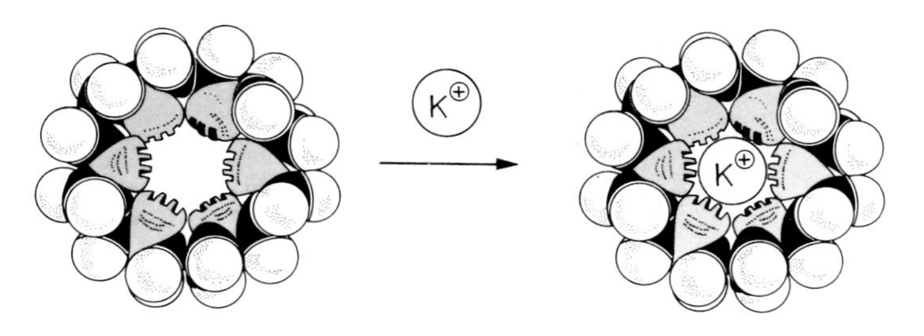

Fig. 20.9. Using a hard-sphere model, it can be shown that a K^+ ion just fits into the cavity of the crown ether [18] crown 6 and forms a complex there. After Vögtle

K^+ ion to fit into it in an optimal manner (Fig. 20.9). A larger alkali-element ion would no longer fit into the cavity, while a smaller one would be less strongly bound. If one wishes to prepare a supermolecule with a different alkali ion in the centre, then another crown ether with a suitable specific cavity size can be chosen; cf. Table 20.1.

Table 20.1. A comparison of the diameters of various alkali-metal cations and crown compounds*

Cation	Diameter [pm]	Crown	Diameter [pm]
Li^+	136	[12] C4 (9)	120–150
Na^+	190	[15] C5 (10)	170–220
K^+	266	[18] C6 (7)	260–320
Cs^+	338	[21] C7 (11)	340–430

* After Vögtle

In recent years, many molecules have been prepared which are effective in the selective binding of substrates having quite a range of shapes, sizes, and structures. They have an eminent value for reactions, catalysis, transport processes, and most especially as models for biochemical reactions. This new supermolecular chemistry will have to be studied by chemists and physicists working together if the great research potential which it promises is to be realized.

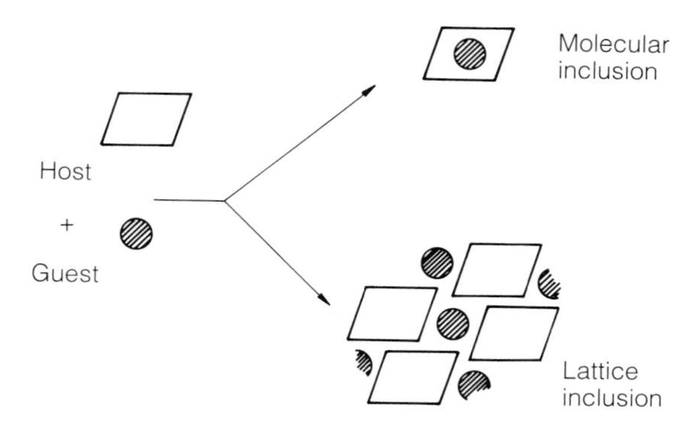

Fig. 20.10. Inclusion compounds can be formed by the molecular enclosure of a guest atom or molecule in a host molecule, or by lattice inclusion within the crystal lattice of the host molecules. After Vögtle

Supermolecules or molecular structures consisting of two or more different molecules have long been known in the form of *inclusion compounds* or clathrates. Such compounds can accept other molecules of suitable size into cavities in their own structures or lattices and bind them there. In this process, binding forces are not decisive; instead, the geometrical

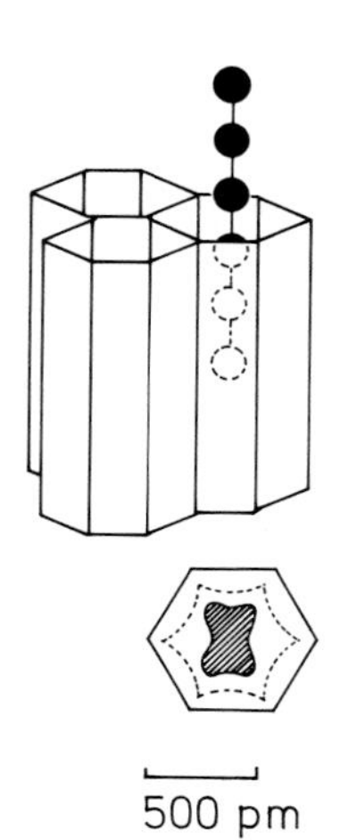

Fig. 20.11. The lattice of the urea molecular crystal permits inclusion of paraffine molecules. After Vögtle

enclosure is important. It can be a molecular enclosure as in the crown ethers, or an inclusion into the crystal lattice, as exemplified by Fig. 20.10. An example of a lattice inclusion is shown in Fig. 20.11, which illustrates the lattice structure of urea, a molecule having the formula

$$\begin{array}{c} H_2N{-}C{-}NH_2 \\ \| \\ O \end{array}$$

The cavities of its lattice can trap for example n-paraffines, n-fatty acids, halogenated hydrocarbons, or indeed even benzene molecules. The channel-like cavities in the urea lattice have a diameter of 520 pm; this limits the size of the molecules which can be bound there.

A number of host substances for clathrates are known, with a great variety of guest molecules. Such molecular complexes have great practical as well as theoretical interest. When inclusion is made possible by the specific lattice structure, then these molecular complexes also bridge the gap to solid state physics.

20.4 Energy Transfer, Sensitisation

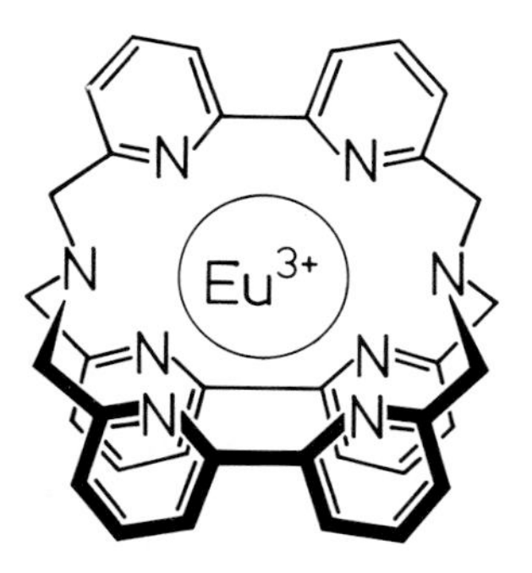

Fig. 20.12. The molecular structure of a europium (III) cryptate with a macrobicyclic polypyridine ligand, enclosing the Eu^{3+} ion. [After J.M. Lehn, Angew. Chem. **100**, 92 (1988)]

Molecular systems can also lead to the conversion of optical radiation along the lines of the following series of steps: light absorption by one partner in the system; transfer of the excitation energy to the other partner; and emission of radiation by the latter. An example from the field of supermolecular chemistry is the europium (III) cryptate of the macrobicyclic ligand polypyridine, shown in Fig. 20.12. The term cryptate is applied to molecular complexes in which one partner (the Eu ion) is almost completely enclosed by the other partner. Although the free europium ion in solution does not fluoresce, the complexes exhibit strong luminescence. As is shown by the excitation spectrum (Fig. 20.13), this luminescence is excited by the absorption of light in the organic ligand, followed by energy transfer to the emitting Eu ion. By enclosing the metal ion in its molecular cavity, this ligand at the same time prevents the supression of the Eu emission by radiationless deactivation processes via the solvent molecules.

Transfer of electronic excitation energy between molecules over somewhat greater distances is an important process in many molecular systems. Such a transfer over distances of up to about 100 Å can be understood classically in the following manner:

A molecule D (for donor), shown in Fig. 20.14, is first optically excited; it could then give up its excitation energy through fluorescence. Regarded as an oscillating dipole, which can radiate electromagnetic waves like an antenna, it could however also excite a molecule A (for acceptor) at a distance r, if this molecule responds to the frequency of molecule D. Using this simple model, one can calculate the distance r_0 at which the probabilities of the direct radiation of light by D and the radiationless transfer from D to A are the same. With the classical oscillator model, one finds the following expression for r_0:

$$r_0 = \left(\frac{C\varepsilon}{n^4 \nu^4} \right)^{1/6} , \qquad (20.1)$$

where C is a numerical factor which takes into account the relative orientation of the two molecules, ε is the absorption coefficient of molecule A at the frequency ν, which is also the oscillation frequency of the oscillator D, and n is the index of refraction of the medium

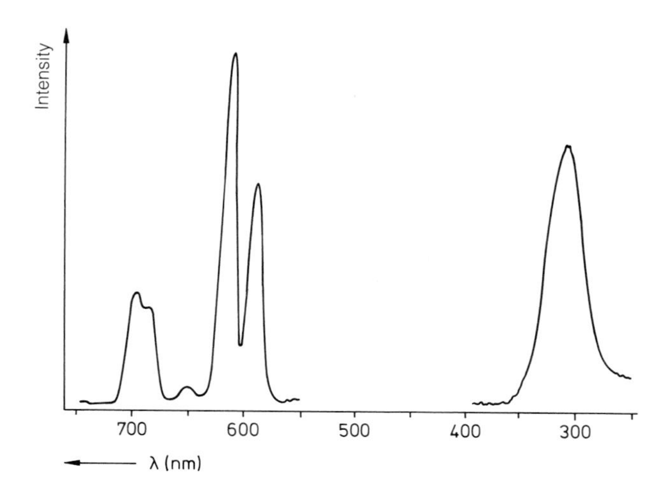

Fig. 20.13. *Right*: an excitation spectrum, that is the spectrum of absorbed light which gives rise to an emission at 700 nm, corresponding to the fluorescence of the molecule shown in Fig. 20.12. *Left*: the emission spectrum; it belongs to Eu^{3+}, while the absorption is accomplished by the organic ligand. This follows from the agreement of the excitation spectrum shown with an absorption band of the organic macrocycle. This is an example of intramolecular energy transfer from excited states of the ligand to excited states of the Eu^{3+} ion. The latter emits with a high yield, owing to its screening from the solvent by the organic partner. The measurements were carried out on a 10^{-6} molar solution at room temperature

between the two molecules. Since for molecules both the emission frequency ν and the absorption ε are spread over a certain frequency range $\Delta\nu$, this bandwidth would have to be taken into account in a more exact treatment.

Such an energy transfer, which was first calculated by Förster and is therefore known as the *Förster process*, can also be formulated quantum mechanically, as well as for the case that the interaction between the donor and the acceptor is not the dipole-dipole interaction but rather the exchange interaction. Typical values of the Förster radii r_0, i.e. for the distances at which this energy transfer is important, are of the order of 50 Å for organic molecules.

An important application of energy transfer in photophysics is *sensitized fluorescence*. It is responsible for the fact that solid anthracene (cf. Fig. 14.21) containing a relative concentration of only 10^{-5} of tetracene as an impurity or "guest"molecule does not fluoresce in the blue-violet as does pure anthracene, but instead in the yellow-green. The energy absorbed by anthracene is transferred with a high efficiency to the tetracene impurities and re-emitted by them; see Fig. 20.15.

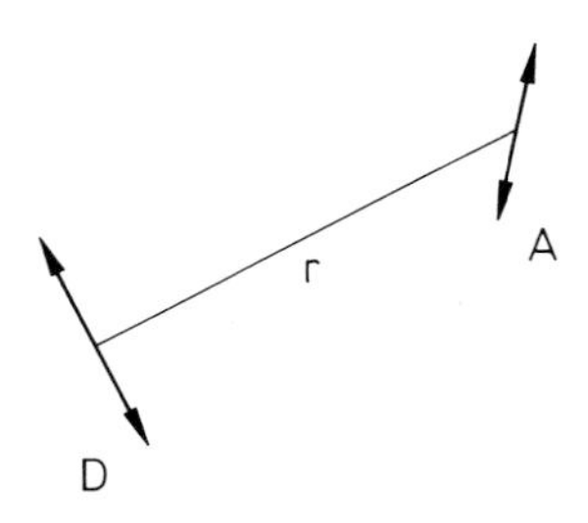

Fig. 20.14. Energy transfer by the Förster process: the light energy absorbed by molecule 1, the donor D, can be transferred to molecule 2, the acceptor A, at a distance r. The probability for this process can be calculated in the simplest case using a classical oscillator model

This can also be expressed in the following way: by dissolving tetracene in anthracene, one can increase the probability of its excitation through light by many orders of magnitude. By using suitable energy transfer systems, one can thus direct the energy preferentially to particular molecules, for example to molecules in which a photochemical reaction is desired.

Such processes play an important role in biological systems, for example in the so-called antenna complexes which collect light for photosynthesis (cf. Sect. 20.6), as well as in technical applications, e.g. in photography.

20.5 Molecules for Photoreactions in Biology

Life on the Earth requires sunlight; without it, the natural lifeforms all around us could not exist. The Sun supplies the energy required for photosynthesis, i.e. for the process which is indispensible to the formation of the material structure and the functions of natural lifeforms. Many other processes which are important to life employ light –we mention here only the phototropic growth of plants, and vision in the realm of animals and human beings.

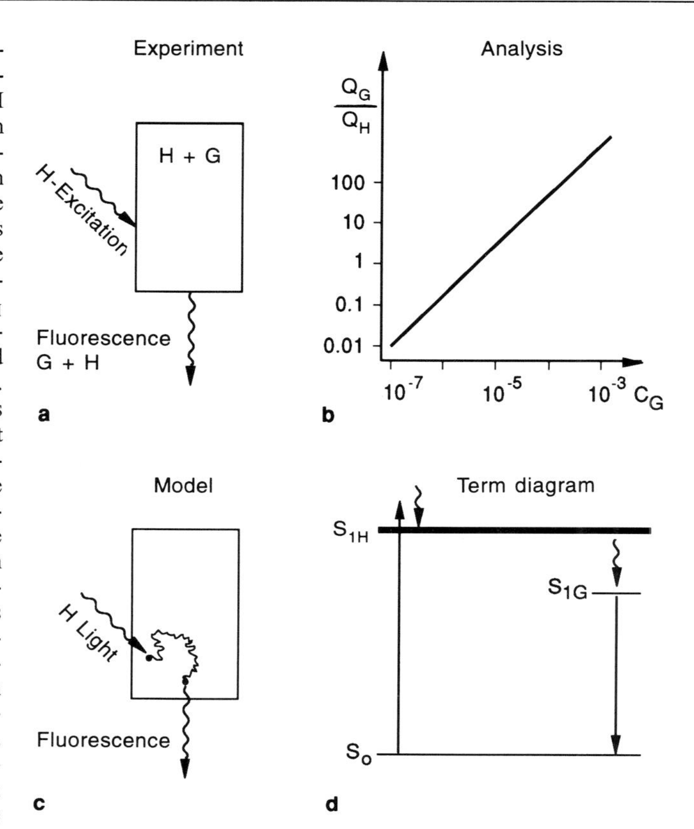

Fig. 20.15a–d. A schematic illustration of sensitized fluorescence. (**a**) The host crystal H contains a small concentration of guest molecules G. The crystal is exposed to light, which is absorbed primarily by the host. The fluorescence consists mainly of guest light. (**b**) The fluorescence light quanta emitted from host and guest, Q_H and Q_G, differ in their spectra, i.e. in their colours, and can thus be detected separately. The quantum ratio Q_G/Q_H is plotted here against the guest concentration C_G. (**c**) The excitation light is absorbed in the crystal by the host; the excitation energy diffuses within the crystal as S_1 excitons. When an exciton reaches a guest molecule G during its lifetime, it is trapped there, and light is emitted by the guest G. (**d**) Excitation energy from the exciton band S_{1H} of the host is transferred to the guest molecule, which is isolated in the host lattice. Emission takes place from the S_{1G} excitation state of the guest. [After H.C. Wolf, *Die Feste Materie*, Umschau Verlag, Frankfurt (1973)]

All these fundamental chemical processes in the biological world take place via photoreceptors. Molecules contained in them are responsible for the primary processes of photochemistry and photophysics in living organisms. They absorb light and transfer the absorbed energy to a site where the primary chemical processes are initiated. In order to understand their functions, one requires the methods of investigation of the molecular physicist.

The most important of these molecules is *chlorophyll*, Fig. 20.16. It consists of a large ring, the macrocycle of the porphyrin ring with delocalized electrons, i.e. conjugated double bonds, responsible for the characteristic light absorption, and a so-called phytol tail, which serves to make the molecules soluble in a membrane and to attach them in the proper way to the associated proteins. The absorption spectra of two somewhat different chlorophyll molecules and of carotene, i.e. of molecules which can selectively absorb light, are shown in Fig. 20.17. The absorption spectrum can be shifted somewhat by making slight modifications in the molecular structure. Living organisms make use of this fact when they exist under conditions where the light has a different spectral composition than that of normal sunlight, for example in swampy water. In the process of photosynthesis, chlorophyll molecules have at least two functions: they serve to absorb radiation energy as light antennae, and they are responsible for the primary process of photosynthesis in the reaction centre, namely the primary charge separation, the transfer of an electron.

Another group of pigment molecules important for photosynthesis in plants and animals are the *carotenoids*. One member of this group, β-carotene, is also shown in Figs. 20.16 and 20.17. In this case, the electronic structure which is responsible for the photoreaction is a linear chain of alternating double and single bonds, a polyene structure (cf. Sect. 14.6). Carotenoids, that is molecules with a basic structure similar to that of β-carotene, are contained in the photosynthesis apparatus of many photosynthetic plants and bacteria.

The absorption spectrum of carotenoids (Fig. 20.17) is complementary to that of chlorophyll and therefore allows a more efficient use of the solar spectrum by the "antennae" which are responsible for light collection in the photosynthesis apparatus. In addition, the lowest-lying excited triplet state of these molecules, in the infra-red at about 8000 cm^{-1}, is also important: it provides for the radiationless removal of excess energy, when more light quanta are absorbed by the chlorophyll than can be processed by the photosynthesis apparatus. It thereby protects the organisms against damage when they are exposed to too much light, and especially against the formation of the extremely reactive excited singlet state of oxygen, $^1O_2^*$ (Fig. 13.9 and 14.1).

This state can, in fact, be formed by a radiationless transition from the triplet ground state, $^3O_2$, according to the reaction scheme:

$$^3O_2 + {}^3\text{chlorophyll}^* \rightarrow {}^1O_2^* + {}^1\text{chlorophyll} \, ,$$

when the deexcitation of the chlorophyll molecule in its triplet ground state ^{3}chlorophyll is not complete, owing to strong irradiation by light. With the reaction

$$^3\text{chlorophyll}^* + {}^1\text{CAR} \rightarrow {}^1\text{chlorophyll} + {}^3\text{CAR}^* \, ,$$

the carotenoid opens an additional deexcitation channel for the removal of excited-state chlorophyll and thereby prevents the abovementioned competing reaction from forming $^1O_2^*$. The carotenoids thus make possible a more rapid removal of excess population from the excited ^{3}chlorophyll* state. (We use an asterisk here to denote an excited electronic state.) Excited singlet oxygen already formed is converted to harmless $^3O_2$ by the reaction

$$^1O_2^* + \text{CAR} \rightarrow {}^3O_2 + {}^3\text{CAR}^* \, .$$

The excited triplet state ^{3}CAR* which results from this reaction rapidly gives up its energy as follows:

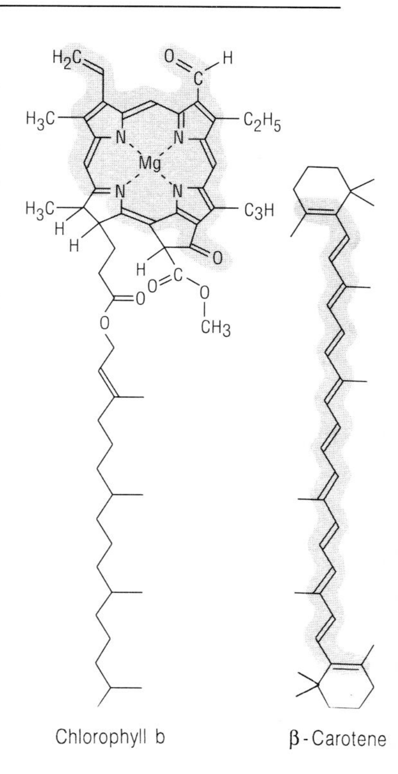

Fig. 20.16. The molecular structures of chlorophyll *b* and of β-carotene. The areas containing conjugated double bonds, which are essential for the absorption of light, are shaded in the figure. The C atoms are not shown in the rings or chains, and the H atoms not at all. Dashes without an atomic symbol indicate CH_3 groups. After Dickerson and Geis

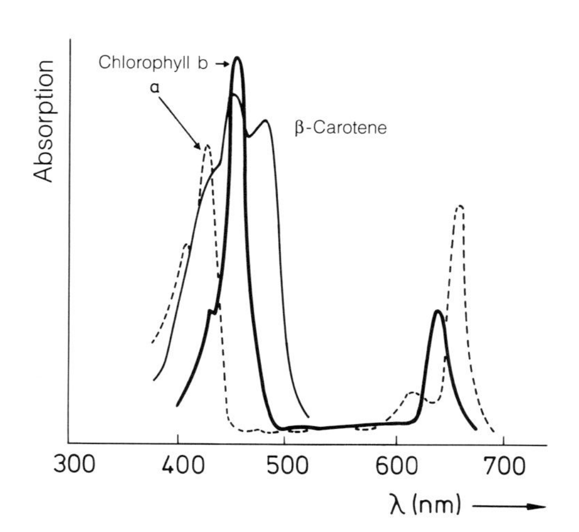

Fig. 20.17. The absorption spectrum in the visible of the two chlorophyll molecules chlorophyll *a* and *b*, and of β-carotene

$$^3CAR^* \rightarrow {}^1CAR + \text{phonons}\ .$$

This is the basis for the protective action of carotenoids against light-sensitized reactions.

The light energy absorbed by the pigment molecules in the antenna system, i.e. the chlorophylls and carotenoids, is conducted within the molecular structure of the photosynthesis apparatus to the reaction centre. There, the photochemical reaction begins.

The molecule responsible for the process of light detection in the eye has, like carotene, a chain of conjugated double bonds as its absorbing structure: it is called *retinal* (Fig. 20.18). When it is bound together with the protein opsin, the resulting complex is named rhodopsin. In this molecule, absorbed light causes a cis-trans isomerisation, that is the spatial reorientation of a bond, as shown in Fig. 20.18. This intramolecular reorientation, the primary process of vision, is then converted into a nerve impulse in the optic nerve by the molecular functional units surrounding the rhodopsin molecules. Such intramolecular reorientations play an important role in many other processes of photobiology, e.g in the regulation of plant growth by light input through so-called phytochromic molecules, which are related

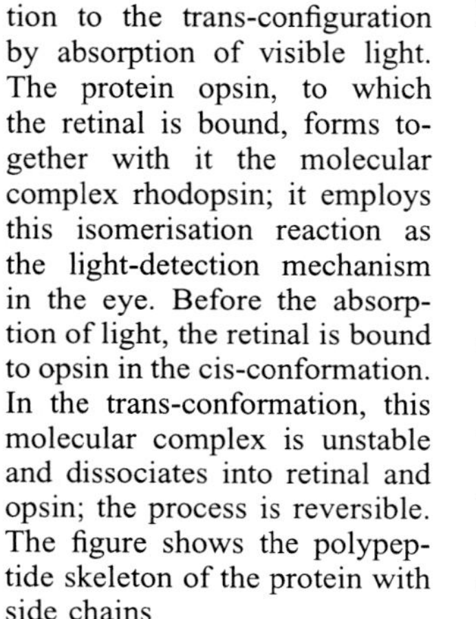

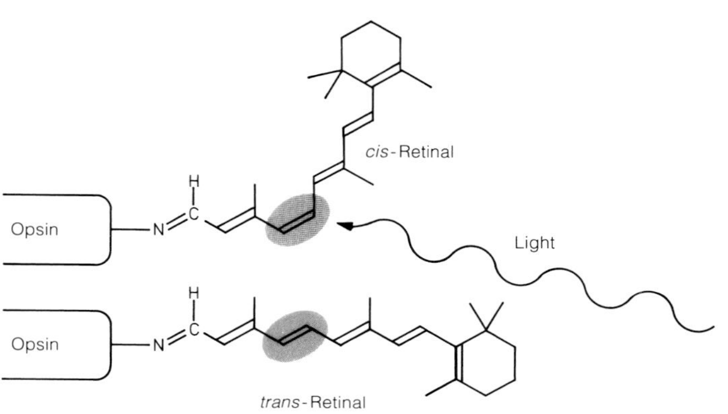

Fig. 20.18. Retinal can be isomerized from the cis-configuration to the trans-configuration by absorption of visible light. The protein opsin, to which the retinal is bound, forms together with it the molecular complex rhodopsin; it employs this isomerisation reaction as the light-detection mechanism in the eye. Before the absorption of light, the retinal is bound to opsin in the cis-conformation. In the trans-conformation, this molecular complex is unstable and dissociates into retinal and opsin; the process is reversible. The figure shows the polypeptide skeleton of the protein with side chains

Fig. 20.19. A schematic representation of the photoisomerisation of 11-cis-rhodopsin. Light is absorbed by the ground state S_0, raising it to the excited state S_1. The colour is blue-green, and the pulse length amounts to 10 fs. From the excited state, a radiationless transition to the alltrans form takes place. After C.V. Shank et al., Science **254**, 412 (1991)

to retinal and in which likewise a cis-trans isomerisation represents the primary step of the photoreaction.

With modern laser-spectroscopic methods, one can even measure the velocities of such isomerisation processes. Figure 20.19 shows a schematic representation of the photoprocess in rhodopsin obtained from bovine eyes. The photoisomerisation is completed after 200 femtoseconds.

20.6 Molecules as the Basic Units of Life

The most important molecular components in the biological realm are the proteins and the nucleic acids. They provide the basic units of the molecular structures from which all lifeforms are constructed. Proteins are the fundamental building blocks of living organisms, while nucleic acids provide for storage and transfer of information.

Proteins are polymers constructed from amino acids. The latter are bifunctional molecules, i.e. molecules with two reactive end groups, an amino group, $-NH_2$, at one end of the molecule, and a carboxylic acid group, $-COOH$, at the other. The general structural formula of an amino acid is:

$$\begin{array}{c} R \\ | \\ H_2N\text{–}C\text{–}COOH \\ | \\ H \end{array} .$$

In this formula, R represents one of many possible side groups, which can have rather diverse chemical properties. If R is simply an H atom, for example, then the resulting amino acid is called glycine; with the side group

$$-CH\begin{array}{l} \diagup CH_3 \\ \diagdown CH_3 \end{array}$$

it is valine. Amino acids can polymerize to give so-called polypeptides (Fig. 20.20) through a

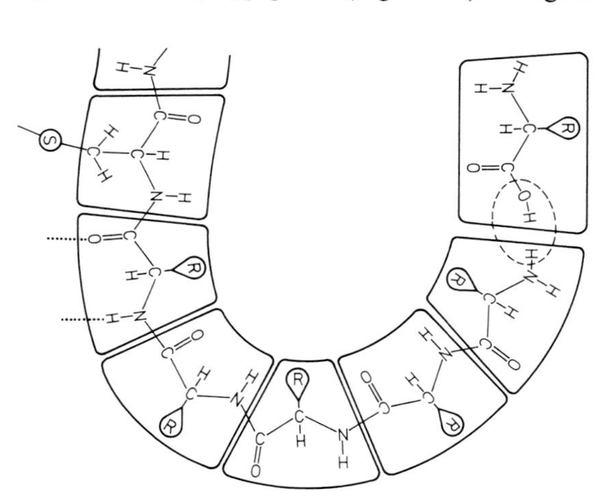

Fig. 20.20. A schematic representation of a peptide chain. Through elimination of a water molecule, amino acids can bind together, forming so-called peptide bonds. The backbone of the chain consists of many identical units. The variable side chains R determine the characteristic properties of each protein chain. The symbol S indicates here a crosslink to another part of the chain in the form of a disulphide bond. A different type of crosslinking is formed by hydrogen bonds (*dashed lines*). After Dickerson and Geis

condensation of their reactive end groups with elimination of H_2O. The sequence of different amino acids, distinguished by their side groups R, is relevant for the biological properties of the polypeptides. These chains with typical lengths of 60 to 600 amino acids coil up and fold to form a three-dimensional protein structure (Fig. 20.21). The molecular masses of such molecules are in the range 10 000 to 200 000. In nature, just 20 different amino acids with differing side groups occur. The sequence of these side groups along the polypeptide chain is determined by the genetic code, i.e. by a structural prescription which is passed on during cell division. The code determines which protein is to be synthesized and how its molecular structure, the primary structure, is to be formed. Certain regions of the polypeptide chain then coil up to give a helical structure, and the helices further fold together to form a three-dimensional molecule; one refers to the secondary and tertiary structures. An example is shown in Fig. 20.21. This structure makes it clear just how such a protein can provide an electronically and geometrically precisely-defined site which can receive another molecule, e.g. as in Fig. 20.21, the iron-containing compound heme.

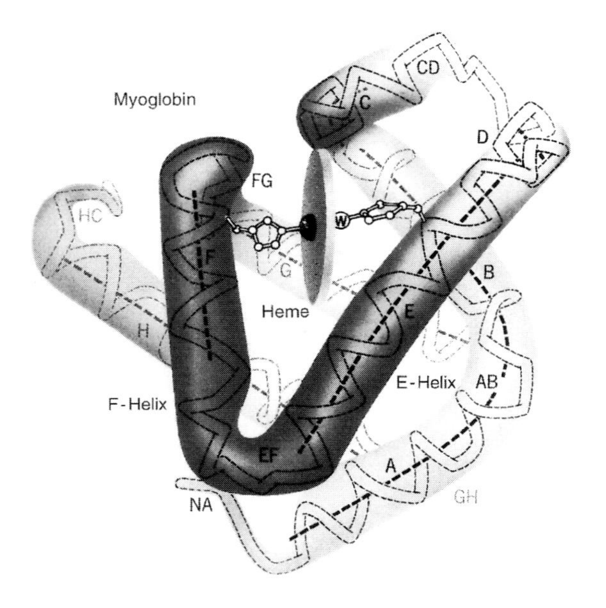

Fig. 20.21. A strongly schematized view of myoglobin. It contains 153 amino acids in its chain and has a molecular mass of 17 000, thus being a relatively small protein molecule. The chain is folded into 8 segments (labelled A through H) by bending of the cylindrical α-helix. The bends between the segments are denoted by the letters of the adjacent two segments, e.g. AB. The pocket formed by E and F holds the heme group, which is an Fe atom in the centre of a porphyrin ring. An O_2 molecule can bind to this central Fe atom, allowing the myoglobin to store oxygen (binding position W). After Dickerson and Geis

A protein is thus a folded polymer made up of amino acids in a specific sequence. There are many possible tertiary structures and functions of proteins. For example, they can serve as carriers for metal atoms or small organic molecules. Such globular proteins with a diameter from 2 to 20 nm can carry out numerous biochemical functions: as enzymes, which show strong catalytic activity; as carriers of oxygen (hemoglobin); as electron transfer agents (cytochrome); and many others. Proteins can however also occur as polypeptide chains which are twisted together to form long cord-like structures. These are important structural and supporting elements; they are called fibre proteins, and form structural parts of hair, skin, and muscles.

The most important carriers of information for organic life are the *nucleic acids*, in particular the DNAs, deoxyribonucleic acids. They make up the central information storage units in the nuclei of cells, from which with the help of ribonucleic acids (RNAs) the stored information can be passed on and made use of in the biosynthesis of proteins.

The primary contribution which can be made by molecular physicists in the investigation of these fundamental building blocks of life are in the areas of structure determinations and the elucidation of intermolecular binding mechanisms.

DNA, deoxyriboneuleic acid, is shown in Figs. 20.22 and 20.23; it is the most important nucleic acid, because it as a rule is used to store and recall genetic information. It is a long-chain polymer made up of molecules of the sugar deoxyribose (the general composition formula of the sugars is $C_x(H_2O)_y$, with x and y between 5 or 6, respectively) together with phsophate groups which alternate with the sugar molecules to form a chain; cf. Fig. 20.22. Each of the individual units is covalently bound to one of the 4 purine or pyrimidine bases, which are symbolized by A (adenine), C (cytosine), G (guanine), and T (thymine); they are all shown in Fig. 20.23. These side groups are thus arranged at regular intervals along the polymer chain.

The genetic information is coded in terms of the sequence of the bases along the DNA strand. A group of three bases – a codon – characterizes a particular amino acid, i.e. it gives the prescription to insert that amino acid into a protein chain being synthesized. For coding the 20 different amino acids from which natural proteins are made, there are thus $4^3 = 64$ distinct codons, since 4 different bases can occur in a given position. A second strand, bound to the first strand to form a double helix (by hydrogen bonding between the bases), contains the same information; cf. Fig. 20.24. This information can be called up by messenger RNA, which has a structure similar to that of DNA, and passed on as a recipe for protein biosynthesis; as mentioned, each triplet of bases, i.e. each codon, specifies a particular amino acid that is to be inserted into the protein strand. The information stored in the codons is transferred to the messenger RNA and brought to the site where protein synthesis is taking place. Figure 20.24 shows again the structure of DNA in a very schematic way. The contributions of physicists to the elucidation of these mechanisms has been considerable. The most important of them was the determination of the double helix structure of DNA using X-ray diffraction by *Crick* and *Watson* in the year 1960. Further significant contributions have been measurements and calculations of the internal motions of biomolecules, of their folding and oscillations, and of their dynamics on being inserted and reformed in living organisms. In this area, there have also been important contributions by means of infra-red and NMR spectroscopies.

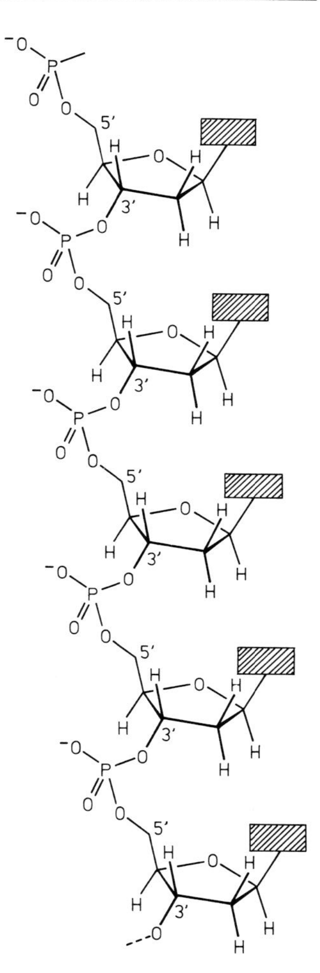

Fig. 20.22. The backbone of deoxyribonucleic acid (DNA) is a long polymer which is made up of alternating phosphate and deoxyribose molecules. They are bound together by an ester bond at the 3′ and 5′ positions of the sugar molecules. The shaded boxes indicate the positions of one of the four possible bases. After Dickerson and Geis

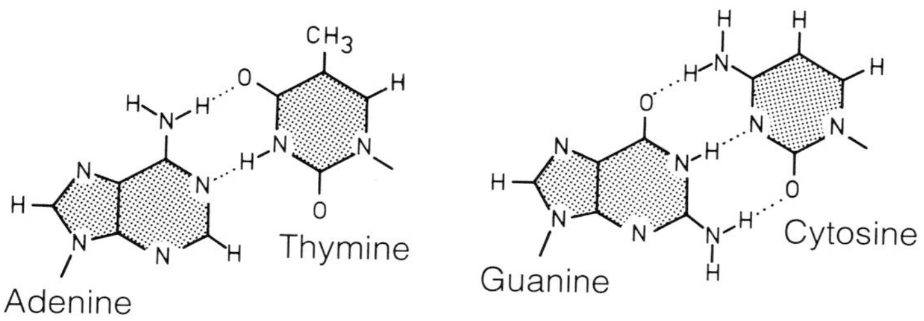

Adenine Thymine Guanine Cytosine

Fig. 20.23. The four bases of DNA, shown joined into pairs by hydrogen bonds

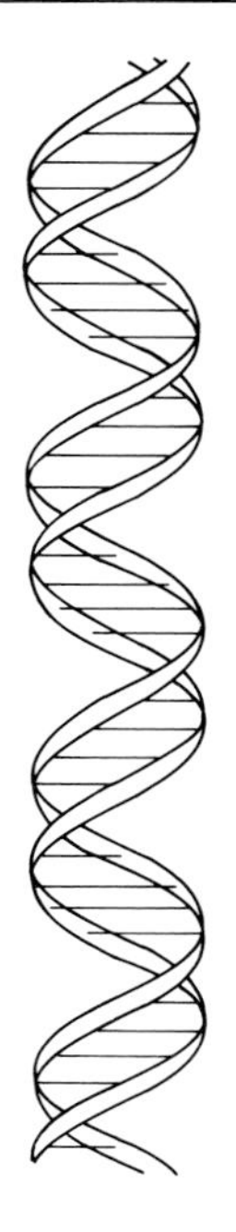

Phosphate			Phosphate
Sugar	Guanine	Cytosine	Sugar
Phosphate			Phosphate
Sugar	Thymine	Adenine	Sugar
Phosphate			Phosphate
Sugar	Adenine	Thymine	Sugar
Phosphate			Phosphate
Sugar	Cytosine	Guanine	Sugar

Fig. 20.24. A schematic representation of DNA. *Left*: a structural model of the deoxyribonucleic acid molecule (DNA). *Right*: some of the structural units of DNA, shown spread out in a two-dimensional manner. The different shapes in the centre indicate how only certain nucleotides can form base pairs by hydrogen bonding

20.7 Molecular Functional Units

The great variety of highly specific reactions which molecules must undergo or initiate, especially in biological systems, is made possible by their being ordered, held fixed, or moved in a controlled fashion at particular distances, with particular orientations and with specific electronic interactions relative to one another or to different molecular partners. One refers to molecular functional units, meaning structures made up of several molecules, which can carry out complex processes such as molecular recognition, information storage or transfer, catalysis, or following specific reaction paths. The elucidation of the structure and functions of such units is currently the object of interdisciplinary research efforts and represents a fascinating application of the capabilities and methods of molecular physics. We have already seen in Sect. 20.5 how proteins can function as receptor–substrate units for transport and reaction processes in the animate world.

Many biologically significant molecular processes can occur only with the aid of molecular *membranes*. These are constructed of long-chain molecules with different end groups on the two ends of the chain, for example a hydrophobic and a hydrophilic group. (The terms hydrophobic and hydrophilic refer to groups which are water-insoluble, like $-CH_3$, or water-soluble, like $-OH$.) Membranes serve as separating partitions between different regions, e.g. in a cell, or as matrices for active molecular units such as proteins, and thus permit oriented and spatially separated reactions and transport processes to take place; see Fig. 20.25.

Membranes can be artificially prepared by placing long-chain amphiphilic molecules onto a liquid surface (Fig. 2.6). This means that the molecules are soluble in differing media at the two ends of their chains. A frequently used group of molecules are the fatty acids, for example stearic acid, chemical formula $C_{17}H_{35}COOH$. They consist of a long zig-zag chain of CH_2 groups with a polar $-COOH$ group at one end and a nonpolar $-CH_3$ group

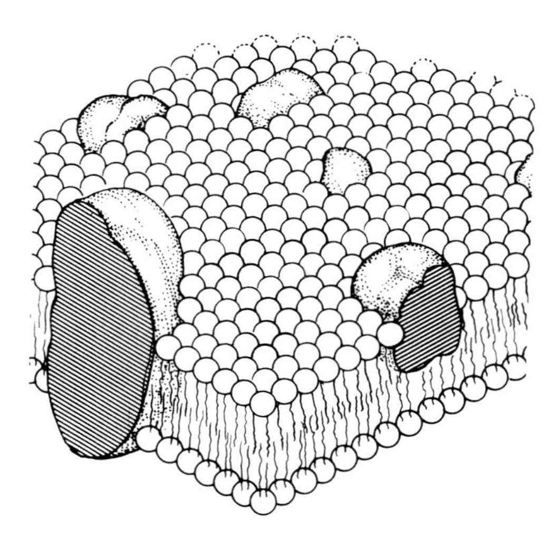

Fig. 20.25. A model of a cell membrane. It consists of a lipid bilayer with embedded protein complexes. The latter can to some extent diffuse between different regions of the membrane and can extend through the membrane on both sides. [After S.J. Singer and G.I. Nicholson, Science **175**, 723 (1972)]

at the other. The –COOH group is water-soluble, while the rest of the molecule is not. As early as 1891, the amateur researcher A. Pockels observed that such molecules dissolve in a two-dimensional manner on a water surface, without entering the volume. Their polar heads point downwards and dip into the water; the hydrocarbon tails point upwards into the air. If the area in which the molecules are two-dimensionally dissolved is reduced by a movable barrier, a continuous monomolecular layer of fatty-acid molecules can be produced on the surface of the water.

Such layers can be drawn out as a monomolecular film onto a glass or metal plate by dipping the plate into the water and pulling it out again, as shown in Fig. 20.26. By repeating this process and carrying it out properly, one can stack up seveal such layers with the same or opposite orientations and thereby prepare mono- or multilayers on a solid substrate. This is the means of obtaining so-called *Langmuir-Blodgett films*, which can serve as models for molecular membranes. The sizes of molecules can also be determined in this way, as we showed in Sect. 2.1. Other molecules can be inserted into the membranes, thus giving an array of molecules with a known separation and orientation for experiments. The technology of Langmuir-Blodgett films opens up many possibilities for the physical investigation of large molecules and for the creation of molecular architectures. Physicists are, for example, interested in questions of the ordering and the possiblities of differentiation in such structures.

One of the most important molecular functional units, without which photosynthesis and therefore most forms of life would not be possible, is the *reaction centre* of photosynthetic systems which we have already mentioned in Sect. 20.5. The structure of this centre in the case of photosynthetic bacteria has been determined recently by X-ray diffraction measurements on single crystals of the reaction centre units. Figure 20.27 shows how the reaction centre complexes are arranged within membranes in the interior of bacterial cells. A small number of molecules is embedded in a surrounding protein with a predetermined orientation and a well-defined spacing. They have the function of carrying out the primary step of photosynthesis, in which the energy of absorbed sunlight is used to create a charge separation across the membrane, i.e. to produce a spatially separated electron-hole pair. The chemical nature of the molecules and their relative orientations must be such that the charge separation occurs rapidly and with a high quantum yield, that recombination of the separated charges does not occur, and that these processes take place in the membrane in such a way

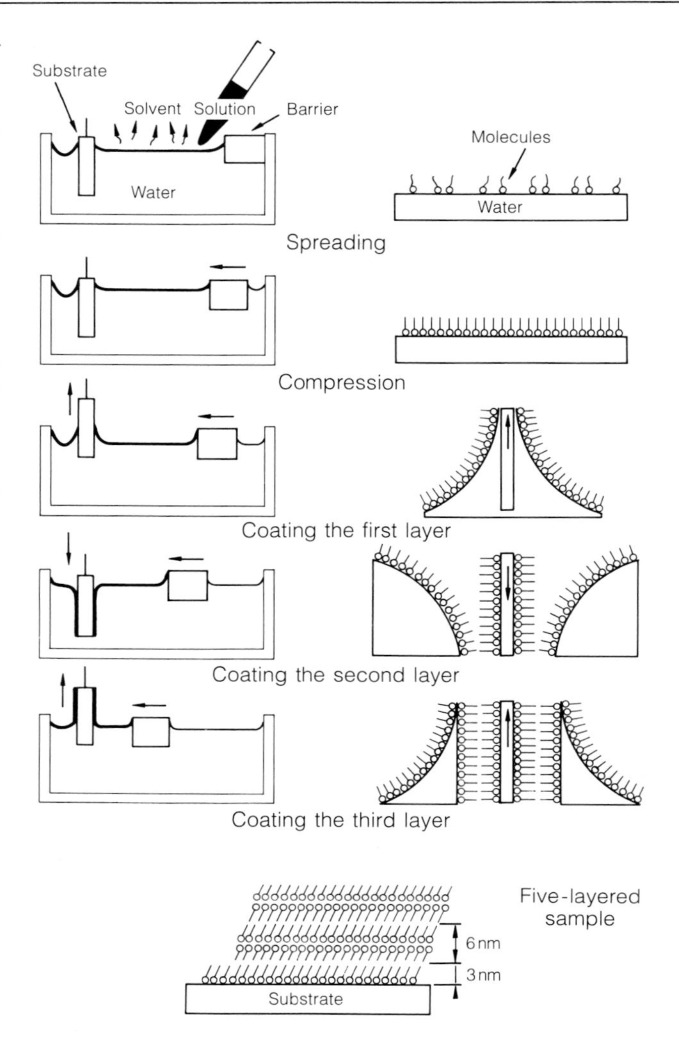

Fig. 20.26. The preparation of Langmuir-Blodgett films. Amphiphilic molecules, that is molecules with a water-soluble and an insoluble end, dissolve two-dimensionally on a water surface, spreading to form a monomolecular layer. They can then be compressed to give a continuous two-dimensional film and can be attached to a substrate by dipping the latter into the water and pulling it out again. This process can be repeated many times. One can thus prepare ordered thin films of a chosen thickness. [see also Nachr. Chem. Techn. **36**, Nr. 10 (1988)]

that the separated charges are available as the starting point for the chemical reactions of photosynthesis. How this is possible will be explained in what follows, referring to Fig. 20.27.

The starting point for the photoreaction in the reaction centre is a pair of chlorophyll molecules, the so-called primary donor; this refers to its function as donor of the electron for the charge separation. It was shown through ESR and ENDOR spectra even before the X-ray structure determination, by the way, that the starting point for photosynthetic charge separation is a chlorophyll dimer in the reaction centre. The light which is absorbed by other nearby chlorophyll molecules, the so-called antennae, and whose energy has been conducted to the primary donor by energy transfer processes, causes an electron in the donor to be released. It is transferred via the two neighbouring molecules, a chlorophyll monomer and a phaeophytin monomer which has a similar structure, to a quinone molecule, called the primary acceptor. An iron complex is clearly also involved in this process, in addition to the quinone. The extremely complex chemical part of the photosynthesis process then begins at

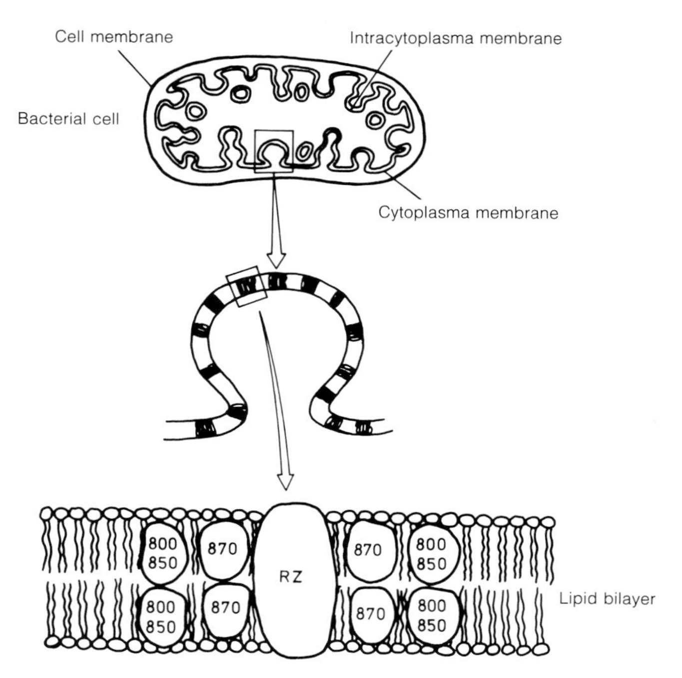

Fig. 20.27. The reaction centre of bacterial photosynthesis is embedded in membrane structures within the intracytoplasmic membrane of the bacteria in the form of a protein complex RZ. Other complexes, here denoted as 800, 850, and 870 according to the wavelength of their optical absorption maxima, contain the chlorophyll molecules which serve as light antennae to absorb incident sunlight. They transfer the absorbed energy to the reaction centre RZ

the primary acceptor. The electron which was removed from the primary donor is replaced via the intermediary of another pigment-protein complex, cytochrome. The entire process of directed charge separation lasts 200 ps and is subdivided into several steps over the intermediate molecules; this evidently serves to prevent a recombination of the separated charges. In Fig. 20.28, the whole reaction centre complex is indicated schematically, with the orientations, but not the intermolecular distances, shown to scale. The left-hand part of the reaction centre, Fig. 20.28, is nearly a copy of the right-hand side; its function is still unclear. The overall reaction which takes place can be summarized in the balance equation

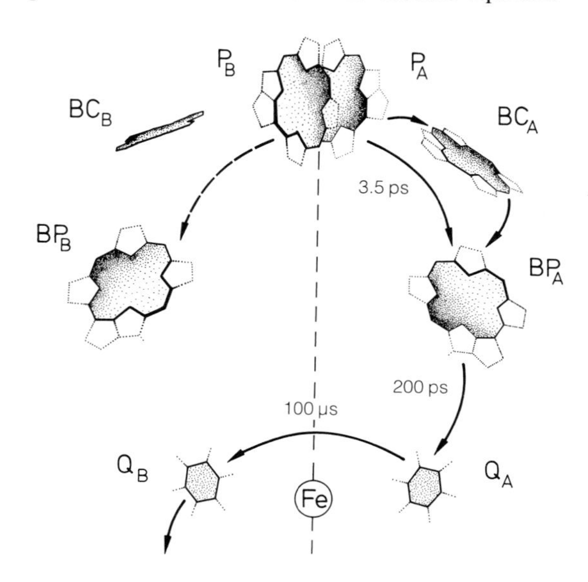

Fig. 20.28. The molecular structure of the reaction centre of bacterial photosynthesis, according to the X-ray structure determination of Michel, Deisenhofer, and Huber. The figure shows the relative spatial arrangement of the molecules, but with somewhat exaggerated intermolecular spacings. Charge separation, i.e. the removal of an electron as a result of light absorption, begins at the primary donor P; it is a chlorophyll dimer. The electron is transferred quite rapidly via a bacteriochlorophyll molecule BC_A and a bacteriophaeophytin molecule BP_A to the quinone molecule Q_A; this process takes less than 200 ps. Then the chemical part of photosynthesis begins. The function of the left-hand branch of the reaction centre, denoted here by a subscript B, is still unclear. Bacteriochlorophyll differs from bacteriophaeophytin mainly through a centrally located Mg atom. The times which are indicated were measured using laser spectroscopy. The first step, from P_A to BP_A (3.5 ps), could be separated using an improved time resolution into two steps across the BC_A molecule. [W. Holzapfel, U. Finkele, W. Kaiser, D. Oestehelt, H. Scheer, H.U. Stilz, and W. Zinth, Chem. Phys. Lett. **160**, 1 (1989)]

$$6CO_2 + 6H_2O \xrightarrow{h\nu} C_6H_{12}O_6 + 6O_2 \ ,$$

with an energy gain of 2870 kJ per mole of glucose, $C_6H_{12}O_6$.

From the point of view of photophysics, the reaction centre is a functional unit that solves the problem of a light-induced charge separation in an efficient and controlled manner, which is still not possible to duplicate in the laboratory using other molecules. The reaction centre will be taken as an example by those who wish to construct artificial molecular units with comparable functions.

The investigation of the structure and the charge-transfer dynamics of the photosynthesis reaction centre represents a great success for the application of molecular-physical methods to solving problems in biological physics.

Only by employing a number of highly developed modern research tools has it been possible to elucidate the structure and dynamics of this complicated functional system. Among them are structure determinations by X-ray diffraction, spectroscopy in all wavelength ranges, and especially also magnetic single and multiple resonance techniques and laser spectroscopy, which gives time resolutions down to the femtosecond range. Here, there are many fascinating objects to be found for physical and chemical research.

21. Molecular Electronics and Other Applications

As a conclusion to this book, we have permitted ourselves to describe the outlook in a field which on the one hand is quite fascinating because of the goals it has set itself, but on the other is still uncertain with regard to its eventual limits and potentials.

21.1 What Is it?

For some years now, the imaginations of molecular and solid-state physicists have been fired by a new idea, or a new catchword, referring to a novel application of their research: *molecular electronics*. What is it? Molecular electronics is the generic term for efforts and speculations which have the goal of complementing present-day conventional electronic devices, particularly microelectronics based on silicon and related semiconductor materials, by electronics which makes use of molecules and molecular functional units, utilizing the specific properties of molecular substances. This goal has been set for the not-too-distant future. The attraction of such ideas is due to the hope of a further significant miniaturisation as compared to silicon-based devices, of being able to use the enormous variety of molecules offered by organic chemistry, and, perhaps, of finding materials which are readily available and easily produced.

In considering the requirements of an electronics based on molecules, one finds that the following functions would be needed:

- Molecules or molecular functional units which can serve as *switches*: molecular systems which are bistable with regard to light or to electric or magnetic fields;
- Molecules or molecular functional units usable as *conductors*: for some years, organic metals (i.e. molecular solids with a metallic electrical conductivity) and even organic superconductors have been known;
- Molecules or molecular functional units which perform as *logic elements*: it must be possible to make such devices by combining switches and conductors;
- Molecules or molecular functional units able to provide *information storage*, i.e. memory devices which permit the writing and readout of information by means of light or electric or magnetic-field pulses.

What, then, is molecular electronics? The concept can perhaps best be defined thus: molecular electronics includes all of the phenomena and processes in which organic molecular materials play an active role in the processing, transmittal, and storage of information.

If we use this definition, then there is at present hardly such a thing as molecular electronics. Whether it will someday exist, and what it would then be like, we do not yet know; but we can already apply ourselves to tasks in basic research which will prepare the ground for its possible development, which foresee its applications in the future and keep them in mind. This is an area requiring the combined efforts of solid-state and molecular physicists and chemists; the concept of molecular electronics may thus serve as a stimulus for varied and interdisciplinary research programmes.

21.2 Molecular Conductors

Are there organic molecules which are suitable for use as electronic conductors ("molecular wires") to connect other molecules together in an electrical circuit? One can immediately think of polyenes (Sect. 14.6), or also of the so-called *organic metals*. The latter are crystalline compounds which are in general composed of two partners, of which the one serves as an electron donor and the other as an electron acceptor. The organic partners are often arranged in stacks, so that they yield a one-dimensional or low-dimensional conductivity when conduction is acheived through overlap of the π-orbitals of neighbouring molecules; cf. Fig. 21.1.

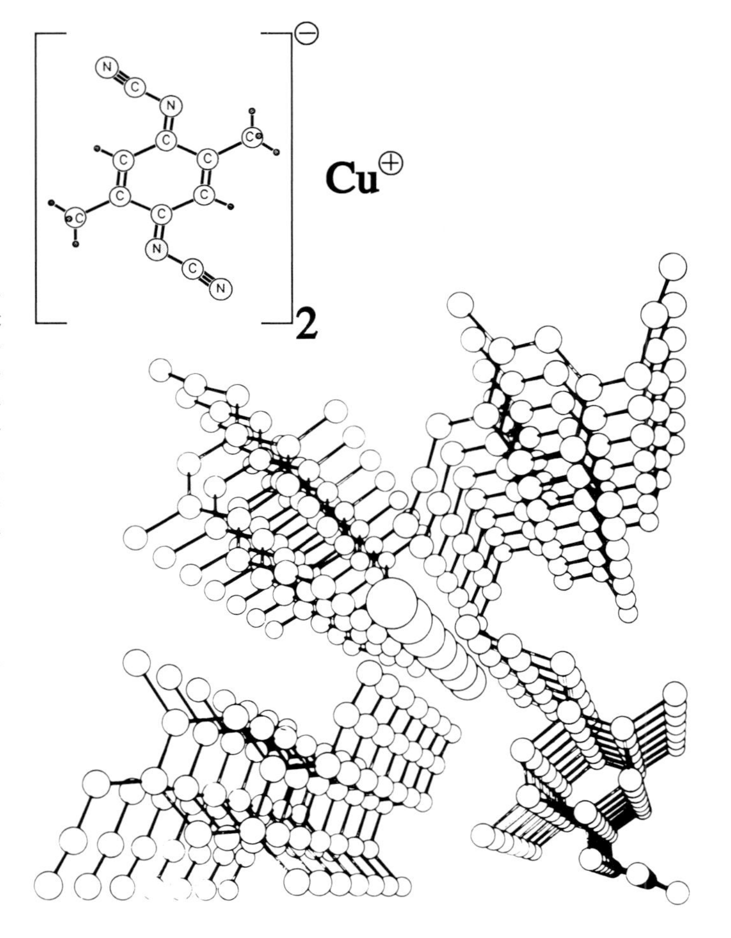

Fig. 21.1. The crystal structure of the radical-anion salt 2,5 dimethyl- dicyanoquinonediimine (DCNQI), with copper as its inorganic partner. In the centre of the picture, a chain of Cu ions can be seen; they are, however, not responsible for the metallic conductivity of the material. Around them are four stacks of the organic partner molecules. Conductivity takes place along these stacks. The stacks are joined together via –CN groups through the central Cu, so that the one-dimensionality is somewhat reduced. [After P. Erk, S. Hünig, J.U. v. Schütz, H.P. Werner, and H.C. Wolf, Angew. Chem. **100**, 286 (1988)]. The molecular structure diagram at the upper left shows the H atoms as dots only

As an example, we consider the radical anion salts of dicyanoquinone-diimine (DCNQI). The structure (Fig. 21.1) illustrates the stacking of the organic molecular units. The possibility of an electronic connexion between the stacks is provided by the central metal ions, i.e. between the central metal and the nitrogen atoms of the –CN groups of DCNQI. The metal counterions of the salt (Cu, Ag, Li and others) as well as the substituent groups of the DCNQI molecules (e.g. $-CH_3$, Cl, Br, I) can be varied over a wide range without essentially changing the crystal structure. The extremely high conductivity and the metallic character even at low temperatures of one of these compounds, the Cu salt of DCNQI substituted with two methyl groups as shown in Fig. 21.1 and in the upper curve of Fig. 21.2, demonstrate why it is referred to as an organic metal. The conductivity of this salt at low temperatures is as high as that of copper at room temperature, and it increases monotonically with decreasing temperature, as is typical of metals. To answer the question of what the mechanism of this conductivity is and why it takes on a metallic character, it is necesary to apply a whole battery of various experimental methods from molecular and solid-state physics. One of those is magnetic resonance spectroscopy; it tells us for example that the high conductivity is directed primarily along the stacks of the organic molecules. This can be seen from the influence of the mobile charge carriers on the relaxation behaviour of the proton spins in the DCNQI molecules.

It is a matter of great interest to compare the conductivities of different salts of this family of compounds which differ either in the identity of the metal ion or in the substituent groups on the DCNQI molecules. For such a comparison, it is important that the different salts have essentially the same crystal structure: only the intermolecular distances are slightly variable, and the molecular orientations show only small differences.

Most of these salts exhibit a behaviour similar to that of the two lowest curves in Fig. 21.2: they have a rather good conductivity at room temperature, which with decreasing temperature first increases slightly, then however shows a strong decrease at lower temperatures. This behaviour is similar to that of a "metallic semiconductor" and is typical of one-dimensional

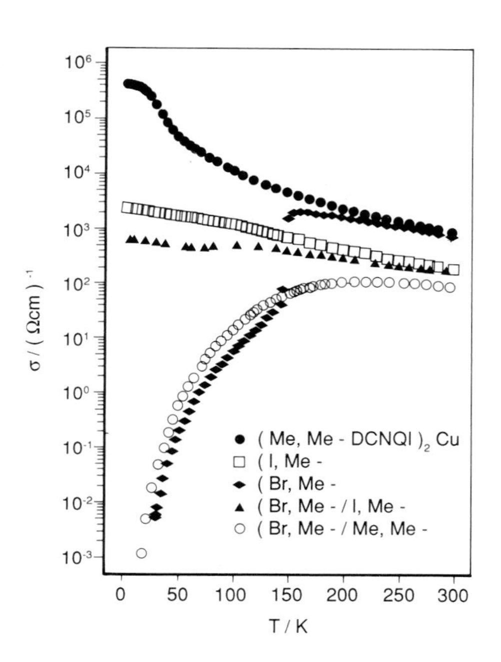

Fig. 21.2. The temperature-dependent electrical conductivity of some Cu salts of DCNQI. The compounds differ in the nature of the substituents on the DCNQI molecules; cf. Fig. 21.1. Me– refers to a $-CH_3$ group (methyl), while I or Br are iodine or bromine atoms as substituents in place of the methyl groups; compare the molecular structure in Fig. 21.1. The crystal structures are very similar in each case. The conductivity ranges from that of an organic metal even down to very low temperatures (*uppermost curve*) to that of a metal-like semiconductor (the two *lowest curves*; one of them is for an alloy). [From H.C. Wolf, Nachr. Chem. Techn, **37**, 350 (1989)]

conductors, in which a phase transition at low temperatures (*Peierls* transition) puts an end to the metal-like state. These compounds are hardly worthy of consideration as "molecular wires".

On the other hand, the slightly altered geometric and electronic configuration of the copper compound belonging to the uppermost curve in Fig. 21.2 are evidently sufficient to cause deviations from one-dimensionality by increasing the stack-to-stack interactions through bonds between the central copper ions and the –CN groups of DCNQI. This supresses the phase transition, and the high metallic conductivity persists down to very low temperatures. A slight modification of the molecular structural units is enough to cause significant changes in the physical behaviour of these salts.

This example clearly shows how cautious one should be with words like "molecular wires". It is precisely the deviation from one-dimensionality which is important here. In terms of molecular electronics, this means that the lateral size of molecular wires must not be too small. From the great multiplicity of chemically possible substances, one must search out those few whose structural properties, and the electronic properties to which they give rise, are suitable for the projected application. A relatively large number of organic metals has been known for some years, including even organic superconductors. It is an engaging task for molecular physics and quantum chemistry to find the structural principles of molecules which show these properties.

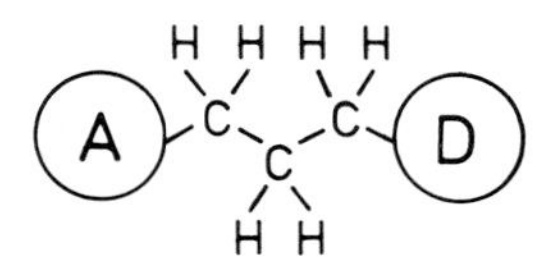

Fig. 21.3. The structural formula of a molecular functional unit which could function as a "molecular rectifier" or diode. The molecule *A* is an electron acceptor, *D* is a donor. See also D. Haarer, Informationstechnik **34**, 4 (1992)

Another hoped-for goal of molecular electronics is the *molecular rectifier*. One is shown in Fig. 21.3: this molecular functional unit has at its one end a molecule *A* which serves as an electron acceptor –for example, a molecule with a –CN substituent. At the other end, there is a molecule *D* with donor properties, e.g. a molecule with amino substituents. The molecule in the centre is insulating, but it can allow the tunnelling of charge carriers from the two ends due to its small size. Since the *A* unit has a tendency to attract electrons, an electron which is added to *D* will tend to move from *D* to *A*; the whole unit has an effectively polar character. There have been attempts for more than 20 years to put this idea into practice and to prepare this type of molecular rectifier or diode. However, some important problems remain to be solved, in particular irreversible changes in parts of the molecules through oxidation and reduction, and the attachment of the functional unit to a metallic or semiconducting interface.

21.3 Molecules as Switching Elements

In order to use molecules as switches which are activated by light, one would like to have a molecule *A* which can be changed to a configuration *B* by absorption of light of wavelength λ_A; this process should be reversed by light of wavelength λ_B. Suitable cases, so-called bistable molecules, are being sought after.

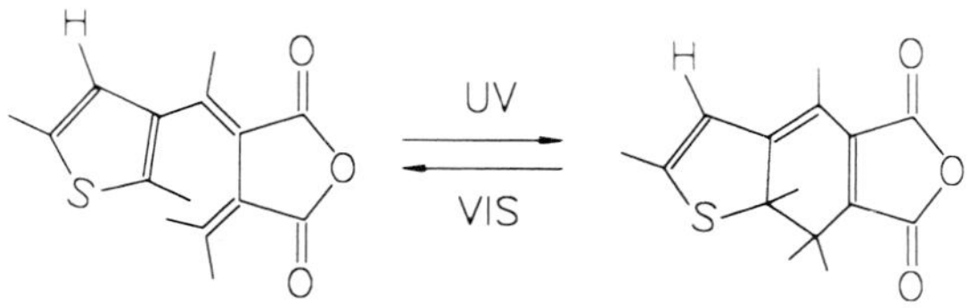

Fig. 21.4. Molecules as switches: the thiophene fulgide molecule is photochromic and can be reversibly switched between the two valence-isomeric configurations shown. (The dashes on C atoms in the figure indicate CH_3 groups)

An example are the *fulgides*, whose photochemistry is based on a valence isomerisation: a π-bond is transformed into a σ-bond, accompanied by ring closure; cf. Fig. 21.4. Such molecules are interesting as switches because it is hoped that they could be used to open and close a conjugated chain of π-bonds, e.g. in a polyene. Figure 21.4 demonstrates that a chain of conjugated double bonds leads through the molecule (perpendicular to its axis) in only one of the two switching states.

The measurements shown in Fig. 21.5 illustrate the spectral behaviour of such a colouration and bleaching cycle; they were carried out on thiophene fulgide. On colouration, the molecules exhibit a decrease in UV absorption and the appearance of a new absorption near 600 nm. Bleaching restores the original state. The process can be repeated many times: it is thus reversible. A possible application could be for example the switching on and off of energy transport along polyene chains, as described in the following section, by inserting the fulgide molecules into the polyenes. The photochromic reaction of the fulgides which is measured in solution is also observed when they are in a solid matrix or in the crystalline form, and is thus a molecular property.

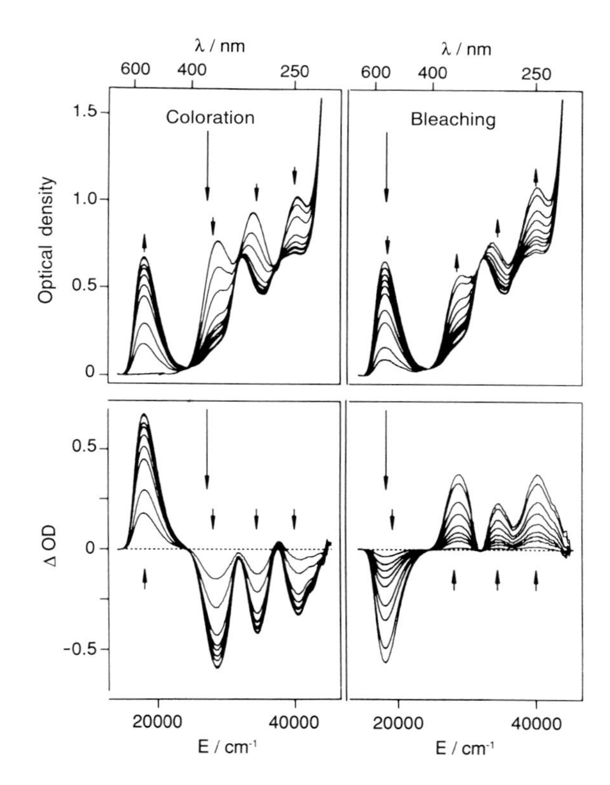

Fig. 21.5. Colouration and bleaching of the thiophene fulgide molecule (in solution), as shown in Fig. 21.4. *Upper part*: the optical density. *Lower part*: changes in the optical density through the effect of light with the wavelength indicated by the arrows. [From H.C. Wolf, Nachr. Chem. Techn. **37**, 350 (1989)]

21.4 Molecules as Energy Conductors

A further topic from molecular physics with a view to molecular electronics is the possibility of transporting excitation energy through molecules. Here, it would be desirable to find molecules in which energy that is absorbed at one end is transported to an end group at the other end of the molecule. In such a supramolecule, the middle portion thus acts like a wire; to be sure, not a wire for electrical conduction, but rather a wire for the transmission of electronic excitation energy without charge transport.

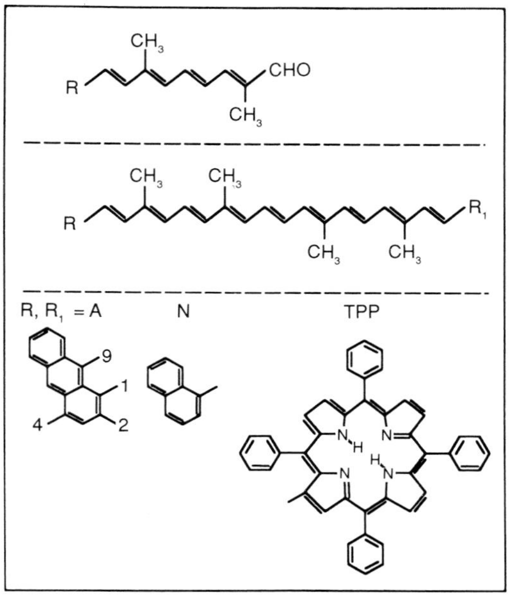

Fig. 21.6. Molecules as energy conductors. Molecular groups R and R_1 are attached to polyene molecules of different lengths, here with 5 or with 9 double bonds. The R group can act as a donor, the R_1 group as an acceptor for electronic excitation energy. The donors here are the anthryl group A, which can be substituted in different positions, or else the napthyl group N. The acceptor is tetraphenyl porphyrin (TPP). In the upper part of the figure, a molecule is shown which has been substituted at only one end; the other end is teminated by a –CHO group. The $-CH_2$ groups serve to stabilize the chains

Measurements on polyene molecules which have been substituted at one end by an excitation group such as anthryl (A) and at the other end by a detector group like tetraphenyl porphyrin (TPP) (see Fig. 21.6) appear to indicate that this kind of energy conduction is in principle possible. Determinations of the absorption, fluorescence, and excitation spectra of these molecules (Figs. 21.7 and 21.8) yield the following results:

- The various molecular partners retain their identities to a high degree even after being joined together to form a supramolecule: their electronic states and vibronic structures remain specific to the structural subunits of the supramolecule. A localized excitation of the aromatic end groups is possible. In Fig. 21.7, one sees for example an excitation band in the absorption spectrum which is characteristic of the anthryl end group (in the 250 nm region), and also absorption bands characteristic of the other substituent, tetraphenyl porphyrin (TPP) (most noticeable is the strong absorption band at ca. 430 nm, the *Soret* band), as well as the absorption bands of the polyene, which shift in a characteristic way towards longer wavelengths as the chain length of the molecules is increased.
- There is, however, a clear modification of the intensities of the bands and the electronic energies of the component molecules, depending on the type of substitution and the length of the polyene. These changes show that the binding exerts a mutual influence on the

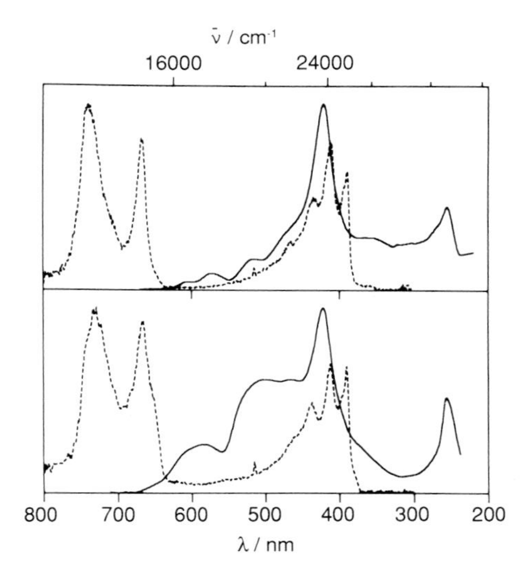

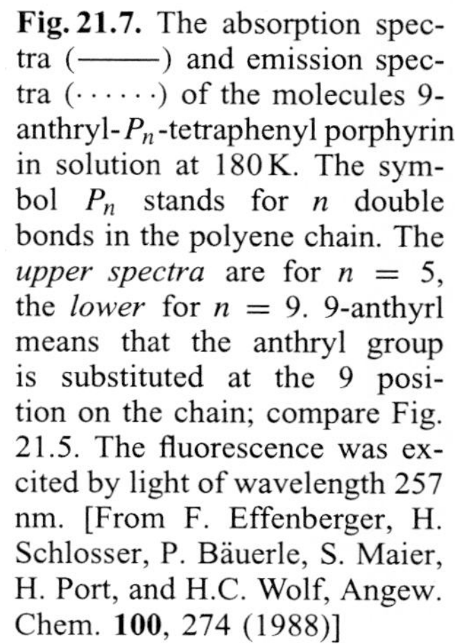
Fig. 21.7. The absorption spectra (———) and emission spectra (······) of the molecules 9-anthryl-P_n-tetraphenyl porphyrin in solution at 180 K. The symbol P_n stands for n double bonds in the polyene chain. The *upper spectra* are for $n = 5$, the *lower* for $n = 9$. 9-anthyrl means that the anthryl group is substituted at the 9 position on the chain; compare Fig. 21.5. The fluorescence was excited by light of wavelength 257 nm. [From F. Effenberger, H. Schlosser, P. Bäuerle, S. Maier, H. Port, and H.C. Wolf, Angew. Chem. **100**, 274 (1988)]

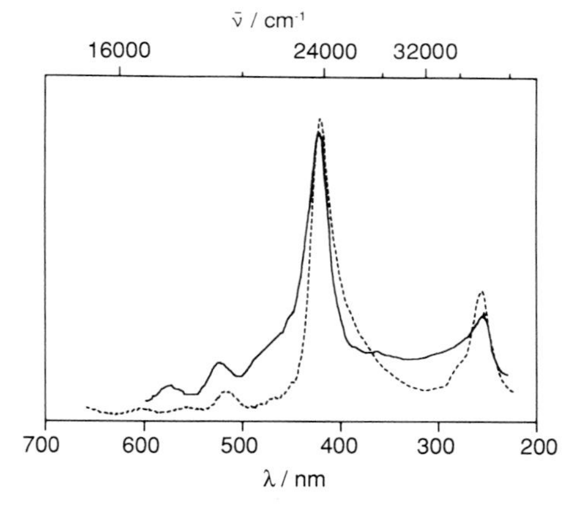

Fig. 21.8. Excitation spectra of 9-anthryl-P_n-tetraphenyl porphyrin at 180 K in solution: $n = 5$ (······), $n = 9$ (———). The intensity of the TPP emission at 665 nm is plotted as a function of the wavelength of the excitation light. The maximum near 250 nm corresponds to an absorption of the anthryl group (S_3 excitation) and shows especially clearly that energy is transported to the TPP. [From S. Maier, H. Port, H.C. Wolf, F. Effenberger, and H. Schlosser, Synth. Met. **29**, E517 (1989)]

electronic orbitals, leading to their partial merging, but without a complete loss of the identity of the partners.

– An intramolecular energy transport from one end, at the anthryl substituent, to the other end, with TPP, is in fact observed. It is mediated by electronic states which evidently belong to the whole supramolecule. This intramolecular energy transport is for example noticeable in Fig. 21.8, which shows the excitation spectrum of the TPP emission. One sees that the TPP emission can be excited by light in the absorption range of the anthryl group, in the range of the TPP absorption itself, and possibly also in the range of the polyene absorption.

These substituted polyene molecules are thus interesting candidates, or at least model substances for the further study of molecules and functional units that might be of use in molecular electronics. In particular, one could imagine that the insertion of molecular switches into the polyene chains would influence this energy transport. It might also be possible to place the polyene chains into monomolecular layers or membranes and to use them for the transport of electronic excitation energy across the layer or the membrane.

A further interesting area of research is light-induced charge separation in molecules and molecular chains; this means that light absorption creates a state in which the molecule carries an unpaired electron at one end and the corresponding hole at the other.

21.5 Molecular Storage Elements

Will it be possible also to use molecules for information storage and retrieval (molecular memories)? Can one for example change the state of a molecular system by the action of light into a new state which is distinguishable from the initial state and which is stable, and then read out this stored information or erase it by a second light quantum? Is it possible to construct a functional unit with a high storage density in this way, i.e. with a large amount of stored information in a small volume?

An interesting approach to this problem is based on the phenomenon of photochemical or photophysical hole-burning, Fig. 21.9. Although an isolated molecule often exhibits a sharply defined absorption spectrum, the emission and absorption spectra of molecules in a solid matrix, e.g. in an inorganic or organic glass, usually consist of broad bands. The individual molecular spectral lines are inhomogeneously broadened, because there are many different local environments for the molecules within the matrix. These various local environments shift the energy levels of the molecules by different amounts (solvent shift). The observed spectrum is then a superposition of many individual sharp lines, which cannot be resolved or distinguished from each other.

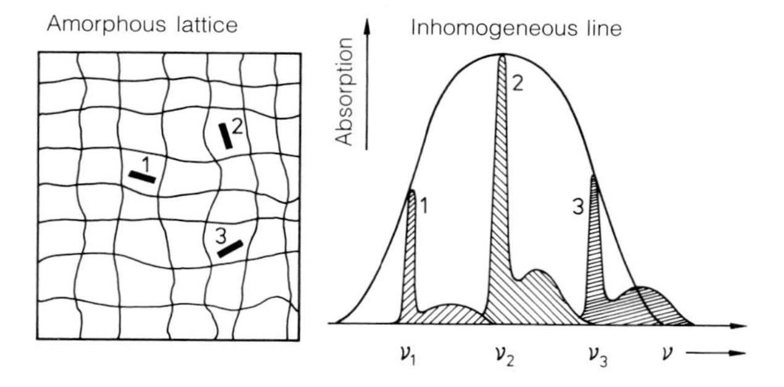

Fig. 21.9. The inhomogeneous broadening of molecular spectral lines in a matrix; this is the basis for the phenomenon of hole-burning. In an amorphous matrix, the electronic excitation energies and thus the transition frequencies of molecules are distributed over a range of energies, because the individual molecules have differing local environments (*left side*). Each individual configuration has a particular absorption spectrum, consisting of a zero-phonon line and phonon sidebands. The overall absorption is a superposition of all these individual absorption spectra (*right side*). [After D. Haarer, Angew. Chem. **96**, 96 (1984)]

If one can now modify a molecule in its structure or its interaction with the environment by causing it to absorb a light quantum, so that it can no longer absorb a light quantum of the same energy at a later time, then this molecule is removed from the ensemble of absorbing molecules within the inhomogeneous line. A "hole" appears in the line at the position of the individual absorption of the bleached molecule. This is the principle of the hole-burning method; compare Fig. 21.10. The first organic molecule to which this method was successfully applied was porphyrin. In its case the absorbed light causes a rearrangement of the central H atoms as shown in Fig. 21.11. This reduces its absorption coefficient at the frequency in the spectrum where the initial absorption occurs (this is hole-burning); the absorption coefficient is correspondingly increased at some other frequency outside the inhomogeneously broadened line shown here. Depending on whether the hole-burning process involves a rearrangement within the molecule itself or in its environment, one refers to photochemical or to non-photochemical hole-burning.

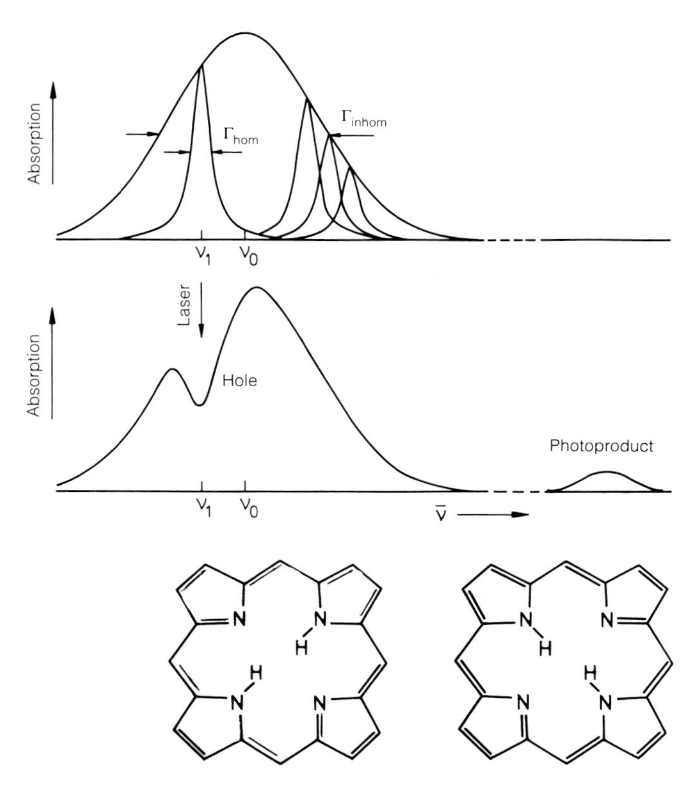

Fig. 21.10. Photochemical hole-burning: laser light of a narrow bandwidth is used to irradiate an inhomogeneously broadened spectral line. This burns a hole into the absorption line, having the width of the laser bandwidth or the homogeneous linewidth Γ_{hom} of the individual molecules. The absorption line of the photoproduct appears at some other point in the spectrum. [After S. Völker, Ann. Rev. Phys. Chem. **40**, 499 (1989)]

Fig. 21.11. In the porphine molecule (free base form), the two central H atoms can be switched back and forth between two configurations by the action of light (photoisomerisation). In solid solutions, this leads to a "hole" in the absorption spectrum. Optical hole-burning with organic molecules was first investigated in this system. The photoisomerisation is observed only at low temperatures; at room temperature, the central protons can move freely between the two configurations, as can be demonstrated by proton spin resonance experiments

One can also obtain sharp molecular absorption lines using narrow-band laser light when the molecules are dissolved in glasses or crystalline matrices, where the absorption lines are strongly inhomogeneously broadened. If the quantum energy of the laser light is changed in small steps, many holes can be burned adjacent to one another in a broad absorption line. In this way, one could attempt to store information in molecular systems. An important problem in molecular physics is the search for such storage media which are insensitive to ageing and radiation damage. The stability and reversibility of the storage system are decisive for its possible practical applications.

In this connexion, the molecule bacteriorhodopsin, which occurs in nature and can be obtained from bacteria, is extremely interesting. It is suitable for the stable and reversible storage of information and thus also for optical holography. This has been demonstrated by *Oesterhelt*, *Bräuchle* and *Hamp* in an impressive series of experiments.

The hole-burning technique can be used not only for information storage, but also for molecular spectroscopy. If a sharp hole can be burned in a broad absorption line, and the width of the hole approaches the limit of the homogeneous linewidth for the observed transition, then the transition energy can be determined much more precisely than with conventional absorption spectroscopy. Molecules, with their sharp absorption lines, can then also be used as probes to analyze the local environment in glasses or in biological matrices. The width of the hole is in the limiting case determined by the lifetime of the excited state, which can thus also be measured in this way.

It has also been shown that spectral hole-burning in systems with inhomogeneously broadened absorption lines can be used for optical holography and as molecular computers.

The properties of the light are converted into photochemical modifications of the irradiated material; the stored patterns can then be processed using logical operations as desired. Applied electric fields are an important external parameter for the dynamics of such systems. [See also *U.P. Wild* et al., Pure Appl. Chem. **64**, 1335 (1992).]

21.6 Spectroscopy of Single Molecules in the Condensed Phase

Using modern experimental methods of laser spectroscopy, it has become possible to investigate *single molecules* in the condensed phase spectroscopically with a high resolution. This is on the one hand interesting for molecular physics: one can study the line positions, line shapes, and linewidths and their modification by the molecular environment, the temperature, or by applied electric and magnetic fields, and thus gain information about the molecules themselves or the matrix in which they are included. But on the other hand, the question of whether or not it is possible to adress single molecules in a controllable way is also very important for the efforts to realize molecular electronics.

The molecules being investigated are included in a matrix at a low concentration. As explained in Sect. 21.5, their absorption lines then give rise to broad, inhomogeneous bands. These arise from the superposition of the absorption spectra of many individual molecules, whose absorption frequencies are shifted relative to one another by the different interactions with their local environments. In an amorphous or glassy matrix, the band profile, i.e. the distribution of the centre frequencies of the individual absorbers, is roughly Gaussian. Its width is a large multiple of the width of the homogeneous (Lorentzian) line profile of a single absorber. In conventional spectroscopy, one observes only the broad inhomogeneous line, because the number of molecules taking part in the absorption is very large.

Figure 21.12 shows the simulated absorption of samples containing 10 000, 1 000, 100, and 10 molecules. While with 10 000 molecules, hardly any of the substructure of the individual components is recognisable, the homogeneous lines of the 10 molecules can be readily distinguished.

By choosing very dilute and extremely small samples for investigation, one can indeed detect the signals from single molecules using the method of fluorescence excitation spectroscopy and an extremely narrow-band tunable laser. An example of such a measurement is given in Fig. 21.13, which shows the fluorescence excitation spectrum of a single pentacene molecule in a host crystal of *p*-terphenyl. The homogeneous linewidth is found to correspond to 7.8 MHz; this value is determined by the lifetime of the excited state, as given by the relation $\Gamma_{\text{hom}} = \Delta\nu = 1/2\pi T_1 = 20.4$ ns. This lifetime was measured directly for a large number of pentacene molecules by means of photon echo experiments, and was found to be 21.7 ns. There is thus good agreement with the value found from the homogeneous linewidth. Another example is given in Fig. 21.14, which shows the spectrum resulting from 3 molecules. It shows clearly how individual molecules can be observed separately from one another with a good signal-to-noise ratio. The interesting experimental goal of manipulating individual molecules has thus been approached very closely not only by using the scanning tunnel microscope (Sect. 2.1), but also optically.

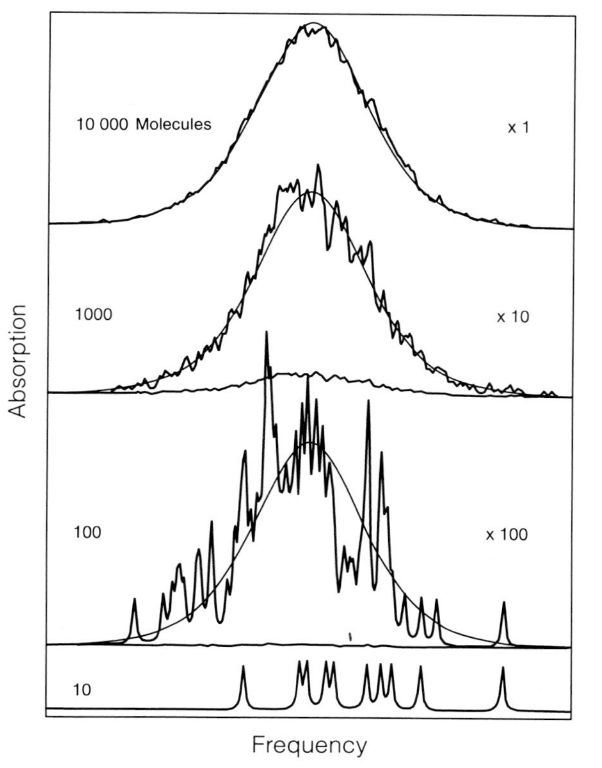

Fig. 21.12. The simulated absorption spectra of N molecules in a disordered matrix. As N decreases from 10 000 to 10, the integral absorption also decreases; this is the reason for the magnification factors shown at the right. The ratio of homogeneous to inhomogeneous linewidths has been assumed here to be 1:40. In reality, for glass matrices it has values of $1:10^4$ to $1:10^6$ at homogeneous linewidths of 10^{-1} to 10^{-3} cm^{-1} and an inhomogeneous linewidth of 10^3 cm^{-1}

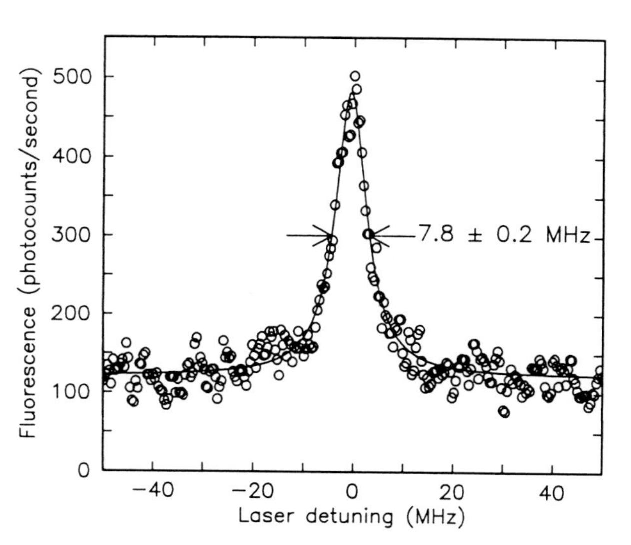

Fig. 21.13. The fluorescence excitation spectrum of a single pentacene molecule in a thin *p*-terphenyl crystal at 1.5 K, concentration $8 \cdot 10^{-9}$ mol/mol = $1.7 \cdot 10^{13}$ cm^{-3}. The centre, at 0 MHz, corresponds to an absorption wavelength of 592.407 nm. From W.P. Amrose, Th. Basché, and W.E. Moerner, J. Chem. Phys. **95**, 7150 (1991). See also W.E. Moerner and Th. Basché, Angew. Chem. **105**, 537 (1993)

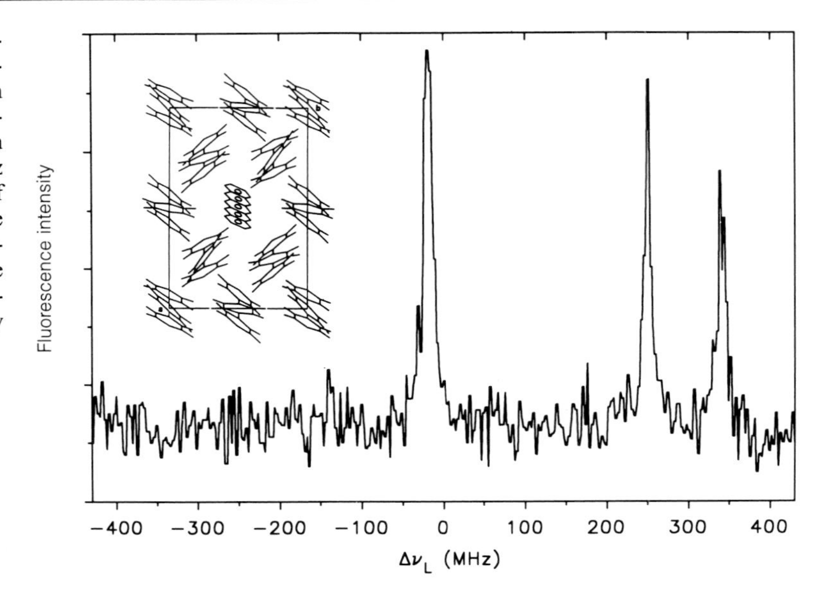

Fig. 21.14. The fluorescence excitation spectrum of three individual pentacene molecules in *p*-terphenyl, showing the crystal structure of the host as an inset. The abscissa value 0 MHz corresponds to a wavelength of 592.362 nm. The fluorescence intensity is plotted as a function of the wavelength of the extremely narrow-band excitation light. Kindly provided by C. Bräuchle and Th. Basché

21.7 Electroluminescence and Light-Emitting Diodes

A light emission from materials which are not excited by light, as in photoluminescence, but rather by the application of an electric field, is called *electroluminescence*. The field can lead to the injection of electrons and holes into the electroluminescent material if suitable electrodes are used. The recombination of these charge carriers produces excited states in the material, which then return to the ground state with accompanying light emission.

Electroluminescent devices have considerable practical applications as *Light-Emitting Diodes* (LEDs). In particular, they are the most important component in integrated optoelectronic devices, where they serve to transform electrical signals into optical ones.

Organic molecules and polymers can also be used for this purpose. If one is thinking of practical applications, the long-term stability of the devices is very important. In this sense, the polymer poly(*p*-phenylene vinylene) (PPV) is an interesting candidate, as was first shown by the research group of R.H. *Friend*: cf. Fig. 21.15.

The figure shows a device which can be used to observe and investigate the electroluminescence. The light-emitting PPV layer is located between one electrode which injects electrons, here a calcium electrode, and a second electrode which injects holes, here a thin, transparent film of indium-tin oxide (ITO) on glass. Such devices are similar to Schottky diodes which are well-known in semiconductor physics. An electric current passed through such a diode, i.e. from the metallized glass surface to the calcium (or aluminium) cover layer, causes excitations in the PPV and gives rise to light emission. This light can emerge through the glass at the bottom. The light emerges through the glass substrate; a potential difference of a few volts is sufficient for this process.

In order to attain high quantum yields and good stability, a number of improvements have been made on the simple device shown in Fig. 21.15; a detailed discussion of these would take us too far afield here, however.

Polymers as active materials for electroluminescence are especially promising because the colours of their luminiscence can be varied between wide limits by making minor changes in

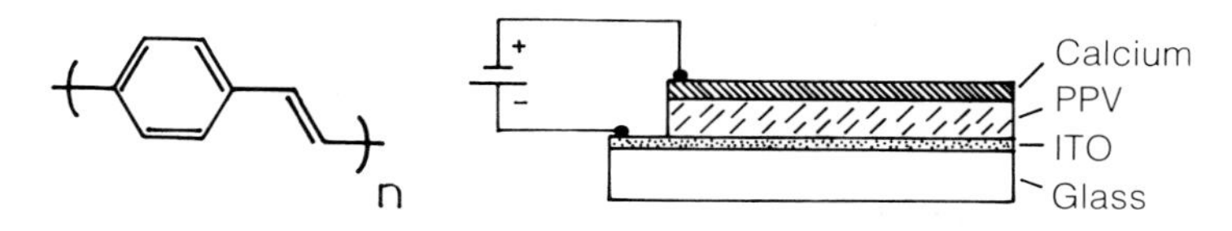

Fig. 21.15. *Right-hand part*: a sketch of a device used to generate electroluminescence. The polymer PPV is placed between two electrodes which have differing work functions for electrons or holes (here Ca and ITO, a glass plate coated with a thin layer of tin oxide). Recombination of the injected charge carriers in the PPV produces light, which can emerge from the device through the glass substrate at the bottom. *Left-hand part*: a monomer unit of the polymer PPV. See also R.H. Friend et al., Nature **347**, 539 (1990), and P.L. Burn et al., Nature **356**, 47 (1992)

the chemical nature of the molecules, and because the polymer materials can be prepared and used as large-area sheets. This has opened up an interesting area of applications of molecular systems and polymers for many fields of technology, in electronics and communications.

21.8 The Future: Intelligent Molecular Materials

These few examples are intended to show how the photophysics and photochemistry of organic molecular systems can contribute to the search for a way to realize molecular electronics. It seems clear that here lies a broad and fascinating area for subtle basic research. It is also clear that this research can not simply concentrate its efforts on individual molecules, but instead has to consider molecules acting together with other molecules or as functional units. It therefore becomes apparent that only a close *cooperation between organic chemists, molecular physicists, and solid-state physicists* can yield useful and interesting results, and that such a cooperation is thus very necessary. Possible applications are an attractive but still distant goal.

Already today, one hears the terms "intelligent materials", "intelligent functional units", or "intelligent supramolecules". They refer to molecular systems with properties like those which we know from biological systems. Self-reproduction, correction of defects and errors, self-organisation and adaptation to external conditions are among the abilities which make a functional unit "intelligent"; they are possessed by molecular systems in the living world, and appear as a far-off goal for molecular physicists and quantum chemists in the laboratory.

Appendix

A1. The Calculation of Expectation Values Using Wavefunctions Represented by Determinants

A1.1 Calculation of Determinants

We begin by reminding the reader briefly of the definition of a determinant, which is given in the theory of linear equations. The determinant is associated with a matrix

$$A = (a_{jk}) \tag{A1.1}$$

where the indices j and k take on integral values $1, 2, \ldots, N$. This determinant is denoted by Det A, and can be represented in the standard form:

$$\mathrm{Det} A = \begin{vmatrix} a_{11} & a_{12} & \ldots & a_{1N} \\ a_{21} & a_{22} & \ldots & \\ \vdots & & & \\ a_{N1} & \ldots & \ldots & a_{NN} \end{vmatrix} . \tag{A1.2}$$

First, we deal with the calculation of the value of such a determinant. We consider a permutation which is derived from the N-tuple

$$(1, 2, \ldots, N) \tag{A1.3}$$

that we denote by

$$(k_1, k_2, \ldots, k_N) . \tag{A1.4}$$

All together, there are $N! = 1 \cdot 2 \cdot 3 \cdot \ldots \cdot (N-1) \cdot N$ such permutations. If $N = 5$, then there are $5! = 1 \cdot 2 \cdot 3 \cdot 4 \cdot 5 = 120$ permutations of the numbers (1,2,3,4,5). Some examples are given in (A1.5):

$$\left.\begin{matrix} (1, 2, 3, 4, 5) \\ (2, 1, 3, 4, 5) \\ (1, 2, 5, 4, 3) \end{matrix}\right\} . \tag{A1.5}$$

Now, the determinant (A1.2) consists of a sum of products of the form

$$a_{1k_1} a_{2k_2} \ldots a_{Nk_N} . \tag{A1.6}$$

Depending on whether the number of steps leading from (A1.3) to (A1.4) is even or odd, one refers to an even or an odd permutation. The number of steps is denoted by P. In the caclulation of the determinant, the individual products (A1.6) are to be multiplied by the factor

$$(-1)^P , \tag{A1.7}$$

i.e. by (+1) if P is even and by (−1) if P is odd. Then the sum is carried out over all pemutations (A1.4), including the original term (A1.3). We thus obtain the following prescription for the calculation of a determinant:

$$\text{Det}\, A = \sum_{P(k)} (-1)^P a_{1k_1} a_{2k_2} \ldots a_{Nk_N} . \tag{A1.8}$$

The index $P(k)$ on the summation means that the sum runs over all the permutations of k.

A1.2 Calculation of Expectation Values

We now turn to the computation of quantum-mechanical expectation values, in which we will represent the wavefunction of a molecule as the determinant of wavefunctions of the individual electrons. We thus replace the matrix elements a_{jk} by wavefunctions, with the index j referring to the electron and the index k to the quantum number(s) of the wavefunction: $a_{jk} \to \chi_k(\boldsymbol{r}_j)$. For simplicity, we assume that the quantum numbers are ordered from 1 through N. Of course, the whole procedure can also be applied to arbitrary quantum numbers. In order to use formula (A1.8), we consider permutations of the quantum numbers k and then employ the substitution

$$a_{jk_j} \to \chi_{k_j}(j) , \tag{A1.9}$$

where we have written $\chi(j)$ instead of $\chi(\boldsymbol{r}_j)$ for brevity.

We thus write the determinant of the overall wavefunction, $\Psi = \frac{1}{\sqrt{N!}} \text{Det}\, \chi$, in the form

$$\text{Det}\, \chi = \sum_{P(k)} (-1)^P \chi_{k_1}(1) \chi_{k_2}(2) \ldots \chi_{k_N}(N) . \tag{A1.10}$$

The normalisation factor will be taken into account later. We now want to compute the expectation value of an operator Ω with respect to this wavefunction:

$$\langle \textstyle\int \text{Det}(\chi^*)\, \Omega\, \text{Det}(\chi)\, dV_1 \ldots dV_N \rangle , \tag{A1.11}$$

where the angular brackets denote the expectation value with respect to the spin variables. We will specify the operator Ω more precisely in the following.

We begin by setting

$$1)\ \Omega = 1 . \tag{A1.12}$$

From the computation of (A1.11) with $\Omega = 1$, we obtain the normalisation factor. Inserting the determinants for Ψ and Ψ^* into (A1.11) with (A1.12) and multiplying out the sums given by (A1.10), we find expressions of the form

$$\langle \textstyle\int \chi^*_{k'_1}(1) \chi_{k_1}(1) dV_1 \rangle \langle \textstyle\int \chi^*_{k'_2}(2) \chi_{k_2}(2) dV_2 \rangle \ldots \langle \textstyle\int \chi^*_{k'_N}(N) \chi_{k_N}(N) dV_N \rangle . \tag{A1.13}$$

We have thus split up the integral over the coordinates of all the electrons into a product of integrals over the coordinates of the individual electrons. The quantum numbers derived from the left-hand determinant in (A1.11) have been denoted by a prime. Since the single-electron wavefunctions are mutually orthogonal and are normalized, only those terms in (A1.13) are nonvanishing for which the quantum numbers are the same, i.e. for which

$$k_1' = k_1,\ k_2' = k_2,\ \ldots\ k_N' = k_N\ . \tag{A1.14}$$

These terms have the value 1. As there are $N!$ permutations of the quantum numbers $k = 1, \ldots N$, we obtain immediately the result

$$\langle \textstyle\int \mathrm{Det}\,\chi^* \,\mathrm{Det}\,\chi \, dV_1 \ldots dV_N \rangle = N!\ . \tag{A1.15}$$

We now consider the case that the operator Ω refers only to a single electron, the one with the index j_0:

$$2)\ \Omega = H(j_0)\ . \tag{A1.16}$$

Here, j_0 is a number from the set of the indices $j = 1, 2, \ldots N$ which enumerate the electrons. We introduce the abbreviation

$$\langle \textstyle\int \chi_{k'}^* \chi_k dV \rangle = \langle k'|k \rangle\ , \tag{A1.17}$$

so that the corresponding term from the expression (A1.13) can be written more compactly:

$$\langle k_1'|k_1\rangle\langle k_2'|k_2\rangle \ldots \langle k_{j_0}'|H(j_0)|k_{j_0}\rangle \ldots \langle k_N'|k_N\rangle\ . \tag{A1.18}$$

Due to the orthogonality of the wavefunctions, the relations

$$k_1' = k_1,\ k_2' = k_2,\ \ldots\ j_0,\ \ldots\ k_N' = k_N \tag{A1.19}$$

must hold for all of the quantum numbers except those which refer to the index j_0. However, since the primed quantum numbers differ from the corresponding unprimed quantum numbers only by a permutation, but all quantum numbers must be the same except for those with the index j_0, it follows automatically that the last pair of quantum numbers must also be the same. Then, in the case that (A1.19) is fulfilled, (A1.18) reduces to

$$\langle k_{j_0}|H(j_0)|k_{j_0}\rangle = \langle \textstyle\int \chi_{j_0}^* \, H(\boldsymbol{r})\, \chi_{j_0}\, dV \rangle\ . \tag{A1.20}$$

Keeping the index j_0 fixed, we find that there remain $(N-1)!$ permutations of the quantum numbers. Then the expression (A1.11) with (A1.16) reduces to

$$(\mathrm{A1.11}) = (N-1)! \sum_{k=1}^{N} \langle k|H(j_0)|k\rangle\ . \tag{A1.21}$$

For later applications, Ω will often consist of a sum of contributions $H(j_0)$, in which the $H(j)$ differ only in terms of the electron coordinates, but not in their form:

$$\sum_{j=1}^{N} H(j) = H\ . \tag{A1.22}$$

As the next case, we consider an operator Ω represented by:

$$3)\ \Omega = \sum_j H(j)\ . \tag{A1.23}$$

Since the indexing of the electronic coordinates must have no influence on the final result of the computation of the expectation value, the replacement of $H(j)$ by (A1.22) means that the result (A1.21) should simply be multiplied by the number of the terms in (A1.22), i.e. by N. We then obtain the expression

$$\langle \int \mathrm{Det}(\chi^*)\, \Omega\, \mathrm{Det}(\chi) dV_1 \ldots dV_N \rangle = N! \sum_k \langle k|H|k\rangle \ . \tag{A1.24}$$

Assuming that the wavefunction Ψ is obtained from the determinant of the χ's with a normalisation factor $\frac{1}{\sqrt{N!}}$, then we obtain as our final result

$$\langle \int \Psi^*\, \Omega\, \Psi\, dV_1 \ldots dV_N \rangle = \sum_k \langle k|H|k\rangle \ . \tag{A1.25}$$

We now consider the computation of expectation values when the operator Ω describes an interaction between an electron with the coordinates $\boldsymbol{r}_l$ and an electron with the coordinates $\boldsymbol{r}_m$:

$$4)\ \Omega = V(l, m) \ . \tag{A1.26}$$

An explicit example would be a Coulomb interaction of the form

$$V(l, m) = \frac{e^2}{4\pi\varepsilon_0 r_{lm}} \ . \tag{A1.27}$$

Due to the symmetry of the interaction, we can assume that

$$l < m \ . \tag{A1.28}$$

Analogously to (A1.13), we again pick out a single element from those which are obtained on multiplying out the two determinants, and, extending (A1.18), we write this element in the form

$$\langle k_1'|k_1\rangle\langle k_2'|k_2\rangle \ldots (l) \ldots (m) \ldots \langle k_N'|k_N\rangle \ , \tag{A1.29}$$

where the factors $\langle k_l'|k_l\rangle$ and $\langle k_m'|k_m\rangle$ are to be left out at the positions marked by (l) and (m), respectively. The matrix element

$$\langle \int \chi_{k_l'}^*(l)\chi_{k_m'}^*(m) V(l, m) \chi_{k_l}(l)\chi_{k_m}(m)\, dV_l dV_m \rangle \tag{A1.30}$$

enters in their place. With the exception of the indices l and m, the orthogonality relations must be fulfilled, leading immediately to the expressions

$$k_1' = k_1, \ \ldots\ (l), \ \ldots\ (m), \ \ldots\ k_N' = k_N \ , \tag{A1.31}$$

where only the positions of the indices (l) and (m) are excepted. Since, again, the primed quantum numbers differ from the unprimed quantum numbers only by a permutation, expressions (A1.31) lead to the result that either the relations

$$1)\ k_l' = k_l, \ k_m' = k_m \tag{A1.32}$$

or else

$$2)\ k_l' = k_m, \ k_m' = k_l \tag{A1.33}$$

must hold. Now, a simple but tedious consideration shows that (A1.33) results from (A1.32) through an odd permutation. Since, for fixed l and m, there are $(N-2)!$ permutations, we obtain for the expectation value of the overall determinant the following expression:

$$\langle \int \mathrm{Det}(\chi^*)\, V(l,m)\, \mathrm{Det}(\chi) dV_1 \ldots dV_N \rangle = (N-2)! \sum_{k \neq k'} (V_{kk'.kk'} - V_{kk'.k'k}) \;, \tag{A1.34}$$

where we have used the abbreviations

$$V_{kk'.kk'} = \langle \int\int \chi_k^*(1)\chi_{k'}^*(2)\, V(1,2)\, \chi_{k'}(2)\chi_k(1)\, dV_1 dV_2 \rangle \tag{A1.35}$$

and

$$V_{kk'.k'k} = \langle \int\int \chi_k^*(1)\chi_{k'}^*(2)\, V(1,2)\, \chi_{k'}(1)\chi_k(2)\, dV_1 dV_2 \rangle \tag{A1.36}$$

in an obvious manner. The minus sign in the second term of the sum comes from the abovementioned odd permutation. The Coulomb interaction of all the electrons (7.5) enters into the Hamiltonian (7.6). We therefore compute also

$$\langle \int \mathrm{Det}(\chi^*)\, \tfrac{1}{2} \sum_{l \neq m} V(l,m)\, \mathrm{Det}(\chi)\, dV_1 \ldots dV_N \rangle \;. \tag{A1.37}$$

Since the integrals in (A1.36) and (A1.37) cannot depend on the indices l and m of the electronic coordinates, we can replace the sum $\sum_{l \neq m}$ by the factor $N \cdot N - N = N(N-1)$, where the subtraction $-N$ is due to the condition that $l \neq m$. Finally, we then obtain

$$(\mathrm{A1.37}) = \tfrac{1}{2} N! \sum_{k \neq k'} (V_{kk'.kk'} - V_{kk'.k'k}) \;. \tag{A1.38}$$

The result (7.15) is obtained as the sum of (A1.24) with (A1.23) and (A1.38) after dividing by the normalisation factor $N!$ [cf. (A1.15)].

A2. Calculation of the Density of Radiation

We first determine the number of modes dN in a volume V and in the frequency or wavenumber interval dk $(d\nu)$. We assume a finite normalisation volume V. Initially, we give a one-dimensional example, and assume that the field strength has the form

$$E = E_0 \sin(kx) \;. \tag{A2.1}$$

We fix the value of k by requiring that E vanish at the boundaries $x = 0$ and $x = L$ (where L is the length of the normalisation "volume"). This is fulfilled if we choose

$$k = \frac{n\pi}{L} \;, \quad n = 1, 2, 3, \ldots \;. \tag{A2.2}$$

The wavenumber of an electromagnetic wave is related to its frequency by the equation

$$\omega = ck \;. \tag{A2.3}$$

On the other hand, we of course have

$$\omega = 2\pi\nu \;. \tag{A2.4}$$

From these two equations, we obtain

$$\nu = \frac{cn}{2L} . \tag{A2.5}$$

Since n is an integer, we can enumerate the waves. Using (A2.5), we then find for the number of allowed waves in the interval $d\nu$

$$dn = \frac{2L}{c} d\nu . \tag{A2.6}$$

In this way, we have solved our problem of finding the number of waves in the frequency interval $d\nu$, at least in one dimension. In reality, we are of course dealing with three dimensions. Then the wavenumbers k_i are given by

$$k_i = \frac{\pi n_i}{L} , \quad n_i = 1, 2, 3, \ldots , \quad i = x, y, z . \tag{A2.7}$$

We define $k = \sqrt{k_x^2 + k_y^2 + k_z^2}$ and $n = \sqrt{n_x^2 + n_y^2 + n_z^2}$. In complete analogy to (A2.5), we obtain the relation

$$\nu = \frac{cn}{2L} . \tag{A2.8}$$

In order to describe the number of allowed waves in the frequency interval $\nu, \nu + d\nu$, we consider a coordinate system in which the axes correspond to the numbers n_x, n_y, and n_z. Each point in this space with integral values of the coordinates represents a possible state of oscillation of the electromagnetic field. If we choose the frequency interval to be sufficiently large, we can assume that the points can be calculated as if they formed a continuum. The number of points in a spherical shell of thickness dn is then given by $4\pi n^2 dn$. Since the numbers n_j must all be positive, this result has to be divided by the number of octants in the sphere, i.e. by 8. And since we are dealing with two polarisation directions (helicity states) of the electromagnetic waves, which must be counted separately, we need to multiply our result by 2. We thus obtain for the number of allowed states in a spherical shell

$$dN = \pi n^2 dn . \tag{A2.9}$$

Using (A2.8) to express n in this formula, we obtain the number of allowed waves in the frequency interval $\nu, \nu + d\nu$:

$$dN = \frac{8\pi L^3}{c^3} \nu^2 d\nu . \tag{A2.10}$$

By applying a similar argument to travelling *plane* waves, whose wavevectors lie within a solid angle $d\Omega$, we obtain

$$dN = \frac{V}{(2\pi)^3} k^2 dk d\Omega \tag{A2.11}$$

and therefore

$$\sum_k \ldots = \frac{V}{(2\pi)^3} \int \ldots k^2 dk d\Omega . \tag{A2.12}$$

Bibliography

of Supplementary and More Specialized Literature

1. Textbooks on Physics and Physical Chemistry

Alonso, M., Finn, E.J.: *Fundamental University Physics* (Addison-Wesley, Reading, MA 1980)

Atkins, P.W.: *Physical Chemistry*, 4th edn. (Oxford Univ. Press, Oxford 1990)

Halliday, D., Resnick, R., Krane, K.S.: *Physics I and II*, 4th edn. (Wiley, New York 1992)
Krane, K.S.: *Modern Physics* (Wiley, New York 1983)

2. Textbooks on Atomic and Molecular Physics

Eisberg, R., Resnick, R.: *Quantum Physics of Atoms, Molecules, Solids, and Particles*, 2nd edn. (Wiley, New York 1985)

Haken, H., Wolf, H.C.: *The Physics of Atoms and Quanta*, 4th edn. (Springer, Berlin, Heidelberg 1994)

March, N.H., Mucci, J.F.: *Chemical Physics of Free Molecules* (Plenum, New York 1992)

Svanberg, S.: *Atomic and Molecular Spectroscopy*, 2nd edn. (Springer, Berlin, Heidelberg 1992)

Weissbluth, M.: *Atoms and Molecules* (Academic, New York 1981)

3. Textbooks on Quantum Mechanics and Quantum Chemistry

Bethe, H.A., Salpeter, E.F.: *The Quantum Mechanics of One- and Two-Electron Atoms* (Plenum, New York 1977)

Cohen-Tannoudji, C., Diu, B., Laloë, F.: *Quantum Mechanics I and II*, 2nd edn. (Wiley, New York 1978)

Landau, L.D., Lifshitz, E.M.: *Quantum Mechanics*, 3rd edn. (Pergamon, Oxford 1982)

Martin, J.L.: *Basic Quantum Mechanics* (Oxford Univ. Press, Oxford 1981)

Merzbacher, E.: *Quantum Mechanics*, 2nd edn. (Wiley, New York 1969)

Messiah, A.: *Quantum Mechanics I and II* (Halsted, New York 1961 and 1962)

Schwabl, F.: *Quantum Mechanics* (Springer, Berlin, Heidelberg 1992)

Shankar, R.: *Principles of Quantum Mechanics* (Plenum, New York 1980)

Wichmann, E.H.: *Quantum Physics* (Vol. IV of *Berkeley Physics Course*) (McGraw-Hill, New York 1971)

4. Specialized Literature

Chapter 1

Heilbron, J.L.: *Elements of Early Modern Physics* (Univ. California Press, Berkeley 1982)
van der Waerden, B.L. (ed.): *Sources of Quantum Mechanics* (Dover, New York 1967)

Chapter 2

Robertson, J.M.: *Organic Crystals and Molecules* (Cornell Univ. Press, Ithaca, NY 1953)

Chapter 3

Ashcroft, N.W., Mermin, N.D.: *Solid State Physics* (Holt, New York 1976)
Jackson, J.D.: *Classical Electrodynamics*, 2nd edn. (Wiley, New York 1975)
Kittel, C.: *Introduction to Solid State Physics* 6th edn. (Wiley, New York 1986)

Chapter 4

Naaman, R., Vager, Z.: *The Structure of Small Molecules and Ions* (Plenum, New York 1988)

Chapter 5

Wagnière, G.: *Introduction to Elementary Molecular Orbital Theory and to Semiempirical Methods*, Lecture Notes Chem., Vol. 1 (Springer, Berlin, Heidelberg 1976)

Chapter 6

Hargittai, I., Hargittai, M.: *Symmetry through the Eyes of a Chemist* (VCH, Weinheim 1987)
Tinkham, M.: *Group Theory and Quantum Mechanics*, (McGraw-Hill, New York 1964)

Chapter 7

Atkins, P.W.: *Solution Manual for Molecular Quantum Mechanics* (Oxford Univ. Press, Oxford 1983)
Christofferson, R.E.: *Basic Principles and Techniques of Molecular Quantum Mechanics* (Springer, Berlin, Heidelberg 1989)
Haken, H.: *Quantenfeldtheorie des Festkörpers*, 2nd edn. (Teubner, Stuttgart 1993)
McWeeny, R.: *Methods of Molecular Quantum Mechanics*, 2nd ed. (Academic, New York 1992)
Szabo, A., Ostlund, N.S.: *Modern Quantum Chemistry* (McGraw-Hill, New York 1989)
Wagnière, G.: *Introduction to Elementary Molecular Orbital Theory and to Semiempirical Methods*, Lecture Notes Chem., Vol. 1 (Springer, Berlin, Heidelberg 1976)
Wilson, S.: *Electron Correlation in Molecules* (Clarendon, Oxford 1984)

Chapters 9 and 10

Banwell, C.N.: *Fundamentals of Molecular Spectroscopy*, 4th edn. (McGraw-Hill, New York 1994)
Demtröder, W.: *Laser Spectroscopy*, Springer Ser. Chem. Phys., Vol. 5 (Springer, Berlin, Heidelberg 1988)
Flygare, W.H.: *Molecular Structure and Dynamics* (Prentice-Hall, Englewood Cliffs, NJ 1978)
Graybeal, J.: *Molecular Spectroscopy* (McGraw-Hill, New York 1993)
Herzberg, G.: *Molecular Spectra and Molecular Structure*, 3 Volumes (van Nostrand, Princeton 1964–1966, reprint 1988–1991)
Hollas, J.M.: *High Resolution Spectroscopy* (Butterworths, London 1982) Steinfeld, J.I.: *Molecules and Radiation* (Harper and Row, New York 1974)
Struve, W.S.: *Fundamentals of Molecular Spectroscopy* (Wiley, New York 1989)
Townes, C.H., Schawlow, A.L.: *Microwave Spectroscopy* (McGraw-Hill, New York 1975)
Wilson, E.B., Decius, J.C., Cross, P.C.: *Molecular Vibrations* (Dover, New York 1980)

Chapter 12

Atkins, P.W.: *Physical Chemistry*, 4th edn. (Oxford Univ. Press, Oxford 1990)
Banwell, C.N.: *Fundamentals of Molecular Spectroscopy*, 4th edn. (McGraw-Hill, New York 1994)
Demtröder, W.: *Laser Spectroscopy*, Springer Ser. Chem. Phys., Vol. 5 (Springer, Berlin, Heidelberg 1988)
Herzberg, G.: *Molecular Spectra and Molecular Structure*, 3 Volumes (van Nostrand, New York 1964–1966, reprint 1988–1991)
Hollas, J.M.: *High Resolution Spectroscopy* (Butterworths, London 1982)
Steinfeld, J.I.: *Molecules and Radiation*, 2nd ed. (MIT Press, Boston 1978)
Struve, W.S.: *Fundamentals of Molecular Spectroscopy* (Wiley, New York 1989)

Chapters 13, 14 and 15

Atkins, P.W.:*Physical Chemistry*, 4th edn. (Oxford Univ. Press, Oxford 1990)
Demtröder, W.: *Laser Spectroscopy*, Springer Ser. Chem. Phys., Vol. 5 (Springer, Berlin, Heidelberg 1988)
Herzberg, G.: *Molecular Spectra and Molecular Structure*, 3 Volumes (van Nostrand, New York 1964–1966)
Hollas, J.M.: *High Resolution Spectroscopy* (Butterworths, London 1982)
Steinfeld, J.I.: *Molecules and Radiation*, 2nd ed. (MIT Press, Boston 1978)
Struve, W.S.: *Fundamentals of Molecular Spectroscopy* (Wiley, New York 1989)
Weissbluth, M.: *Atoms and Molecules* (Academic, New York 1981)

Chapter 16

Haken, H.: *Licht und Materie I, Elemente der Quantenoptik*, 2nd edn. (BI Wissenschaftsverlag, Mannheim 1989)
Haken, H.: *Quantenfeldtheorie des Festkörpers*, 2nd edn. (Teubner, Stuttgart 1993)

Chapters 18 and 19

Abragam, A.: *The Principles of Nuclear Magnetism* (Oxford Univ. Press, Oxford 1989)
Atkins, P.W.: *Physical Chemistry*, 4th edn. (Oxford Univ. Press, Oxford 1990)
Banwell, C.N.: *Fundamentals of Molecular Spectroscopy*, 4th edn. (McGraw-Hill, New York 1994)
Carrington, A., McLachlan, A.D.: *Introduction to Magnetic Resonance* (Harper and Row, New York 1974)
Ernst, R.R., Bodenhausen, G., Wokaun, A.: *Principles of Nuclear Magnetic Resonance in one and two Dimensions* (Oxford Univ. Press, Oxford 1990)
Freyman, R.: *A Handbook of Nuclear Magnetic Resonance* (Longman, London 1988)
Harris, R.K.: *Nuclear Resonance Spectroscopy* (Longman, London 1994)
Slichter, C.P.: *Principles of Magnetic Resonance*, 3rd edn., Springer Ser. Solid-State Sci., Vol. 1 (Springer, Berlin, Heidelberg 1992)

Chapter 20

Vögtle, F.: *Supramolekulare Chemie*, 2nd edn. (Teubner, Stuttgart 1992)

Subject Index